Optical Networking
& WDM

ABOUT THE AUTHOR

Walter Goralski has more than 30 years of experience in the data communications field, including 14 years with AT&T. He is currently a Senior Course Developer with Juniper Networks and is the bestselling author of *SONET, Second Edition*, and several books on ADSL, VoIP, ATM, and frame relay. He has also written numerous articles on data communications and other related issues.

ABOUT THE REVIEWER

Gary Kessler is an Assistant Professor & Program Coordinator for the Computer Networking Program at Champlain College in Burlington, Vermont, as well as a Networking consultant, frequent contributor to Network magazine, and author of the bestseller *ISDN : Concepts, Facilities, and Services*.

Optical Networking
& WDM

WALTER **GORALSKI**

Osborne/**McGraw-Hill**

New York   Chicago   San Francisco
Lisbon   London Madrid   Mexico City
Milan   New Delhi   San Juan
Seoul   Singapore   Sydney   Toronto

Osborne/**McGraw-Hill**
2600 Tenth Street
Berkeley, California 94710
U.S.A.

To arrange bulk purchase discounts for sales promotions, premiums, or fund-raisers, please contact Osborne/**McGraw-Hill** at the above address. For information on translations or book distributors outside the U.S.A., please see the International Contact Information page immediately following the index of this book.

Optical Networking & WDM

1234567890 FGR FGR 01987654321

ISBN 0-07-213078-4

Publisher
Brandon A. Nordin
Vice President & Associate Publisher
Scott Rogers
Executive Acquisitions Editor
Steven Elliot
Project Editor
Jennifer Malnick
Acquisitions Coordinator
Alex Corona
Technical Editor
Gary Kessler

Copy Editor
Robert Campbell
Proofreader
Marian Selig
Indexer
Claire Splan
Computer Designer
Peter F. Hancik
Dick Schwartz
Illustrator
Beth E. Young
Series Design
Peter F. Hancik

This book was composed with Corel VENTURA™ Publisher.

AT A GLANCE

CONTENTS

Part II

Optical Fiber and Electrical Nodes

Part III

Wide-Area Optical Networking

Part IV

Local Area Networks

Part VII

Reference

PREFACE

What's going on here? Ethernet, the venerable workhorse of LANs everywhere, has moved far beyond 10 Mbps speed and 100 Mbps. New flavors of Gigabit Ethernet (GBE) have invaded the MAN and WAN at speeds of 1 Gbps, and soon 10 Gbps. Qwest, which started life as a company with many miles of fiber routes and little else to boast of, has merged with (or acquired, depending on one's sensitivities) US West, a proud segment of the once mighty Bell System. Enron, not too long ago solely a power company, has earned a name for itself as a communications giant. New carrier Yipes Communications plans to offer connectivity for customer LANs at speeds from 1 Mbps to 80 Mbps, and in 1 Mbps increments. Sun Microsystems plans to send traffic directly onto fiber links because traditional SONET fiber links were pricey, took too long to deploy, and Sun personnel were much more familiar with the typical LAN architecture of IP packet inside Ethernet frame over fiber optic cable. And SONET equipment vendors Lucent and Nortel, not wanting to be left on the sidelines, have boosted SONET (and the international equivalent, SDH) to 40 Gbps in the form of OC-768. Lucent already markets the Wavestar-Opticair free space laser system that can operate up to 10 Gbps over 5 km—without any fiber at all. Students everywhere download gigabytes of free music over the Internet and Web in a few minutes at schools where a megabyte of information would have swamped the whole network for an hour not too long ago. And this is only a survey of the changes in networks and networking wrought in the recent past.

What has fueled all of this change? The rise of optical networking and the promise of DWDM. All of the examples just mentioned have been made possible by recent advances in fiber optic transmission and the network elements that handle the traffic on the links. When seen in this light, it seems less astonishing, at least from the technological perspective. GBE has shown that modern fiber systems can carry LAN traffic 40 km or more without the need for conversion to or the cooperation of traditional "long haul" carriers and networks technologies. Qwest has shown that whoever has the most bandwidth (and is not limited in action by a history of voice regulation) wins. Enron has confirmed that fiber systems have leveled the communications playing field as never before. Yipes has shown that the emerging optic technologies have made the communications field more open than ever. Sun has demonstrated that there is little need for voice-oriented carrier systems like SONET when new optical systems can easily accommodate IP directly. Nortel has shown that there might just be a place for SONET in this new optical world. Lucent has shown that optical networking need not be constrained by former ways of doing things. And people everywhere, not just students, have proven that there will always be a use for increased bandwidth—at least for the foreseeable future.

In a very real sense, this book is all about bandwidth. But it is about a particular form of bandwidth, the bandwidth delivered by new optical networking techniques coupled with the multiplexing of lightwaves onto a single strand of fiber made possible more efficiently than ever before with DWDM. Service providers see the pressure for increased bandwidth from customers everywhere. Competition is reduced in many cases to who can provide the most bits for the lowest price. Equipment vendors face a world where many observers see SONET and SDH equipment demand over the next few years as a 7 percent and 14 percent combined annual growth rate (CAGR) respectively, while DWDM equipment demand is expected to have a 45 percent CAGR. The writing is on the wall. Add bandwidth quickly, or be left behind.

Optical Networking & WDM explores all aspects of this new optical networking and DWDM phenomena. While intended for readers with at least some background in basic computer and communication technologies, this book is also intended to be of interest to general readers and specialists alike. There is always something new to be learned in the field of optical networking. This book begins by examining the fundamental differences between electrical and optical transmission systems; it continues by exploring how early optical networking employed fiber optic links but continued to employ entirely electrical network devices at the ends of these links—and even in the middle. This book next investigates wide area optical networking, with special emphasis on optimizing fiber optic systems for IP packet transport; it continues by extending the investigation to local area optical networks, where some of the most exciting developments in this field have begun; it also considers the use of fiber to the home in the form of residential networks. Finally, it closes with a look at the future, especially important as new non-optical wireless systems promise more and more bandwidth themselves.

Along the way, this book will offer a detailed look at just how light travels in a fiber; how fiber optic cable is constructed; how optical transmitters, receivers, and amplifiers function; how DWDM works; how optical cross-connects and switches function; and much, much more. There should never be a dull moment. This is the future of networking.

ACKNOWLEDGMENTS

No one works or lives in a vacuum. There are always people and groups who should be thanked for their assistance in the making of a book.

I would like to thank my former employer, Hill Associates, Inc., for creating the type of intellectual work environment where personal growth is always encouraged, and for nurturing a climate that makes the knowledge needed to write books like this in the first place. I owe special thanks to Dave Hill, not only for hiring me, but also for embracing optical networking and never failing to bring new developments to my attention.

I also must thank my current employer, Juniper Networks, for encouraging me to continue my writing. The atmosphere at Juniper has always been one of great excitement and a continuing sense of discovery and exploration for all new technologies.

I owe a great deal to individuals I have had contact with over the years who took the time to show me their work and correct my numerous misunderstandings of what at times seemed beyond my comprehension. I must especially mention Dave Beebe of Fujitsu, Dewayne Bridges and Randy Ellis of Alcatel, and Eric Spurrier of Lucent Technologies.

On the publishing side, Steve Elliot encouraged the production of this text. McGraw-Hill supported the process from start to finish with their usual efficiency.

My children, Chris, Alex, and Ari, have accepted my role as writer, and have even mentioned this fact to others, a sure sign that I have arrived. Finally, I owe more than I could ever express to Camille, who at long last showed me what life is all about. Thank you one and all.

PART I

Electrical and Optical

The first part of this book opens with a chapter that is intended to be a more or less historical overview of wide area and local area networks. Although the emphasis will be on traditional, electrical transmission systems, starting with the electrical telegraph, a brief analysis of early optical networks in the form of the "optical telegraph" will be offered. These early systems were "free space" optical networks in the sense that no optical fiber cable was needed (or yet invented!) to carry optical signals. In fact, these optical network were an elaborate form of semaphore networks.

The main emphasis in Chapter 1 will be electrical telegraphs and telephone transmissions. However, carrier systems, which multiplex many communications sources such as voice conversations onto a single transport facility, will be fully explored. Then the differing needs of local area networks will be examined, keeping in mind that all electrical systems have real limitations in terms of performance and efficiency. There is also the issue of cost effectiveness at very high bit rates; this will be examined as well.

This leads to Chapter 2, which covers the introduction of fiber optic cable itself. How such cable is manufactured is explained, since the extremely small fiber cores pose challenges in both fabrication and performance testing today. The fundamental steps involved in pushing light through glass are investigated, and all of the major fiber types are introduced. This is followed by an explanation of fiber applications, which are not limited to communications. Light can be distributed on fiber optic cable not only to carry information, but to serve other purposes as well.

This is followed by Chapter 3, which focuses on fiber optic transmitters and receivers, with the emphasis on semiconductor laser transmitters, although light emitting diodes (LEDs) are also treated. The receiver section emphasizes avalanche photo-detectors (APDs), but others forms of fiber optic receivers are also explored. Then mid-span regenerative repeaters are introduced, along with the key functions they must perform on long fiber spans.

This part of the book closes with a look at basic optical transmission systems. Chapter 4 combines the concepts of all three previous chapters. Network components, both electrical and optical, are assembled for WAN and LAN alike, showing the differences and similarities between the needs of the two. Specialty systems are also discussed, such as the restricted WAN environment of the metropolitan area network and the unique needs of a true global area network.

Since the emphasis here is mainly historical, not all of the "nifty new stuff" is explored right up front. The first part of this book basically ends with the state of optical communications around 1988. But this examination of fundamentals is necessary to appreciate the revolutionary changes taking place in optical networking today.

CHAPTER 1

Electrical Transmission Systems

The use of optical transmission in modern networking is sometimes positioned as an obvious improvement on electrical transmission systems. But this should not be taken to mean that there was no form of optical networking before electrical systems began carrying all forms of messages, public and private. In fact, before there was any form of electrical networking, a primitive form of optical networking did exist, and it flourished for a time. This is in fact a good place to start, since many of the same features and functions of modern networks began way back then.

THE FIRST OPTICAL NETWORKS

In the beginning, there was optical networking. Optical networking does not include isolated instances of signal fires or other forms of prearranged messaging. White smoke meaning "the battle is won" worked once, but not with general messages whose content could not be anticipated and coordinated ahead of time. Even links of this type covering long distances using relay bonfires barely fit the modern definition of a "network," which is usually defined as a permanent set of interconnected links set up expressly for the purpose of general communications. But such forms of visual communications, as opposed to drums, were undeniably optical in nature.

The roots of true optical networking go back to early 1791, when Frenchman Claude Chappe set up pivoting five-foot-high wooden panels painted black on one side and white on the other. By coordinating panel flips with his brother stationed ten miles away armed with a telescope and code book of meanings, Chappe managed to optically (eyeball) transmit a message of nine French words in about four minutes. The message, IF YOU SUCCEED, YOU WILL SOON BASK IN GLORY in English, ultimately proved false, but the local officials present were reportedly impressed. Not many forms of transport could cover ten miles as quickly, although even a slow horse represented much more bandwidth in terms of information that could be carried. The trade-off for loss of bandwidth was the lower delay for spontaneous and short messages.

This bandwidth restriction in the form of slow signaling may have been the reason that Chappe was discouraged from calling his invention the "fast writer," or *tachygraphe* in French. A friend and government official, Miot de Melito, suggested "far writer" or *telegraphe,* and Chappe agreed. And so the *optical telegraph* was born. In actual fact, this was more of a *visual telegraph,* but the term "optical telegraph" has stuck for better or worse.

It was by all accounts a painful journey from successful demo to functioning network. In that sense, not much seems to have changed in more than 200 years. Chappe had even experimented in 1789 with banging on copper pots and sending messages acoustically. No doubt the neighbors were happier when the Chappe family turned to optics. In 1791, France was deeply mired in the turmoil of the French Revolution, and in spite of Chappe's successful demonstration in his northern home town of Brulon, people were understandably wary of too much change too quickly. So more was required than a simple demonstration to the locals.

People were also suspicious of anything that could remotely be interpreted as a display of royalist sentiments. A second demonstration in 1792, this time in Belleville, near Paris, was disrupted when rioters suspecting a royalist plot destroyed the equipment and almost the Chappe brothers as well. Perhaps placing the test near the line of sight of Temple Prison, where royalist prisoners were held, was not such a great idea.

Like all technologies, the optical telegraph was evolving even as it was being developed into a working system. The first systems demonstrated required synchronized clocks to coordinate between sender and receiver. This prevented letters from being repeated or dropped when the link "desynchronized." For the 1792 demonstration, the black and white panels were replaced by a long rotating bar having two smaller rotating arms at the ends. The long bar, which Chappe called the "regulator," could be positioned horizontally, vertically, or at a 45-degree angle. The smaller arms, called "indicators," could take on one of seven positions 45 degrees apart. Altogether there were 98 possible combinations of regulator and indicators, which of course yielded a 98-position "alphabet" for communications purposes. Six of these "letters" were reserved for special uses.

The remaining 92 positions, or *codes,* were used to represent letters, numbers, and even the most common syllables in the French language. A system with even 92 codes was deemed inadequate, so the codes were always conveyed in pairs. The first code referred the operator at the receiving side of the link to a page in a special 92-page code book. The second code referred the operator to one of 92 numbered meanings on that particular page. This 92×92 system contained a total of 8,464 words and phrases. The slowdown in signaling speed because of the necessity of sending in pairs was offset by the enhanced information content of each individual code. This is an early example of increasing the number of "bits per baud" on a communications link. The use of the term *baud,* short for Baudot, who was an early telecommunications pioneer, is anachronistic here and is used merely for illustrative purposes. A *baud* is just any change of line condition on a link that can be used to convey information (bits). If a baud is used to convey only one letter at a time, a link must have a lot of bauds to send whole sentences quickly. But if each baud can represent a word, then fewer bauds can convey sentences much faster, even if the overall baud rate of the second link is slower.

The optical telegraph as a system was evolving physically as well. Chappe had attracted the interest of many technically inclined people, and a noted clock maker named Abraham-Louis Breguet built a new control mechanism for the new regulator-indicator design. Older models relied on the operator at the sending side to set the position of the arms manually and directly, exposed to the elements. The Breguet system used a fairly sophisticated system of pulleys to allow the sender to manipulate the arms of what was basically a scale model of the entire apparatus. Any change to the arms' positions on the smaller model was mirrored through the pulley system by the larger signaling arms. The control model could then be placed in a small hut below the signaling tower itself. This arrangement is shown in Figure 1-1. There were 36 arm positions used to send letters and numbers, since words and names not listed in the code book had to be spelled out letter by letter.

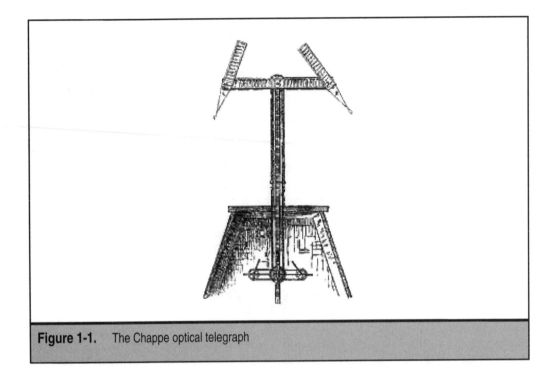

Figure 1-1.   The Chappe optical telegraph

Chappe had visions of much more than just a simple link a few miles long between single sender and receiver towers. He realized that messages could easily be relayed over great distances by building a network of towers within sight of each other. The National Convention, the ruling body of France at the time, received a detailed proposal from Chappe in 1793. Charles-Gilbert Romme, the president of the Committee of Public Instruction, immediately saw the potential of the optical telegraph system for military and other uses and proposed the Convention supply funds for further experimentation.

A committee of three, including scientist Joseph Lakanal, immediately set up a small network of three telegraph stations spanning twenty miles. What worked on this small scale could conceivably be used to cover the nation. As a sign of Chappe's new respectability, the mayors of the three towns were made responsible for the continued safety of the telegraph equipment and their operators. Construction went quickly, and a test conducted on July 12, 1793, was deemed a rousing success, although by today's standards the system was abysmally slow.

There were two operators at each tower: one to watch the upstream station through a telescope, and the other to manipulate the arms signaling the downstream station as quickly as possible. Each signal thus needed to be held for only a few seconds. In spite of this tandem operation, it took 11 minutes to send an official message of 27 words from source to destination, for a rate of about two and a half words per minute. A shorter reply took nine minutes to return to the source. But 20 miles in twenty minutes is a mile a minute, or 60 miles per hour, an unheard-of speed for an ad hoc conversation in that day and age. All three committee members were quite impressed, Lakanal in particular.

Lakanal became the champion of the new technology in the central government. There was still opposition to the new government of the Revolution, especially in the far provinces, and perhaps the new speedy communication system could be used to give a further advantage to the central government authorities in Paris. Lakanal, also being a politician, emphasized that Chappe's invention obviously showed the salutary effect that the new environment of the French Republic was having on the population of France. Be that as it may, a new fifteen-tower line was proposed to connect Paris to Lille, a city about 130 miles to the north. Chappe was rewarded with a government salary and the use of a public horse.

The military value of the optical telegraph was demonstrated almost immediately. The French State Telegraph to Lille, which started operations in May 1794, was used on August 15 of that year to report the retaking of a town from the Austrians and Prussians less than an hour after it happened. The French army's successful campaign in the north into Holland that year was reported on the Paris-Lille line, and the central government came to appreciate the relatively new concept of the "time value of information." Within a few years the line had been extended from Lille to Dunkirk, and a second line snaked across France to Strasbourg in the east.

One military leader who embraced the optical telegraph was Napoleon Bonaparte. Once Napoleon took over in 1799, the French telegraph system was expanded in many directions. One important line went to Boulogne on the coast to prepare for the anticipated invasion of England. The younger Chappe brother, Abraham, was enlisted to create an optical telegraph capable of signaling farther than ever before: all the way across the English Channel. The project actually succeeded between Belleville and Saint-Martin-du-Tertre, roughly the same distance needed. And although the French tower was eventually installed at Boulogne, the slight glitch arose that the English tower actually had to be built in England before the link could be used. Since the invasion of England never came off as planned, the special "long distance" optical telegraph remained a curiosity. But Napoleon's other military plans in the southeast were more fruitful, and the telegraph eventually wound its way from Paris through Dijon, Lyons, and Turin all the way to Milan.

Napoleon's use of optical networking soon was emulated by other European countries, notably Great Britain and Sweden. The other governments either blatantly copied Chappe's system or created various "improvements" on the original. In Britain, for example, a form of optical telegraph was used by the Admiralty as early as 1795 to connect London with the southern ports. In 1797, the British system used six wooden shutters that could be open or closed. This yielded 64 possible code combinations (2 multiplied by itself 6 times), more than enough for military use.

By 1810, the French system extended from Paris to the borders of Italy, Germany, Belgium, and the Atlantic Coast. Even as early as 1799, the code books had grown to three volumes and had 25,392 entries. This was because the usual response to adding more words and phrases for business and other purposes was just to add more codes. It never occurred to anyone to just go back to the simpler form of purely alphabetic representation for all message content. Also, there seems to have been some sensitivity to the privacy issue. Optical telegraphs were essentially broadcast media, where anyone could see the message as it was sent, not just the receiver. But only the receiver had easy access to the code books, of course, which were far too bulky to memorize (thirty-six or so alphanumeric codes, however, could easily be committed to memory).

Even with this form of maximum information per signal state, the highest speed for information transfer was about 20 characters per minute, or about 160 bits per second. Until the electrical telegraph began to make its impact felt, at its 1852 peak the French system linked 29 cities through a system of 229 towers, forming a network with some 3,000 route miles. Chappe semaphores sprouted from many rooftops, as shown in Figure 1-2. The British and Swedish systems were also quite large and elaborate.

Chappe does not seem to have been overly disturbed by the predominantly military use of his system, but he still had dreams of using the optical telegraph for general business purposes. He wanted to see a pan-European system doing things like conveying commodity prices from Amsterdam to Paris to Cadiz. He wanted daily national news updates, as long as the content was approved by the state, of course. Napoleon was not in favor of either idea but did approve the sending of the winning numbers of the weekly lottery. This cut down on a lot of fraud, since the numbers were now known nationally on the very day they were drawn.

Figure 1-2. Chappe optical telegraphs mounted on rooftops

Other problems with the optical telegraph continued to plague Chappe and his successors. These had nothing to do with the technology but everything to do with the ownership and originality of the idea behind the invention. Challenges to a pioneer were also to play a part in most breakthroughs since Chappe, and challenges to a communications invention were not unknown previously. But the success of national networks raised the stakes dramatically and made these issues all the more important. Rival inventors came out of the woodwork claiming to have had the idea prior to Chappe or to have a better version of the same basic device. Even Breguet, the control pulley inventor, claimed to have a much larger stake in the overall scheme. Things got so bad that Chappe suspected his rivals of poisoning him and eventually committed suicide in 1805.

Other than the initial French optical telegraph, the most documented system belonged to Great Britain. The Admiralty had chosen the six-shutter system, but from the start they were besieged with proposals to make the British system faster, or cheaper to operate, or both. The proposals ranged from bigger and better code books to bigger and better towers.

Besides France, Great Britain, and Sweden, there were also optical telegraphs in Denmark, Finland, Germany, and Russia. Several important concepts were first established in these early versions of what were essentially not only optical networks, but data networks as well. Code books were an early form of data compression, and in a sense a crude form of encryption also. Error recovery was a vital principle as well, both in the form of error detection ("that's not in the code book!") and error correction ("send that over, please!"). Even the idea of flow control (senders must never overwhelm receivers) had its roots in these early networks.

From the very beginning of the optical telegraph, some annoying problems became increasingly obvious. Most seriously, the system only functioned in clear and bright weather conditions. Darkness terminated service until the morrow, and rain and fog limited the usefulness of the device. Some tried to operate after dark using colored lanterns on the arms, but this experiment had mixed success at best. The optical telegraph was also very expensive to construct and operate. The towers were costly and numerous. The operators had to be trained, worked in shifts, and were highly skilled. Only governments could really afford the optical telegraph systems, and so most of the traffic was restricted to official government correspondence. People noticed the towers, of course, since they were hard to miss. But the optical telegraph hardly transformed the lives of ordinary citizens.

The demise of the optical telegraph came with the rise of the use of the electrical telegraph. Optical networking in the 1800s ended with the abandonment of the last three Swedish towers in 1881. The towers still dotted the landscape in many places, but these funny-looking "windmills" no longer carried the vital news of the day. That role was now played by the electrical telegraph.

ELECTRICAL TELEGRAPH TRANSMISSION

The United States did not build an optical network based on either the French or British model. There were many reasons for this. The new federal government in the United States did not have a lot of money to spend on anything but the essentials. And just as important, the vast size of the United States compared to more compact European nations meant that a series of links placed 10 miles apart spanning the 1,100 mile distance from Boston to Atlanta would require the construction of 111 towers. Yet the same distance considerations would ultimately require the United States to build some type of rapid national communications network.

Strangely, even as the Chappe optical networking was busy flapping its arms all over France, and the British shutter system was clicking open and closed, research into using electricity to send information was underway. Ultimately, the electrical telegraph came into being from the cross-fertilization of British research and American development. These early electrical telegraphs just borrowed the coding system of the optical systems. Semaphore towers gave way to closely spaced poles with iron wire (iron was stronger and much cheaper than copper at the time). But the roots of electrical transmission went back a long way.

Electricity, mostly in the form of static, was known since early times. In more modern days, Michael Faraday (the last major scientist without a university degree) showed that electricity could be harnessed and made to flow in wires. Danish physicist Hans Christian Oersted showed in 1820 that electricity flowing in a wire causes a magnetic field to appear, thus discovering the whole phenomenon of electromagnetism. This discovery was soon followed by the invention of the galvanometer (showing the flow of electric current), the electromagnet (some metals act as permanent magnets when electricity is flowing around them), and better forms of batteries (which stored electricity). The electrical telegraph wedded all three into a system.

This took some doing to get right. For instance, in 1824, a British physicist and mathematician named Peter Barlow used an arrangement of these three elements to produce clicks capable of conveying information. But now the wire was the problem: even 200 feet of wire caused the clicks to diminish in volume. Clearly work needed to be done on sending electricity over long distances before the electrical telegraph could be of much use to anyone. This was a job for the engineers. Ultimately, they were up to the challenge.

Many of these early proposals for a telegraph based on electricity went before the British Admiralty as keeper of the optical telegraph system. One of the earliest proposals was by William Sturgeon, the inventor of the practical electromagnet using copper wire. In 1822, Francis Ronalds approached the British Admiralty with a design for an "electric telegraph" based on Sturgeon's electromagnet. Electrodes were to be used to point to letters or numbers at each end. Ronalds ran eight miles of wire around his estate at Hammersmith for demonstrations. Ultimately, however, the Admiralty decided that the electric telegraph was "wholly unnecessary," since Britain was well served by the existing optical system, which had been around since 1796.

But as annoying as these proposals were to the Admiralty, an important side effect of all this activity was to expose people to these new possibilities. One of these people happened to be a New York University art professor named Samuel Finley Breese Morse, born in 1791, the year of Chappe's first demonstration of the optical telegraph. In 1824, Morse attended a lecture on electromagnetism, which got him thinking about using electricity for communications. Morse soon had ample reason to spend a lot of time pondering the problem of slow communications over long distances. The limitations of communication were made painfully obvious to Morse in the following year, 1825. His wife died when he was out of town painting portraits of the powerful in Washington, D.C., a very profitable business Morse had tried to enter for some time. He learned of her death on the day of her funeral.

While Morse was still thinking, in 1831 Joseph Henry set up a one-mile-long electromagnetic telegraph in Albany, New York. Henry's device used clicks and bell rings to show that a sender was present, but he never filed for a patent or explored more practical uses for his device. In fact, Henry never even bothered to encode the signals sent by his device. Remote bell ringing remained its sole accomplishment. Henry seems to have regarded it as little more than a parlor trick.

Although Morse's career still focused on painting, he was always exploring new ideas, not only related to electricity and communications. During three years in Europe, starting in 1829, Morse not only studied painting in France and Italy but also had a scheme to repaint miniatures of the art treasures of the Louvre for United States audiences. Morse also experimented with a sculpture-copying marble cutter and had actually sold a new type of water pump in 1817 to the local fire brigade. But much of what Morse pursued seems to have come to naught as he bounced from one great plan to another.

However, Morse's thoughts on telegraph systems became clearer and more determined in 1832 when he was returning from Europe. The shipboard talk revolved more and more around communication by means of electric signals over wire, and Morse was inspired to pursue his ideas in a more systematic fashion. Morse spoke to shipmate Dr. Charles Jackson from Boston, who even had brought some electromagnetic trinkets along to delight the passengers. The questions of how far? and how fast? came up. Jackson suggested that the speed of the electrical signal was instantaneous (the speed was close to the speed of light, so for all terrestrial intents and purposes, electricity was instantaneous) and no one really knew the longest distance practical (also true enough). But wire by this time had improved in structure and electrical performance.

Morse decided to give this idea of fast remote communications another shot. By the time the ship docked, Morse had filled his notebook with ideas for a simple coding scheme ("bi-signals," which evolved into dots and dashes), a code book (numeric bi-signals were the index), and a way to make a record of the message (pencil marks on moving paper, suggested Jackson).

By 1837, Morse's notebook doodle ideas had reached the patentable stage. He had strung 700 feet of wire around his room at New York University (where he lived in a dorm). That same year, he staged the first public demonstration of his device. But longer wires were still a problem.

Meanwhile in England, scientist Charles Wheatstone and anatomy student William Cooke had combined their independent efforts and set up a British electric telegraph. Cooke had invited the more scientifically inclined Wheatstone into his telegraph venture. With Joseph Henry as a visitor, the two Englishmen developed what came to be known as the "ABC telegraph." It employed five "needles" to point to letters and numbers on cards under the needles. This synchronized rotation limited the speed at which the device could operate, and this soon became the most serious obstacle to large-scale deployment.

Morse also had grappled with this problem in the United States. His associate and assistant, Alfred Vail, soon hit on an ideal solution. Discarding the idea of transferring coded letters that had to be looked up in voluminous code books, Vail proposed sending simple text by means of dots and dashes, where a dash was three times the duration of a dot. A spool of paper at the receiving end printed out the dots and dashes as they were sent. This Morse code telegraph seriously challenged the British five-needle telegraphic system in terms of both speed and ease of use. By 1845, as the academic Wheatstone and the amateur Cooke quarreled over who had invented what, the American Morse code became the de facto universal standard for electrical telegraph systems.

Having now demonstrated a practical digital communications system, where the dots and dashes could be easily thought of as the 0's and 1's of modern binary codes, Morse took the next logical step. He approached the federal government, naturally assuming that such a system of nearly instantaneous communication would be of interest to a country that was rapidly expanding to continental dimensions. He was right.

This meant that the electrical telegraph had to work over long distances. Ironically, Joseph Henry in Albany had already solved the problem. He had found that instead of using a single larger battery to try to force the current over longer distances, it was actually better to use a series of small batteries. To go farther, just add more batteries to the series. Henry was using more than 1,000 feet of wire in his doorbell toy. Henry's work came to the direct attention of Wheatstone and Cooke (Wheatstone knew of Henry's work), and to the attention of Morse through a fellow NYU professor, chemist Leonard Gale.

In 1838, Morse demonstrated a working telegraph to a congressional committee in the Capitol building in Washington, D.C. By this time, the telegraph was working over 10 miles of wire filling the room. In March 1843, after some delay and bickering, including seriously proposed amendments for funding to study hypnotism and the possibility of the end of the world in 1844, Congress approved $30,000 for a telegraph line to be run between Washington, D.C., and Baltimore, Maryland. This 40-mile run would be a true test of the technology's capabilities.

The project did not proceed smoothly. Morse tried to lay the insulated telegraph wire inside a pipe run alongside the right-of-way of the Baltimore and Ohio railroad, but the spring rains and defective wire conspired to end the project with only a few miles installed. Down to his last $7,000, and aware that the future of his invention depended on the success of the project, Morse reluctantly pulled up the wire and ran it on poles alongside the tracks instead.

Even before the project was completed, Morse had the opportunity to demonstrate the telegraph's "killer application," a use so unique that the telegraph became a requirement and not merely an expensive luxury to most Americans. In the spring of 1844, both major political parties in the United States at the time, the Whigs and the Democrats, had their nominating conventions for the presidential elections that year in Baltimore. When the Whigs gathered in May, the telegraph had only reached Annapolis Junction, about halfway from Washington. Morse, desperate to show the utility of his invention, arranged for Alfred Vail to meet the train from Baltimore back to Washington there. As the train passed by, Vail asked who had been nominated. "Clay and Frelinghuysen!" one of the delegates shouted back.

Henry Clay for president was no surprise, but the nomination of Theodore Frelinghuysen for vice president was a real shock. Vail tapped out the news to Morse at the Capitol. When the delegates arrived with their astonishing news, they were the ones that were surprised instead. Everyone had already heard the news. It was a brilliant maneuver. This little application of "electronic mail" made a lasting impression on many influential people.

After this demonstration, things began to happen quickly. The first official "telegram" was sent on May 24, 1844, between Vail at the Baltimore and Ohio Mount Clare railroad station and Morse in the Supreme Court chamber of the Capitol. The famous message "What hath God wrought?" was not a Morse inspiration. To prevent possible collusion, the assembled dignitaries in Washington decided to go along with Morse's suggestion that a spur-of-the-moment message be sent and returned. The expression was selected by Annie Ellsworth, daughter of a government official who was a longtime friend of Morse. Vail immediately echoed it back, and the witnesses cheered.

If there were any doubts as to the utility of the telegraph, they were laid to rest when the Democrats met in Baltimore that summer. This time, the vice-presidential nominee was not even in Baltimore but remained in Washington. When the telegraph clicked out that Senator Silas Wright was to be the running mate of James K. Polk, Wright astonished one and all by refusing. The telegraph was still so new that when word of Wright's decision was telegraphed back to Baltimore, the convention halted while a delegation was dispatched by train to verify the message's content. No one doubted the capabilities of the telegraph after that. The two national conventions of 1844 became the defining moment of the telegraph as a transforming technology. After this, things would never be the same again. An early telegraph is shown in Figure 1-3.

So far, Morse's telegraph had proved to be both a viable technology and socially acceptable. The question of economic viability, however, was still an issue. The Baltimore-Washington line was paid for, true enough, and the batteries added only minimally to operating expenses. The main expense was in personnel to operate the telegraph equipment at each end.

Morse tried to sell the whole thing, including his patents, to the government for a fixed price of $100,000, but the deal fell through. The Washington end of the line was moved to the city post office building, and it operated for free to the public for almost a year. Finally, on April 1, 1845, a charge of one cent for every four characters sent was imposed by Morse's Magnetic Telegraph Company. The public, which had been playing chess games over the wire, immediately found alternate means of communication adequate. Total revenue for the first four business days was one cent. One person came in and just offered a penny to see the system work. "You get four letters," he was told. He just shrugged and told the operator to make something up.

But business picked up, and by the end of April, the telegraph had taken in a sum total of $21.23. At that rate, the original investment would be recovered in 118 years. In fact, by

Figure 1-3. The early Morse telegraph

October, the total revenue of $413 barely cut into the $3,925 of operating expenses. Cave Johnson, the postmaster general, in his annual report, was under the impression that whatever the rate, the telegraph system would never be profitable.

Morse was not discouraged. By May 1845, he had extended the line to Philadelphia. It cost about $50 per mile to build a telegraph line, so expanding service was not an enormous financial burden. Other telegraph companies even sprung up, linking Buffalo, New York, and Boston, Massachusetts, by 1846. In 1847, the wires stretched all the way to Missouri. New York and Chicago were linked the same year. Newspapers in both cities decided to use the line to share information, and so the Associated Press was formed. Many newspapers took on the designation *Daily Telegraph* to emphasize the origin of their news. Business use of the telegraph far outstripped personal use. People still wrote and sent letters for much less than the cost of a telegram.

Rates in the United States remained based on message size. In England, by contrast, two networks had been set up by September 1847. The rate structure of the Electric Telegraph Company was based on distance, and this proved too expensive for most potential customers. By 1850, a maximum rate of 10 shillings was imposed, and this was dropped to 2 shillings by 1860.

Telegraph message transfer remained slow, mostly due to the laborious task of interpreting the paper-tape dots and dashes into letters and words. But in 1848, a 15-year-old boy in Louisville, Kentucky, became a celebrity of sorts when he demonstrated the odd ability to interpret Morse code directly by ear alone. Soon this became common, and speeds of 25 to 30 words per minute were achievable. Figuring a five-character word, this rate of almost 20 "bits" per second is impressive for its day. It compares very favorably with the 2.67-bits-per-second rate of optical telegraphs. By 1858, newer mechanical senders and receivers boosted the rate on the telegraph lines up to 267 bits per second.

Data compression was used on the telegraph lines as well. There was no systematic code use, but an ad hoc abbreviated writing taken from the newspaper industry was widely used. Known as *Phillips code,* after the Associated Press's Walter P. Phillips, it enabled operators to tap out "Wr u ben?" for "Where have you been?" and even "gx" for "great excitement." The code was only used internally, and customers were still charged by the full words they used.

There were exceptions to the basic rates, especially on "special" lines. When the first transcontinental link was opened in 1861 between New York and San Francisco (on Telegraph Hill, of course), the government subsidized the link to keep the cost down to $3 for 10 words. But the company immediately charged $1 a word, claiming cost overruns.

By 1873, Western Union, an amalgam of many smaller telegraph companies, was handling 90 percent of all telegrams. Their assets were close to $40 million, an enormous amount for the day. Western Union had increased its capital by 11,000 percent in the preceding 10 years. They were adding about 19,000 miles of wire per year to the network, at a cost now of between $100 and $150 per mile. So in many ways, the analogy of the telegraph as an early form of Internet is a very good one.

Like the Internet, the telegraph spawned whole new ways of doing business as well. In 1886, a young telegraph operator found himself in possession of a shipment of watches refused by a local jeweler. Working with his telegraph, he soon sold them all to his fellow operators and railroad employees. In six months, he had made $5,000, quit his job, and founded the company that eventually became Sears, Roebuck and Company. So e-commerce was born with the telegraph as well.

A lot has been made recently of the telegraph as a kind of "Victorian Internet." A very successful book by Tom Standage bearing that title appeared in 1998, although the analogy had been made before. The story told is fascinating, and it is refreshing to see such a distinctly American invention as the Morse telegraph (the vast United States territories even in the 1800s made other forms of communications painfully slow) placed in a British and global context. Oddly, Morse tried in vain to patent the telegraph in Britain, where opposition from native sons Cooke and Wheatstone proved insurmountable. The Cooke and Wheatstone "ABC telegraph" was indisputably an electrical telegraph, but it used ten wires and five rotating needles to point at letters, a slow and costly arrangement eventually abandoned. The system had other quirks, such as no way to indicate the letter "Q" ("KW" was usually substituted). The fact that Cooke and Wheatstone barely got along did not help the situation. The British telegraph system also employed ideas that were quickly abandoned or never tried in the United States, such as a distance-sensitive charge for messages. Even in much smaller England, where a 10-shilling maximum message charge in 1850 dropped to a 2-shilling maximum by 1860, messages were too expensive to send over long distances with any billing system other than message length.

There is no doubt that the telegraph was treated as a form of *transforming technology:* one that showed life would and could never be quite the same again. By 1865 the telegraph had been run across entire continents and beneath oceans. Romance flourished on the wires, but so did crime and fraud. Codes were added for security, and the first hackers and crackers appeared. The new networks had their unabashed supporters and unflagging critics. Overall the people loved the new telegraph technology. There were local celebrations whenever a telegraph office opened in town. The parallels between telegraph and Internet are astonishing.

But the telegraph had a window of only thirty years or so to flourish before a newer, more exciting invention came along and swept the public imagination. In many ways, this wave of new development also mimics the situation in modern times. This new technology was the telephone.

ELECTRICAL TELEPHONE TRANSMISSION

Even as the telegraph was becoming an indispensable part of all citizens' lives, the public began to look beyond the telegraph. If a telegraph could send messages by clicks and taps, which were just sounds after all, why not just send the whole voice? As early as 1854, suggestions were offered to make this possible. Perhaps silver and zinc plates could be held in the mouth and hooked up to a telegraph line. Then the receiver might produce a voice instead of dots and dashes.

The excitement in the public imagination was illustrated in 1866, when a series of articles chronicling the exploits of one Joshua Coppersmith appeared in many New York City newspapers. Supposedly, this gentleman was convicted of fraud for exhibiting a device he called a "telephone," which adapted the principles of the electric telegraph to send the human voice through a wire. The problem was that no one named Joshua Coppersmith was ever proved to have existed, and no record of such a court action was ever found. This whole incident was a legend, but one that served the same function as legends today: It outlined a story that, even if not true, people firmly believed and desired to be true.

Thus the idea of voice communication was established long before a young teacher of the deaf named Alexander Graham Bell sat down in an office in Boston and tried to solve a problem that had been vexing the telegraph for years. The problem was that telegraph lines were inherently unchannelized. That is, only one telegraph message could be transmitted at a time. The others had to wait until the operator finished tapping out the current message. If some way could be found to multiplex the telegraph signals on the wire, then many messages could be sent simultaneously. This would make more effective use of the single strand of telegraph wire, and short messages would not have to be delayed until longer ones were finished.

This problem of "serial delay" with short but important messages held up by longer but less vital ones still plagues all forms of serial communication. Many ways to divide the total line speed into separate frequencies and time slots have been used. In a sense, the dense wavelength division multiplexing (DWDM) used in optical networking is just an application of frequency division multiplexing (FDM) applied to light waves.

In spite of the promise of multiplexed telegraph messages, Bell failed in his attempt. He tried tuning forks, which seemed promising, but in order for this plan to work for Morse Code, the bandwidth available on the wire had to be greatly expanded in order for the tuning forks to properly respond to the different frequencies. Bell's efforts to expand the bandwidth on the telegraph wire led him to accidentally tighten a single screw too far. This caused electricity to be present on the wire all the time, not just during a pulse for a dot or dash. This made all the difference, and the apparatus started to pick up voice, much to the delight of Bell and his assistant Watson. So when Bell also accidentally spilled sulfuric battery acid on his lap (Bell seems somewhat clumsy by modern laboratory standards), Watson heard his typically restrained cry of "Watson, come here, I want you!" (Few modern inventors could or would be as calm with a lap full of sulfuric acid.)

But Bell's invention of the "harmonic telegraph," as the 1876 patent called it, hardly measured up to Joshua Coppersmith's urban legend. Only Bell could speak loudly enough be heard at any distance, and only the enthusiasm of Emperor Dom Pedro of Brazil at Bell's exhibit at the Centennial Exposition in Philadelphia that July convinced Bell's backers to continue their financial support. (Legend has it that the demonstration was originally set up for a famous soldier named George Armstrong Custer. But Col. Custer had one last stand to make and missed the steamboat back east.) Better transmitters, the best one built by Thomas Edison, soon helped the volume problem.

Very early in the history of the telephone, in 1877, Bell's backers, trying to make good on their investment, offered the Bell telephone to the Western Union Telegraph Company for $100,000. The Western Union president refused. Later that year, Bell replaced his Patent Association with his own Bell Telephone Company. By 1878, the whole system had grown to 1,000 phones, and the first switch (switchboard) was installed in Hartford, Connecticut.

Connections were made by hand by the switchboard operators. Young men, mostly college students, were used as operators until their tobacco-chewing, foul-mouthed ways and relaxed work ethic forced their replacement by milder young women. For years, the only respectable work a young, unmarried woman without a college education could find was at the local Bell System office. In the switching office, connections were requested by the subscribers, and by name. It soon grew difficult to keep track of the growing client list, so the connectors on the operators' switchboards were numbered sequentially. At first some customers resisted, but once the advantage of fewer wrong connections became obvious, telephone numbers were accepted by the public in 1879.

The economics of the telephone worked differently than for the telegraph. The telegraph was never intended by its developers to extend to residential use. Everyone could not be expected to learn Morse Code, and there was no real equivalent of the switchboard in the telegraph world. There was no easy way to take a telegraph message from one source and send it out to one of several possible destinations. This message distribution was all done by manually relaying messages through the telegraph system.

Everyone could talk, however. Bell at first targeted the telephone for business use. He saw it as an easy way for retailers to contact suppliers, much as the telegraph was used. (The general public wrote letters to communicate. The telegraph was only needed for situations of a particular urgency. In those days, most areas had two mail deliveries per day, one in the morning and one again in the afternoon, just in case someone had written a letter in the meantime.) But others saw much more potential for the telephone than just for businesses.

Telephones were not sold to users, since they were useless without the network. Bell's company leased the telephones to users. They would have otherwise been prohibitively expensive, it was felt, and this emphasized the company's "service only" position. The charge was set at $40 per year for business use and $20 per year for home. The distinction was based on anticipated utility and ability to pay. A residence had few users, but a business had many more potential users of the telephone.

At first the telephone hardly seemed much of a threat at all to the telegraph. In general, only the well-to-do and technically curious, including Mark Twain of Hartford, bought telephone service, and it still seemed that most communications situations were best handled by going down to the local Western Union office and sending a telegram. At the other end, local messengers rushed off to the home of the recipient, usually on bicycles, another new form of technology.

The real potential for the telephone grew apparent in January 1878. A serious train wreck occurred near Tariffville, Connecticut, just outside Hartford. The local druggist had phone lines available directly to all the local doctors, in the business use model that Bell had championed. In this case, the druggist was able to summon all the doctors to the crash site almost immediately. The obvious advantage of direct voice communication over a messaging service was dramatically pointed out.

Some 50,000 telephones were in use in the United States by 1880, although nagging technical problems threatened to limit telephone use solely to local areas. Early telephone systems used rented telegraph wires. These were of a single strand of copper or iron wire with what was known as a *ground return*. While quite efficient for the intermittent dots and dashes of the telegraph, these circuits formed long antennas when converted to telephone use. Every rumble of thunder within 30 miles caused the voice on the circuit to dissolve in a burst of static. When adjacent wires were chattering with telegraph signals, the phone users heard this too, through a phenomenon known as *cross talk* caused by electromagnetic induction. The closer the wires, the worse the problem. When telephone wires were run across the new Brooklyn Bridge in New York, the cross talk drowned out the intended conversation.

The static problem was solved in 1881 when a technician named John Carty, exasperated by the poor sound quality on a wire running the 45 miles from Providence to Boston, accidentally hooked up two wires between the telephones. By not disconnecting the first telegraph wire before trying the second to check for a better voice connection, Carty invented the *isolated metallic ground* circuit that dramatically reduced noise.

Another real problem was the use of iron wire in telephone circuits. Much cheaper and stronger than copper wire, iron wire was used extensively in the telegraph network. Although quickly covered with a layer of rust, "open" (uninsulated) iron wire remained the medium of choice for many telegraph lines. However, copper wire proved to be much better for two-wire "paired" voice circuits. Fortunately, in 1884, a new method of hard-drawing copper resulted in copper wire that was as strong as iron and yet cheap enough to string for voice circuits. The use of copper soon made longer runs, such as from New York to Boston (292 miles), possible.

The cross talk problem was solved not long afterward. It was discovered that twisting the two wires in a pair virtually eliminated the annoying cross talk among closely packed wires. The newer twisted pair wire made cables (specially constructed packages with many wire pairs) possible. And cables made deployment of multiple-voice paths more efficient and therefore less expensive. Cables were more protected from environmental damage as well.

By 1890, the number of telephones had grown to 250,000. New York and Chicago were linked by 1892. Nebraska, Minnesota, and Texas had service by 1897. The nation was crossed in 1915. Once again, as with the telegraph, San Francisco and New York were linked. Telephone use exploded with the expiration of Bell's original patents in the early 1900s. Now anyone could build and operate a phone company, and many people did. The telephone had arrived as a technology.

The first calls from New York to San Francisco took 23 minutes to switch by hand from switchboard to switchboard and cost $20.70. It seems expensive, even now, and it was. The Bell system really had little idea how much to charge for phone service, and local rates varied widely. At late as 1923, the Bell system followed 206 different local rate structures, with some places being charged three times what others were for the same basic service. Rates were set according to the community's ability to pay, for historical reasons, and sometimes out of simple greed. Candidly, many executives admitted that telephone service was overpriced, but the vast profits did fuel further network expansion and research into improved equipment.

The technology of the telephone network continued to advance. Dial phones replaced hand cranks in the 1920s, and switches became automated to break strikes by the unionized operators. A shortage of readily available replacement workers during World War I had already accelerated the pace of mechanical switch deployment. An early switchboard is shown in Figure 1-4.

By the 1930s, the situation of the telegraph and telephone companies had completely reversed itself. The national network was now based on the telephone transmission and switching system instead of the telegraph. Ownership passed to telephone operating

Figure 1-4.   An early telephone switchboard

companies, and instead of phones over rented telegraph lines, telegrams now rode over rented voice lines from the phone companies.

The national network was now analog, not digital, as it was with the telegraph network. Analog signals vary continuously between maximum and minimum. A receiver must distinguish not just between a 0 (dot) and a 1 (dash) but between closely spaced values such as 0.31245 and 0.31246 (for example).

CARRIER SYSTEMS

The preceding sections, as fascinating as they may or may not be historically speaking, have real importance when it comes to optical networking and DWDM. It is always good to realize that information, be it text or voice, can be carried across a network in two major forms. The information can be represented by a line code (sometimes called *signaling*) that is analog or digital. Text was easy to express as digital 0's and 1's. Voice was easier to express as continuously varying tones or sounds. So the telegraph network was digital and the voice network was analog.

Efforts were made in the 1960s and 1970s to digitize voice, mainly for sound quality reasons ("you can hear a pin drop"). Digital transmission systems use 0's and 1's, employing *regenerative repeaters* to "clean up" a weakened (attenuated) digital signal on long links. The repeaters perform what are known as the "3 R's": the *restoring* (strengthening) of the weakened pulses, the *reshaping* of the signal, and *retiming* of the signal to enable downstream receivers to more easily distinguish 0's and 1's. Analog transmission systems make use of *amplifiers,* which strengthen the received signal but cannot distinguish between noise and the original signal strength (since all allowed signal levels are valid in analog systems). Repeaters can "clean up" a digital line and remove noise. Amplifiers on analog links will amplify noise right along with the original signal. With amplifiers, noise is *cumulative.*

These efforts at voice digitization were successful, if expensive. But optical networks have much more in common with analog transmission systems. That is, the optical links are mainly analog line coding and employ amplifiers in which noise will accumulate. Analog links multiplex most efficiently using frequency division multiplexing (FDM), while digital links multiplex most efficiently using time division multiplexing (TDM) and time slots. Oddly enough, wavelength division multiplexing (WDM) is a kind of blend of these two methods.

This section will briefly outline some of the analog carrier (multiplexing) systems used on the analog telephone network. There are two reasons to explore analog transmission systems a little more thoroughly as an introduction to optical network multiplexing. First is the reason just mentioned: Optical networking has much in common with analog carrier systems. Also, the emphasis on digital telephony transmission in the past 30 years or so has made this information rather hard to find.

Analog Carrier Systems

All of the voice circuits in the global public switched telephone network (PSTN) typically originated and terminated at individual access lines. At the end of the access lines there was usually only one telephone handset. There was only one telephone because, until the late 1960s, the telephone itself could not belong to the customer. The telephone was telephone company property and leased on a monthly basis from the local telephone company. If anyone wanted a second telephone, the second telephone had to be paid for on a monthly basis also (people cheated all the time). And recurring costs always dominate over time compared to one-time, nonrecurring charges.

In any case, only one telephone call at a time could be carried on one twisted pair access line, this being the essence of circuit-switching. But once the voice circuit hit the central office, the usual term for the local switching telephone office, it made no sense to have a separate wire pair for every voice call that the central office was designed to carry. So the calls in progress were routinely multiplexed onto higher-capacity trunks. Trunks are just transmission lines that carry more than one voice cell. Since FDM was more efficient for analog voice conversations, the voice channels were combined into an FDM analog multiplexing hierarchy. Lower levels of the hierarchy used multiple pairs of ordinary twisted pair copper wire, while higher levels used coaxial cable or microwave towers to provide the huge bandwidths that these trunking carrier systems required.

The role that trunking plays in the telephone network is shown in Figure 1-5.

This section describes the analog multiplexing systems used in the Bell System. Other telephone companies could have their own methods, but most followed the Bell hierarchy, simply because what the Bell System used became a standard way of doing things. This assured that the equipment used was in plentiful supply as well.

The analog multiplexing hierarchy began with N-carrier, a multipair system for local trunking needs. The N1 (introduced in 1950) and N2 equipment multiplexed 12 voice channels onto two pairs of twisted pair wire identical to the wires used on the local loop, or the access line used to carry one conversation from customer to central office. N-carrier links spanned anywhere from 5 to 250 miles, but most were used to link a central office to its nearby neighbors, or to the nearest long distance office. Later, N3 in 1964 and N4 in 1979 doubled the capacity to 24 voice channels while at the same time taking up less space and consuming less power, although everything was still analog.

There were also two distinct short-haul microwave multiplexing systems. Microwave systems are line-of-sight, radio frequency transmission systems that require repeater towers placed along the path of the link. Short-haul microwave systems were commonly used to aggregate a lot of voice traffic between tandems or toll offices where running landline wire would be very expensive, such as across major rivers. In spite of the *short-haul* name, which applied to both the primary use and the fact that the repeater towers were often only 5 miles or so apart, these systems could span 10 repeater hops and have spacing up to 25 miles apart. So a "short-haul" microwave system could be anywhere from 5 to 250 miles long.

These 6 GHz systems included the TM-2, which carried 8,400 voice channels, and the TM-2A, which carried 12,600 voice channels. This reflects the enormous bandwidths available on these microwave systems. The 11 GHz systems included the TL-A2, which carried 6,000 voice channels, and the TN-1, which could have either 19,800 or 25,200 voice channels. By the early 1980s, deployment of both 11 GHz systems had ceased because the bandwidth consumed was far more than actually needed for the number of voice channels carried.

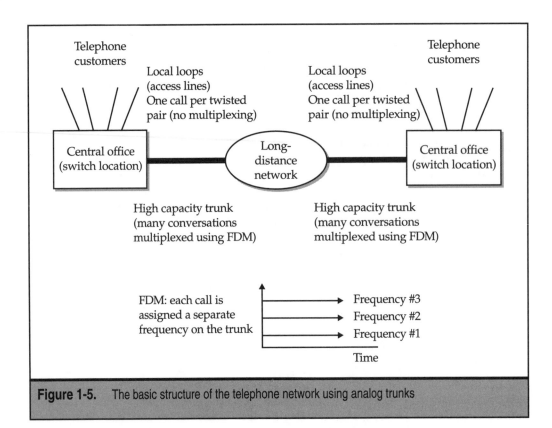

Figure 1-5. The basic structure of the telephone network using analog trunks

Most notable for the purposes of this section were the landline analog multiplexing methods based on coaxial cable, most notably the L-carrier system. The coaxial cable was similar to the cable used today in cable TV systems and Ethernet LANs, but quite different electrically and in terms of frequencies used. L-carrier dates all the way back to 1941, but it was in 1946 that deployment began in earnest. In L-carrier, several coaxial cables could be grouped together as what were called "tubes" to multiplex even greater numbers of voice channels together. The early L1 system carried only 1,800 voice channels, but the L5E system of 1978 carried 132,000 voice channels on 22 "tubes." Ten were for the voice channels in each direction, and two cables were for protection purposes. So each coaxial cable carried 13,200 voice channels. L-carrier links could be as short as one mile or as long as 150 miles.

Table 1-1 shows the L-carrier family in terms of date introduced into service and total number of voice channels in a single cable sheath. Each cable sheath could have many coaxial cables inside. Note the absence of an L2 designation.

For international voice communications, undersea coaxial cables were used. This family began with 24 voice channels on two coaxial cables from Key West, Florida, to Havana, Cuba, in 1950. Transatlantic cables from the United States to Europe were installed every six

L-Carrier Level	Year Placed in Service	Number of Voice Channels
L1	1946 (1941)	1,800
L3	1953	9,300
L4	1967	32,400
L5	1974	108,000
L5E	1978	132,000

Table 1-1. The Analog L-Carrier Family

years or so to keep up with the growing voice traffic between the United States and Europe. By 1976, analog coaxial cables could carry 4,000 voice channels under the North Atlantic.

The next major category of analog carrier system is long-haul analog microwave. Again, the name is more intended to describe the purpose of the system, not the length of the spans, although few long-haul microwave repeaters were less than 20 miles apart, and 30 miles was more normal. There were 4 GHz systems (the TD family) and 6 GHz systems (the TH family and AR6A) developed between 1950 and 1983. The TD systems carried from 2,400 to 19,800 voice channels, depending on version. The TH-1 system carried 10,800 voice channels, the TH-2 system had 12,600 voice channels, the TH-3 system had 16,800 voice channels, and the AR6A had 42,000 voice channels.

Finally, satellite communications were often used instead of undersea coaxial cable for international communications. Satellite systems basically point huge microwave dishes at the sky. The proof that this would work was provided by Echo I in 1961. Echo I was a big, shiny balloon inflated in orbit and used to bounce signals uplink and downlink between earth stations. In 1962, Telstar carried all of 12 voice channels in a low earth orbit, but at least this satellite was an active system that received and retransmitted the voice channels through a system of *transponders*. Telstar moved in the sky also, limiting its use for long telephone calls, unlike today's *geosynchronous* satellites, which are stationary over the earth's equator. Today, of course, international and private satellites carry 21,000 voice channels or more as well as television channels.

These descriptions have been of the *physical* multiplexing systems. But for multiplexing purposes, the analog hierarchy is divided into conceptual levels that are observed regardless of the physical medium and multiplexing capacity that the conceptual level is deployed on. This is a big help in international trunking situations where different national carrier systems must interconnect.

The basic unit of analog voice multiplexing is the *group* of 12 voice channels. The ITU-T defines ten groups as making up a *supergroup* of 120 voice channels, but in the United States a supergroup has only 60 voice channels. Five supergroups make up a *mastergroup* of 600 voice channels. In the United States, the next level is the *jumbogroup* of

Name	VCs	U.S. Equipment Level	ITU-T Equipment Level
Group	12	N-Carrier	
U.S. Supergroup	60		
ITU-T Supergroup	120		
Mastergroup	600	L1	G.341 (5 supergroups)
	1,200	TD	
ITU-T Supermastergroup	1,800	TH/L3	G.343 (15 supergroups)
	2,400		G.344 (20 supergroups)
U.S. Jumbogroup	3,600	L4	
	5,400		G.332/G.345 (45 supergroups)
	6,000	AR6A	
	7,200		G.344/G.346 (60 supergroups)
	10,800	L5	
	13,200	L5E	
	21,600		G.333 (180 supergroups)

Table 1-2. The Analog Group Hierarchy

3,600 voice channels (6 mastergroups), while internationally the highest level is called the supermastergroup of 1,800 voice channels (3 mastergroups). So no matter how the voice channels are physically carried, they are always organized into these logical groupings. And although the United States and international versions don't quite match up, the groupings are much the same. Table 1-2 shows the levels of the analog hierarchy.

The term "VC" stands for the number of voice channels supported at each level. Note that in actual practice the levels of the group hierarchy are "mapped" to the actual physical multiplexing carrier systems. In other words, no real multiplexing hardware carries a single 60–voice channel supergroup on coaxial cable, microwave, or anything else. Supergroups are called *intermediate multiplex levels* because of their conceptual existence, and hardware-based multiplexing levels are called *system levels*. Not very many levels even carry the same number of voice channels, making international connectivity that much more interesting.

Digital Carrier Systems

Although optical networking has much in common with analog carrier systems, a brief historical introduction to digital carrier systems is in order. This is because in many cases the new optical networks are intended mainly to carry existing digital transmission systems' traffic, and sometimes they carry the digital signal itself transparently across the fiber link.

Analog multiplexing in the PSTN is still around, if only because regulated telephone companies are only allowed to replace a certain amount of their equipment per year and the useful lifetime in any case is generally set at anywhere from thirty to forty years. So analog multiplexing systems installed in 1980 could be around for some time to come.

Most analog carrier and multiplexing systems have been converted to digital multiplexing systems, and all new multiplexing systems installed in the PSTN carry digital signals. Digital multiplexing has been around since 1962, with the introduction by AT&T of the highly successful T1 multiplexing system. T1 is the first level of what became a whole digital hierarchy that multiplexes voice channels not as analog voice passbands at 4 KHz using frequency division multiplexing (FDM) but as digitized voice channels at 64 Kbps using time division multiplexing (TDM). Both are still circuit-switching "all the bandwidth all the time" systems, but the basic unit of analog multiplexing is the voice channel and the basic unit of digital multiplexing is 64 Kbps serial bit streams known as *DS-0s*. Digital multiplexing terminals (transmitters and receivers) cost less than their analog counterparts even in 1962 and had better performance characteristics (i.e. the voice sounded better to human listeners) than analog systems that amplified noise right along with the voice signal. As installation costs for wire pairs increased and the cost of digital multiplexing equipment fell with the price of digital components, T1 became cost effective at shorter and shorter distances compared to analog trunks.

Technically, T1 is a *physical* trunk network consisting of two pairs of unshielded twisted pair copper wire, transmitter and receiver specifications, and repeaters usually spaced every 6,000 feet or so in between. The organization of the bits sent on a physical T1 system is defined as the *DS-1 frame structure*. A DS-1 carries twenty-four 64 Kbps DS-0s on the two pairs of copper wire. This gives a T1 digital transfer rate of 1.536 Mbps (24 × 64 Kbps), and there are 8 Kbps of overhead for a total line rate on a T1 of 1.544 Mbps, usually rounded off to 1.5 Mbps.

Another way of looking at the T1 line rate is to realize that 24 digital voice channels generate 8 bits and are sampled 8,000 times per second (this is where the 64 Kbps comes from). The 24 channels therefore pack 192 bits (24 × 8 bits) in each T1 frame. A 193rd bit known as the *framing bit* is added to each basic frame to allow receivers to synchronize on the incoming frame boundaries. Since there are 8,000 samples per second in each digital voice channel, the T1 link must operate at 1.536 Mbps (192 bits × 8,000 per second) or actually 1.544 Mbps (193 × 8,000 per second) just to keep up.

Each 193-bit T1 frame is sent in 1/8,000th of a second, or every 125 microseconds. This is below the threshold of human perception, and much faster than the blink of an eye, which takes place in 1/10th second or 100 milliseconds or 100,000 microseconds. Oddly enough, golfers might actually have the best perspective on T1 frame speeds. The brief "click" sound of the driver club hitting the golf ball, a sound at about the limit of human perception, is about 1/2 of a millisecond long, or 500 microseconds. So during that click, a T1 transmitter has sent four frames out on a T1 link.

A T1 link can be used to carry digitized voice channels, or serial bit streams from a computer in any combination. For instance, a T1 multiplexer can carry 12 voice conversations from a corporate PBX to another PBX at another site within the same corporation and at the same time link 12 routers throughout the building to their counterparts at the other site at 64 Kbps.

Although voice channels traverse a T1 link at 64 Kbps for the most part, serial data is often limited to 56 Kbps, or 7/8ths of the 64 Kbps DS-0 channel rate. This is because sometimes the eighth bit of each DS-0 channel is used by the T1 equipment for functions other than bit transfer. For voice applications, this intermittent "bit robbing" is not audible, but of course computers rely on all the bits they send in a DS-0 channel to be delivered correctly. If bits are "robbed," the true content of the eighth bit is obliterated and cannot be restored. Now, not every eighth bit of every channel in every T1 frame is robbed, but since only the carrier knows for sure when the bit is used or "robbed," the customer has no choice but to avoid using the eighth bit in each DS-0 channel altogether.

There are actually four uses for this eighth "robbed" bit. However, one of these uses is only required when the DS-0 channel is used for voice trunks ("robbed bit signaling") and so does not apply when the DS-0 channel carries serial computer bits. The other three uses of the eighth bit are:

1. To maintain *1s density* on the line. This makes sure that there are enough 1 bits appearing on the link.

2. To signal *yellow alarm* to a transmitter. This indicates a problem at the receiver.

3. For internal carrier *maintenance* messages.

Each use should be explained in a few words. The T1 line code, bipolar AMI, represents 0 bits by no voltage at all on the line. The term AMI means Alternate Mark Inversion and just means that successive 1 bits are represented by alternating voltages. If too many 0 bits in a row appear, as might be the case when adjacent DS-0 channels carry computer bits, the receivers can lose *frame synchronization* ("was that 30 bits of nothing, or only 29? Or maybe 31?"). The 1 bits, on the other hand, are represented by voltage pulses. So when a T1 using pure AMI encounters 16 consecutive 0 bits to be sent on adjacent DS-0 channels, the equipment flips one of the 0 bits (the eighth bit in the first channel) to a 1 bit to maintain a minimal *1s density* on the line. Yellow alarm occurs when a T1 receiver loses the incoming signal entirely. Bits can still be sent, perhaps, but there is no sense in continuing operation if one pair is no longer functional. So yellow alarm sets every eighth bit in every channel upstream to a 1 bit in order to tell the sender to stop sending downstream (and potentially switch to a backup link if available). So computer bits that otherwise would arrive okay are obliterated by the equipment function. Finally, technicians can use the eighth bit in the channels to signal T1 equipment to perform diagnostics, gather network management information, and the like. None of these methods are used to penalize data users, of course. They just began to be used when T1 was exclusively a voice trunking method and continued to be used when T1 began to be sold for data purposes in the early 1980s.

If a T1 link allows data users to employ the full 64 Kbps DS-0 channel rate, known as *clear channel capability*, two things must be done by the T1 service provider. First, instead of using straight bipolar AMI, the transmitter, receiver, and all repeaters and other equipment in between must use something called binary 8 zero substitution (B8ZS) line coding. B8ZS uses a special string of 8 bits that is substituted for a string of eight zeros in a row. Sometimes B8ZS is positioned as a distinct line code from AMI, but B8ZS is best thought of as a *feature* of AMI. This solves the 1s density problem. Second, another form of T1 framing known as the T1 extended superframe (ESF) is used to remove yellow alarm and maintenance message from the channels themselves and place them in the 193rd framing bit pattern in 24 consecutive T1 frames.

Strangely enough, the international equivalent of T1, known as E1, never suffered from this 56 Kbps limitation. The E1 frame structure is different, so yellow alarm and mainte-nance messaging never became an issue. A special signaling channel in E1 avoids robbed-bit signaling altogether. The E1 line code is not bipolar AMI, and a special synchronization channel ("word") avoids 1s density concerns, so B8SZ is not required either. All E1s support 64 Kbps transmission, and all circuits that begin in one country (as defined by the United Nations) and terminate in another must use E1 instead of T1. T1 can only be used within the borders of the United States (Canada also uses T1, but also internally).

E1 multiplexes 30, not 24, digital voice channels into the basic E1 frame. Frames are still grouped, but not for the same reason as in T1. E1 channels still run at 64 Kbps but use a slightly different (and incompatible) form of voice digitization. The basic structures of the T1 and E1 frames are shown in Figure 1-6.

With either T1 or E1 the basic unit remains the 64 Kbps DS-0 circuit-switched voice channel. But as the voice traffic loads offered to telephone networks increased, some method had to be found to multiplex more and more digital voice channels between the same pair of voice switches on the same physical facilities, the better to increase capacity on the trunk.

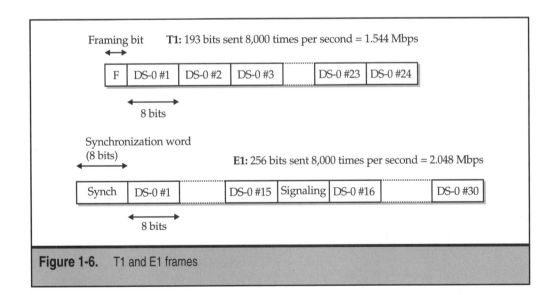

Figure 1-6.   T1 and E1 frames

LOCAL AREA NETWORKS

The emphasis on optical networking in this chapter so far has been on wide area networks (WANs). But optical networking in this volume will also be explored as it applies to local area networks (LANs) as well, in particular Gigabit Ethernet (GBE) and 10 Gigabit Ethernet (10GBE). Optical networking in LANs can be used to ease traffic loads and carry information far beyond the boundaries of early LANs. In the past, speedy LANs were severely limited in the distances they could span, but you can always go farther by going slower.

In the beginning, all networks were wide area networks. Computers were huge, expensive "mainframe" computers, and few organizations could afford more than a handful. Mainframes cost millions and required special environments and cooling to work at all, so mainframes were usually placed inside special computer rooms, many with glass walls, the better to impress potential customers of the organization. Even rich organizations had little need to place more than one computer at a site.

Things changed with the rise of the PC. This step toward individual, or personal, computing began at several research institutions, most notably Xerox's Palo Alto Research Center (PARC) in California. In a frenzy of creativity rarely seen anywhere in history, a few short years in the late 1970s and into the early 1980s saw the birth of the personal computer (PC), "windows" on monitors, graphical user interfaces (GUIs), the mouse, the LAN, and the first network protocols and operating systems designed for LAN instead of WAN use.

Most of these ideas grew out of a previous one in a natural progression. The early PCs, which could be built for about $10,000 by the early 1980s, were as powerful as a mainframe that could run 16 programs all at once. But there was only one monitor to watch all these programs execute. So add a "window" for each one. But how to get simple text-based monitors to display windows of varying sizes? Shrink or expand the graphics (invent the GUI) to fit the window instead of fixing the letter "A" to always be the same size and appearance. But there is only one keyboard. How can the keyboard quickly be associated with one window or another? (Today we say "give the window keyboard focus.") Add a mouse that can be used to point at a window and click to say essentially "I want the keyboard to be associated with *this* window for now." (Special keystroke sequences were tried, but they needed a window of focus of their own, and if anyone terminated the keystroke program, the computer was effectively hung.) PC users were isolated from other users, even though they were only a cubicle away. Invent the LAN to tie them all together!

Ethernet was invented by a team led by Dr. Robert Metcalfe at Xerox PARC in the late 1970s. His task was to link about 100 computers spread around the complex. His development team ultimately came up with a one-kilometer (.625 miles) length of coaxial cable that ran at the considerable speed of 2.94 Mbps. Since many computers needed to share access to this "bus" cable, they came up with a scheme known as Carrier Sense Multiple Access with Collision Detection (CSMA/CD), a way of accessing the cable pioneered by the University of Hawaii for a packet radio network linking the island campuses.

Metcalfe, after considering a number of names such as "Xerox Wire," came up with the name "Ethernet" from the "luminiferous ether" that nineteenth-century scientists believed permeated the universe and allowed light waves to travel in a vacuum. Light waves needed no such ether, it turned out, but Metcalfe's Ethernet was very much tied to the cable needed to carry the 0's and 1's his computers generated.

When the Institute for Electrical and Electronics Engineers (IEEE) began to entertain thoughts of standardizing a way of implementing a local area network, one of the first proposals was made by Digital Equipment Corporation (DEC), Intel, and Xerox. These "DIX" companies had built and used Ethernet as a proprietary LAN method and now approached the IEEE to standardize their method. IBM proposed Token Ring, and General Motors proposed something called "Token Bus" around the same time. The IEEE eventually rubber-stamped Token Ring as 802.5 and Token Bus as 802.4, but for some reason modified Ethernet to the point of incompatibility as 802.3. In fact, the 802.3 committee invented a specific version of Ethernet known as "10Base5" (10 Mbps, baseband signaling, and a 500-meter segment). To this day, when people say "Ethernet" unless they work for DEC or Xerox, they probably mean some form of 802.3 LAN. But according to the IEEE, 802.3 is a committee. People should really say "CSMA/CD" LANs, since this form of operation is something all types of Ethernet employ, even GBE and 10GBE.

CSMA/CD was needed to address the fact that stations on an Ethernet could essentially choose to send as long as another station wasn't already sending (the "carrier sense"). All stations shared the media (the "multiple access"), so there were times when two stations almost simultaneously ("almost" due to the finite speed of electricity in a cable: about 2/3 of the speed of light) started sending. These stations generated a collision. The "collision detection" mechanism meant that an Ethernet station listened to its own transmission. If it heard exactly what it sent, then the transmission was good. If it heard "garbage," or even one bad bit out of thousands, that indicated a collision had occurred and the entire frame had to be retransmitted. The presence of collision effectively limited total Ethernet throughput to a small fraction of the total bandwidth available.

This was not at first a major limitation. When the first 10 Mbps Ethernet LANs began to appear in the early 1980s, what PCs and microprocessors existed were used in an overwhelmingly stand-alone fashion. LANs were used to connect minicomputers for the most part (DEC equipment, mostly), with PCs only participating as terminal emulation devices on the LAN. That is, the PCs became text-based teletype-equivalent terminals that depended on the locally attached minicomputer for all processing. PCs and workstations had their own operating systems and applications, but they were not generally used in the client/server environment initially.

This was probably just as well. Early PCs and even workstations were built around processor chips that were extremely modest in power and speed. Early PC chipsets ran at 8 MHz and executed about 4 million instructions per second (Mips) and had only 64 K of memory. The common rule of thumb is that a microprocessor chip can generate bits for a network interface at about half the rate of the instructions per second it can execute internally. So an 8 MHz machine with 4 Mips of power could generate about 2 Mbps of data for a network, at the peak rate. Usually, the applications that the machine ran had much less of a bandwidth requirement.

So an Ethernet LAN running at 10 Mbps could easily accommodate 10, 50, or 200 of these early PCs when the time came to hook them up to LANs directly. This was done as the applications these PCs and workstations ran easily overtook in power and sophistication the applications that were earlier reserved for the mainframe and minicomputer environment. The need for PCs and workstations to share information and resources (generally, resources are considered physical entities, such as color printers, scanners, and other equipment an organization felt it could not afford to buy for each and every desktop) meant that they came to be deployed directly on the Ethernet and Token Ring LANs that were being built in the mid-1980s.

The PCs and workstations shared the bandwidth provided by the LAN. Since data applications were inherently "bursty," with long periods of silence (called "think time") punctuated by a demand for high speeds and throughput, it was a very good match. Applications got all 10 Mbps of bandwidth on the Ethernet, but when the application was silent, other stations could use it, on a contention basis. LANs could efficiently parcel out the bandwidth available, and PCs and workstations had a very high speed network available whenever they needed it. And since the stations could only generate about 2 Mbps of data in one burst anyway, there was no danger at all that *any* station on the LAN would monopolize the bandwidth or demand higher speeds for any application.

But by the late 1980s there were signs that this situation would not hold true much longer. The newer 386 chipsets ran at 32 MHz and executed about 12 Mips, so they could churn out bits at about 6 Mbps, which exceeded the capacities of at least one of the more common LAN architectures available: Token Ring. When a LAN suddenly became swamped with traffic from these higher-speed machines—and the applications that ran on them—the only choice a network manager had was to split the LAN into two parts. The need for formerly connected LANs to still communicate (that was the whole point in the first place) was handled by adding an internetworking device—a bridge or router—to the now separated LANs. These LANs more and more frequently came to be built around hubs (Ethernet-based—really 802.3 CSMA/CD, the IEEE standard—network devices) instead of Token Ring solutions, mostly due to cost considerations.

By the 1990s, even this concept was running into problems in large LAN situations. The processors were now Pentium-type machines with 66 MHz and 100 Mips. These could pump out 50 Mbps and easily overwhelm not only a 10 Mbps Ethernet, but a 16 Mbps Token Ring. Network managers found that their LANs were starting to show 30 percent utilization during busy hours and 50 percent during peaks. The idea of dividing the LANs up into smaller and smaller units with fewer and fewer stations connected by bridges or routers did not appeal to many of these network managers. This reluctance was mainly due to the uncertainty of the practice (it was not always clear just what the net effect would be), but also because the practice seemed so ad hoc.

So pressure began to mount on standards organizations, especially the IEEE as the source of all earlier LAN specifications, to invent the "second generation" of LANs. These would preferably run at 100 Mbps or higher and be similar to Ethernet and Token Ring LANs that most organizations were familiar with. The users generally did not want *new* technology, just *faster* technology.

But new technology was what they initially got. First, they got a 100 Mbps LAN technology—sort of. Around 1989, a technology called Fiber Distributed Data Interface

(FDDI) was developed by the American National Standards Institute (ANSI). FDDI ran at the required speed, but on fiber optic cable, a medium most organizations were not familiar with that was generally more expensive than other media. It was designed to run over distances as great as 100 Km, a capability that did not really appeal to users who were concerned with the clients in the next office and the server down the hall. Early FDDI networks were not so much built as engineered: There was so much design-dependent variation that it required a regular engineering team to design and implement even a modest FDDI network. Eventually FDDI was modified to run off copper-based media like Unshielded Twisted Pair (Ethernet and Token Ring had taken the plunge into UTP from Coax and Shielded Twisted Pair [STP] long before), but the damage was done. Users stayed away in droves.

Another possibility was being explored, and users liked this idea much better, for one very good reason. They had essentially proposed it. Users—really the network managers and implementers, but they are still users too—did not like the limitations in terms of size and weight of the coaxial cable Ethernet required. Cable TV networks used much smaller and cheaper cable. It was more flexible than "thick" Ethernet coax, and so cheaper to have installed.

So when a vendor offered a version of Ethernet that could use this cable TV coaxial cable, users voted with their dollars. Soon the scramble was on to introduce "thin coax" products, which created such anarchy that the IEEE was forced to standardize this "Cheapernet" as 10Base2 to form a companion to the original 10Base5. It was still Ethernet in terms of speed, but the physical layout and distances were highly restricted (200 meter segments) compared to classic Ethernet-based LANs.

Once users started building larger and larger Cheapernet networks, however, it became apparent that the new standard was totally unsuited for a general office environment. The stations had to be "daisy-chained" together from one to the other, making it next to impossible to insert or remove stations without recabling the LAN. If one station was powered off, the whole network essentially halted. There were other complaints as well, but all focused on the impracticality of wiring a building efficiently with thin-wire Ethernet.

What users ideally wanted was not something that looked like special data cabling at all. They wanted data cable for LANs to look like UTP phone wire for telephones, for a number of reasons. First of all, it was very cheap. Second, installers had been working with it for years and were familiar with it. Third, it was a star-wired system radiating out from a central point, not a complicated bus (Ethernet 10Base5) or daisy-chain (Cheapernet 10Base2) arrangement.

Of course, for the new "star-wired UTP" the same Cheapernet scenario happened again, with one vendor offering 10 Mbps Ethernet-based LANs over UTP wiring. This time, however, instead of scrambling with different flavors, the other vendors got smart. They went directly to the IEEE and insisted on a "10BaseT" standard right away. These vendors argued they could not wait years for a new standard to be researched and debated, while one vendor dominated the market. They needed something immediately, so that all vendors who wanted to sell 10BaseT solutions could do so on an even footing with the same technology. So in a mere 18 months—an astonishing speed—the 10BaseT (802.3i) specification became the overwhelming favorite LAN technology. This star-wired UTP LAN created the 10BaseT hub market almost overnight. More than half of all LANs today are 10BaseT LANs.

Today, most PCs are attached to LANs that can handle a full 10, 100, or even 1,000 Mbps. But gigabit speeds (1,000 Mbps) can be carried on coaxial cable or twisted pair wire only for short distances. To go farther with the bandwidth needed by modern LANs today, fiber and some form of optical networking is needed. The time has come to explore some aspects of the fiber optic cables used in newer WANs and LANs.

CHAPTER 2

Fiber Optic Cable Fundamentals

In contrast to the "optical network" (or "visual network") described in Chapter 1, modern optical networking depends for the most part on the deployment of fiber optic cable. The only exception to this light-on-cable architecture is "free space" transmission between an optical transmitter and a receiver, separated only by air. However promising newer free space optical systems might be, the vast majority of all fiber optic systems retain the use of fiber optic cable between transmitter and receiver.

This chapter will explore all aspects of the fiber optic cable that runs between transmitters and receivers. All fiber optic cable is not the same. There are many variations on the basic theme of sending light down a fiber, and all of the useful ones will be investigated. There will be coverage along the way of the manufacture, the use, and even the basic performance characteristics of all major fiber types.

It is probably best to begin with a brief discussion of the light that shines down the cable itself. Light is an odd sort of phenomenon, behaving sometimes as a wave (usually when light travels) and sometimes as a particle (usually when it stops traveling). It turns out almost everything shares this wave/particle aspect of physical nature to some degree, including you and me. But it is with light that this dual nature of reality is most noticeable, and most important.

OPTICAL NETWORKING DEFINED

Since the concept of *optical networking* is central to this book, it is important to give rather full definition to the term right up front. The term optical networking has two main parts: the *optical* aspect and the *networking* aspect. Networking has been around for a long time and has a more or less accepted definition or series of characteristics that all networks share.

First of all, networks exchange information between *end users*. The end users can be human beings, or the applications they run on computers attached to the network one way or another, or even physical devices such as a printer or a temperature sensor. In most cases, the less knowledge end users need about the inner workings or detailed operation of the network, the better. Any user can in theory communicate with any other user attached to the network, although some form of permission or configuration process might be required before this happens. Today, it is more or less assumed that the network functions in a *digital* fashion. That is, the network deals only with bits, 0's and 1's. The information exchanged might be *analog*, voice being the main example, but the analog information is always represented digitally as a stream of 0's and 1's on the network.

Networks exist to enable end users to share information or resources, both of which can be represented by hardware or software. The resources of the network are often shared as well, typically the *network nodes* that shuttle bits back and forth between end users and the *links* that tie the users and network nodes together. Shared links and network nodes allow a minimum of network hardware to connect the maximum number of users. Common examples of network nodes are voice circuit switches such as central offices (the term most often used in the United States) or local exchanges (the international term for the same thing). Other examples would be routers (connectionless network nodes), packet switches (connection-oriented network nodes), and so on. Networks also usually have some mechanism for controlling and managing the entire network, although usually the control aspect is just lumped into an overall network management procedure. The key is that network nodes function in the electrical realm, even if the links are fiber optic cable.

Network nodes vary widely in operational detail. But almost all just take a piece of information in on an input port, look something up in a table, and send it out on an output port. The lookup can be as simple as the port number ("all bits arriving on port 22 go out on port 43") or as complex as processing a 20-byte IP packet header and changing the values of some of the fields. Network nodes all carry out a sequence of steps to accomplish their task of connecting input to output.

Optical networking is therefore here defined as involving a network in which at least some of the operational steps in a network node take place in the optical realm, not just in the electrical realm. If this sounds simple enough, it is. But it is the unique nature of light that makes optical networking so interesting.

GUIDING LIGHT

No technology springs to life fully formed without a long and sometimes painful trip to the marketplace. The road to fiber optic transmission using a long, twisting and turning piece of "glass" to carry light is no exception. Although silica-based fiber optic cables almost always are said to be made out of "glass," the term is somewhat misleading. Fiber is not the same as window glass, although both are classified as glasses. Fiber optic cable and other glasses belongs to a special class of materials that do not form a regular solid crystal lattice when they cool. They just get thicker and thicker, but they remain "liquids" at heart. And in spite of legends about old cathedral stained glass windows being thicker at the bottom than at the top due to liquid flow, most glasses are as thick as solids and even more brittle.

A few words about the development of fiber optic cable are appropriate, if only because many of the impairments encountered in optical networking today (such as attenuation) are more easily understood when placed in historical perspective. Several other key concepts such as *critical angles* and *decibels* are introduced here as well.

A Swiss physicist named Daniel Collodon and a French physicist named Jacques Babinet showed in the 1840s that light could be bent along jets of water streaming from a fountain. Physicist John Tyndall in England showed much the same thing in 1854 by guiding light inside a jet of water flowing from a tank. In 1870, Tyndall showed the Royal Society that light would follow a curved stream of water. Once inside the water, it seemed, the light became somehow "trapped" and could not escape. This demonstration was repeated many times and in many ways, mostly to amuse. It was just a very neat phenomenon to observe. When coupled with the new electric light, multicolored fountains could be made to put on light shows to dazzle and amaze the public.

Light inside water was not the only way that light could become "trapped." By the turn of the century, scientists had figured out that light would also travel inside bent rods made of quartz. These were even patented as "dental illuminators" to carry light inside the human body. As late as the 1940s, many doctors used plexiglass tongue depressors that were illuminated.

Today, of course, we would observe that water has a different *index of refraction* or *refractive index* (RI, but both terms are used) than air. By definition, the RI of air is set to 1. Most translucent (semitransparent) materials have RIs higher than 1, some much higher. The materials used in fiber optic system typically have RIs of about 1.5. When faced with media having different refractive indices, light is either reflected or bent (refracted) depending on the angle at which the light encounters the differing RIs. Reflected light bounces off the material, but refracted light will enter the new medium as long as the new material is transparent enough.

Angles turn out to be very important in any consideration of reflection or refraction. The amount of reflection and bending depends on the angle that the light wave encounters zthe change in refractive index. At some angles, reflection dominates and little light enters the new medium. At other angles, refraction dominates and a lot of light enters. Seldom is light entirely reflected or refracted, however. Even modern fiber optic cables that supposedly function through "total internal reflection" are not quite that perfect.

This introduces the concept of a *critical angle*. Within the critical angle, light will be reflected back into the medium with the higher refractive index when facing the choice of entering a material with a lower refractive index. Above the critical angle, the light is refracted right out of the region of higher refractive index. The concept of a critical angle is shown in Figure 2-1.

Whenever light encounters a boundary at an angle greater than the critical angle, the light is reflected. A simple example of this involves what an underwater swimmer sees when looking upward at the surface of the water. When looking more or less straight up toward the surface, the swimmer will see what is above the surface of the water. But if the angle is sufficiently large when looking toward the surface, only a reflection of whatever is below the water appears. Fish experience exactly the same effect and can only see over a range of angles quite near the perpendicular to the surface of the water. So a fish contemplating an attractive lure cannot see the fishing rod to which the lure is attached. Thus, even this simple phenomenon is good news to fish catchers everywhere.

When applied to fiber optic cable, the concept of a critical angle means that the material having the higher refractive index acts as a waveguide for the light inside. Since the angle extends to three dimensions, and not just the two dimensions shown in the figure, the critical angle forms the basis for the *numeric aperture (NA)* of the structure as a whole. The NA is just a measure of the total light capturing capability of the system. The larger the NA, the easier it is to inject light into the system. Light travels down the material, be it water bounded by air or some other materials altogether, as waves.

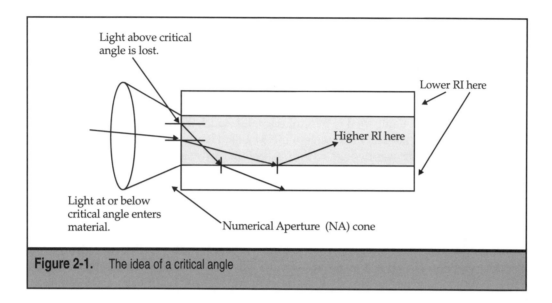

Figure 2-1.   The idea of a critical angle

Much of the following discussion in this chapter revolves around the fact that light is neither wave nor particle, but both. Furthermore, this dual nature of light cannot really be visualized, a fact that caused a great deal of distress among physicists, including minds as great as Einstein's. Light just does not behave in an intuitive fashion in many cases, especially when light is confined in small spaces. The equations all work, but the result cannot be visualized or easily reconciled with how scientists imagine the universe must function. Einstein, in spite of his pioneering work in light effects and relativity, went so far as to conclude that something was seriously wrong with human intuition or the universe itself. The details are not of concern, except to note that light is a very strange thing to work with.

In the time of Newton, scientists imagined that light consisted of tiny "rays," and this image persists to this day. Light is usually represented as a line with a little arrow at one end. In most illustrations, this image of light as ray is not harmful, but it is not accurate either. Light is not a little arrow. Light might behave as a wave or as a particle, but never as a ray. This is good to keep in mind, since many of the figures in this book feature little arrows representing light as Newtonian rays, just as in all other books that concern light.

It was only a short step from the early quartz rods to true optical fibers. Fibers were just transparent rods of glass or plastic made very long and very thin so that they were also very flexible. Optical fibers were applied to all sorts of imaging applications from the start. For example, in the 1920s, American Clarence W. Hansell and Briton John Logie Baird patented the idea of using arrays of either transparent rods or just hollow pipes to carry images for faxing or television. Their work was theoretical, but Munich medical student Heinrich Lamm reported in 1930 that he had used a bundle of fibers to send the image of a light bulb a short distance. The short distance was unimportant to Lamm, who was trying to extend the medical application of optical fiber to illuminate the insides of the human body.

These early fibers were just unclad pieces of glass or plastic. They transmitted light poorly, and the fact that Lamm was a Jew in Nazi Germany did not help popularize his work, especially after Lamm fled to the United States. Much the same territory was covered in Denmark by Holger Moeller Hansen in 1951. Moeller Hansen applied for a patent on fiber optic imaging, but this was denied on the basis of the "prior art" of Hansell and Baird. So the interesting properties of fiber bundles were discovered, considered, and all but abandoned several times. In 1954, two reports appeared on fiber bundles all over again. The first was from Holland, the work of Abraham van Heel of the Technical University of Delft, and the second was from England, the work of Harold H. Hopkins and Narinder Kapany of Imperial College in London.

This time the 1954 reports on the properties of fiber bundles had an impact on scientists everywhere. The reports were widely read, although the biggest issue was the short range of transmission. An optical physicist from America, Brian O'Brien, made a crucial observation to van Heel, who instantly realized the implications. All early fiber bundles were unclad (bare) and relied on the air to provide the lower refractive index needed for internal reflection. But the *crosstalk* or interference between fibers in the bundle was a real limitation once the unclad fiber surface became contaminated in any way, even through handling. What van Heel did was to enclose the fiber rod itself inside a transparent outer coating of glass with a lower refractive index, called the *cladding*.

Losses were still high, however, high enough to limit practical fiber applications to medical use, in particular endoscopy (looking inside the human body through either end). For example, University of Michigan undergraduate Lawrence Curtiss developed an endoscope for the stomach with Basil Hirshowitz, a physician, and C. Wilbur Peters, a physicist. By then bitter patent battles were raging over glass-clad fibers, such as the one waged and lost by Will Hicks and his group at the American Optical Company. Medical fibers had attenuation (signal loss ratio) of about one decibel (dB) per meter by 1960, rendering them useless for communication purposes. This was 1,000 dB per kilometer. Since decibels are a logarithmic scale, where a 30 dB loss represents an input-to-output power ratio of 1,000 to 1, and 60 dB loss is one million to one, medical fibers were impossible to use in networking applications.

Engineers routinely use decibels instead of percentages to measure power ratios, and all other types of ratios as well. This is because decibels are easily added, while percentages are not. Consider a simple network where the output power on one link is piped directly to the input on a second link. The signal power loss in watts on each link is 90 percent, meaning 1/10th of the input power arrives at the output. What is the end-to-end signal power loss on the whole system? Adding percentages gives 90% + 90% = 180%, which is nonsensical. Using decibels, the loss on each link is 10 dB, since an input-to-output power ratio of 10 to 1 translates to 10 dB. Ten to one is 10 raised to the first power (10^{-1}), which is then multiplied by 10 to give decibels. In decibels, the total end-to-end loss of the system is 10 dB + 10 dB = 20 dB, which is correct, since 1/10th of 1/10th is 1/100th or 10 raised to the second power (10^{-2}). Naturally, not all ratios work out to neat whole numbers. For example, an 80 percent power loss in watts translates to a decibel loss of 6.99 dB (1/5th power received is expressed as 10 to the –0.699 power). The original unit was often so small that engineers had to add up too many fractions. So the whole unit was multiplied by ten to eliminate some of the fractions. Decibels can just as easily be used to measure signal power loss (attenuation) or a power boost (gain) or almost any other ratio.

A small table giving the relationship between ratios, powers of ten, and decibels is shown in Table 2-1. Note that a 60 dB loss is not *twice* as bad as a 30 dB loss, but a thousand times worse!

Why consider fiber optics for communications anyway? Simply because by around 1960, the United States and other countries such as Great Britain had pushed their telecommunications infrastructure to the limit. More bandwidth was needed to carry telephone calls, and even radio and microwave frequencies were pressed into service in large numbers. Higher and higher frequencies were used, and many researchers were working on *millimeter waveguide* systems. These were just very smooth hollow pipes used as waveguides for frequencies in the tens of gigahertz (gigahertz frequencies have millimeter wavelengths). These waveguides had lower attenuation than pure air, but they still had problems going around corners and the losses were still rather high.

Some far-sighted people proposed fiber optic cable as a waveguide for light. Alec Reeves, who had invented digital telephony before World War II mainly to combat analog noise, looked to optical systems to increase telephone network capacity in 1958. At optical frequencies, the shorter pulses that could be used translated to very high bandwidths, if only the attenuation issues could be addressed. Still, most interest was focused on millimeter waveguides. At Bell Telephone Laboratories in the United States,

Ratio (P_{out}/P_{in})	Power of Ten (Negative when P_{out} is smaller (loss), positive when P_{in} is smaller (gain))	Decibels
1/10	10^{-1}	–10 dB (1 × 10)
1/100	10^{-2}	–20 dB (2 × 10)
1/1,000	10^{-3}	–30 dB (3 × 10)
1/10,000	10^{-4}	–40 dB (4 × 10)
1/100,000	10^{-5}	–50 dB (5 × 10)
1/1,000,000	10^{-6}	–60 dB (6 × 10)

Table 2-1. Decibels

Stewart Miller led a group in 1965 that was working on gas lenses to focus light from Bell Labs' newly invented laser (1960) into the hollow waveguides. The large attenuation on fiber did not encourage researchers to couple lasers to fibers—yet.

Research on fiber continued, however. Elias Snitzer at American Optical Company did some work with Will Hicks (now at Mosaic Fabrications) in 1961 and managed to make fibers so small that the *core* (the light-carrying structure inside the cladding) carried only one *waveguide mode* of light. That is, there was only one path for light to take through the fiber. This also helped to cut down on signal loss, but the overall attenuation was still too high for long distance communication over fiber.

The big fiber breakthrough came in July 1966 in the form of a paper by Charles Kao and George A. Hockham. Both worked, as Reeves did, at Standard Telecommunications Laboratories Ltd. (STL) in Harlow, England. Kao had been born in Shanghai, China, and came to work with Antoni E. Karbowski at STL under Reeves. When Karbowski left in 1964 to join the University of New South Wales in Australia as the chair of the electrical engineering department, Hockham replaced Karbowski at STL and worked with Kao.

Kao in particular realized that the problem was basically that glass is not particularly clear, cladded, or uncladded. The paper pointed out that glass fiber attenuation was largely the result of impurities in the form of metal ions, with chromium, copper, iron, and vanadium ions being the most common sources of trouble. Nickel and manganese and other *transition metals* play a role too. Iron, the worst source of signal loss, was common in glass (iron makes many glasses have a greenish tinge and makes Coke bottles green). If these could be eliminated from the fiber during the manufacturing process, fibers with attenuation of 20 dB per kilometer (dB/km) carrying thousands of telephone calls were possible, Kao claimed. With 20 dB power loss, one percent of the input power reaches the end of a fiber one kilometer long. With a powerful enough transmitter, and a sensitive enough receiver, such fiber optic cable could be used for communications purposes.

Kao was in contact with several people at the British Post Office, who then ran the British telephone system. One was F. F. Roberts, an engineering manager at the Post Office Research Laboratory in London. Roberts was interested in the increased telephone traffic capacity that fiber optic cable promised, and he persuaded his boss, Jack Tillman, to fund a study into ways to decrease fiber attenuation.

Several Corning Glass Works scientists visiting England from Corning, New York, heard about Kao's ideas through the Post Office. Corning was willing to work on the problem, and so with the encouragement of the British Post Office, Robert D. Maurer began work on modern fiber optic cable in 1966. There was no pressure, and work was intermittent. Still, by 1970, Maurer, along with Corning Glass Works' Donald Keck, Peter Schultz, and Frank Zimar (who ran the glassworks itself), had managed to manufacture cladded glass fiber with an attenuation of 20 dB/km. This made the first fiber optical systems of the 1970s possible.

Corning used silica glass to achieve their magic. Fused silica glass can be made very pure but is hard to work with. Even at 2,000 degrees, silica does not flow freely like a liquid, and silica has a low refractive index. Silica has to be drawn, almost like copper wire. By intentionally adding just the right impurities to the glass (a process called *doping*),

Corning produced a core and cladding with the right optical properties. The early fibers broke easily, mostly from bending, so strengthening fibers had to be added under the jacket, a procedure still routinely used, although moderns fibers have a cross-section tensile strength greater than steel wire of the same diameter. But as fragile as the fiber was, British Post Office Research Laboratories tests showed that the attenuation goal had finally been reached.

There are many impairments in optical networks that limit speed and distance, but attenuation is first and foremost among these. Fiber attenuation performance improved quickly. By 1975, it was down to 4 dB/km. This dropped to 0.5 dB/km in 1976 and 0.2 dB/km in 1979. At 0.2 dB/km, a fiber 10 kilometers long delivers about 63 percent of the signal power to the receiver. Such a fiber would make a very clear window, although windows 10 km thick would have limited application!

PUSHING LIGHT THROUGH GLASS

The first thing to realize is that pushing light through glass is not at all like driving electricity through a wire. Networks based on electronics are very different from networks based on fiber optics. In an electrical circuit, most of the electricity travels on the outer surface of the wire. In a fiber optic cable, the fiber acts as a *waveguide* for the light inside it. This makes even the simple action of connecting two fibers quite different than the even simpler act of joining two electrical wires together.

This is not to say that electronics and electricity do not play an important role in even the simplest point-to-point fiber optic links. There are no purely optical computers or chipsets to speak of yet, although some intriguing prototypes promise to enable devices to perform many computing tasks using optical technology. The very concepts of bits and bytes (or octets: groups of eight bits) used for communications purposes have no real counterparts in the purely optical domain yet.

In order to appreciate some of the more profound differences between electrical and optical transmission, as well as the relationship between the two domains, this might be a good time to examine the basic parts of a simple fiber optic link. Naturally, this book will go on to talk about more complex optical networks and DWDM systems, but this is a good place to start.

All of the essential components of a basic fiber optic link are shown in Figure 2-2. There are many important concepts to be explored here, so some time will be spent discussing the operation of this system. The discussion is somewhat simplified but accurate enough for a basic introduction to fiber optic transmission.

There is a lot going on in this figure, so some detailed explanation of all the piece parts is necessary. Keep in mind that this is a simple link, not a network. The figure shows one-way transmission across a fiber, which is the most common arrangement. Two-way transmission on the same fiber is possible, but typically a second fiber would be used to send bits back to the source in the figure.

First of all, the fiber optic link exists to transfer bits (0's and 1's) from one side of the link to the other. The bits might come from and go to a voice switch, a router, or any one of

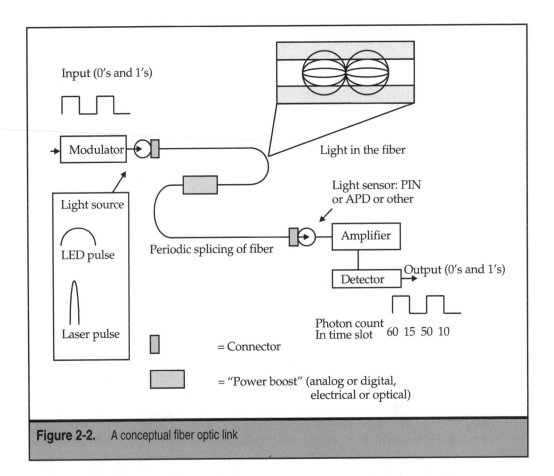

Figure 2-2.   A conceptual fiber optic link

many types of device. All that is required is that the receiver understand the meaning of the bits that the sender is transmitting. There are even fiber links that allow analog signals to be sent, such as on the fiber links in many cable TV networks, but only digital information is considered here. It is important to note that the bits originate from and end up in the electrical realm.

The input bit stream is *modulated* onto the fiber. Modulation just means that some form of a transition state of the fiber link is used to represent a 0 or 1 or groups of 0's and 1's. The link has an unmodulated "ground state" called the *carrier*. Modulation is some form or pattern of deviation from the carrier used to indicate bit patterns. One of the simplest forms of modulation is to change the amplitude of the carrier to indicate a 0 (low amplitude, for instance) or a 1 (high amplitude). Amplitude modulation (AM) can be used in all forms of communication, such as AM radio, not just fiber optic links. When used on fiber links, AM is commonly called *intensity modulation,* since the light source is never really off, but just at one of the allowable intensities. Other types of modulation are possible on fiber optic links, but intensity modulation remains the simplest and most cost effective in most cases. The input 0's and 1's will produce a series of pulses, with the timing or duration of the pulses being proportional to the bit rate of the fiber optic link. The higher the serial bit rate, the shorter the pulse duration.

The modulated input is used to drive the light source itself. This is where the transition from the electrical realm to the optical realm occurs. Two major types of light source are used on fiber optic links. The first is the light emitting diode (LED). The pulses characteristic of an LED light source are shown in the figure. LED pulses are useful, but they are spread out in time (and across wavelengths) and not of a very high amplitude. LED sources can be quite cost effective, however, for shorter spans, for lower bit rates, and for some types of fiber optic cables. For longer spans, for higher bit rates, and for some types of fiber optic cables, lasers are used as light sources. Lasers generate a sharper (in time and across wavelengths) and more powerful pulse than LEDs. However, since all lasers in common use today for optical networking include an LED light source as well, lasers are more costly to use in fiber optic systems.

Both LEDs and lasers today are implemented as chipsets on a common card or board, the same as used to occupy almost any computer bus slot. The LEDs and lasers are just forms of the *diode,* a common electronic, semiconductor-based component. In the early days of fiber optic transmission, the laser was a bulky, desktop device itself. Today the *semiconductor laser diode* is the light source for most long-distance, high-speed fiber optic links. The light source card typically has a short "pigtail" ending in a fiber connector, also shown in Figure 2-2. As it turns out, this initial connector is a considerable source of signal loss end to end over the fiber link. More details on fiber optic light sources will be discussed later on in this book.

Light is thus coupled onto the fiber optic span itself. On long spans, the fiber optic cable might need to be spliced, since splices have lower losses than connectors. On the other hand, connectors allow for the easy replacement of a failed light source, easy moving of a light source to another fiber link, and so on. There could be a long way between splices, since reels of fiber optic cable can be 10,000 feet (about 3 km) long.

Within the fiber, the light pulses travel as light waves. The light waves are spread out in time and across a range of wavelengths (even from lasers), as shown in detail in Figure 2-2. Most of the light travels in the core of the fiber, confined by the concept of total internal reflection. However, an appreciable percentage of the light also travels in the cladding of the fiber, where the light can either disappear (adding to attenuation) or interfere (interact) with the core light at the receiver. As a wave, light can be split, reflected, or refracted. Light waves also interacts with themselves and/or other light waves, a process called *interference.* There is both *destructive interference* (usually bad) and *constructive interference* (which can be good), so the term *interaction* is actually more descriptive. But "interference" is the standard term used.

On very long spans, over tens of miles or kilometers, it might be necessary to counter the effects of attenuation with a periodic "power boost," as shown in Figure 2-2. The power boost can come from one of several methods, either analog or digital, and the device can operate either electrically or optically. Today, electrical, digital devices known as *regenerators* are commonly found on most long fiber spans. Fiber regenerators are actually small receivers and transmitters placed back-to-back in one compact device. They recover the 0's and 1's from the light waves on the fiber on the input side, and then resend the bit stream as light waves on the output side.

All regenerators, not just the ones in use on fiber optic links, perform three valuable functions. Regenerators 1) *Restore* the signal power level of the digital signals, making higher amplitude output, 2) *Reshape* the signals, which always spread out in time, and 3) *Retime* the signals, centering them inside their characteristic bit interval. Regenerators are said to perform the "3 R's" of restoring, reshaping, and retiming the signals.

Fiber links today, especially those in use with optical networks such as DWDM systems, do not employ digital, electrical regenerators. These systems use fiber optic *amplifiers* instead, which operate in an analog fashion, but in the purely optical realm. All amplifiers are analog devices, and fiber optic amplifiers only see the light waves, not the individual bits making up the digital information stream. So fiber optic amplifiers can only do "1 R" restoring of power levels, since reshaping and retiming require the device to be aware of the bit stream. There are other ways of making sure light signals do not spread out too much or arrive outside their proper times, so fiber amplifiers are actually an improvement over regenerators on a fiber optic link. More details on the operation of fiber amplifiers will be discussed later in this book.

At the receiving end of the fiber link is the light sensor or detector. These devices fall into one of two main categories. There are P-Intrinsic-N diodes (PIN diodes) and avalanche photodiodes (APDs). Details on the precise operation of both are given later in this book. For now, it is enough to point out that PINs are less expensive than APDs, but PINs are limited in terms of sensitivity. APDs are much more sensitive devices, since they amplify as well as detect light, but are more expensive for this very reason. Like light sources, light sensors are card-mounted semiconductor devices and are attached to the incoming fiber through a connector and short pigtail fiber.

Once through the light sensor, the signal reenters the electrical realm. When a PIN light sensor is used, it is common to use a separate amplifier at the receiving end. The output from this amplifier is fed directly into more detector circuitry, usually on the same board as the other electronics. This time, the detector is more concerned with bit timing and the output bit stream is the same as the input bit stream (less bit errors).

Note that PINs or APDs essentially count photons to determine bit output. The light is firmly in the particle world here. More light at the input generates a higher photon count at the output side, but not *different* types of photons. A photon can have higher or lower energy, but it still counts as one photon. What the receivers basically do is count photons arriving within a time slot. If the photon count exceeds a certain threshold level, a 1 bit is registered for that time slot. Below the threshold, a 0 bit is registered. But all time slots contain photons. In fact, fiber optic cables without any transmitters at all generate photons (and 0's and 1's) when a receiver is hooked up! This is called *dark current* and must be compensated for once the fiber link is operational. There are several sources of dark current, but all are related to the wave nature of light.

The wave nature of light always complicates even the simplest fiber optic link. For example, since light waves usually spread out in time, a phenomenon that is called *dispersion*, it is possible for the light sensor to register photons at the receiver *before* the light source has been turned on! This *phase velocity* exceeds the speed of light in the material but is the result of the light pulse "pushing" some photons forward in time. No information can be sent this way, so the Theory of Relativity is not violated. (Actually, relativity does not forbid things to travel faster than light, and many things do. Relativity forbids anything being *accelerated* to the speed of light or beyond, and the use of super-luminal things to send information.)

Receivers on fiber optic links work because there are enough photons in the proper time slots to register the bits correctly.

Advantages of Fiber Optic Cable

Optical networking starts with the use of fiber optic cable. While not a perfect transmission medium, fiber optic cable enjoys a number of advantages over other forms of media. These advantages are briefly outlined here. The fact is that fiber optic cable offers more advantages and fewer disadvantages than any other medium in use today. Even the cost of fiber optic cable in terms of cost per bit/second is the least of any cable or wireless medium. It is also true that no other cable medium is even being researched today. Research is confined to more effective and efficient use of fiber optic cable such as the use of DWDM. The general guideline in the fiber industry is that speeds achievable on fiber optic cable quadruple every two years, and that the distance achievable without digital repeaters also quadruples every two years.

The main fiber optic cable advantages are discussed as follows.

Higher Bandwidth Capacity

Fiber optic cable is unsurpassed when it comes to high bandwidth and information capacity. The bandwidth of a transmission medium is usually measured in terms of the analog frequency range (bandwidth is more properly an analog term). This is appropriate for fiber optics, where the signals are inherently analog in nature (light waves). Analog bandwidth is measured in Hertz (cycles per second, or Hz). The symbol for cycles per second is f, and the rule for figuring out the bandwidth of a medium is given by the formula $f = c / \lambda$, where c is the speed of light and λ is the characteristic wavelength of the medium.

Now, c is 186,000 miles per second or 300,000 kilometers per second, and λ is about 1 micron or 1,000 nanometers (billionths of a meter) for visible light (lasers are typically infrared, but this is close enough). These numbers give a characteristic bandwidth for fiber optic cable of about 300 terahertz, or 300,000 gigahertz. This is only a simple computation, which does not take into account digital modulation techniques for sending digital signals on the fiber.

To offer a contrast to fiber, Figure 2-3 shows the typical bandwidths of other common media. The scale is logarithmic, which means that each step on the vertical axis actually represents 1,000 times the bandwidth of the previous one.

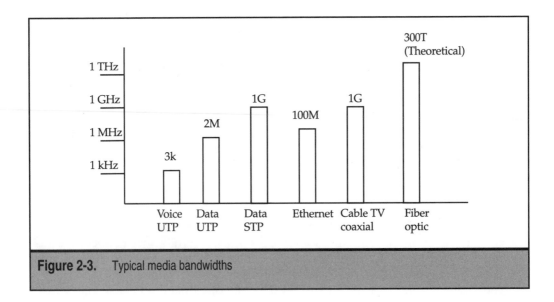

Figure 2-3.   Typical media bandwidths

Lower Bit Error Rates (BERs)

Fiber optic cable has lower bit error rates (BERs) than any other transmission medium. Lower bit error rates (BER) are fine, but as link transmission speeds grow, the error rate issue becomes crucial. Consider a copper link (coaxial cable or twisted pair with a BER of around 10^{-6}. This means one bit in error for every million bits sent (10^6 is 1 million). At the DS-3 rate of 45 Mbps, a bit error will occur every 22.0 milliseconds on average, or about 45 bits in error every second. (Actually, copper errors are usually in bursts, but the average BER is still a valid concept.)

The significance of the BER for high-speed networks is twofold. First, at higher speeds only fiber optic cable can give acceptable average bit error intervals. Second, at lower bit rates, fiber can give astonishingly low bit error intervals, leading to the common observation that "with fiber, bit errors just go away." This is not really an exaggeration. For example, some service providers offer 64 Kbps links on fiber with a BER of 10^{-13} (1 per 10 trillion bits). At this rate, the average bit error interval at 64 Kbps will be one bit error every five years. A decrease in the BER from 10^{-6} to 10^{-9} is a thousandfold decrease. In other words, the number of bit errors a network has in one day at 10^{-6} on copper is the same as the number of errors it has in three years at 10^{-9} on fiber!

Fiber optic links today run at 10^{-10} or better. So the number of errors a copper network had in one day are seen over a 30-year period on fiber. Hence the thought that with fiber, errors go away. Table 2-2 shows some average bit time intervals (the time between bits in error) for a variety of line speeds.

Bit Error Rate	64 Kbps	256 Kbps	1.5 Mbps	45 Mbps	135 Mbps
10^{-6} BER	16.0 seconds	3.9 seconds	0.7 seconds	22.0 millisec.	7.4 millisec.
10^{-9} BER	4.3 hours	65.0 minutes	11.0 minutes	22.0 seconds	7.4 seconds
10^{-12} BER	6.0 months	1.5 months	7.7 days	6.2 hours	2.1 hours

Table 2-2. Average Bit Error Time Intervals

Longer Distances Between Repeaters

Digital repeaters, which detect and repeat the string of 0's and 1's onto another fiber link, require power and add potential points of failure. A given length of fiber will need fewer repeaters than the same length of cable composed of another medium. Repeaters are a significant cost factor in any transmission system, of course.

Fiber optic cable performs so well over long distances because of the low attenuation on fiber optic cable. The distance between repeaters on a digital link is purely a function of the signal loss over the distance. Figure 2-4 shows the typical signal losses over one kilometer

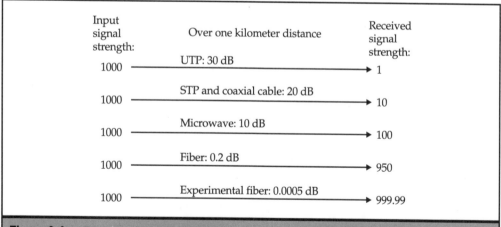

Figure 2-4. Typical media signal losses

(about 5/8 of a mile) for various transmission media. In each case, the signal strength is rated an arbitrary 1000 "units." The typical attenuation in terms of dB per kilometer are shown in the figure. There is even experimental fiber that receives 999.99 of the 1000 transmitted signal strength over a kilometer.

Many fiber optic cable systems are engineered for repeater spacings of 25 miles (40 km). This compares to about 7.5 miles (12 km) for coaxial cable systems. Newer fiber optic cable systems have repeater spacings of about 75 miles (120 km). And DWDM systems using newer optical amplifiers can go even farther.

No Electromagnetic Interference

Most electromagnetic interference in the form of noise comes from the fact that copper cables are long antennas. Fiber will not pick up electromagnetic noise from the environment. In contrast, local loops, the long runs of copper cable used to connect telephone central office to customer premises, must be electrically bonded and grounded to shield them from radio noise.

Fiber optic cable is not only noise resistant, fiber is basically impervious to noise. This is because fiber operates in the optical range of the electromagnetic spectrum. This is one reason that fiber optic cable error rates are so low. While stray electrical signals may be picked up by copper media, there is really no such thing as "stray light." Even more simply put, while the "bandwidths" of storms, power supplies, and many wireless transmitters overlap with those of copper media, they do not overlap with optical bandwidths.

Fiber optic cable has been used for years where noise is a factor in transmission systems, for example, many factories with heavy power equipment or power companies that run communications and control lines along high-voltage power transmission lines. With noise no longer an issue, route selection is a much easier process for service providers.

No Electrical Connectivity Needed

The issue of electrical connectivity is related to the previous point about noise. The need to couple links electrically is a real issue. *Ground loops* can cause problems over an electrical link even a few miles or kilometers in length. And while grounding is essential to make an electrical link function at all, grounding *both* ends of an electrical link can be a disaster.

Also, electrical links sometimes can carry high voltages. Outside plant personnel must be very careful working near some points in an all-electrical network. Also, lightning can easily strike electrical cables and travel long distances, posing another source of shock hazard. While these dangers are not entirely removed in optical networking with fiber optic cable, they are minimized.

High Security

Fiber networks can be constantly monitored for increases in signal loss, which may indicate the presence of taps. It is a myth that fiber is more secure than copper because fiber is difficult to tap into. All that is needed is a simple optical coupler. Any fiber link can be tapped by bending the fiber, scraping off the outer layers of cladding, and strapping the optical coupler device to the cable at that point. The fiber is then tapped, and any signal that travels down the fiber is picked up by the eavesdropping device. However, there are fewer points where such devices can be inserted, and it is much more difficult to inject false signals *onto* a fiber link.

And the inserted coupler is very easy to detect. The inserted optical coupler will add at least 0.5 dB of signal loss (attenuation) to the fiber link. In normal situations, the attenuation on any link, even a fiber one, may increase with time (especially if the link is not maintained properly). But a jump of 0.5 dB very quickly is not normal under any circumstances.

So security is handled on fiber networks by closely monitoring the end-to-end attenuation on a link. If this rapidly increases by a half dB or so, then some physical damage has occurred. It could be innocent (e.g., environmental damage), or it could be an intruder. There are even ways to find out exactly how far along the fiber the loss was injected, focusing detection efforts at that area. The this reason, the federal government and military prefer fiber to any other type of link in many cases. Even banks and many business have come to recognize the enhanced security that fiber optic networks can provide.

Bandwidth Upgrades

With DWDM, it is possible to expand the bandwidth on a fiber optic link. Upgrading a digital link with copper networks has been difficult, to say the least. For instance, going from a 1.5 Mbps link (a DS-1 or T1) to a 45 Mbps link (a DS-3 or T3) requires the running of new media in most cases. T1 is defined on unshielded twisted pair (UTP) copper wire, and T3 is defined on coaxial cable, fiber, or microwave.

Many organizations are asking for 45 Mbps links to support new interactive multimedia or other high bandwidth, broadband applications. The price to the customer for the T3 must offset the cost of running the new media to the service provider in a reasonable payback period. This fact has kept the cost of T3 links high, and it forms yet another incentive for installing fiber optic cable in the first place.

Once run, the fiber optic cable need not be a limiting factor at all. And because a lot of fiber optic cable is buried below ground level, the cost of burial is the most significant cost factor in fiber optic installations. It often costs tens of thousands of dollars per mile just to dig a hole in the ground. Obviously, the cost component is to be avoided if at all possible.

It is important to note that this upgrade capability is not inherent to fiber optic networks. In many cases, it must be carefully planned for, and there is plenty of older fiber optic cable that cannot be used with DWDM. But the fact that fiber links can be upgraded for DWDM at all is significant and a vast improvement over older digital media.

Low Cost

Fiber optic cable costs less per foot than almost any other cable type. (It is the electro-optical conversion costs that keep working fiber optic links costs relatively high.)

Small Size and Weight

The same size crew can install much more fiber in one day than any other cable medium. Fiber also offers numerous advantages in terms of size and weight. Even 10,000-foot fiber reels weigh only a couple hundred pounds. Smaller reels can be carried under a person's arm and weight about 60 pounds or so. Fiber optic cable can be run above hung ceilings without worry, and attached to a building slab without concern about an engineering report to determine if the building floor load factor might be exceeded. Fiber easily snakes its way through crowded conduits. In fact, one of the earliest incentives to run fiber optic telephone cable in metropolitan areas was to relieve the stress on crowded trunk cable conduits.

Also, a single fiber optic cable may contain 12 to 36 strands of fiber. Each can be spliced in about 10 to 20 minutes, depending on the type of fiber and immediate conditions. So service can be restored very quickly.

Drawbacks of Fiber Optic Cable

Fiber, in spite of its numerous advantages, is far from perfect. The most serious drawbacks of fiber optic cable are outlined in this section.

Problems Bending Fiber

This is probably the most serious drawback of fiber. Most of the light in fiber optic cable travels in the core and is reflected back into the core at the boundary of the cladding and core. The amount of reflection, and thus the signal loss, is determined by the angle of the light striking the core/cladding boundary. Only when the angle is low does the light stay in the core. Bend the fiber too much, and too much light escapes into the cladding and is lost for information carrying purposes.

This is one reason why early fiber optic cables were seen to be fragile and difficult to work with. Fibers today still routinely add strengthening materials such as Kevlar and are covered far beyond the cladding with additional layers, usually of plastic, called *buffering*. By the time even a single fiber strand is completely jacketed, it can be as thick as twisted pair cable. Not only does this additional layering make the fiber easier to handle or even to *see* (unbuffered fibers can be thinner than human hairs), the added layers help to prevent the fiber optic cable from being bent too tightly around corners when installed. The fiber can always be kinked, however, often from something as simple as stepping on it the right way. The point is that the strengthening is not so much to prevent breakage when the fiber optic cable is pulled (fiber has stronger tensile strength than steel wire of the same cross section), but to prevent kinks and tight bends.

Fiber optic cable types differ in the amount of bending allowed. It all depends on the refractive index difference between core and cladding. The higher the difference, the tighter the allowable bend. However, there is a penalty from making the refractive index too high, and for many fiber optic cables, it is best to keep the refractive index differential between core and cladding as small as possible.

Problems Connecting Fiber

Another drawback of fiber optic cable concerns the ways that fiber optic cable sections can be connected. Connecting should be done in a fashion that minimizes signal loss or attenuation, naturally. The best way to join fiber optic cables is with a *fusion splice*. A fusion splice actually melts part of the glass in the two fibers together, and if done correctly and with care it can yield a connection point with very little signal loss. One problem with fusion splices is that they are difficult to do well outdoors when the temperature is low, such as during a North American or European winter. Not too long ago, fusion splicing was a complex procedure requiring highly skilled personnel. Some technicians had reputations as good fusion splicers, and some were shunned. One bad fusion splice could doom a link many miles long. Now "fusion kits" have taken much of the human factor, as well as the excitement and drama, out of fusion splicing.

Fusion splicing is fine for permanent installation of fiber optic cable. But there are always places on a link where fusion splicing is not practical. In particular, the connection to the transmitter or receiver should not be a fusion splice, or else changing either end of the fiber link when a transmitter or receiver has failed or needs to be upgraded can be difficult. There are also purely physical fiber optic connections, such as fiber patch panels that can be rearranged as needed. In these cases, the fiber join is best accomplished with a fiber optic connector. Most modern fiber optic transmitters and receivers come with short connectorized fibers ready-made to plug into the fiber link itself. This is called a *pigtail*. A lot of the signal loss on a fiber link comes from the pigtail connectors.

Early fiber connectors were not very good and had large signal losses (up to 3 dB per connector). Large cores are easier to connectorize with acceptable signal losses, of course. Fiber optic LANs, which employ large core fibers for the most part, can use connectors all over the place, and the short distances involved keep signal losses low. A typical connector loss today is about 0.3 dB, good enough for LANs, but not for long WAN links.

Limitations of Electrical Components

This is a more subtle drawback. Until more of the network nodes function in a purely optical fashion, fiber optic cables will start and end at electrical devices.

Potential Damage to the Glass

If the glass that a fiber optic cable is composed of changes its optical properties during operation or permanently, the fiber link will stop working. Here are the most common environmental threats to fiber optic cable:

▼ **Water**   Water gets into everything if enough time passes. Some fiber optic cables, especially older ones, have tiny voids or "air bubbles" in the glass. If water penetrates the glass of the fiber optic cable and fills the bubbles, the fiber stops working. This is one reason that undersea fiber optic cables are sheathed with layers and layers of water-resistant materials.

■ **Ultraviolet (UV) radiation**   UV radiation is very penetrating. This is why people can get a sunburn even on a cloudy day. Aerial fibers must be protected from solar UV radiation, since this could break down the boundary between the core and the cladding on the fiber.

■ **Gamma radiation**   Gamma radiation darkens many types of glass. It can also cause the glass to emit light, or "glow," and is thus a source of noise. Nuclear power plants are not friendly to fiber.

■ **High-voltage electrical fields**   Voltages above 30,000 volts tend to make fiber optic cables emit light and discolor as well. Ironically, fiber has to be shielded when used along high-voltage transmission lines.

▲ **Chomping**   In spite of many reports, shark bites on undersea fibers appear to be a legend. Fibers are no more appetizing to sharks than any other cable laying on the ocean floor. But squirrels, gophers, and termites will merrily chew through fiber optic cable (and almost anything else in their path that does not get out of the way).

FIBER OPTIC CONNECTORS

Many types of fiber optic connectors are in use today, although some have fallen out of fashion and others remain wildly popular. Generally, the trend is now away from bulky connectors to smaller types, and connectors that join multiple fibers, especially for LAN applications. It is important to note that the types of *fiber* being connected are much more important than the type of connector used. In most cases, simple adapters will allow one type of connector to interface successfully with another. Some of the latest and most popular types of connectors are illustrated in this section.

In no particular order, the fiber optic connector types are:

▼ **SC connector** The Subscriber Connector (sometimes called the Square Connector) has been the workhorse of the fiber optic connector world for a long time. A variation of the SC connector is the *duplex SC* connector, sometimes called the *568SC* connector. The variation is just two SC connectors in one plastic housing. The connector is "pull proof," meaning that the fiber will not pull out of the connector when someone trips over it: the connector gives first. SC connectors can be terminated with the same tools as some older connectors discussed later in this section (the ST in particular). The biggest drawback of the SC connectors is that they are relatively wide, making port densities in vendor equipment and faceplates a real issue. The typical loss for the SC connector is about 0.15 dB, which is very good. The SC connector is shown in Figure 2-5.

■ **LC and LC duplex** LC seems to stand for "Lucent connector" because Lucent pioneered it. This connector is only one-half inch wide, as opposed to older inch-wide connectors. The duplex version fits into a standard copper RJ-45 connector "footprint," which is nice when upgrading a LAN from copper to fiber. This connector, like many others, features an audible "click" when latched and is pull-proof. The typical loss is only 0.1 dB. LC connectors can be used in LANs, WANs, and cable TV applications. IBM is using LC connectors in newer Fibre Channel products. The LC connector is shown in Figure 2-6.

■ **MT connector** The Media Termination connector is used to terminate fiber optic *ribbon cable.* Ribbon cables are groups of fiber optic cables manufactured side-by-side, forming a fat, flat ribbon structure. MT connectors terminate anywhere from 2 to 12 fiber optic cables all at once. This is good for some forms of fiber, mostly used for LANs. The typical loss is about 0.25 dB.

Figure 2-5. The SC connector

Figure 2-6.   The LC connector

■ **MT-RJ connector**   The Media Termination-Recommended Jack connector is another "half sized" connector. This is the product of a consortium of leading companies (HP (Agilent), AMP, Siecor, Fujikura, and US Conec) seeking to develop a connector compatible with copper RJ faceplates. Many equipment vendors have embraced the MT-RJ, among them Cisco Systems. Suitable for LANs and WANs, these connectors have a typical loss between 0.2 and 0.4 dB. The MT-RJ connector is shown in Figure 2-7.

■ **MP connector**   The MultiPurpose (meaning not certain) connector is a variation on the standard SC connector adapted for multiple fiber connections, from 2 to 12 per connector, similar in plan to the MT connector. It is the same size as the SC connector, and MT plug compatible. MP connectors are mostly used for LANs but can also be used as board-to-board "jumpers" and for mainframes. The typical loss is about 0.4 dB. The MP connector is shown in Figure 2-8.

Figure 2-7.   The MT-RJ connector

Figure 2-8. The MP connector

- **E2000 connector**   Also known as the FLSH connector, this connector is used in LANs and WANs, and cable TV systems as well. The nice thing is that the connector allows the fiber to be positioned at an angle, in 60 degree increments. The E2000 connector has a typical loss of 0.1 dB. The E2000 connector is shown in Figure 2-9.

- **ST connector**   The Straight Tip connector has fallen out of favor somewhat but is still widely used, especially in WANs. Pioneered by AT&T, the ST also comes in a duplex flavor. Unlike modern fiber connectors, the ST was quite large and twist-locked onto the receptacle, just like a coaxial cable TV connector. The problem was that if someone tripped over the cable, the fiber pulled out before the connector gave way! So the fiber had to be reterminated. The ST connector is still sometimes seen in WANs, and LANs and cable TV systems as well. The typical loss is about 0.25 dB. The ST connector is shown in Figure 2-10.

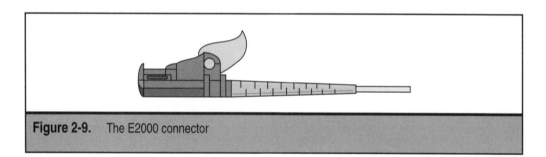

Figure 2-9. The E2000 connector

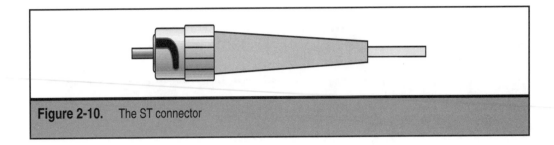

Figure 2-10. The ST connector

■ **FC connector**   The Fibre Channel connector is quite similar to the ST connector. A variation of the FC connector is the *D4 connector,* which adds a hood over the fiber tip for protection. The same tools used for SC and ST termination can also be used for the FC connector. Typical loss is 0.15 dB. These connectors are used for Fibre Channels LANs, although in some cases FC links can reach WAN distances. The FC connector is shown in Figure 2-11.

■ **Biconic connector**   The biconic connector was also pioneered by AT&T and has also fallen out of favor. Seen almost exclusively on WANs, the screw-type connection was difficult to remove easily. Typical loss is 0.35 dB, relatively high for today's optical networking needs. The biconic connector is shown in Figure 2-12.

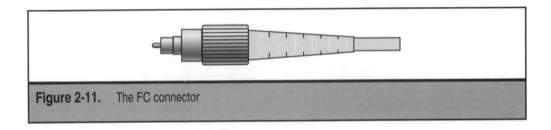

Figure 2-11. The FC connector

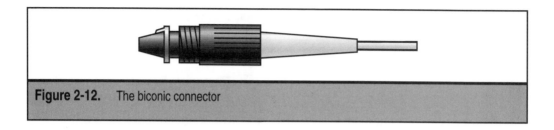

Figure 2-12.   The biconic connector

■ **FDDI connector**   The Fiber Distributed Data Interface connector (also known as the FDDI *MIC [media interface connector]*) is used exclusively with FDDI networks. FDDI is an early (1990) fiber ring architecture, very similar to a token ring LAN running at 100 Mbps on fiber up to 100 km (about 63 miles). The connector terminated both inbound and outbound fibers in one housing. Today, since FDDI is becoming scarce, it is common to adapt FDDI connectors to something else such as an ST connector. The typical loss is 0.2 dB. The FDDI connector is shown in Figure 2-13.

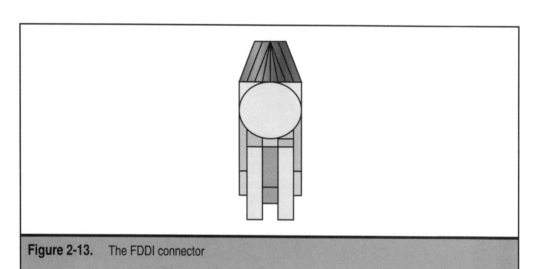

Figure 2-13.   The FDDI connector

▲ **ESCON connector**   The Enterprise System Connector is used exclusively with IBM's ESCON architecture. ESCON is a way of linking IBM computers over fiber, both LANs and WANs. Typical loss is 0.2 dB. The connector is quite similar to the FDDI connector.

There are even other types of fiber optic cable connectors for specialized use, such as the Quantum QX used for ribbon cables. AMP Corporation had the SMA connector, which was also popular and still is for military use. But this section has detailed the most common types of fiber optic cables in use today.

MAKING FIBER OPTIC CABLE

Fiber optic cable is very small. Most fibers are only 125 microns (micrometers, or millionths of a meter) in diameter including the cladding. Buffering, which is the process of placing layers of protective plastic coating around the cladding, can double this diameter to 250 microns, but the fiber is still quite small. Cores can be only 9 or 62.5 microns in diameter, which is smaller yet. To put these sizes in perspective, the human hair has a diameter of about 125 microns (to be strictly accurate, human hair ranges from about 100 to 140 microns in diameter). This is one reason fibers are buffered with layers of white plastic: otherwise, fiber cables would be quite hard to see and pick up!

The small size of fiber optic cable raises an important issue. How are structures so small fabricated in the first place? This section will take a quick look at how fiber optic cable is manufactured.

The goal of all of the fiber optic cable manufacturing methods is to create a solid rod known as the *preform*. The preform has all of the optical properties of the finished fiber optic cable, but on a very large scale compared to the finished fiber. A preform can be up to several inches in diameter and several feet long. To make the finished fiber, the preform is stood on end in a *drawing tower* and the lower end of the preform is uniformly heated. The miracle of fiber optic cable manufacture is that the preform melts in the exact optical proportions—core and cladding—laid down in the preform. The hot fiber is cooled, buffered, cooled again, and put onto a take reel all in one operation. This process is shown in Figure 2-14.

So where does the preform come from? That's relatively easy. There are four main ways to make the preform:

1. Outside vapor deposition (OVD, used in the United States and Europe)
2. Vapor axial deposition (VAD, used mainly in Japan)
3. Modified chemical vapor deposition (MCVD, also used in the United States and Europe)
4. Plasma-activated chemical vapor deposition (PCVD, a faster and improved MCVD method)

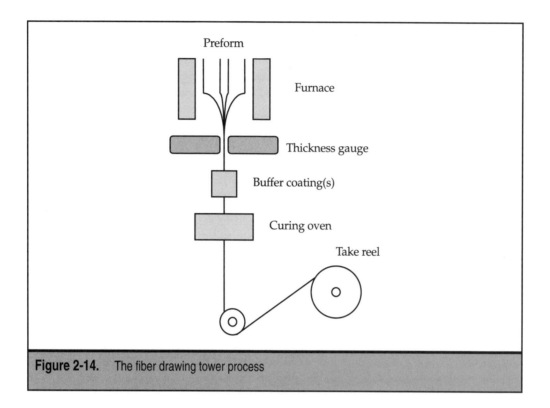

Figure 2-14.   The fiber drawing tower process

The first two methods are closely related, as are the last two. The first two methods essentially build the fiber optic cable from the inside out, and the last two basically build the cable from the outside in. Only these two main methods need be detailed to understand the basics of fiber optic cable manufacture.

Fiber optic cable is very pure silica glass. Purity is important to produce the low attenuation that characterizes modern fiber optic cable. The complicating factors in fiber optic manufacture are the high temperatures involved and the precise amounts of dopants (impurities) that must be added in just the right quantities to achieve the optical properties desired in the fiber optic cable. The final production of fiber optic cable is actually 1,000 times purer than the materials used as input to the manufacturing process.

The fiber optic glass itself is made by the high-temperature reaction of gases. There are many reactions that can be used, but the following reactions are typically used in the MCVD production process.

$$SiCl_4 \ + \ O_2 \ \rightarrow \ SiO_2 \ + \ 2Cl_2$$

$$GeCl_4 \ + \ O_2 \ \rightarrow \ GeO_2 \ + \ 2Cl_2$$

$$4POCl_4 \ + \ 3O_2 \ \rightarrow \ 2P_2O_5 \ + \ 6Cl_2$$

The fluoride forms are sometimes used, but here the substances are silicon chloride, germanium tetrachloride, and phosphorous oxychloride, or "sickle," "geckle," and "pockle" to those who work with these materials every day. These are oxidation reactions, but OVD and VAD use hydrolysis reactions employing both oxygen and hydrogen. The dopants introduce the desired characteristics into the glass, such as the refractive index or expansion coefficient.

Whichever specific ingredients and reactions are used, all modern fiber optic cable manufacturing techniques employ some form of vapor deposition. The difference in the four manufacturing forms lies mainly in the details.

In VAD and OVD, the fiber is built basically from the inside out, core first and then cladding, around a carbon rod in the case of OVD. In OVD, the preform is made horizontally, and in VAD the preform takes shape vertically. In MCVD and PCVD, the fiber optic cable is essentially built from the outside in, cladding first and then core, inside a tube. The refractive indices are controlled in all cases by simply changing the composition of the input vapors used. It is the "soot" gathering on the rod or inside the tube that slowly builds up the fiber itself. The main ideas behind these methods are shown in Figure 2-15.

Each of these methods is important enough to discuss more fully.

Outside Vapor Deposition (OVD)

The OVD method employs a technique known as *flame hydrolysis.* The main reaction takes the "sickle" ($SiCl_4$) and produces silica (SiO_2) with HCl as a by-product. This happens inside an oxy-hydrogen flame, producing small particles of molten silica (the "soot"). A carbon or metal rod rotates on a lathe inside the furnace, and the silica flame is moved back and forth so that the silica is deposited evenly. This happens because the rod is much cooler than the air inside the oven. Dopants like germanium are added to the flame to control the refractive index as first core and then cladding are laid down.

Naturally, once the process is stopped and the rod is withdrawn, there is a rather large hole down the middle of the core. So the preform is heated again (sintered) to collapse the hole. Only a small part of cladding is built up in this way. The rest of the cladding is added by inserting the collapsed preform into a silica tube with the proper cladding optical properties. Then the whole assembly is heated and collapsed yet again.

If there seem to be a lot of steps and heating involved in OVD, that is because OVD was the first practical method of making fiber optic cable. OVD also had problems with water (really OH radical groups) inside the fiber. The OH groups absorb light at a wavelength of 1,385 nanometers (nm), making it imperative to avoid this region of operation on the finished fiber cable. OVD fiber tends to have a depressed refractive index along the axis as well. And finally, the preforms are necessarily limited in size and another step with other equipment is needed to draw the cable. There can be a lot of handling in between these *batch* processes, making it an expensive process and harder to control quality end to end.

The batch processing limitations of OVD, along with the high *water peak,* both eventually led to the abandonment of OVD as a commercial manufacturing method.

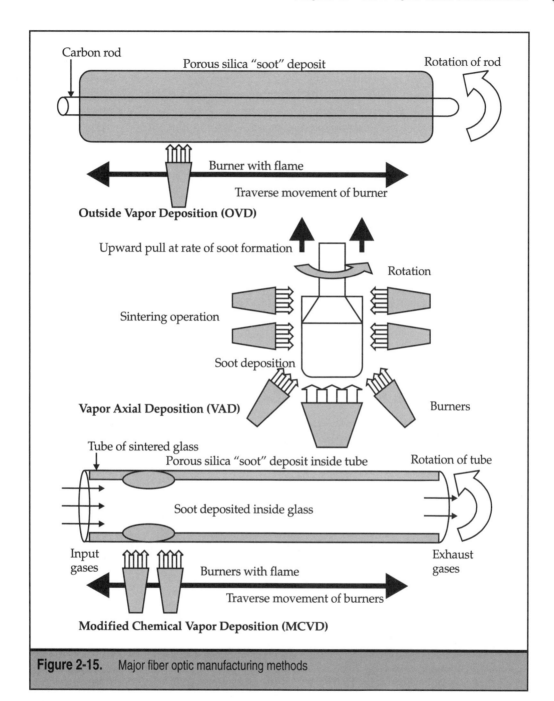

Figure 2-15.   Major fiber optic manufacturing methods

Vapor-Phase Axial Deposition (VAD)

Think of VAD as contemporary OVD. A large percentage of all of the world's fiber optic cable is made with VAD. The basic idea in VAD is the same as OVD, but the whole apparatus is stood on end, making for larger preforms (up to 250 km of fiber can be drawn from VAD preforms). VAD can also be used as a continuous process, which can be very cost effective, but most VAD is still a batch process.

Instead of moving the flame, soot vapor deposition and the refractive index are controlled by slowly pulling the preform upward during rotation. The sintering process occurs further up the preform. Water contamination can be controlled better this way, and the preform rod can be fed directly into a drawing process.

Modified Chemical Vapor Deposition (MCVD)

MCVD is fundamentally different from OVD or VAD. MCVD grew out of the inside vapor deposition (IVD) method, and MCVD is used today to produce most of the world's supply of fiber optic cable. MCVD forms silica inside a premade, rotating tube of silica. The gases used are introduced into one end of the tube, and the flame burners travel the length of the tube. The soot is deposited internally, and exhaust gases leave the other end of the tube. At the end of the process, there is a hole in the center of the preform in this method as well. To seal this hole, the heat is turned up and the tube collapses as the burners pass slowly from end to end.

The MCVD method has a number of advantages. First, no water is produced, so the OH absorption peak can be eliminated in many cases. Next, the refractive index profile can be very closely controlled. Also, the process is fast and can be integrated into a complete preform-to-draw process. Finally, MCVD handles the germanium core dopant much better than other methods, since these other methods tend to evaporate the germanium when closing the central "hole" in the fiber core. There is always a risk then that the refractive index profile will dip along the axis of the fiber core, which can be a problem in some types of graded index multimode fiber optic cable (see the section "Fiber Optic Cable Types" for more).

Plasma-Activated Chemical Vapor Deposition (PCVD)

Think of PCVD as an MCVD variation. The variant is that instead of heating the outside of a silica tube as in MCVD, the PCVD method uses microwaves to heat the inside of the tube in a high-powered version of a microwave oven. Since the microwaves are a field, they can be moved quickly along the tube, and the silica tube does not heat up itself. So the process can employ thousands of passes instead of hundreds, and the thinner layers can be controlled very precisely in terms of refractive index.

This is not to say that the tube in not heated. But the temperature is a modest 1,000 degrees Celsius, not 1,600 or more degrees as in other processes. Another nice feature of PCVD is that there is no need to rotate the tube. The reaction is very efficient and is faster than MCVD. Preforms can be made that can yield several hundred kilometers of fiber.

FIBER OPTIC CABLE TYPES

Three major types of fiber optic cables are used in optical networking today. In addition, two variations on one of the major types form distinctive subfamilies. These newer fiber variations are of great interest in optical networks that stretch more than a few miles or kilometers. The major types are:

▼ Multimode step index (fine for LANs)

■ Multimode graded index (GRIN) (even better for LANs)

▲ Single mode (step index) (best for WANs)

In addition, single-mode fibers are divided into two major categories today:

▼ Dispersion shifted and dispersion flattened (good for WDM)

▲ Non-zero dispersion shifted fiber (NZ-DSF: best for WDM)

Some newer fibers seem to break all the rules for optical network operation. These will be discussed at the end of this section.

The three types of fiber differ in the way that light travels down them. Multimode fibers allow multiple paths, called *modes,* for the light to travel in them. The various modes tend to interact ("interfere" is the usual term, but not all light interactions are destructive) with each other in these fibers, so *modal dispersion* can be a problem, limiting speeds and distance. The term *dispersion,* often used as a generic term in fiber optics, usually means the same as the term "noise" used in electrical networks, although purists might cringe at such a gross comparison. Whatever the definition, dispersion is almost always bad in optical networking. Although the light in the fiber is anything but a ray, representing light in fiber as a line with an arrow attached to it is the norm, and that is the convention used here. In reality, light more or less spirals its way helically down the fiber, dipping into the cladding surrounding the core before being reflected back along the axis of the core.

Single-mode fiber allows only one mode of light to propagate down the fiber. In single-mode fiber, all of the other modes introduced at the sender interact and cancel each other out ("destructively interfere") and only one mode arrives at the receiver. Multimode fibers are limited in terms of speed and distance because the multiple modes arriving make the job of the receiver more difficult. So optical networks needing the highest speeds and spanning the greatest distances always use some form of single-mode fiber.

The overall structure of the three main fiber optic cable types is shown in Figure 2-16.

The conventions used in naming these fibers are just about universal. It is common to give the diameter of the core in microns or micrometers (same thing), followed by the outside diameter of the core and cladding together. The most common outside diameter is 125 microns, and although during buffering layers of plastic are added to the fiber, these do not seem to count (the buffers make no difference optically in any case).

So we have 50/125 MMF (multimode fiber) and 62.5/125 MMF. There are also 9/125 SMF (single-mode fiber) and so on, although these are the main types seen. The

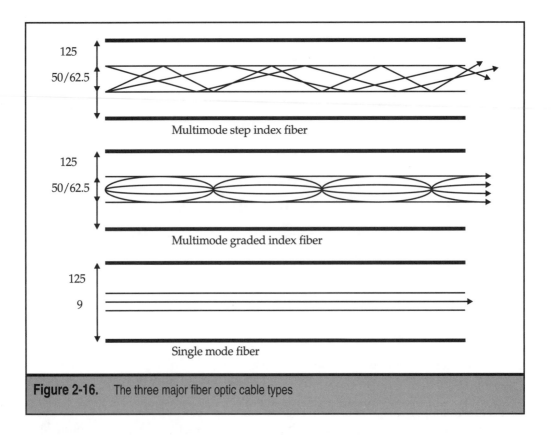

125

50/62.5

Multimode step index fiber

125

50/62.5

Multimode graded index fiber

125

9

Single mode fiber

Figure 2-16. The three major fiber optic cable types

core always has a higher RI than the cladding, which is why the light is reflected back into the core at the boundary: the cladding looks like a mirror to light in the core.

Multimode Step Index Fiber

Light does not enter fiber at an infinite number of angles. The angles are determined by the numeric aperture (NA) of the fiber, as previously discussed. Nor does light travel in the fiber over an infinite number of paths to emerge haphazardly at the receiver. There are only a finite number of paths that the light can take down a fiber. These are called *modes*. All of the other possible paths will interact with each other and disappear within the fiber itself in the process known as destructive interference.

The number of possible modes through a fiber optic cable is determined by three factors. These are the core diameter, the wavelength used, and the difference between the refractive index in the core and that in the cladding. A simple formula relates the three factors. The larger the core, the more modes that are possible at a given wavelength. Whenever the core diameter is large enough to propagate multiple modes down the fiber, this is called *multimode* fiber. Fiber optic cables with core diameters of 62.5 microns (very common) have about 400 modes at the 1,300 nm wavelength.

The term *step index* refers to the fact that there is an abrupt transition between the refractive index used in the core and the lower RI used in the cladding. Multimode step index fibers are commonly used today in LAN applications. These fibers are limited in distance and bandwidth, but this seldom becomes a factor at the LAN networking scale. Most fiber optic LANs still offer much higher bandwidths and much longer runs than other media. The limitations of multimode step index fiber are due to the phenomenon of modal dispersion.

The bad thing about modal dispersion is that the different modes (paths) through the fiber are of different lengths. So light arrives at the end of the fiber spread out in time ("dispersed"). Naturally, this limits the characteristic bit time used to represent a 0 or 1 bit on the fiber, which in turn limits the useful bandwidth. The longer the fiber, the more the paths differ in length, so distance is limited as well. To go faster, go shorter. To go farther, go slower. Those are the main tradeoffs for multimode step index fiber, although the wavelength(s) used also play a role today.

Sometimes this multimode fiber relationship between bandwidth and distance is expressed as the "bandwidth.distance" (read as the "bandwidth dot distance product") of the fiber. Bandwidth on fiber is usually expressed in MHz, since light is an analog signal form. As long as the bandwidth used for information transfer times the distance the fiber spans does not exceed the bandwidth.distance product established for the particular multimode fiber, the fiber link should work as advertised. Typical figures today are about 1,000 MHz.km for 62.5/125 multimode fiber. In other words, if the distance spanned is 100 km, then 10 MHz can be used for information transfer. If 200 MHz is needed for information transfer, then only 5 km can be spanned, and so forth. (The bandwidth.distance product only properly applies to multimode fibers, not single-mode fibers that have no modal dispersion.)

The major limitation is the modal dispersion in all multimode fibers. This contributes to what is called *intersymbol interference* at the receiver. Generally, all forms of dispersion, not just modal dispersion, contribute to the total intersymbol interference at the receiver.

Multimode Graded Index (GRIN) Fiber

One of the biggest contributors to modal dispersion in multimode fibers is the abrupt difference in the higher core RI and the lower cladding RI. Multimode graded index (GRIN) fibers gradually change the RI from the center to the edge of the core. Note that the RI in the cladding is *not* graded, only in the core itself. This limits the number of modes propagated and so cuts down on intersymbol interference.

The reason GRIN fiber is so effective at cutting down on modal dispersion is that light near the center of the core experiences a higher RI than light traveling near the cladding. This slows the light traveling on these physically longer paths. So light follows a more curved path in a GRIN fiber. The whole idea is to keep the light traveling at about the same speed through the fiber. So a light pulse, even though the pulse still has many modes, stays together better as it travels down the fiber. This allows for higher bandwidths and longer runs. However, the GRIN fiber manufacturing process must be very precisely controlled, a factor which can make GRIN fiber quite expensive compared to other fiber types.

It is sometimes said that GRIN fiber "gives multimode fiber some of the advantages of single-mode fiber" in many LAN applications. To appreciate this statement, it is necessary to examine single-mode fiber in some detail.

Single-Mode Fiber

The trick of single-mode fiber is that the core is very small, usually about 9 microns instead of the 50 or 62.5 micron cores used in multimode fibers. The smaller core makes it possible for only a single mode to propagate down the fiber, although this also depends on the wavelength used. The single-mode fiber must be carefully "tuned" to the anticipated wavelength used. This can be an issue when multiple wavelengths are in use, as with DWDM. So in a very real sense, it is somewhat misleading to speak of just "multimode" or "single-mode" fiber without specifying wavelength, although people talk like this all the time. *All fibers are multimode or single-mode only within a certain range of wavelengths.* At very long wavelengths, even "multimode" fibers will propagate only a single mode. And at 600 nm (for example), even single-mode fiber will exhibit multiple modes, although perhaps only three or four modes. So all single-mode fibers have what is called the *cutoff wavelength.* Below the cutoff wavelength, a "single-mode" fiber is really multimode. Typically, the cutoff wavelength for 9 micron single-mode fiber is 1,100 nm.

Assuming the single-mode fiber is used at the intended wavelength, there is no longer any reflection at all between core and cladding as light travels down the single-mode fiber. Light essentially travels right down the center of the core. This sounds simple enough, but it is difficult to execute in practice. It is hard to inject light into a small core, since the NA is very small. The *injection loss* is high. Also, a large portion of the light (up to 20 percent) actually travels in the cladding. This means that the *apparent diameter* of the core where the light travels is larger than the core itself. For this reason, single-mode fibers are often characterized by the *mode field* or *mode field diameter* instead of the actual core diameter. The value of the mode field diameter depends on the difference in the RI between the core and the cladding, which is much less in single-mode fiber than in multimode fibers. Light traveling in the cladding is the major source of one type of dispersion found in fiber optic cables, called *waveguide dispersion.*

The single-mode fibers used today are the result of a series of tradeoffs in the optical fiber design, as should come as no surprise. Smaller cores have higher losses when the fiber is bent around corners. If the minimum bend radius of single-mode fiber is exceeded, the light escapes from the core into the cladding and is lost at the receiver. Three factors are important in the design of single-mode fiber. First, the core must be small enough, but not too small. Second, the difference between RI in core and cladding must be small, but not too small. Finally, the wavelength used must be long enough, but not too long (or else no transmitter exists at that wavelength).

Single-mode fibers have very low attenuation, typically half that of multimode fibers. This actually has little to do with modes or structure. The major source of attenuation in fiber is signal loss due to the presence of the dopants that are used to separate the RIs between

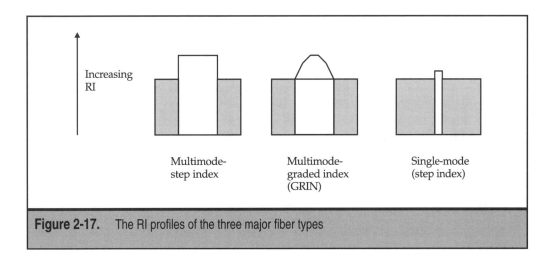

Figure 2-17.   The RI profiles of the three major fiber types

core and cladding. Since single-mode fiber RIs differ much less between core and cladding than multimode fibers, less dopant is needed. This automatically gives single-mode fibers less attenuation and makes single-mode fiber the best choice for WAN optical networks. The RI profiles of all three major fiber types are shown in Figure 2-17.

DEALING WITH CHROMATIC DISPERSION
AND WAVEGUIDE DISPERSION

Single-mode fibers do away with modal dispersion entirely. However, there are many more types of dispersion that conspire to limit the useful bandwidth and distance on a fiber optic span. Two of the more basic types of dispersion are known as *chromatic dispersion* and *waveguide dispersion*. Sometimes chromatic dispersion is called *material dispersion*, and the term "chromatic dispersion" is used to describe the sum of both types. The book will just refer to the sum of chromatic (material) dispersion and waveguide dispersion as *total dispersion*. Since many modern single-mode fibers have been designed to manipulate these two major forms of dispersion, a few words about each are in order.

Waveguide dispersion occurs because light on a single-mode fiber also travels in the cladding. Since light travels faster at lower RIs, and slower at higher RIs, the light in the cladding arrives slightly before the light traveling in the core. In the cladding, the light is at a slightly longer wavelength than the light in the core and travels quickly in the lower RI material. Therefore the shorter wavelengths in the core, where the RI is higher, arrive "late" at the receiver. So the pulses are dispersed.

Chromatic dispersion, which is just another term for material dispersion, occurs because the RI in a fiber core varies slightly with wavelength. This is called the *optical Kerr effect* and is one of the facts of optical networking. All light pulses, even those from laser light sources, generate more than one wavelength in most cases. The usual spread is from about 0.2 to 5 nm and is called the *spectral linewidth* of the pulse. So even though single-mode fiber has only one mode, the different speeds of these slightly different wavelengths in the spectral linewidth cause chromatic dispersion.

Oddly, these two types of dispersion tend to operate in *opposite* directions. In other words, at many wavelengths, one form of dispersion tends to smear the light pulses out while the other form tends to "squish" the pulses together. Neither form of dispersion is really good, however: it makes little difference at the receiver if the light pulse disappears because it has spread out too much over time or if it has been compressed down to nothing!

Fiber dispersion is often measured in terms of picoseconds per nanometer per kilometer (ps/nm/km). Sometimes dispersion is also measured in picoseconds per kilometer per nanometer (ps/km/nm), or even picoseconds per nanometer-kilometer (ps/nm-km), but the effect is more important than the units used. The effects of the two forms of dispersion are called *chirp*. Positive chirp occurs when the longer wavelengths arrive before the shorter ones. Negative chirp occurs when the shorter wavelengths arrive before the longer ones.

What can be done about dispersion with single-mode fiber is to carefully design the fiber optic cable so that the waveguide dispersion and chromatic dispersion cancel each other out at a particular wavelength. Since the total dispersion is just the simple sum of the chromatic and waveguide dispersion effects, it is possible for the total dispersion to be exactly zero. In most single-mode fibers, this *zero dispersion point* occurs at about 1,310 nm. At this wavelength, light pulses will not disperse, although they will still attenuate, of course. The zero dispersion point is shown in Figure 2-18.

One of the reasons that the synchronous optical network (SONET) and synchronous digital hierarchy (SDH) operate at 1,310 nm is to take advantage of this zero dispersion point. In Recommendation G.652, in fact, the ITU-T has defined the standard single-mode fiber to have:

▼ A cladding diameter of 125 microns

■ A mode field diameter of 9–10 microns at 1,300 nm

■ A cutoff wavelength for single-mode operation at 1,110-1,280 nm

■ A bend loss at 1,550 nm of less than 1 dB, when wound 100 times on a spool 7.5 cm in diameter

■ Dispersion less than 3.5 ps/nm/km from 1,285 to 1,330 nm, and less than 20 ps/nm/km about 1,550 nm

▲ A rate of dispersion change with wavelength that must be less than 0.095 $ps/nm^2/km$ (the *dispersion slope*)

This is the type of single-mode fiber used for most Sonet/SDH systems today.

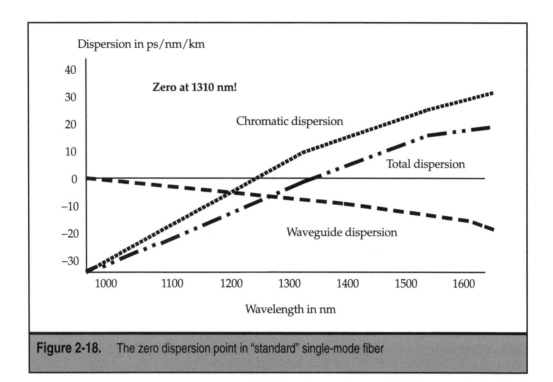

Figure 2-18.   The zero dispersion point in "standard" single-mode fiber

DISPERSION SHIFTED FIBERS

However, for modern optical networking, it would be better to shift the zero dispersion point to the neighborhood of 1,550 nm. This is where most DWDM equipment operates today. This operational range has been chosen because overall attenuation is lower at 1,550 nm and most fiber amplifiers operate much more efficiently around the 1,550 nm wavelength. Standard 1,310 nm zero point single-mode fibers can be used for DWDM systems, of course, but the dispersion on these fibers at 1,550 nm is rather large, around 17 ps/nm/km. Lasers with very small spectral linewidths will help, but such lasers can be expensive. It would be more cost effective to have single-mode fiber with a zero point around 1,550 nm.

Fortunately, manipulating the interplay between chromatic and waveguide dispersion is relatively straightforward. This can be done during fiber optic cable manufacture by carefully controlling the profile of the RI in core and cladding. Basically, the idea is to add a larger waveguide dispersion effect. Waveguide dispersion is caused by light traveling in the cladding, so the RI core profile is more complex to compensate for this effect. A dispersion-shifted fiber with a zero dispersion point is shown in Figure 2-19.

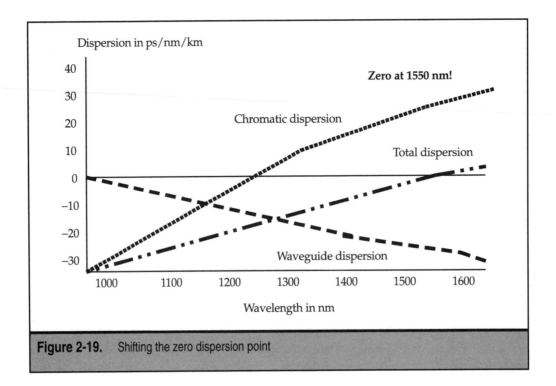

Figure 2-19. Shifting the zero dispersion point

There is also dispersion-flattened fiber, which has relatively low dispersion (about 3 ps/nm/km) in the whole range from 1,300 to 1,700 nm. However, these fibers have proved to have very high attenuation and are seldom used for WAN applications.

In Recommendation G.653, the ITU-T has defined a standard dispersion-shifted single-mode fiber designed to operate in the 1,550 nm range to have:

▼ A cladding diameter of 125 microns

■ A mode field diameter of 7.8–8.5 microns at 1,550 nm

■ A cutoff wavelength for single-mode operation at 1,270 nm

■ A bend loss at 1,550 nm of less than 0.5 dB, and less than 1 dB around 1,580 nm, when wound 100 times on a spool 7.5 cm in diameter

- Dispersion less than 3.5 ps/nm/km from 1,525 to 1,575 nm
- A rate of dispersion change with wavelength that must be less than 0.085 ps/nm^2/km (the *dispersion slope*)

Dispersion-shifted fiber might seem ideal for DWDM systems. But there are still problems. For highly technical reasons, light signals on dispersion-shifted fibers break up more often than on standard fiber optic cable. And ironically, having a number of wavelengths traveling nearly all at the same speed on the same fiber maximizes another impairment effect known as *four wave mixing (FWM)*. When many wavelengths are in phase over long distances, FWM is a real problem. FWM is the optical equivalent of interference between channels, a kind of optical near-end cross talk (NEXT).

So it is actually better to have a small amount of dispersion near the 1,550 nm DWDM operation point. These fibers exist, and are called non–zero dispersion–shifted fibers (NZ-DSF). These fibers have a complex core RI profile and so can be very expensive, but they are optimized for DWDM systems. The NZ-DSF dispersion is about 4 ps/nm/km in the 1,530 to 1,570 nm range. This small positive dispersion minimizes the effects of FWM in DWDM systems, although it is possible to use other fiber types.

In the United States, the two main sources of fiber optic cable are Lucent (formerly AT&T) and Corning. Lucent calls their NZ-DSF Tru-Wave, and Corning uses the designation SMF-LS. The core RI profiles of dispersion-shifted fiber, dispersion-flattened fiber, and NZ-DSF are shown in Figure 2-20.

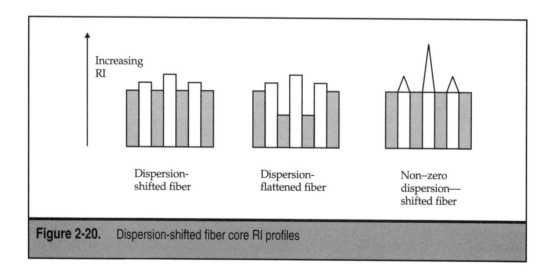

Figure 2-20.   Dispersion-shifted fiber core RI profiles

In Recommendation G.655, the ITU-T has defined a standard NZ-DSF fiber designed to operate in the 1,550 nm range and to minimize WFM effects to have:

▼ A cladding diameter of 125 microns

■ A mode field diameter of 8–11 microns at 1,550 nm

■ A cutoff wavelength for single-mode operation at 1,470 nm

■ A bend loss from 1,550 nm to 1,580 nm of less than 0.5 dB when wound 100 times on a spool 7.5 cm in diameter

■ Dispersion less than 0.1 to 6.0 ps/nm/km from 1,530 to 1,565 nm (final figures still under study)

▲ Rate of dispersion change with wavelength is still under study

Individual fibers are often gathered into larger cable structures. The details vary depending on the purpose of the cable. Some of the more common types are discussed more fully in Appendix A.

CHAPTER 3

Transmitters and Receivers

The previous chapter introduced some of the basic concepts of fiber optic cable. The cable runs between a fiber optic transmitter and a fiber optic receiver. A fiber optic regenerator (or repeater) is just a receiver and transmitter hookup back-to-back on a single fiber span. This chapter, which builds on the foundation of the basic terms and ideas examined in the preceding chapter, explores some of the details of the fiber optic transmitters and receivers. It also investigates the basic operation of fiber optic regenerators. In keeping with the introductory nature of these early chapters, this chapter starts out with a discussion of single-wavelength transmitters and receivers and moves on to consider the needs of transmitters and receivers for multiple wavelength use in WDM and DWDM systems. But the details of a fully functional, multiple wavelength optical system are left for a later chapter.

The ideas presented in this chapter are some of the most interesting and vital ideas in networking today. Only with the proper transmitter, fiber optic cable, and receiver can a fiber optic link function at all. These concepts form the foundation for the idea of combining multiple wavelengths of light in the same fiber to produce wavelength division multiplexing (WDM), and finally of literally jamming these wavelengths as close together as technically possible to produce a "denser" version of WDM known simply as DWDM. This is the challenge for transmitters and receivers and regenerators. All three components change when optical networking is introduced. In fact, the regenerator must become an *optical amplifier* as soon as multiple wavelengths are introduced, for reasons explained later on in this chapter.

A SIMPLE FIBER OPTIC LINK

This section emphasizes the components used on a simple fiber optic link; it is not intended to duplicate the operational aspects of a fiber link covered in Chapter 2. The simplest fiber optic link consists of an optical transmitter, a length of fiber optic cable, and an optical receiver. The optical transmitter itself is not the source of the bits being sent over the link, often called the *span*, but the optical transmitter must *modulate* the electrical bit stream onto the fiber. For example, light intensity above a certain threshold can represent a 1 bit and light intensity below this threshold can represent a 0 bit. So some form of electrical-to-optical (E-O) conversion must happen at the transmitter, and a complementary optical-to-electrical (O-E) conversion takes place at the receiver. This complementary process is usually just covered by the term *E-O conversion*, although it should always be kept in mind that the process must be performed in both directions. The basic architecture of a simple fiber optic link (span) is shown in Figure 3-1. The electrical source and optical transmitter have been bundled into an E-O converter on the sending side. On the receiving side, the optical receiver, or detector, has been bundled with the electrical receiver as another E-O converter, although it is understood that the conversion is actually optical to electrical at the receiver.

There is only one more element that could be added to the figure. If the fiber span is long enough, and not too long ago this meant more than a few miles or kilometers, it would be necessary to place a *regenerative repeater* somewhere along the link. The term

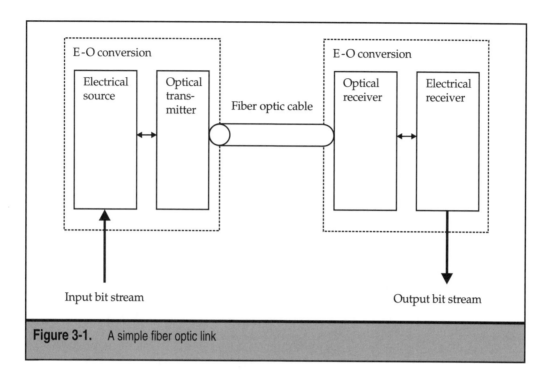

Figure 3-1.   A simple fiber optic link

"regenerative repeater" is a mouthful, and many people just say either *repeater* or *regenerator* and usually mean the same device when using either term. A regenerative repeater must be used due to the attenuation on the link, which reduces the received optical power on the link. The end points that form the source and the destination of the electrical bit stream are just too far apart for the receiver to operate reliably and within the bit error rate that has been established for the link.

How does anyone know just when a repeater is needed on a link? There are some standard span lengths, of course, but the key concept here is the idea of a *power budget.* (Power budgets will be discussed in full in the next chapter.) Standard span lengths rely on standard power budgets, but in many network designs, standards can be sacrificed in favor of a custom solution that works. And even the standards have a number of options and variations that must be considered when deciding just when and where a repeater is needed on the fiber link.

Moreover, regenerative repeaters require electrical power. This power might not be available locally at the standard span distance. Power can always be brought in remotely, sometimes over the sheath of the fiber cable itself. But it might be better to tweak the specification standards or go back to the power budget when designing a fiber optic link. Figure 3-2 shows a fiber span with not only one, but two repeaters. When repeaters are in place on the fiber optic link, it is common to refer to the entire string of devices from input bit source to output bit destination as the *link.* The link consists of one or more *spans,* as determined by the repeaters, but this terminology is by no means universal.

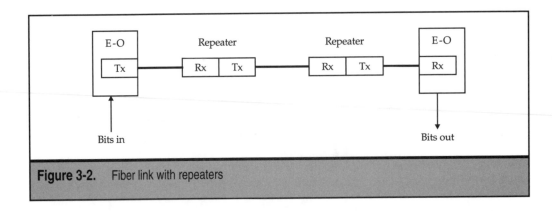

Figure 3-2.   Fiber link with repeaters

Repeaters on a fiber optic link are nothing more than a simple O-E and E-O converter placed back-to-back in the same device. (These are labeled Rx and Tx in the figure.) So the repeater recovers the electrical bit stream from the fiber span entering the repeater, but it then just turns around and remodulates the electrical bit stream onto the fiber span leading out of the repeater. It is not necessary to do any processing on the bits when they are in the repeater, although all practical systems use some of these bits not just for user information transfer, but also for some internal housekeeping and network management purposes.

Actually, the figure is not a particularly good representation of a fiber optic link. Bits can go from left to right in the figure, but how do bits flow in the other direction? It is possible to try to make the same fiber carry bits in both directions. After all, light beams do not interfere with each other when two flashlights are pointed at each other. True enough. The main problem lies in the fact that fiber optic transmitters and receivers usually do just that: transmit or receive. They don't do both.

So it is much more common to have separate transmitters and receivers (Tx and Rx) units in the E-O converters and the repeaters as well. Of course, this requires a second fiber optic cable to be run from E-O device to E-O device, but this turned out to be much less expensive in the 1970s and 1960s than trying to make bidirectional fibers and components operational on a large scale. (Things have changed somewhat today, and optical amplifiers have made bidirectional fibers practical in many cases.)

So a more complete component-level picture of a simple, single-wavelength fiber optic link would be more like the one shown in Figure 3-3.

The fibers in the figure are labeled "outbound" and "inbound" with regard to the E-O converter on the left of the figure. From the perspective of the E-O converter on the right, the top fiber is "inbound" and the bottom fiber is "outbound." Note that the repeaters have now become coupled Tx/Rx units, but there are still four distinct components in the repeater and the optical signals always become bits within the covers of the repeater.

Since regenerative repeaters occupy such a key position on the fiber optic link, it might be a good idea to look at these components of a single-wavelength fiber optic system in some detail. This is also a good idea because the dependable but limited regenerative repeater must be done away with in a WDM or DWDM environment.

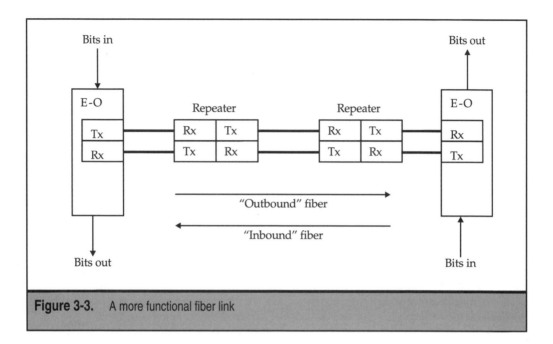

Figure 3-3. A more functional fiber link

REGENERATIVE REPEATERS

Regenerative repeaters can be placed at the end of each span on a fiber optic link wherever they are needed, and an almost unlimited number of repeaters can be placed along the link. This might sound odd. What is it about repeaters that allows them to be cascaded almost without limit along a fiber link?

The key is the fact that repeaters completely recover the bit stream through the receiver E-O converter before sending them out again through the transmitter E-O converter. Repeaters perform three essential services on the span. Repeaters

▼ *Restore* the output signal level.

■ *Reshape* the pulses that represent the bits.

▲ *Retime* the signal pulses.

Repeaters are sometimes called *3R devices* because these three functions are so essential. Each of these restore/reshape/retime repeater actions is worthy of some further explanation.

A regenerative repeater will *restore* the power level of the signal received in the sense that the output power level will be what the transmitter at the repeater is set to generate. This restored power level is usually just as high as the power level at the sending E-O device end of the link. Why not? The repeater transmitter is typically the same as the transmitter in the sending E-O device.

It is important to note that the signal strength restoration that is performed at a regenerative repeater is *not* the same as signal amplification. There is a key difference between restoring signal power levels in a repeater and amplifying signal power levels in an amplifier.

And the difference will prove crucial when we consider the operation of a true *optical amplifier* in a later chapter.

A regenerative repeater is inherently a *digital* device, one that deals with 0's and 1's just like the bit source on the optical link or a computer. An amplifier is inherently an *analog* device, one that deals only with the modulated signal in its native state. The problem is that amplifiers amplify any noise right along with the modulated signal. So amplifier noise is said to be *cumulative* along a link. Sooner or later, the amplified noise swamps the modulated signal and a practical limit is reached. The signal is now so distorted by noise that operation is impossible. Repeaters do not accumulate noise along a link. As long as the noise along a span to a repeater is kept below a threshold, there will be no errors in the bit stream and noise will be removed from the reconstituted signal. This is why cascading repeaters are seldom a problem but cascading amplifiers will cause higher and higher bit error rates at the receiving end of the link.

Regenerative repeaters work by linking E-O conversion steps back to back. So the repeater is essentially an O-E-O device, with optics on each end and electrical bits in the middle. The main point here is that the bit stream is always completely recovered in a repeater device. There might still be bit errors, of course, but a good optical link design will place repeaters as necessary to compensate for these errors. The whole purpose of the repeater "E" stage is to give the repeater optical transmitter a "clean" set of 0's and 1's to work with at the end of a span. So the signal-restore process is just that: the optical signal is converted to a bit stream, and reconverted to an optical signal outbound.

Any signal is distorted by attenuation and noise as it travels. Attenuation weakens the signal, and noise distorts the signal in a number of ways, making the sharp edge of a pulse appear ragged and curved (purists will cringe at this cavalier description of noise effects on a digital signal, but there is a point to be made). Attenuation is handled by the restoration aspect of the repeater. Noise effects are handled by the second function of the repeater.

The second function of a repeater is to *reshape* the pulse that represents the bit, be it a 0 bit or a 1 bit. Even if a 0 bit is represented by a totally flat signal level, noise can make this look like a 1 bit to the receiver. So some form of pulse reshaping is needed for all the bits. Pulse reshaping is pretty much exactly what it sounds like: ragged pulses distorted by noise are reshaped into sharply defined pulses by the repeater. This is all made possible by the complete recovery of the bit stream.

The third function of a repeater is to *retime* the pulse representing the bits. Retiming is required because a serial (one by one) digital signal is always constrained by a characteristic *bit time*. This is sometimes also called the *bit interval*, but the meaning is the same. The bit time is always determined by the speed of the link. For example, a link operating at 1 Kbps will have a characteristic bit time of 1/1,000th of a second, or 1 millisecond. This just means that every millisecond, a new bit is sent out on the link. A link operating at 1 Mbps will have a bit time of 1 microsecond (1/1,000,000th of a second) and so on.

Now, the pulse that represents the bit does not usually occupy the entire bit time. There are some very technical reasons why this is so, but these are well beyond the scope of this section and involve the way that signal modulation has to occur. For instance, a DS-1 link operating at 1.544 Mbps has a characteristic bit time of about 0.65 microseconds. But the pulses that represent the 1 bits on a DS-1 link only last half of this time interval (this is called a "50 percent duty cycle" in the specifications).

The main point here is that attenuation and noise and other distortion effects on a link will spread out the pulse not only in space but also in time. So pulses arriving at a repeater are not necessarily positioned where they are supposed to be in their bit time. The repeater will reposition the regenerated pulses within the bit time. This is the retiming function of the repeater.

Restoring, reshaping, and retiming is what regenerative repeaters do. Note that amplifiers, being analog devices, can restore signal levels (along with any noise present), but they cannot reshape or retime pulses, because amplifiers do not even see any of the bits that might be present in the modulated signal. Amplifiers are "1R" devices.

The major differences between digital repeaters and analog amplifiers are shown in Figure 3-4.

A digital signal differs from an analog one, even light, in that digital signals can be sent without any distortion at all. Think of the receiver at the end of the link as performing a repeater function to recover the original bit stream. The bit stream should match up bit by bit with what was sent, as long as noise is kept within limits that allow the signal to be recognized at all.

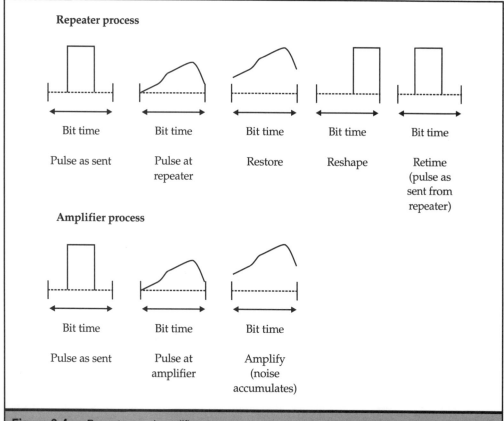

Figure 3-4.   Repeaters and amplifiers

It is somewhat ironic that optical networks, especially those using WDM and DWDM, function by replacing digital repeaters with optical amplifiers. An optical amplifier is just as much an analog device as any other amplifier. So noise (optical noise, of course, but noise nonetheless) accumulates through a cascade of optical amplifiers. What saves optical networking is the fact that inherent optical noise levels are so low that noise accumulation is not so much of a problem as other optical effects.

The biggest problem with digital repeaters is that they are *code and timing sensitive*. This just means that in order to reshape a pulse, a repeater must know what line code is being used to represent the 0's and 1's. And to retime the signal, the repeater must know the bit rate of the link in order to derive the proper bit interval. So a repeater that works with a 10 Mbps link will not work if the transmitter and receiver are changed to operate at 100 Mbps.

A repeater will not function on a fiber optic link if multiple wavelengths are present, either. The transmitter in a simple repeater, designed to be an inexpensive and yet robust device for field operation, can operate at only one wavelength. Multiwavelength repeaters could be made, but these are experimental devices at present and anything but inexpensive and robust.

WAVELENGTHS

Wavelengths are always important in optical networking, especially when discussing transmitters and receivers. Today many wavelengths and ranges of wavelengths are used for different purposes. This is a good place to take a quick look at the wavelengths that will be referred to in the rest of this book. These are shown and explained in Figure 3-5.

The best perspective on the wavelengths used for optical networks is linked to history. The range from 800–900 nm can be called the *1970s window,* or first window (this term originates with Harry Dutton of IBM). Early optical links in the 1970s all used this range. The band around 1,310 nm is the 1980s window (second window), and this is where the standard fiber optic systems SONET/SDH, which matured in the early 1980s, operate. Today, the range from 1,510 nm to 1,600 nm is the 1990s window (third window) and is used because it is at the lowest point of attenuation on a fiber optic cable link. This window is used in all modern WDM and DWDM systems.

However, the opening of the third window was not easy. Light sources are expensive in this range, and there is more to the fiber story than simple attenuation. More was needed to enable fibers to function well in the 1990s window. This work addressed the degradation effects mainly due to *dispersion.*

There are many reasons why a pulse of light sent along a fiber will *degrade,* which is just another way of saying *change.* The light will be weakened, a pulse, lengthened in time, and so on. There are five major reasons for the degrading of a pulse of light sent along a fiber optic cable.

▼ **Attenuation** This is always the major concern. A light pulse will always weaken because the fiber absorbs light. The glass itself does not absorb the light, but any and all impurities in the glass will. And if the fiber is not manufactured with precise uniformity in diameter and so on, these flaws cause scattering of the light. Scattering effects are the biggest cause of light loss in fibers today.

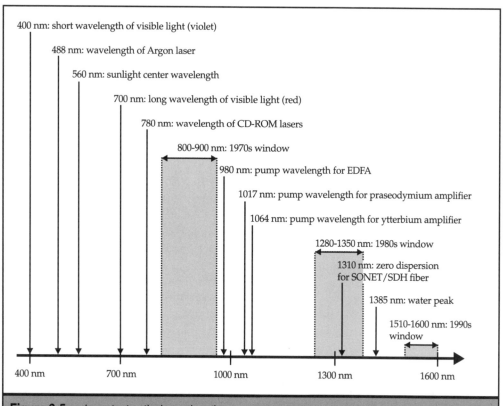

Figure 3-5. Important optical wavelengths

■ **Dispersion**   This is exactly what it sounds like: a pulse of light becomes spread out (disperses) in time as it travels down a fiber. Over long enough distances, all of the pulses eventually merge together and the peaks needed at the receiver to distinguish 1's all disappear. There are many different types of dispersion, some of which can even be used to advantage in optical networking. Three types of dispersion are of primary concern. *Material dispersion* (also called chromatic dispersion) results from the fact that fibers present slightly different RIs at different wavelengths, which therefore travel at slightly different speeds. Even the best lasers produce more than one wavelength of light (although in a very narrow range). *Waveguide dispersion* is caused by the shape of the core and RI profile across the core. Waveguide dispersion can actually cause the light pulse to "bunch up" in time and disappear. This also means that waveguide dispersion, if carefully planned, can counteract the effects of material dispersion. (It should noted that in some references chromatic dispersion is defined as the sum of material dispersion and waveguide dispersion.) This will be explained more fully later on in this section. *Modal dispersion* occurs in multimode fibers because the distance traveled, and hence the arrival time, is slightly different for each mode.

■ **Power limits**   One way of overcoming attenuation effects is just to pump up the input power. But the amount of optical power coupled onto a fiber is limited by nonlinear effects. This just means that doubling the input does not double the output but in fact might decrease the output to lower levels than before. Optical networks must function in the linear power region to be useful. Most single-mode fiber systems are limited to about half a watt of input power to remain in the linear region. The nonlinear effects in a fiber are mainly caused by the intense electromagnetic fields in the core when light is present.

■ **Polarization effects**   The core of fiber optic cable should be perfectly round and symmetrical. But there are always imperfections, either in manufacture or during installation, which distort the core in places. As a result, light in the fiber is changed in polarization. Currently, few if any optical networks rely on polarization of the light to carry information. But in future systems, polarization effects could be a concern.

▲ **Noise**   It might seem odd to speak of noise in an optical network, since immunity from electrical interference is one of the big advantages of fiber. It is true that optical networks will not pick up noise from *outside* the network. But noise is just an unwanted signal detected by the receiver. In optical networks, *modal noise* is a complex effect present on multimode fibers, and *mode partition noise* is a problem with single-mode fibers as power "hops around" (is partitioned) among many close wavelengths.

None of these optical impairments are fatal, fortunately, and most can be overcome through a variety of methods. Generally, it is desirable to make the fiber core as narrow as possible, and to allow only one mode of propagation at a particular wavelength. This is the idea behind single-mode fiber, of course. Next, the wavelength(s) chosen for operation must be selected to match the characteristics of the fiber itself. Cost is a factor as well, since the shorter the wavelength used in terms of nanometers (nm), the lower the cost. Also, dispersion effects that depend on wavelength can be minimized by narrowing the *spectral width* of the light source. The spectral width is a measure of the range of wavelengths generated by the light source. Even lasers do not produce all of their power at one particular wavelength, although lasers are much better than LEDs. Modern semiconductor lasers can have spectral widths of around 1 nm to 5 nm, while some more sophisticated (and expensive) lasers can have spectral widths as low as 0.01 nm. LEDs, on the other hand, will have spectral widths between 30 nm and 150 nm. Finally, material dispersion and waveguide dispersion act to spread out and narrow waves, respectively, and so can be used together to counteract each other to the benefit of the optical system.

All in all, modern optical network design depends on the interplay of two factors, with a third factor added in some cases. The two most important considerations are signal strength and dispersion effects. The third factor is noise, but noise is rarely a problem in simpler optical systems.

FIBERS AND OPERATIONAL WAVELENGTHS

As described in the previous chapter, pure fused silica, or silicon dioxide, is the primary ingredient in optical fiber cables. But using silica requires high temperatures, and the RI of silica is not particularly well suited for fiber optic systems. So *doping* is used to change the RI of the silica core, or cladding, or both. Doping is just the process of intentionally adding impurities to a substance during fabrication. Even chandeliers can have lead oxide added in to produce the sparkle of "lead crystal" glass. In optical fiber, 4 percent to 10 percent germanium oxide is added to the silica to increase the RI. This is a huge amount of dopant, considering that semiconductor fabrication processes dope materials with only one part in ten million of so. Boron trioxide will decrease the RI. Other substances are also used, but they all tend to also increase the attenuation of the fiber.

When all is said and done, the typical attenuation characteristics of modern fibers are fairly consistent. It should be noted that the infrared range starts at about 730 nm, so none of these wavelengths are visible. Also, the typical fiber attenuation in 1970 was about 20 dB/km, which is quite high for any medium. By the 1980s, this was reduced to about 1 dB/km, a figure widely used by vendors and writers. In the mid-1990s, attenuation on fiber of around 0.2 dB/km was common.

Fiber attenuation varies dramatically with wavelength. Figure 3-6 shows the variation in some detail. There is a "water peak" at around 1,400 nm caused by water ions that are always present in the fiber.

So the battle against fiber attenuation can be fought and won by opening the 1990s window wavelengths for operation. However, dispersion effects have to be taken into consideration as well. One of the reasons that the 1970s window was not the best window was that the fiber *zero dispersion point,* where material dispersion and waveguide dispersion exactly canceled each other out, was around 1,310 nm. So pulses at 1,310 nm stayed nice and tight as they made their way along many kilometers of single-mode fiber, although they still weakened from attenuation, of course.

The standard single-mode fibers used in SONET/SDH had their zero dispersion point at 1,310 nm. With changed fiber fabrication methods (doping) and manipulation of the RI profiles across the fiber core, a whole generation of *dispersion-shifted fibers* appeared in the 1990s that made operation in the even lower attenuation 1990s window attractive. So in the 1990s newer fibers moved the zero-dispersion point into the 1990s window, centered around 1,550 nm. This interplay between material dispersion and waveguide dispersion is shown in Figure 3-7.

The story of special fibers for optical networks might stop here, if not for a few other important points. First, SONET/SDH still is defined to operate at 1,310 nm, yet obviously

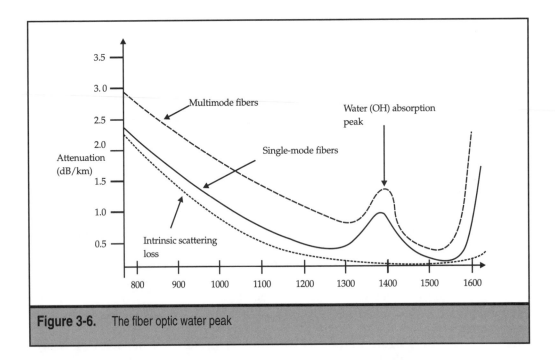

Figure 3-6. The fiber optic water peak

the current way of building optical networks is firmly in the 1,510 nm to 1,600 nm range. Second, this is not to say that wavelengths in the 1,310 nm range cannot be used with dispersion-shifted fibers (sometimes called DSF). However, the dispersion (and attenuation)

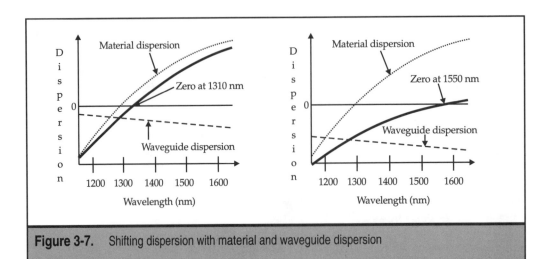

Figure 3-7. Shifting dispersion with material and waveguide dispersion

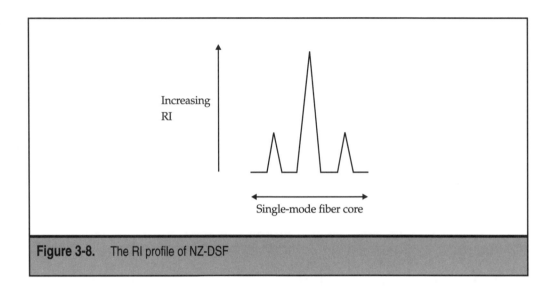

Figure 3-8. The RI profile of NZ-DSF

will be higher at 1,310 nm (the characteristic SONET/SDH wavelength) than at, say, 1,550 nm. Third, optical networks today rely more on WDM and DWDM than a single serial stream of bits at a single wavelength. So a single zero dispersion point right in the middle of the transmission band is not always desirable. For WDM and DWDM, it would be better to have a range between 1,510 and 1,600 nm where dispersion is low, but still nonzero. Again, clever doping and RI profiles made the manufacture of such *non–zero disper-sion–shifted fiber (NZ-DSF)* possible.

The RI profile of NZ-DSF is shown in Figure 3-8. These fibers are recommended for WDM and DWDM systems, although "regular" 1,310 nm zero dispersion-shifted fibers can be used as well (as long as the higher dispersion is accounted for during system design).

There are many other interesting developments in the field of fiber optic cables. These can only be mentioned here. There are large effective-area fibers (LEAF) that will accept more optical power, dispersion-flattened fibers with low dispersion in the 1990s window (but high attenuation), and dispersion-compensating fibers that can be spliced in periodically to combat dispersion effects on long NZ-DSF runs.

LASERS FROM LEDS

It was all well and good that clear optical fibers with low attenuation could be fabricated to operate in the three wavelength windows. But without a transmitter and a receiver that could operate in the same ranges, optical networking would remain a dream. And technical feasibility was only one part of the equation, of course. The technology had to be economically feasible and socially acceptable as well. The transmitters and receivers not only had to be affordable, but someone still had to make money selling them as well. Interestingly, social acceptability is not an idle parameter when it comes to lasers. Social acceptability comes into play when using lasers in general office spaces. Lasers can be dangerous to vision in many cases, and if the general public were convinced that lasers were too dangerous, this could be a serious impairment to optical networking (similar health concerns sometimes plague cellular telephony discussions).

Getting a light to shine down a piece of fiber optic cable is not an easy thing to do. The small core of the fiber is only one aspect of the problem. Light bulbs and other simple light sources are not "clean" enough for fiber optic transmission anyway. Light is normally too "noisy" and spread across much too large a spectrum to allow for efficient transmission on fiber. What was needed was a signal oscillator that could be modulated and that generated light instead of radio waves or microwaves. When the answer came, it came from the arena of microwave research.

In 1951, Charles H. Townes, a physics professor at Columbia University, came up with the maser: microwave amplification by stimulated emission of radiation. It took him three years to build a working maser, but the general principle of exciting a group of molecules with more energy than usual, then stimulating them to release the energy all at once, was a sound one. Townes realized that the same thing could be done with optical wavelengths, and the race was on to create a laser: *light* amplification by stimulated emission of radiation.

Many researchers took up the laser challenge. On May 16, 1960, Theodore Maiman got red light pulses to emerge from a ruby cylinder at Hughes Research Laboratories in Malibu, California. After some initial skepticism, by October scientists were firing laser pulses over 25 miles of air, not fiber. Lasers based on helium-neon gas soon followed, and an early semiconductor version of the laser was demonstrated in 1962, but most engineers kept trying to make laser communications work through the air. But fog, rain, snow, sleet, hail, and many other atmospheric conditions made these early "free-space lasers" useless for communications. Not until 1965 did Charles Kao and other visionaries succeed in convincing the industry that the laser should be coupled with better fibers for communications purposes.

But the real promise of lasers applied to fiber optic cable had to wait for the development of the semiconductor laser, a process that took until 1977 to become acceptable. In the early 1960s, primitive laser diodes, which is what semiconductor lasers really are, operated at the temperature of liquid nitrogen, generated a pulse that lasted 10 milliseconds (about a thousand times longer than desired), and required an enormous amount of power to fire each pulse. The early semiconductor lasers burned out often, too.

The whole idea of a semiconductor laser was intimately tied up with the state of the semiconductor technology of the times, of course. Semiconductors, as the name implies, fall between materials that are conductors and carry electricity easily and materials that are insulators and do not carry electricity easily at all. Electrons in metallic conductors such as copper or aluminum move around freely with a little energy push. Electrons in insulators such as glass or rubber are tightly bound and don't move much at all. Semiconductors have some electrons that can move around in a *conduction band* and some electrons that are bound in a *valence band*. It's all or nothing, so electrons in a semiconductor need enough of an energy boost to jump this *band gap* before they can be used to carry an electric current.

It turns out that semiconductors can function either by using the negatively charged electrons to do the work (an *n-type* semiconductor) or by using the positively charged "holes" the electrons leave behind to carry current (a *p-type* semiconductor). When n-type and p-type semiconductors are put together, the simple two-terminal (positive and negative) device that results is called a *diode*.

When a voltage is applied across a semiconductor diode, it only carries electricity in one direction. This is when a positive voltage is applied to the p-type material, and a negative voltage is applied to the n-type material. The n-type electrons move toward the positive terminal, and the p-type holes move toward that negative terminal. At the *junction layer* between the n-type and p-type materials, the electrons and holes essentially mix together, and electrons "fall" into the holes, releasing energy as they perform this process of *recombination*. In silicon semiconductors, the released energy is heat. But when the semiconductor material is gallium arsenide (GaAs) or indium phosphide or other materials, some of this energy is also released as light. The end result is the light emitting diode, or LED, as shown in Figure 3-9.

It took many more years for semiconductor LEDs to operate at room temperatures, last a long time, and consume a reasonably small amount of power. But making a laser out of an LED proved deceptively simple. Any time light is confined in a small enough space, it will "lase." Polishing the ends of the n-type and p-type semiconductor materials would do the trick, since the semiconductor LED was only some 400 microns long. But this mirroring step required the addition of *a heterojunction* to increase the efficiency of

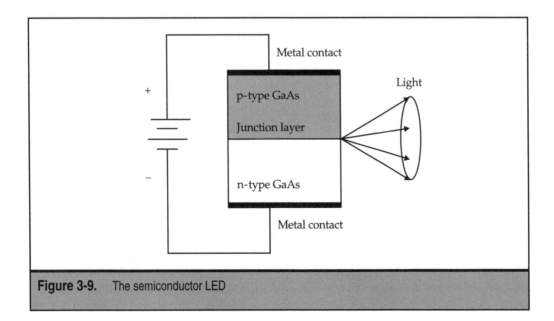

Figure 3-9. The semiconductor LED

the device. It took a lot of effort to make this happen. But by 1977, the semiconductor laser had entered the world of optical networking.

Further work throughout the 1980s resulted in semiconductor lasers that not only operated in the 1,550 nm window, but also proved to be *tunable*. That is, these laser diodes could be adjusted to generate light at not just one wavelength, but many.

The Tunable Laser Diode Operating at 1550 nm

Another key piece of the optical networking puzzle, in addition to the creation of special fibers that could be used with confidence for WDM and DWDM, was the creation of the tunable laser diode operating at around 1,550 nm, right in the middle of the 1990s optical communications window. There are actually three pieces to this invention. The first part is the *laser diode* itself. The second part is the *tunable* nature of the laser, which enables the same physical device to be used to generate different wavelengths of light. Finally, the third part is operation at around 1,550 nm. All are equally important for optical networking today, and each aspect will be discussed in some detail.

In the early days of fiber optic systems, lasers were bulky affairs that were very expensive and hard to work with in spite of their numerous advantages over LED-driven fiber systems. Not only was the cost often prohibitive, but lasers often had to be locked in a closet with a warning sign on the door and anyone entering had to wear protective eyewear. Today, both LEDs and lasers are semiconductor devices mounted on a simple computer card assembly. LEDs remain simpler than lasers when implemented in a chipset, but they share many features with lasers. So it makes sense to start with a quick look at modern semiconductor LEDs.

An LED is just a *forward biased p-n junction*, but this is not the place for a full discussion of semiconductor electronics. All that is needed is an appreciation that when the proper electric potential (voltage) is applied to a wafer of p-type semiconductor material and n-type semiconductor material, light will be emitted at the junction between the materials. That's the easy part. The hard part is getting the LED light at the desired wavelength out of the junction and onto the fiber. In practical LEDs, sometimes the light emerges from the surface of the device (SLEDs) or from the edges of the device (ELEDs). In most SLEDs, the n-type semiconductor has a "hole" in the middle to allow the light to emerge from the junction. A very simple diagram of a SLED (sometimes called a "Burrus LED") is shown in Figure 3-10.

The characteristics of LEDs as compared to lasers as light sources (transmitters) in optical networks are important. LEDs have been lower in cost than lasers, although this price differential has all but disappeared in recent years. Much of the price differential was based on production volume. Certainly the use of lasers in simple CD-ROM data and audio devices has leveled the cost playing field drastically. LEDs have had lower power outputs than lasers (100 microwatts or so) and produce light not at a single wavelength, but in a large band or "spectral width" of about 5 percent of the wavelength (50 to 100 nm). The spectral width can be reduced through the use of filters, but this filters out part of the signal power also, of course. LED light is not coherent as is a laser's light, and so the LED light must be focused with a lens onto the fiber (this is why LEDs are not used with single-mode

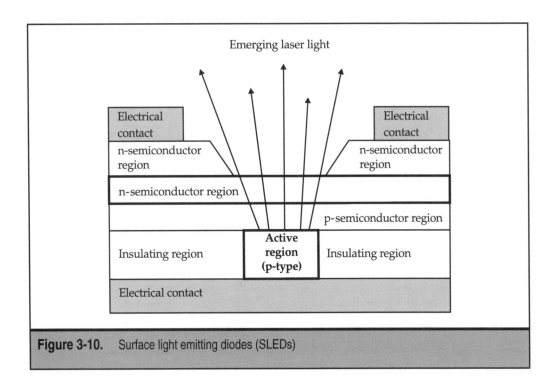

Figure 3-10. Surface light emitting diodes (SLEDs)

fibers: it is too hard to get much light into the small core). LEDs cannot be turned on and off fast enough to drive gigabit per second systems as lasers can. Most LEDs top out at about 300 Mbps. On the other hand, LEDs can easily be used with analog modulation techniques. This can be done with lasers, but only with much more difficulty.

In contrast to LEDs, lasers are much better to use in optical networks as light sources. Laser light in theory is "coherent" and consists of a single wavelength. This is not strictly true in practice, as all lasers used for communications exhibit a characteristic *linewidth*. But lasers can be turned on and off very quickly, in some cases on the order of femtoseconds (one millionth of a nanosecond, or 10^{-15} seconds), and produce higher powers than LEDs (about 20 milliwatts for communications systems, and up to 250 milliwatts for optical amplifier pump lasers). Laser light, which emerges highly parallel from the device, can inject 50 percent to 80 percent of its power onto a fiber.

But however close lasers and LEDs come in price, the fact remains that temperature control and power control is needed in lasers. These controls require electronic circuits to function. And lasers are generally nonlinear with respect to input power.

In spite of some claims to the contrary, most semiconductor lasers used for communications do *not* produce a single wavelength of light. Just like LEDs, they produce a range of wavelengths, and the spectral width is the measurement of this. This turns out to be important if a *tunable* laser is desired. Usually, the spectral width of the laser is about 8 nm, scattered across eight peaks or "modes" (not the same as fiber modes). Not all of these wavelength peaks are of equal strength. The laser first outputs power at a "dominant mode," then at another mode and another, switching back and forth unpredictably. Thus power is distributed in a bell-shaped curve, with ill-defined edges. So it is common to quote spectral width as between the points on the curve where the power drops to one half the maximum "mode" power. This is called the full width half maximum (FWHM) and is a valid measure for LED spectral widths as well.

Spectral width is the crucial laser parameter for a number of reasons. First, the wider the spectrum, the greater the dispersion on the fiber. When using WDM or DWDM, the closer the wavelengths can be packed (i.e., the smaller the spectral width), the more channels can be used on a single fiber. Also, sophisticated modulation techniques such as phase modulation cannot be used unless the spectral width is very narrow. On the other hand, very narrow spectral widths, usually desirable, can actually cause unwanted non-linear effects in the fiber (such as Stimulated Brillouin Scattering [SBS], discussed more fully later in this book).

In addition to spectral width, semiconductor laser output has a characteristic *linewidth*. So not only are the "peaks" spread across the spectral width, but each individual peak is slightly different in width. The relationship between spectral width and linewidth is shown in Figure 3-11.

There are many other characteristics of semiconductor lasers, such as coherence length and time, power, operational range in terms of wavelength, wavelength stability, and so on. However, these are beyond the scope of this discussion.

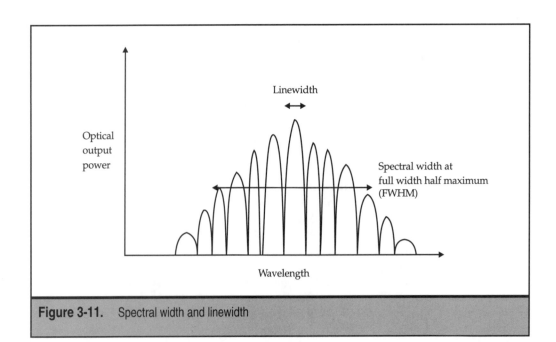

Figure 3-11. Spectral width and linewidth

OPTICAL RECEIVERS

As fascinating as LEDs and laser diodes are as transmitters, they would be useless without reliable optical receivers or detectors. Sometimes the word *detector* is used for the optical device and the word *receiver* is reserved for the destination of the bits, but in the world of optical networking there is less of a need to distinguish between the two terms. In this section, the terms *optical receiver, optical detector,* and *photodetector* are used more or less interchangeably.

The vast majority of optical receivers used in communications networks today are based on the idea that light causes ionization and consequently a flow of electricity in semiconductor materials. So instead of electricity causing light, as in the semiconductor laser, the optical detector uses light to produce electricity. This electric current production in an optical receiver can be done in a number of different ways, but they are all variations on this basic theme.

The different types of light receivers vary mainly in one or more of four major parameters: First, detector efficiency, which is a measure of the ratio of output current power to input optical power. Second, the range of optical wavelengths over which the receiver will operate. Third, response time, which is a measure of how quickly the receiver can react to changes in the input optical power and reset itself. Fourth, the noise level of the receiver itself, which must be considered as a lower bound to the operational range of the device.

The optical receiver is based on the *photoconductor*. This is just a piece of semiconductor material with a voltage applied across it by means of electrical contacts. Any arriving photon is absorbed by the semiconductor material and an electron and "hole" pair is created. Recall that the electron itself has a negative charge and the hole left behind in the semiconductor crystal has a positive charge. The pair move through the applied field to the metal contacts. So the amount of electrical resistance through the device depends on the number of photons received per unit time. The general principle of the photoconductor is shown in Figure 3-12.

Photoconductors are not normally used as optical receivers. But the principle is sound, and the modern *photodiode* converts light directly to electric current on an almost one-to-one basis. One photon, one electron. This is the idea behind the PIN photodiode. PIN stands for p-intrinsic-n diode, which just means that there are p-type and n-type semiconductors separated by an insulating layer called the *intrinsic zone*. A straight "p-n diode" would work as a type of LED in reverse, but these devices are very slow in terms of response time, and the junction is too thin to convert many photons to electric current (so they are not very efficient either).

PIN Diodes

PIN diodes are faster and more efficient than p-n diodes. They are so efficient that any photon having energy higher than the "band gap energy" will be detected. This is not necessarily a bad thing, since the same device can now respond to a wide range of wavelengths. This responsiveness is an advantage in optical networks with many wavelengths, since the same *wideband* receiver can be used for many different wavelengths. However, if many wavelengths are combined on a single fiber, then some type of fiber

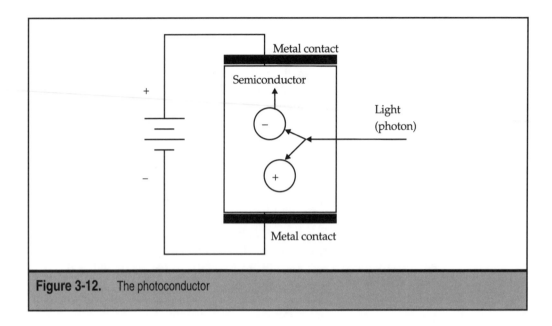

Figure 3-12. The photoconductor

grating or other arrangement must be used to separate the wavelengths and keep them from all registering at the same receiver.

It is common with PIN diodes to speak of the *cutoff wavelength*. PIN diodes will respond to any photon having a wavelength shorter than this cutoff wavelength, since photons have higher energy at shorter wavelengths. The general structure of the PIN diode is shown in Figure 3-13. Note that the I layer actually has a slight n-doping to it.

However, the lower the band gap energy of the PIN diode, and so the higher the cutoff wavelength, the higher the internal noise of the device. Internal noise results in *dark current* being detected at the receiver. Dark current is a flow of electricity that is present in the optical receiver even when no light source is hooked up at the sending end of the link. Obviously, dark current must be kept to a minimum value so that the receiver can operate at maximum efficiency. So band gap energy is usually kept slightly lower than the energy of the longest wavelength that needs to be detected.

Optical receivers are manufactured to operate specifically in each of the three optical transmission windows. Silicon PIN diodes are used between 500 and 1,200 nm. These devices are very inexpensive but not very sensitive. Around 1,300 nm (really 1,250 to 1,400 nm), indium gallium arsenic phosphide (InGaAsP) or germanium can be used, but these devices are more expensive as well as more sensitive. Around 1,550 nm, indium gallium arsenide (InGaAs) is used.

Avalanche Photodiodes (APDs)

PIN diodes are one major class of fiber optic receiver devices. But another type of light detector device is used as a fiber optic receiver as well, known as the *avalanche photodiode,* or APD.

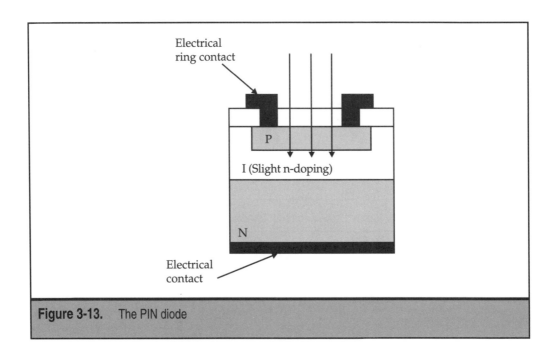

Figure 3-13.   The PIN diode

The nice thing about APDs is that they amplify the signal as well as detect it. In other words, one input photon results in *more* than one electron being freed to move about the semiconductor material and cause an electrical current to flow. PIN diodes only convert one input photon to one electron at best.

An APD works on the same basic principle as a *photomultiplier tube.* A photomultiplier tube has an electrical field that causes a cascade of electrons to flow whenever a single photon is detected. This relies on the presence of a *multiplication region* in the semiconductor device that is characteristic of the APD.

The APD starts out in the same way as a PIN diode. That is, a photon striking the detector releases an electron. But in the APD, the electron is accelerated by an electric field until it hits an appropriate target material. This impact causes more electrons to be released. The new electrons are further accelerated and strike another target, which produces more electrons yet. The process can be repeated even more times until all of the electrons reach a collector and are registered. So one photon causes a current of many electrons.

APDs can result in internal amplification of between 10 and 100 times. So one photon basically produces between 10 and 100 electrons. Technically, the APD is just a PIN diode with a 50-volt reverse bias, which is very high. This is actually an improvement on older APDs, which required a reverse bias of 100 volts or more. A PIN diode has a reverse bias of about 3 volts, and sometimes less.

The APD includes a lightly doped p-region renamed the π-region. This region is sandwiched between two p-regions and capped by an n-region where the light enters the APD. The structure of a basic APD is shown in Figure 3-14.

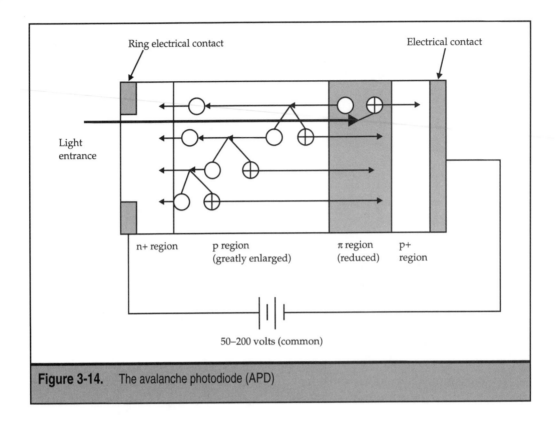

Figure 3-14.   The avalanche photodiode (APD)

The figure shows an avalanche in progress. One photon entering on the left has triggered the formation of only four electron-hole pairs. The freed electrons are shown as plain circles. The "+" circles are the "holes"; some APDs can count these instead of electrons. In the figure, the p-region has been greatly enlarged just to make room to show the avalanches, and the π-region has been reduced, again just to make room in the figure. Note that one electron can cause another electron-hole pair to appear, which is the whole idea behind the APD. Indeed, the APD process is so efficient that sometimes the main design issue is not to encourage avalanches, but to make sure that the avalanches stop in time for the next serial bit (in the form of arriving photons) to register.

There are a few related points to make about APDs. First, the APD has a very strong electrical field. So uncontrolled avalanches can be a real problem. Next, the voltage needed to produce this field is very high for a semiconductor device, which usually deals with 12 volts or less. This makes APDs expensive. Finally, APDs can be very sensitive, but they often miss many photons. This is not as paradoxical as it sounds. A weak photon can trigger an avalanche, but there is also a good chance that a photon might not trigger anything at all. Fortunately, APDs work very well as receivers across all three windows.

Like PIN diodes, APD optical receivers are manufactured to operate specifically in each of the three optical transmission windows. Silicon and germanium APD diodes are used between 800 and 1,000 nm. The germanium devices are not very sensitive, however, due to internal noise levels. Also, APDs are expensive to use in this window, and PINs are usually used instead. Around 1,300 nm (really 1,250 to 1,400 nm), germanium APDs can be used, but other materials such as those used at 1,550 nm are used as well due to the noise levels in germanium-based APDs. Around 1,550 nm, indium gallium arsenide (InGaAs) or indium phosphide (InP) are used.

ADPs are distinguished by four major operational characteristics. These are sensitivity, operating speed, gain-bandwidth product, and noise level.

ADPs are used whenever extreme receiver sensitivity is required. Until recently, this more or less mandated that APDs be used on WAN optical links, and indeed they were.

All operating speed concerns with PIN diodes also apply to APDs. But with APDs, there is an additional limiting factor. This is often called the *avalanche buildup time.* An avalanche just takes time to develop and calm down. This factor limits the characteristic serial bit time or interval that the APD can deal with. The serial bit limit to date is about 40 Gbps, giving a bit time of 1/40 of a nanosecond. (Not so long ago, the limit was 1/10 of a nanosecond, limiting SONET/SDH to about 10 Gbps.)

The main measure of optical receiver desirability is the product of the gain in dB and the highest bandwidth detectable expressed in GHz (Gigahertz). Good modern APDs have a gain-bandwidth product of about 150 GHz, but this is improving all the time.

When it comes to noise, APD avalanches just make internal noise worse. This is because any electrons wandering around free in the material, such as those caused by ambient operating temperature, can trigger an avalanche. The longer the operational wavelength of the APD, the worse the noise level.

Once operation at 1,550 nm becomes necessary, as in DWDM systems, APDs are not always the best way to go. With cascaded optical amplifiers instead of digital repeaters, the internal noise of the APD is often the limiting factor. So in many DWDM systems, PIN diodes are used with a *preamplifier.* This gives a configuration with greater sensitivity than even APDs can offer and much lower noise levels.

Other Types of Receivers

Silicon is a fine APD material, but only below 1,000 nm. In the first optical window, silicon APDs have good gain, fast response times, and low noise. But above 1,000 nm, silicon cannot absorb light. Silicon is highly transparent to 1,550 nm light, although the RI bends the light, of course.

So to use APDs above 1,000 nm, it is better to use a *hetero-interface photodetector.* This type of APD replaces the p-region semiconductor with something that can absorb light at the desired wavelength. InGaAs is a good material to use here. The device can detect in InGaAs but multiply in silicon. The biggest issue in fabrication is getting the silicon and InGaAs crystal lattice structures to line up, but this has been done with the process of *wafer fusion* in devices known as Silicon Hetero-Interface Photodetectors (SHIPs). The gain-bandwidth product is about 350 GHz, and there is very little noise.

Other research work has focused on the PIN diode all over again. As mentioned, with DWDM a preamped PIN can be more sensitive than an APD and have lower noise levels as well.

But at higher speeds (and lower bit times), PINs all have a real problem. It takes a lot of time for the electrons to diffuse through the regions, especially the I layer, and the high APD electrical field is not there to drive the electrons. To speed up a PIN, it is possible to make the I layer smaller and smaller. But if the p and n layers are too close, the PIN will not work at all. One strategy is to reduce the size of the device, but this just makes the PIN diode that much less efficient. A real dilemma.

This brings up an important related point about optical detectors in general. Whenever the serial bit rate on an optical link increases, the bit time shrinks. But it always takes a given number of photons to register a bit. It is not possible to just crank up the power of a transmitter, since semiconductors, lasers or not, have inherent power limits. Thus there are essentially the same number of photons available, just fewer photons available per unit time (bit time). So whenever a link increases speed by a factor of two, the number of photons available drops by half. Go from 10 Gbps to 40 Gbps, and the number of photons available to register a bit is only 1/4th of what there was at 10 Gbps.

The problem is even worse with PIN diodes at high speeds. Not only are there fewer photons, but PIN devices suffer from a loss of *quantum efficiency*, which just means that more arriving photons fall through the cracks of the PIN, so to speak. PIN diodes can be built as *traveling wave devices* to try to counteract the effects of this loss of quantum efficiency at higher speeds.

Basically, a PIN diode built as a traveling wave device hooks up one PIN after another so that any photons missed by one PIN have a chance of being absorbed by the next PIN in the array. The outputs line up too, so that if any electric current is produced by one PIN, all of the electric currents will be added together at the end of the PIN array. The real trick is to assure that both electricity and light pass through the traveling wave device in phase and on time.

Another attempt to deal with the limited number of photons available at very high speeds is the *resonant-cavity photodetector (RECAP)*. The idea here is to make the PIN diode I layer very thin to minimize the time needed to register a photon, but also essentially to bounce the photons back and forth across the I layer to maximize the chance for the photon to be absorbed by the device.

One issue with a RECAP is that these devices are highly wavelength specific, since only certain wavelengths can be made to "bounce" in one device. This is not the best idea for receivers, which ideally should respond over a wide range of wavelength to be most cost effective and useful for WDM and DWDM purposes. However, RECAP devices have a future as wavelength splitters and/or isolators in DWDM systems.

Finally, there is the class of optical receiver devices known as *phototransistors*. Light has always been a source of noise to even the simplest transistor: they are all basically light sensitive. So phototransistors are just transistors designed to be used as optical receivers. Transistors are inherently amplifiers, so a phototransistor is quite efficient.

Phototransistors are lower in noise than APDs and have even higher output. But overall, phototransistors are much less responsive than either PIN diodes or APDs. This limits the usefulness of the phototransistor for optical communications in the second and third windows.

This chapter has taken a look at working fiber optic links, but only those in which one wavelength is used. All of the components needed have been examined, from transmitter to regenerative repeaters to receivers. These systems are a form of "optical networking" to be sure, but not what optical networking is considered to be today. But these optical links were good enough to be used in a number of networks, both LAN and WAN, for many years. The next chapter takes a closer look at LANs and WANs built around the types of fiber optic links introduced in this chapter.

CHAPTER 4

Optical Transmission Systems

It is all well and good to examine fiber optic cables, and transmitters, and receivers in isolation. That is how an understanding of these key optical network components can best be introduced and built upon. But up to this point there has been a lot of emphasis on *optical* and little mention of *networking* as such. When it comes to optical networking, there is even a higher level of understanding to be gained from an examination of how fiber optics have been used and are still used in networks as a whole.

This understanding is necessary in preparation for the introduction of WDM and DWDM. Only with an appreciation of some of the limitations of fiber optics in local area and wide area networks up to the present can the breakthroughs that optical networking represents be fully appreciated. So this chapter examines some of the ways that fiber optics have been used in LANs and WANs before the recent introduction of more complete methods of optical networking.

This chapter is meant as an overall introduction to optical transmission systems. Many of these LAN and WAN architectures will be treated more fully later on in this book in chapters of their own. But it might be helpful to investigate some of the overall aspects of optical transmission systems all in one place.

FIBER OPTICS IN THE LAN

Until relatively recently, fiber optic cable was not used very often in a local area networking situation. One compelling reason not to use fiber in a LAN environment was the cost of running and using fiber compared with other media. In many cases it was not so much the cost of the fiber optic cable itself that was the disincentive, but the cost of the E-O conversion devices required to use the fiber. In most cabling schemes, purchasing and running the fiber was not the expensive part of the installation. The fiber still had to be terminated (just the formal term for "connectorized"), and this was sometimes an expensive operation as well. But even when fiber connectors became almost as cost effective as terminating other forms of media, the big expense was the devices needed to convert electricity to light and back again.

One of fiber's big advantages over other forms of media is distance, and distance was not much of a concern when it came to LANs. Most LANs, in spite of the capability for many standard LAN architectures to span a few kilometers or miles, were and are quite small. Often a LAN covered only a single building, or even a single floor. For short distances, other media, usually some form of twisted pair copper, was just as good as fiber.

A powerful incentive to use fiber in the LAN came from the need for increased bandwidth on the LAN, especially for the LAN backbone. Actually, it was the combination of distance and bandwidth that made fiber attractive for the LAN backbone.

A backbone LAN is a special type of LAN. A typical LAN will connect client and server computers to each other, of course, but LANs today all consist of other network components such as hubs (hubs also perform a *repeater* function on a LAN), switches, bridges (LAN bridges today are often called *switching hubs*), and routers. A backbone LAN is a LAN that connects only these network devices. No clients or servers are on the backbone LAN. Backbone LANs connect only hubs or routers or some other forms of network devices.

The problem is that a backbone LAN has to link all of the smaller LANs together. For instance, an office campus with several buildings occupied by the same organization might have a LAN in each building, or more likely on each floor. But each of the LAN hubs spread throughout the complex might only have a single large router through which all traffic to and from the Internet must flow. There could always be multiple routers, but too many routers raises control and management issues within the organization, and since each router usually requires its own link to an ISP, there is an expense issue as well.

But backbone LANs do more than just supply connectivity to network devices such as hubs and routers. Because there are many clients and servers on each LAN, traffic is concentrated onto the backbone LAN, which usually connects only a handful of very busy devices. So high bandwidth on the backbone LAN is not a luxury, but an absolute necessity. It does not help matters that the backbone LAN might have to span greater distances as well. Links can always go farther by going slower, but this option is not available when backbone LAN capacity is an issue.

So one of the earliest and simplest applications of fiber optics to the LAN was known as the FOIRL, or fiber optic inter-repeater link. Repeaters were used in Ethernet LANs to connect the segments of the Ethernet. Each Ethernet segment could span 500 meters, or about 1,500 feet. Beyond that distance, an Ethernet digital repeater unit was required to perform the traditional "3 Rs" of restoring, reshaping, and retiming. The number of segments and repeaters in Ethernet was strictly controlled, since Ethernet segments based on the original coaxial cable could not carry signals too far before another repeater was needed. (There were also limitations arising from the time it took for a signal to travel across all of the segments.)

Not only could fiber handle all of the bandwidth needs of Ethernet repeaters (no matter how many clients and servers were on the segments), its use extended the reach of segments connecting them. These Ethernet repeaters later came to be housed in hubs, and the use of fiber also extended the reach of the segment connecting the repeaters up to 15 kilometers (almost ten miles).

When Ethernet hubs were connected with an FOIRL backbone link, the maximum distance between the hubs was determined by the mode of operation of the Ethernet segments within a hub. Ethernet can work in either half duplex mode or full duplex mode. In half duplex mode, Ethernet devices must listen for collisions as they transmit. This need to listen while sending limits the total span of all the Ethernet segments to one *collision domain*. In full duplex mode, Ethernet devices can send and receive at the same time. This is the normal operating mode for modern switching hubs, and since there are no collisions in full duplex Ethernet, the Ethernet hubs can be farther apart.

Half duplex Ethernet operation limits FOIRL links to 2 km (a little more than a mile) regardless of whether multimode or single-mode fiber is used. The signal propagation delay for collisions is fixed, which limits the span drastically. In full duplex Ethernets, pretty much the rule for all business Ethernet LANs today, the FOIRL span limits are 2.5 km for multimode fiber and 15 km when single-mode fiber is used. At that distance, the signals take 75 microseconds (millionths of a second) to traverse the fiber link. This would be a real problem for half duplex Ethernet collision domains, but not much of a problem for full duplex operation. The 2.5-km distance limit for multimode fiber in full duplex Ethernet mode is more an issue of signal attenuation than anything else. Two Ethernet switching hubs connected by a 15-km single-mode fiber FOIRL link are shown in Figure 4-1.

Admittedly, a point-to-point FOIRL span between two Ethernet hubs is not much of a backbone LAN. But the FOIRL arrangement does fulfill all of the requirements of a backbone LAN. Ethernet frames go back and forth on the FOIRL span, and there are only the hubs as the network devices on the backbone. FOIRL was standardized in 1987 and extended to become 10Base-FL in 1993.

FOIRL spans are not often used for backbone LANs today. The point-to-point nature of the link made it difficult to add a third hub to the configuration. A hub with *two* FOIRL boards in the middle was needed to link three hubs, and two boards in each of the hubs were required to fully mesh the hub's "backbone." Creating backbones with FOIRL spans quickly became quite expensive. Today, more complete backbone LANs such as the Fiber Distributed Data Interface (FDDI) are used.

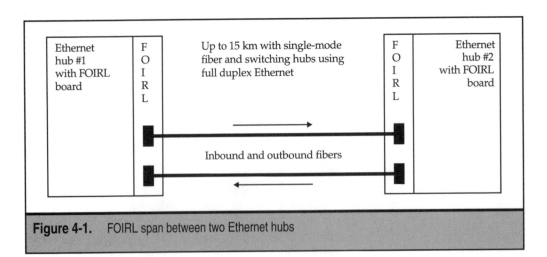

Figure 4-1. FOIRL span between two Ethernet hubs

FDDI

The FDDI standard appeared in 1990; FDDI was, in many ways, ahead of its time and, in some ways, an oddball technology with little hope of widespread acceptance, especially for the desktop. FDDI was essentially a "stretched" version of the Token Ring LAN technology. FDDI stretched Token Ring in terms of distance (100 km for FDDI, making FDDI technically a metropolitan area network [MAN] technology), speed (FDDI runs at 100 Mbps), and frame size. There were significant differences between Token Ring and FDDI, of course. But there was still a token circulating on a ring, in FDDI a physical ring of dual-fiber spans, and without the token an FDDI station could not send. The token is just a special frame structure that allows the station holding the token to send.

The presence of the token immediately limited the throughput on an FDDI ring. A station with something to send had to wait for the token, fill the essentially empty token frame with the information, send the frame to the destination station, and then wait for the frame to complete the journey around the entire ring. If all went well, the sending station "drained" the ring. In an effort to speed the process, FDDI added a method of *fast token release* so that the sender could generate a token frame and send it while still waiting for the information frame to make its way around the ring. However, the price of avoiding Ethernet-type collisions at higher loads was the perception of limiting traffic at lighter loads on a token-passing network.

FDDI is normally deployed with a backbone of what are called *dual counter-rotating rings* of fiber. One ring is active and carries the token and data between attached nodes, and the other ring is a standby ring that can take over through a process known as "ring wrap" if a node or fiber span fails. This is exactly the same concept employed for SONET and SDH rings today, applied on a smaller scale and with much less complexity than full SONET/SDH rings.

Now, the reason for the rings was due to the risk to the fibers running outside of a building. Nodes on the backbone, since they have two pairs of fiber connecting each node, are called *dual attached stations (DASs)*. There are also *dual attached concentrators (DACs)* that have multiple ports for linking devices to the dual ring. FDDI also supports *single attached stations (SASs)* with two fiber connections and *single attached concentrators (SACs)* with multiple ports.

Within a building, the dual backbone rings were overkill, and just as important, the need for *four* E-O converters at every DAS would be very expensive. So FDDI also allowed SASs to link to the backbone DACs. The SACs supported multiple ports in this two-fiber configuration. There was no ring wrapping with single attachments, but there was less risk of a fiber cut on a span that ran down the hall to a telecommunications closet.

However, running fiber to the SASs was still an expensive proposition at the time. So FDDI was quickly amended to allow short spans of twisted pair copper to carry signals at 100 Mbps. This was sometimes called CDDI (for Copper Distributed Data Interface), but the main feature of FDDI was always the dual rings between DASs. The basic FDDI architecture is shown in Figure 4-2.

Even with the development of CDDI and concentrators to cut down on the amount of fiber and E-O converters required in an FDDI ring, FDDI remained much too expensive to deploy with fiber to all desktops. Then again, in 1990, the only standard technology that ran reliably at 100 Mbps was FDDI, which was ten times faster than normal 10 Mbps Ethernet. There was talk at the time of extending Ethernet speeds to 100 Mbps, but no one knew how far off 100 Mbps Ethernet was (not far at all, as it turned out). So FDDI found a

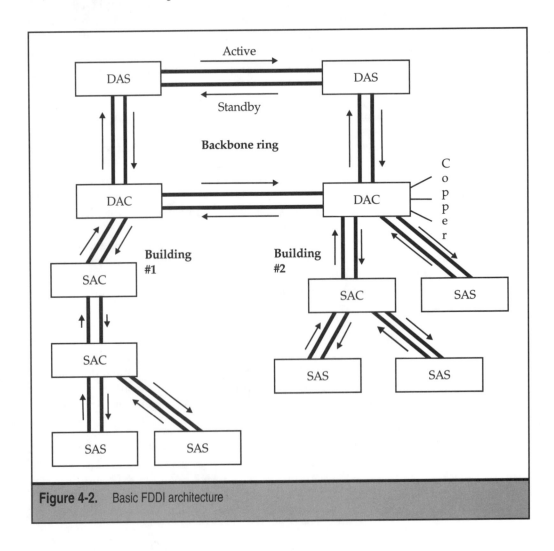

Figure 4-2. Basic FDDI architecture

home as a LAN backbone technology, and FDDI still flourishes to some extent there today. FDDI as a backbone LAN connecting hubs and routers is shown in Figure 4-3.

The "dual attached stations" are now just boards in the hubs and routers and not stand-alone devices as they once were. But there are still fast release tokens capable of handling very high loads, dual counter-rotating rings, and ring wrapping on failures.

FDDI usually uses multimode fiber with at least 500 MHz-km modal bandwidth at 1,300 nm. Attenuation should not exceed 11 dB between nodes, or in some cases 7 dB. LEDs are used as transmitters, and PIN photodiodes are used as receivers. The bit error rate must be better than 1 in 2.5×10^{10} bits (no problem for fiber), and minimum receiver sensitivity must be no higher than –27 dBm. FDDI variations allow for the use of single-mode fiber, and four distinct combinations of lasers and receivers, depending on the distance spanned.

In spite of attempts to tinker with FDDI over the years, such as FDDI-II (which added voice channel support [!] to FDDI) and FDDI Follow On LAN (FFOL, a way to optimize FDDI operation for LAN backbones), interest in FDDI has waned. By and large, the position of FDDI as a backbone LAN has been supplanted by other architectures.

Ethernet as 10Base-F and 100Base-FX

The presence of FOIRL spans in an Ethernet environment led to the creation of a full-blown definition of 10 Mbps Ethernet over fiber in 1993. This specification is known generally as 10Base-F, but it consists of three main pieces: 10Base-FL (FOIRL at heart), 10Base-FB, and 10Base-FP. These three main segment types serve different purposes but are compatible at the fiber level. However, 10Base-FL will not interface with 10Base-FB or 10Base-FP, and so on.

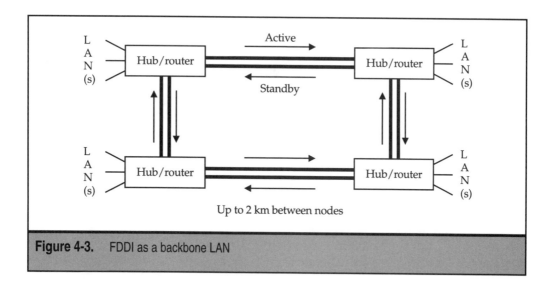

Figure 4-3.   FDDI as a backbone LAN

It is important to realize that 10Base-F does not increase the speed of Ethernet in any way. In spite of the enormous bandwidths available on the fiber links, 10Base-F is firmly 10 Mbps Ethernet transmission. It was intended for all-fiber environments, a day when all Ethernet desktops were linked by fiber and there was no copper at all to be found in the building. This explains the relative scarcity of 10Base-F today. There are just very few places where 10Base-F is wanted or needed.

It is easiest to start with 10Base-FL ("fiber link"), since it both updates and expands FOIRL. This type allows 10 Mbps transmission on two fiber links connecting two devices up to 2 km (a little more than a mile) apart. The devices can be computers or hubs, or even one of each. Like FOIRL links, 10Base-FL links are always point-to-point. There have to be appropriate transceivers on each end of the link, and if the end device is a computer, there is usually an external transceiver. So a computer will have a regular 10Base-T NIC card. The fiber optic connectors are usually ST connectors, but they are officially known as "BFOC/2.5" connectors, where BFOC stands for Baseband Fiber Optic Connector. The 10Base-FL fiber is usually 62.5/125 multimode fiber (MMF), but others MMFs can be used, such as 50/125 MMF, 85/125 MMF, or 100/140 MMF. However, these other types of MMF might not achieve the full 2-km span. The light wavelength used is 850 nm, which makes 10Base-FL quite affordable.

The 2-km span is for half-duplex Ethernet. The fact that there are two fibers in 10Base-FL, one for transmit and the other for receive, makes it more likely that full duplex Ethernet is used, but this is entirely optional. Full duplex operation, free from collision timing restrictions, can support 5-km spans on MMF, and even longer when single-mode fiber (SMF) is used. A 10Base-FL hub connecting two computers with external fiber optic (FO) transceivers is shown in Figure 4-4. Note that the hub can still have nonfiber ports supporting other media such as unshielded twisted pair (UTP). The connection from the NIC card to the transceiver could be a standard Ethernet AUI (attachment unit interface) cable or UTP.

The second 10Base-F type, 10Base-FB (where "FB" stands for Fiber Backbone), is strictly for repeater (hub) connections. This 10Base-F component runs a special *synchronous signaling protocol* that allows the number of repeaters (or hubs with repeaters) to be extended beyond the number of repeaters allowed in the native Ethernet specification. As with 10Base-FL, the length of the fiber between 10Base-FB hubs can be 2 km with regular MMF.

The special 10Base-FB protocol is needed to overcome the limitations that Ethernet encounters when many repeaters are strung together to connect Ethernet segments. The first problem is the fact that repeaters add delay to the collision domain when half duplex Ethernet is used. Collisions must be heard in a certain amount of time by the sender. This delay requirement is countered by using full duplex Ethernet on the 10Base-FL fiber links, of course. But this option is not open to 10Base-FB because of the *second* problem with cascading Ethernet repeaters.

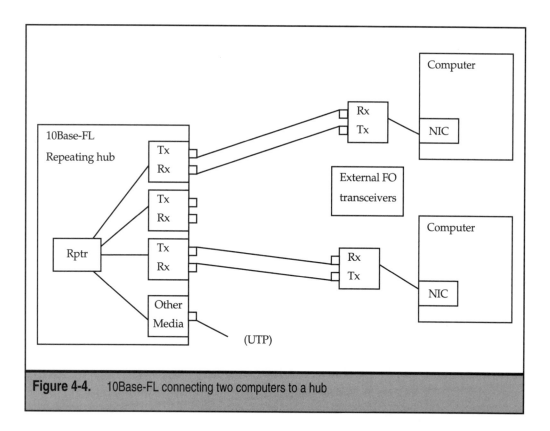

Figure 4-4. 10Base-FL connecting two computers to a hub

The second problem is that repeaters will add a random bit loss into the *preamble* of an Ethernet frame. The preamble is a bit pattern 56 bits long that basically tells a device to "get ready to receive an Ethernet frame" by freeing up buffer space and so on. When preamble bits are dropped by cascaded repeaters, the interframe gap between otherwise valid Ethernet frames can fall below the 10 Mbps Ethernet limit of 9.6 microseconds. Now, repeaters are required to regenerate the preamble bits when the frame is retransmitted, but the damage has been done on the receiving side. The Ethernet specification allows some "shrinkage" of the interframe gap due to preamble bit loss, but the receiver might still lose one or two frames if the shrinkage is too much. The more repeaters, the greater the chance of frame loss. So Ethernet restricts the number of repeaters for the reason of interframe gap shrinkage also.

But 10Base-FB reduces this interframe gap shrinkage by using the synchronous signaling protocol to synchronize the transmission of frames between the two hubs. With normal Ethernet, a preamble shrinkage of eight bits is not unusual. The synchronous signaling protocol reduces preamble bit loss to only two bits, and usually no bits at all are lost. The 10Base-FB receiver is locked to the 10Base-FB transmitter, but now the link must be point-to-point (no surprise) and 10Base-FB cannot be used with 10Base-FL.

Both ends of the link must be designed to use 10Base-FB, so special ports are the rule. Also, 10Base-FB is not to be used to connect a computer to a hub, since that's what 10Base-FL is for. The 10Base-FB type supports the same fiber optic cable and connector types as 10Base-FL. However, 10Base-FB does not support the use of full duplex Ethernet.

The third component of the 10Base-F family, 10Base-FP (Fiber Passive), addresses the point-to-point limitation of both 10Base-FL and 10Base-FB. This type allows for the inclusion of what is called a *passive star* coupler into the Ethernet LAN. This requires the deployment of a 10Base-FP Star hub device in place of the hub. Unlike a normal Ethernet hub, the passive star requires no electrical power, which is why the device is "passive." All of the power is in the NIC cards in the end devices. Up to 33 devices can be connected by the 10Base-FP Star, each with two fibers. The hub just takes in the optical signals from an active device and passively couples them onto all 33 output fiber ports. True collision domains are established by the 10Base-FP Star, so no full duplex Ethernet operation is possible. This lack of full duplex Ethernet support has limited acceptance of 10Base-FP. Also, passive optical couplers have enormous attenuation because the power of a single optical input must be multiplied 33 times without any optical power compensation.

Most 10Base-F implementations today (and there are few) use 10Base-FL to link computers to a 10Base-F hub, and then 10Base-FB to link these hubs together.

The 100Base-T Ethernet specification allows operation of the LAN at 100 Mbps rather than 10 Mbps. It defines two physical media types: unshielded twisted pair copper (100Base-TX) and fiber optic cable (100Base-FX). The 100Base-TX specification is not examined in this section, but 100Base-FX is.

When 100Base-FX is used to link computers to a 100Base-T hub, two fiber optic cables are used, as might be expected. When half duplex Ethernet is used, the maximum distance is 417 meters (1,350 feet); this figure grows to 2 km or more when full duplex Ethernet is used. In general, 100Base-FX is very similar to 100Base-TX, right down to the line coding used. Usually 62.5/125 MMF is used, although 50/125, 85/125, or 100/140 MMF can be used as well over shorter distances. The light wavelength to be used is 1,300 nm, not 850 nm as with 10Base-F. The recommended connector types are SC connectors, but most implementations use ST connectors or even FDDI connectors.

NICs for 100Base-FX usually have optical transceivers built in, but external transceivers can be used, as with 10Base-F, but with a special 40-pin NIC interface known as the Media Independent Interface (MII). Full duplex Ethernet mode is supported; it increases the link length from 412 meters to the full 2 km. Single-mode fiber can extend this even farther.

Gigabit and 10 Gigabit Ethernet

Fiber optics play a key role in the two latest additions to the Ethernet family. These are Gigabit Ethernet (GBE) and 10 Gigabit Ethernet (10 GBE). GBE, to some extent, and 10 GBE, to a large extent, break the boundaries of the traditional LAN and even MAN (100 km) fiber applications and send Ethernet frames structures over a WAN reach. In a very real sense, GBE and 10 GBE represent nothing less than an invasion of the LAN into the WAN domain. Traditional WAN technologies might even be supplanted in many applications in favor of GBE and especially 10 GBE.

Gigabit Ethernet was standardized in a very short period of time compared with other ambitious networking schemes. One reason for this speed to market was that the groups behind GBE were quite willing to incorporate work that had already been done on high-speed fiber optic transmission into the GBE standard. So some of the key pieces of GBE were not newly developed for GBE but freely borrowed from a technology called Fibre Channel (sometimes seen abbreviated as FC). The components of GBE based on Fibre Channel were originally developed for 1 Gbps transmission using Fibre Channel devices. But it is important to realize that GBE is not in and of itself a component of Fibre Channel. Fibre Channel remains a separate technology, which is widely used for creating Storage Area Networks (SANs) with fiber optic cable.

GBE is a large and sprawling specification. This section emphasizes the support for fiber in GBE. GBE allows for the use of UTP in what is known as 1000Base-T, but only for very short, 25-meter (about 75 feet) runs for patch cables and the like. There is also 1000Base-CX, defined on coaxial cable with the same distance restriction. The real promise of GBE and meaningful distances are fulfilled only with fiber optic cable. 1000Base-CX is not an optimal solution, and 1000Base-T is really just an add-on to GBE.

GBE defines two physical media interfaces for fiber: 1000Base-LX and 1000Base-SX. Both share a common coding technique and other aspects with Fibre Channel.

The 1000Base-X specification defines a Gigabit Media Independent Interface (GMII) that plays the same role as the Attachment Unit Interface (AUI) in 10 Mbps Ethernet and the Media Independent Interface (MII) in 100 Mbps Ethernet. But no connector is defined for the GMII, so the optical transceiver (the E-O conversion device) cannot be an external device. This restriction was made because of the parallel nature of GBE data transfers. External cables would be hard to control with such high frequencies and bit rates.

The 1000Base-LX (the L is for "long") GBE interface uses long-wavelength lasers, operating in the 1,270 to 1,355 nm range. Multimode (62.5/125 or 50/125) and single-mode fibers can be used. Half duplex mode with collisions is supported, but most GBE operation is full duplex from switched hubs. Half duplex mode supports a 316-meter (1,036 foot) segment, and full duplex supports 550 meters (1,804 feet) on MMF and 5 km (about 3 miles) on SMF. SC connectors are used.

The 1000Base-SX (the S is for "short") GBE interface use shorter-wavelength lasers, operating in the 770 to 860 nm range. These lasers are less expensive than longer-wavelength transmitters, but only MMF is supported. Half and full duplex operating modes are supported. Half duplex segments can be up to 275 meters (902 feet) on 62.5/125 MMF and up to 316 meters (1,036 feet) on 50/125 MMF. Full duplex segments can be up to 275 meters on 62.5/125 MMF and 550 meters (1,804 feet) on 50/125 MMF.

Still very much a work in progress, 10 GBE extends Ethernet into the reaches of state-of-the-art fiber optic speeds for serial transmission, which currently tops out at 40 Gbps. The whole idea behind 10 GBE comes from the realization that the vast majority of LANs linked by WANs today run Ethernet in some shape, manner, or form. So it would be more efficient to minimize the amount of reframing, conversion, encapsulation, and other forms of overhead processing that must take place as a "low speed" Ethernet frame makes its way from LAN to MAN to WAN and back again.

Using 10 GBE, service providers can aggregate many GBE bit streams into one fat pipe for WAN transport. This role for 10 GBE is shown in Figure 4-5.

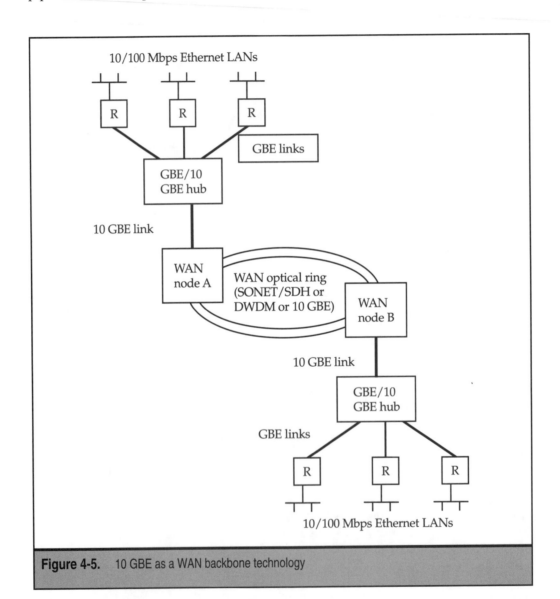

Figure 4-5.   10 GBE as a WAN backbone technology

Work on 10 GBE has just begun. Much of the initiative on the IEEE 802.3ae committee working on 10 GBE has been taken by a group of vendors and service providers organized as the 10 Gigabit Ethernet Alliance (10GEA).

Fibre Channel

Fibre Channel was developed by the American National Standards Institute (ANSI), and so the use of the British spelling of "fibre" has been something of a mystery. Early Fibre Channel documentation always spelled it "fiber." But the technology was developed at a time of rather testy exchanges between standards organizations over the proper use of English documentation meant to be adopted as international standards. So to better Fibre Channel's chances of becoming an international standard at some point in time, ANSI switched to the British spelling to impress the International Organization for Standards (usually abbreviated ISO), or so the story goes.

Fibre Channel (FC) is not a traditional networking protocol like Ethernet or even SONET/SDH, but more like the small computer system interface (SCSI) for connecting scanners or hard drives to a computer. FC runs at 100 Mbps, 200 Mbps, 400 Mbps, 1 Gbps, and even 2 Gbps speeds. FC is defined on coaxial cable, shielded twisted pair (STP), and multimode and single-mode fiber. Supported lengths range from 50 meters (about 150 feet) for links on STP to 10 km (about 6 miles) on single-mode fiber. Delays are very low, about 10 to 30 microseconds, due to the simple buffer-to-buffer transfer mechanisms used by FC.

Several topologies are possible for an FC "network." These include a simple point-to-point link, a topology using a central switching hub or FC "fabric," and an odd type of ring called an FC "loop" because there is only one fiber between all of the nodes.

Today, FC is most often employed to solve an annoying problem with large Web sites. It is not unusual for a large company to have perhaps 30 or more servers all acting as a single Web site, all appearing to users as www.example.com. This is done to spread the load generated by millions of hits from people all over the world. However, the servers must still be coordinated to the extent that all of the servers look and feel the same in terms of Web content and capabilities. Users that begin a transaction such as buying a sweater have to make sure that the Web site does not drop the session if a page request goes to a different server in midstream. And if a hard drive fails, the server that has the hard drive is out of commission.

The term *storage area network (SAN)* has been used to describe this arrangement. In a simple SAN, servers can be connected to the router on an FDDI ring (this is only an example). But on the back end of the servers, the redundant arrays of inexpensive disks (RAIDs) make sure that users see a consistent Web site. Web content updates are done once, since all servers draw from a single (but redundant) pool of pages. FC as a SAN is shown in Figure 4-6.

FC is the most popular technology for this SAN arrangement, due to the high speed and higher layer protocol transparency of FC. There are many other uses for FC, of course. The SAN application is just a very popular one.

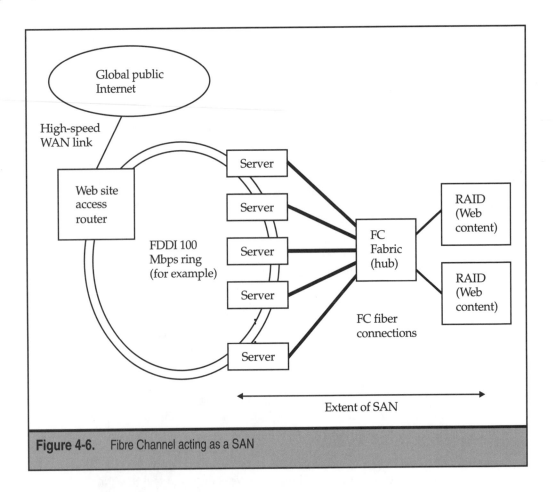

Figure 4-6. Fibre Channel acting as a SAN

FIBER OPTICS IN THE WAN

One of the major attractions of using fiber optics in the LAN is the enormous bandwidth available on fiber optic cable. The fact that fiber spans greater distances without repeaters is a related plus. These same qualities make fiber optic cable attractive for WAN use as well.

Just as in LANs, however, fiber optics have been applied as a series of optical links connecting electrical devices. There is nothing wrong with this, but it is not yet optical networking. This section will briefly outline some of the traditional methods that have been used when fiber optics are applied to WANs.

Telephone Trunking Systems

The major attraction of fiber optics for telephony has been and is the enormous bandwidth capacity of fiber optic cable. Voice is a relatively low bandwidth–consuming application (64 Kbps or less), but usually voice circuits require that whatever bandwidth used for voice be dedicated for the duration of the connection. This is the essence of circuit switching, of course. The need to support circuits even when there is no meaningful information exchange going on in one direction (more than 50 percent of every telephone call is silence) is one reason that many low-bandwidth voice circuits require huge aggregate bandwidths.

Fiber optic cable was first used to support voice service in Chicago in 1967. But such systems remained more of a curiosity for quite some time. In the Bell System, where Bells Labs invented the laser and AT&T was the largest telephone company in the world, fiber optics played only a small role for many years. Only 5 out of 884 pages of the 1984 Bell System engineering guide mentioned "lightwave" systems and "lightguide" cable. The word "fiber" did not even have an index entry.

The most common speed for these early fiber systems was 44.736 Mbps, which was universally rounded off to 45 Mbps. The Bell System called this a T-3, which carried a DS-3 frame structure and supported up to 672 simultaneous voice call circuits. When a T-3 was provisioned on one of the new fiber systems, it became an FT-3.

There was plenty of bandwidth to spare even on these early fibers, however. The reason for the support of the pedestrian T-3 speed was mainly for backward compatibility with the enormous quantity of 45 Mbps equipment in place. So multiple T-3s could be multiplexed onto a single fiber to support even more voice calls. This was simple time division multiplexing, not wavelength division multiplexing. When 12 bit streams from T-3 were combined onto the single fiber supporting the T-3s, there was still a single stream of serial bits on the single fiber. But each bit time was one-twelfth of what it had been on the T-3 (actually, there was always some additional bit overhead).

Vendor-specific systems combined anywhere from 9 to 14 T-3s onto a single strand of fiber. Aggregate cables could carry many such fibers on a national or international backbone. These were known generically as Fiber Optic Transmission Systems (FOTS). Every major telephone company and equipment vendor had its own proprietary FOTS configuration. No vendor interoperability was provided or expected.

FOTS capacity for voice channels was enormous. A system with 12 T-3s carried 8,064 calls on a single slim, low-weight fiber. In contrast, as mentioned earlier in this book, a 3,600-pair copper cable, where each pair carried one call, was four inches in diameter and weighted seven pounds per foot. A 400-foot reel weighed over a ton.

A fiber cable with 32 FOTS fibers in it could carry more than a quarter million (258,048) voice calls at the same time. AT&T had fiber bundles by the mid-1980s that would allow everyone in the United States east of the Mississippi to call everyone west of the Mississippi at the same time and carry all of those calls on a single fiber cable. However, the vendor-specific nature of FOTS links made it impossible to supply details in terms of wavelengths, speeds, or even spans.

Fiber was primarily used by the carriers at the time to support voice calls. There could be megabits going back and forth on the fiber, but all a user could ever get was access to 64 Kbps in most cases, because of the channelized nature of the fiber link. Obviously, if one serial bit stream was able to use all of the bandwidth capacity of the fiber, then the fiber would have to be *unchannelized*. An unchannelized FOTS with 12 T-3s represents about 540 Mbps of bandwidth that would now be available to a single source. That single source could now be a multiplexed stream of data packets, not digital voice.

But voice support could not be neglected. It might still be necessary to support voice in the form of *virtual circuits* on the otherwise unchannelized fiber cables. By 1998 an international standard was available that allowed virtual circuit voice traffic to mix with data packets on a switching network where the switching nodes were connected by fiber optic links. This was *asynchronous transfer mode,* or ATM.

ATM was the middle layer of a three-layer architecture. Below ATM were the unchannelized fiber links, and there were international standards for speeds and the like at the lower layer. The lower layer defined SONET/SDH links (later rings). ATM provided the switching and multiplexing for the traffic on these links. ATM employed neither standard variable-length data packets nor standard voice channels, but instead defined a standard fixed-length *cell* structure. The short cell made it possible to mix voice and data cells, since a delay-sensitive voice would not be held up for long periods of time behind a long data packet.

Above the ATM layer was the broadband services layer. There was an official definition for just what "broadband" was, but essentially a broadband service was one that could not be delivered over a low-bandwidth voice channel or a small number of voice channels bundled together to appear as one. Broadband services were things like broadcast-quality video, online shopping, and so on.

The entire three-layered architecture was officially known as Broadband Integrated Services Digital Network (B-ISDN). ISDN itself was a series of telephony-based services with some video thrown in for good measure, but it had no cells. B-ISDN extended the basic capabilities of ISDN in many ways. Figure 4-7 shows the basic architecture of a B-ISDN network.

Shortly after B-ISDN was defined, the major players were split between service providers and equipment vendors looking for technologies that would fit their immediate needs. There was not enough SONET/SDH fiber around in 1988 to base an ATM network delivering B-ISDN services on. ATM was quickly redefined to use nonfiber links, but soon after the Web had hit town and there was no turning back. Today, broadband services are available over the Internet. Voice is going packet as voice over IP (VoIP). SONET/SDH links provide either large numbers of voice channels or huge unchannelized bandwidths for LAN and router connectivity. ATM has been more or less left as the orphan layer in between.

Access Networks

FOTS links were used on the carrier backbone to aggregate huge numbers of calls for long-distance transport. But once the cost of the fiber itself, and the transmitters and receiver, came down in the 1990s, fiber was positioned not only for national and regional long-haul backbones but also for access networks to businesses and even residential neighborhoods.

Access networks are often called subscriber networks, or local loops, or even more exotic things like the "last mile." But whatever its name, the access network links customer premises equipment (CPE) to an appropriate counterpart in the service provider's office

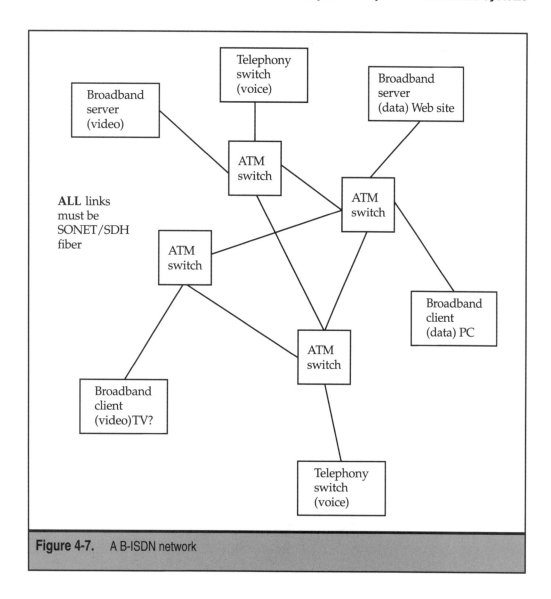

Figure 4-7.   A B-ISDN network

location. Using fiber in this access network means adding or replacing existing copper technologies with some technology based on fiber.

In the case of a business, fiber access is relatively easy to cost-justify and implement. Businesses know they need more bandwidth, and they can pass the costs of running the business on to their customers. But there is a limit to how deep a business's pockets are. And the cost of running the fiber cannot be totally offset by doubling or even tripling the cost of the service delivered on the fiber. In fact, economies of scale seem to dictate that the cost of a service in terms of cost per bit be *less* when delivered on fiber. So out of some 750,000 commercial office buildings in the United States today, only about three percent have fiber running directly to the building for local access, according to the U.S. Department of Commerce. And most of these are concentrated in the top 50 metropolitan areas.

Fiber has been even slower to reach out to residential areas, except for some isolated and controlled experiments. A lot of this slowness has to do with scale. According to the federal government, in the United States alone, there are some 200 million access lines, and 70 percent of these—140 million—run to people's homes. And in most cases, local residential services are tightly regulated, and price increases are not often received happily by customers or regulators unless something really new can be done with the fiber. Delivering the same old voice on fiber might be exciting to some, but not to most customers. Residential customers have no place to pass support service costs along to someone else.

The general process of provisioning local access for residential areas on fiber is known loosely as *fiber in the loop* or FITL. The biggest issue to be faced today is just how close to the subscriber premises the fiber should come. There is the conservative fiber to the neighborhood (FTTN) approach, which has actually been used quite successfully, especially in the densely populated northeastern part of the United States. The FTTN architecture basically uses fiber instead of copper to support anywhere from 500 to 1,000 or even more subscribers in what is often called a *carrier service area (CSA)*. Again, vendor specifics make it impossible to offer any details of these access schemes.

The two more ambitious residential fiber architectures are fiber to the curb (FTTC) and fiber to the home (FTTH). FTTC serves small clusters of anywhere from 10 to 100 subscribers, depending on the vendor. FTTH serves one home per fiber. Naturally, once fiber reaches the home, the task is complete. It must be pointed out and appreciated, however, that it is not the cost of buying and installing the fiber that is the major hurdle for FTTC and FTTH to overcome. Most service providers would much rather run fiber than wrestle with multipair cable bundles. The problem is that there is nothing on the customer premises that can support a fiber interface. So the E-O conversion must be done by the service provider. With FTTN, this is not some much of a problem, since E-O conversion for 1,000 homes can be very efficient. FTTC can be done as well, but at greater cost per home.

FTTH was first suggested in 1972 by John Fulenwider of GTE. Japan started trials in 1978, but all of the work done in the past 20-plus years has barely scratched the surface. And E-O converters for homes, which must usually be mounted outside in the heat and snow and rain, are not yet environmentally hardened enough to be both reliable and at the same time affordable.

The basic architecture of FTTC is shown in Figure 4-8. The optical network unit (ONU) houses the E-O converters and related service functions such as multiplexing. FTTC architectures that service many homes approach FTTN in scope, while others that service only a handful of locations approach FTTH.

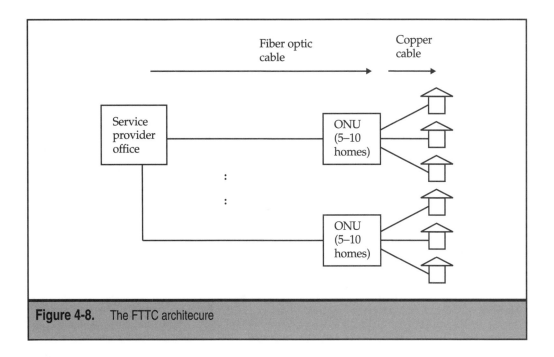

Figure 4-8. The FTTC architecure

Cable TV Systems

Terms like "FTTC" and "FTTH" are basically telephone company words. They describe the process of moving fiber closer to the customer in a telephony environment. But many residential customers have not one but two major types of network services today. Most people have telephone service from a telephone company and cable television service from a cable TV service provider. It is possible today to get voice over cable TV, or video over a telephone line, but this is not often done for a variety of reasons.

Naturally, cable TV companies are as anxious to take advantage of the benefits of fiber as the telephone companies. But instead of terms like FTTC and FTTH, the cable TV industry tends to use terms like *hybrid fiber coax (HFC)* and *passive optical network (PON)*. When the concepts of HFC and PONs are put together, the end result often looks very much like FTTC, with the same services, and often at lower prices. But there are enough differences to make things interesting.

The concept of HFC is an interesting one. Traditionally, cable TV networks used coaxial cable to reach subscribers. The topology used is called the *random branching bus* or *arbitrary branching bus*, meaning that a single coaxial cable with all the channels leaves the cable TV *headend* office and is split among the branches wherever the system designer deems necessary. The headend is where cable TV service providers get most of their programming from satellites and local programs by other means to distribute to their customers over the cable TV network. There could be up to 20,000 "homes passed" (meaning not all houses "passed" by the cable will subscribe to the service) on a single coaxial cable. There are many amplifiers, large and small, that amplify (and add noise to) the analog video signal. The general architecture of a purely coax cable TV network is shown in Figure 4-9.

The numerous amplifiers used in coaxial cable TV systems were a real bother. They not only added noise, there were lots of them (remote subscribers were 30 or 40 amplifiers away from the headend), and they all needed quite a bit of electrical power. Amplifiers also disrupted service for all those downstream from a failure. And if upstream services such as telephony or cable modems were to be supported, the downstream pointing amplifiers often prevented such services from functioning at all. Also, upstream traffic was aggregated onto a single coax cable in many cases.

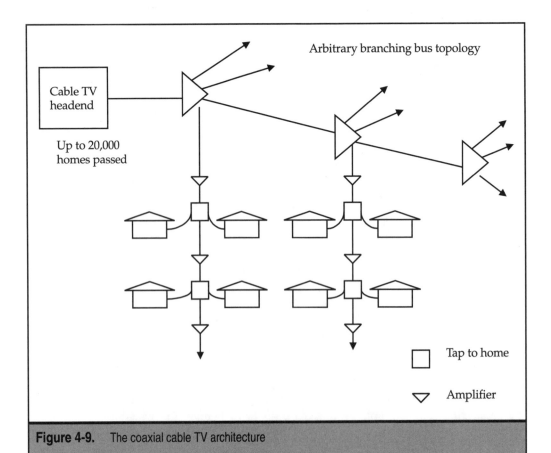

Figure 4-9.   The coaxial cable TV architecture

The solution to many of these problems with analog amplifiers was to replace at least some of the coaxial cable with fiber optic cable. The result was not a *digital* cable TV network, but a better analog cable TV network. Amplifiers could be bypassed (so there was less noise), fiber components were more reliable, upstream traffic was not consolidated to as great a degree, and less powerful upstream transmitters were required in the home.

When fiber is combined with coaxial cable in a cable TV network (in a way that only serves to preserve the customer's nonfiber interface), the result is called a *hybrid fiber coax* network, or HFC. All in all, fiber improves the situation greatly. The general architecture of an HFC is shown in Figure 4-10.

Again, there are few standards for these newer cable TV architectures. So it is not possible to detail numbers of channels supported, upstream characteristics, type of fiber used, or distance specifics. In the United States alone, there are currently some ten vendor

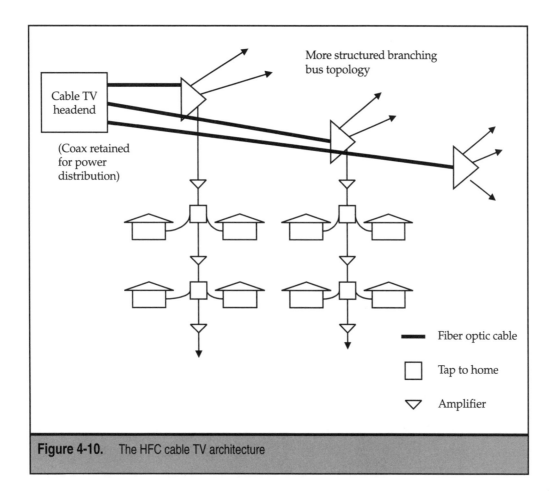

Figure 4-10.   The HFC cable TV architecture

variations with competing and distinctive architectures and schemes for adding fiber to a coax cable TV network.

Just as with FTTC and FTTH, many of these HFC architectures are distinguished by how close to the customer premises the fiber actually comes. Terms like FTTC and ONU are mainly telephony service provider terms, so the cable TV industry has its own terminology to describe these alternate architectures. Thus in cable TV HFC environments, the key terms are *fiber to the feeder (FTF)* and the idea of a *passive optical network (PON)*. In an HFC cable TV network, the role of the ONU is played by the *fiber node*.

Figure 4-11 shows how all of these cable TV terms fit together. The main factor is how many analog amplifiers the fiber nodes displace. An all-coax network could have 20,000 homes passed on a single coax feeder. Fiber can replace many of these large backbone amplifiers simply by connecting a few thousand customers on a fiber node (so there would be ten or so fiber feeders in total).

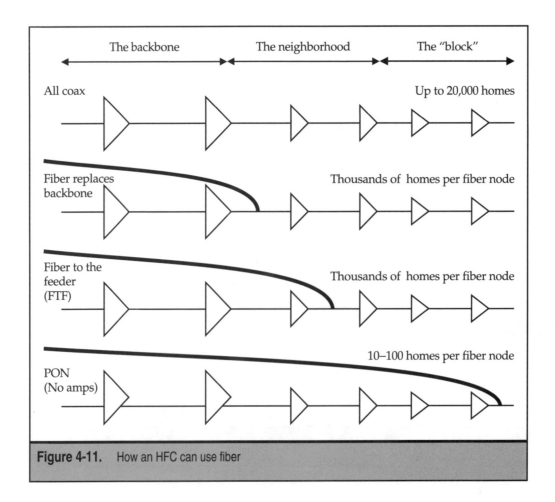

Figure 4-11.   How an HFC can use fiber

In fiber to the feeder (FTF), some of the smaller branch amplifiers are also bypassed, leaving up to 500 or so customers services on the fiber node. This requires more fiber, of course, but bypasses all but the smallest amplifiers.

Why not just bypass each and every powered amplifier with the fiber? If the problem is the analog amplifiers, just get rid of all of them. This is the idea behind the passive optical network (PON). It is passive in the sense that the fiber node just passively splits and/or combines downstream and upstream signals without the need for local electrical power. PONs require no additional power between cable TV headend and customer premises, an ideal state of affairs. Because the truth is, most HFCs retain the original coax cable plant to distribute power to the remote fiber nodes! Fiber will do many things, but it won't carry electricity through the glass. PONs do away with any power concerns in the field.

However, PONs are tricky to design and implement. This is because of a key optical design concept known as the *power budget*. Since the idea of a power budget becomes even more important in DWDM, this chapter will close with a quick look at the optical power budget in action.

MILITARY AND INDUSTRIAL USES OF FIBER

When fiber optic networks are considered, most space is usually devoted to applications of fiber to LANs and WANs. But other applications of fiber optics do not always fall neatly into these two categories. Among these applications are fiber optic applications for remote control and guidance of missiles, tanks, and helicopters for the military. Industrial uses include submersible robots, and internal control and monitoring of aircraft and ships. There is also a lot of interest in applying fiber optics to automotive use as well. As more and more components of a car become computerized, from ignition to windows, communications and control of all these chipsets makes fiber an attractive alternative to noisy electrical systems.

Only brief mention of these applications is needed here. When used within a vehicle, the fiber resembles LAN applications, and when used for remote control and monitoring, the fiber resembles WAN applications. However, there are few standards in place, and customization and vendor-specific implementations are the rule of the day. But as these systems mature, standardization will come slowly but surely.

MANs AND THE POWER BUDGET

By now it should be clear that fiber optic cable is used in WANs and LANs for a number of reasons. And one of the most compelling reasons to use fiber is the fact that fiber spans can be much longer than the spans supported with copper, coaxial cable, or radio waves. Here the term "span" just means "the distance traveled without needing amplifiers or repeaters." The low attenuation (signal loss) and noise on fiber optic spans translates directly to longer spans.

But just how long can a fiber span be? The answer is always tied up in the concept of the optical *power budget*. This concept largely determines the distance a fiber optic span can be used over. Power budgets are usually computed in decibels (dB). The power budget determines if enough light is present at the receiver to cover all the losses on the span. The receiver can be an optical amplifier or an optical detector. If the receiver is an optical amplifier, then the power budget computation is somewhat complicated, but valid nonetheless. The simplest case connects an optical transmitter with an optical receiver at the endpoint of a span. This is the configuration discussed in this section.

Fiber can also be used productively in metropolitan area networks (MANs), which occupy an overlapping position between LANs and WANs. That is, MANs can be seen as "stretched LANs" or "small WANs." Whenever fiber optic links are used in MANs, the power budget becomes very important. This does not mean that LANs and WANs using fiber do not care about power budgets.

It should never be thought that, when standards are followed and fiber span distances are spelled out in detail, power budgets never enter the picture. All this means is that the standardizing body has considered the performance and operational characteristics of all of the span's components and then translated these into maximum speeds and distances. For example, if a given specification recommends a semiconductor laser of a given power as the transmitter, and an APD of a given sensitivity as the receiver, and a given fiber with so many splices and connectors in between, this does not mean there are no power budget concerns. What standards for spans do mean is that the organization issuing the standard has, unilaterally or with other entities such as equipment vendors, carefully computed worst-case power budgets and knows, unless something unforeseen comes up, that this configuration of transmitter, span, and receiver will work.

However, there are always cases where the power budget needs to be understood in detail or even computed from scratch. There are times when standard configurations are pushed to the limit, or configurations where no standard yet exists. For these situations, which arise quite often when fiber optic links are used in LANs, a firm understanding of power budgets is required.

But the overwhelming reason that the power budget today is so important is *dark fiber*. Dark fiber is fiber that has been run but is not yet hooked up to the transmitter and receiver electronics. Now, if the whole reason that the dark fiber has been installed is to implement one of the standard fiber technologies, then the power budget consideration is less critical.

It is when the dark fiber can be used for a number of different purposes that the power budget is so important. Almost every company with right-of-way, from telecommunications companies to cable TV companies to local municipalities, has been busily running single-mode fiber as fast as it can. Never mind that there is no immediate use or customer for the fiber, the bandwidth will not go to waste. In many cases, less than 10 percent of the total fiber capacity is "lit" or in use.

It is usually in the metropolitan area network (MAN) fiber arena that an understanding of the power budget is absolutely essential. MANs are typically defined as less than 100 km (about 60 miles) total distance end to end, but many MANs are much smaller. Typical MANs will cover no more than 20 or 30 miles. When linking buildings or other facilities that are not so far apart, dark fiber is an attractive choice. A full service offering including all E-O conversion, transmitters, and receivers in many cases grows more cost effective as the distance end to end increases. This makes sense, because much of the expense of a fiber link is not the fiber optic cable, but the end electronics. But this also makes short fiber links quite expensive compared with their longer cousins. So full service offerings are often expensive for shorter distances and organizations with several facilities in a concentrated area. Lower bandwidth will save cost but also decrease the effectiveness of the network. Dark fiber is attractive for all of these reasons.

Now, dark fiber has no transmitter and receiver attached. So the range of options open to the eventual user of the fiber is considerable, not only in terms of supported services and protocols, but in terms of the equipment. If users can get the right to use dark fiber, then users can supply their own E-O conversion, transmitter, and receiver. However, the more powerful the transmitter and the more sensitive the receiver, the more expensive the devices become. And even if a competitive service provider is used to provide the link, in many cases this service provider will rely on dark fiber from a third party.

The dilemma of the MAN solution is shown in Figure 4-12. A full service offering is too expensive, but if dark fiber is used, what transmitter and receiver should be used on the span? With a proper understanding of the power budget, dark fiber can be used with the proper transmitters and receivers for the span.

COMPUTING THE OPTICAL POWER BUDGET

The basic idea behind computing an optical power budget is easy. If there is enough light left over, the span will work. If there is not enough light left over, the span will not work. As important as they are, optical power budgets are not difficult to compute. All that is needed is some simple skills like adding, subtracting, and multiplication. Power budget computations do not even require an understanding of decibels, but few practitioners would feel comfortable leaving power budget calculations up to such naive implementers.

So the first step in computing the overall optical power budget is to figure out just how much light there is for the electronic devices on each end of the span to use. Two numbers are of paramount importance here. The first is the minimum transmitter power, and the second is the minimum receiver sensitivity. Note the use of minimums here, just to be on the safe side. These numbers are usually expressed in decibels with reference to one milliwatt of power, or dBm.

Both of these numbers are available only from the equipment vendor. Some equipment vendors try to be coy about these numbers, while others are quite open. Saying the equipment complies with "all relevant international and national standards" just forces interested parties to do more work. In this case, mystery is never an attraction.

The minimum transmit power is just what it says. This is the worst-case performance of the device. The transmitter is absolutely guaranteed to deliver that amount of power, and might give more. Some of the more coy vendors are fond of listing an "average transmit power" for the device, but this number is useless for power budget considerations

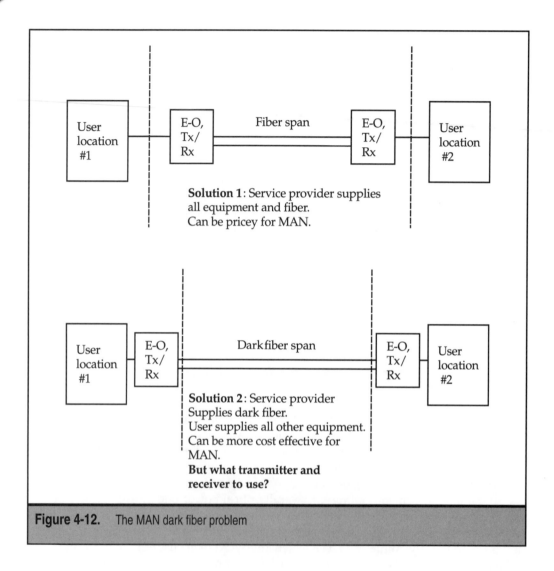

Figure 4-12. The MAN dark fiber problem

and meaningless by definition. A power average means that sometimes the power is more, and sometimes the power is less. There is no "floor" to an average. If a power budget calculation is performed using the average power of the transmitter, what happens on the span when the power generated is considerably below the average listed? It most likely will not work.

The minimum receiver sensitivity is just as important. This is the minimum amount of light required to make the device operate properly. No average allowed here either.

If there were nothing more to the optical power budget, simple subtraction should give a valid result for the amount of light available. That is:

Light available = minimum transmit power – minimum receiver sensitivity

The only thing to remember is that the minimum transmit power and minimum receiver sensitivity are often supplied as negative numbers already. A typical minimum transmit power might be –11 dBm, and a typical minimum receiver sensitivity might be –33 dBm. That's okay, since minus a negative gives a positive, or

Light available = –11 dBm – (–33 dBm) = –11 dBm + 33 dBm = 22 dB
(the difference is always just dB)

There is one small complication with this simple, overall power budget example. Fiber spans almost universally operate in *both* directions. The fibers are different, but probably of the same type, and the transmitter-receiver pairs on both ends can also be different. This can happen when one organization links up to another. Each organization makes independent equipment choices, but of course the fibers still have to work in both directions.

So the rule for bidirectional operation is simple. Figure out the overall power budget in both directions, and then use whichever number is *smaller* for further calculations. It makes sense that if the smaller power budget works in one direction, then the larger budget will work in the opposite direction. However, this will only be absolutely true if the inbound and outbound fibers follow the same route between sites and are of the same construction. Fortunately, this is almost always the case.

For example, consider Company A and Company B. They wish to connect sites over a dark fiber link. Company A has chosen Vendor A with minimum transmit power of –1 dBm and minimum receiver sensitivity of –31 dBm. Company B has chosen Vendor B with minimum transmit power of –3 dBm and minimum receiver sensitivity of –32 dBm. Which available power numbers should be used?

Table 4-1 shows the answer. The power budget is computed twice, once from Transmitter A to Receiver B and then from Transmitter B to Receiver A (the "crossover" is critical for proper results). The lower number should be used, in this case 28 dB, from Vendor B to Vendor A.

If this is all there were to computing power budgets, the topic would not require a section on its own. However, the power budget calculation process has really just begun. All that has been done so far is to compute the *overall* available power on the span end to end. Light is not delivered direct from transmitter to receiver without additional losses. The next step is to consider these additional causes of signal loss (attenuation) in the light of the power budget numbers derived so far.

Light is lost on a span due to a number of things. Any fiber, of course, has a characteristic attenuation per kilometer or mile. Then there are losses due to splices or connectors on the span. No fiber span is a straight shot of continuous fiber from transmitter to receiver.

Vendor A		Vendor B	
Minimum Tx power	–1 dBm	Minimum Tx power	–3 dBm
Minimum Rx sensitivity	–31 dBm	Minimum Rx sensitivity	–32 dBm
Power Budget #1		**Power Budget #2**	
Vendor A Tx	–1 dBm	Vendor B Tx	–3 dBm
Vendor B Rx	–32 dBm	Vendor A Rx	–31 dBm
Power available	–1 dBm –(–32 dBm) = 31 dB	Power available	–3 dBm –(–31 dBm) = 28 dB

Table 4-1. Using the Lower Overall Power Budget

There are always connectors at the equipment ends and at any patch panels along the way, and simple splices on the fiber span in between. This situation is shown in Figure 4-13.

In the figure, sites A and B are to be connected by a fiber optic link. The E-O converter shown also includes the transmitters and receivers needed, although only one of the two fibers needed for simple two-way transmission is shown. This just makes the figure less cluttered. At each site, the E-O unit has a connector on the fiber tail end leaving the device. This mates with another connector that takes the signal down the hall or to another floor on user fiber to a fiber optic patch panel. The patch panel, which is really just a "manual cross-connect," is not strictly necessary but often used. The patch panel just makes rearrangements (known as "adds, moves, and changes") of the building fiber cabling that much easier to do. Naturally, there are connectors on both sides of the patch panel as well. There is also a short length of fiber jumper cable connecting the patch panel ports, but this is not shown in the figure. The patch panel is just a passive housing and adds no loss to the system. The network side of the patch panel at the entrance facility interfaces with the dark fiber itself. This arrangement is mirrored at site B. There could be additional connector pairs, but this scheme with three pairs on each end of the fiber is quite common.

The most important point about connector loss is that connector losses are always figured in connector *pairs*. There are 6 connector *pairs* in the figure, but 12 connectors total. But it makes no sense to compute the loss for *one* connector, since light will not flow end to end unless all of the connectors are mated. So it is connector pairs that count.

Once out of the building, the dark fiber is not usually connectorized. Rather, the fiber on the span itself is spliced. Splices are much more permanent than connectors, are less bulky, and have lower losses than connectors as well. The figure shows four splices on the span between the two sites. There are usually three good reasons to splice fiber optic cable. All three are illustrated in the figure. First, fiber optic cable does not come on a 40-km (25-mile) reel.

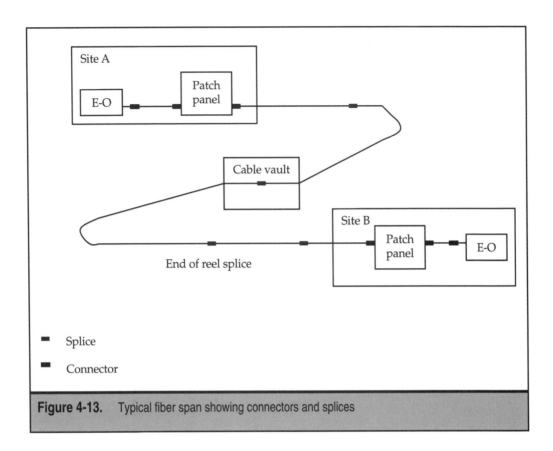

Figure 4-13. Typical fiber span showing connectors and splices

Most fiber reels are only a few kilometers or miles long. At the end of reel, a splice is needed when installing the fiber. Second, there are places along the span where it is easier to splice fiber than other places. If there is an environmentally hardened cable vault made out of steel or concrete, this is a good place to put a splice even if the reel is not quite finished. Splices in a vault can last much longer than splices in the middle of a plastic conduit or between utility poles. Finally, fiber optic cable can be dug up ("backhoe fade") after installation or damaged during installation by related construction activities. These problems usually occur nearer to the connected sites, since construction activity is more intense in the immediate vicinity of a building. To fix the damage, splice the fiber.

There are many variations on this simple end-to-end theme. But all of the basics for power budget considerations are shown in the figure.

So the additional losses from the fiber itself, the splices, and the connectors have to be subtracted from the end-to-end power budget. Fiber optic cable attenuation across the span is the most serious cause of signal loss. Then come connectors. Splices contribute the least to power budget losses.

The typical numbers used in optical power budget computations are shown in Table 4-2. Keep in mind that these are example values and should never be used instead of the actual numbers from vendors.

The standard value for connector pair loss is set by the Telecommunication Industry Association (TIA) at 0.75 dB per connector pair. The table also shows typical fiber cable attenuation for major types and wavelengths. The fiber cable vendor is the only reliable source for actual figures, and sometimes these numbers are stamped right on the cable jacket itself. Actual figures for single-mode fiber range from 0.22 to 0.5 dB/km. Typical splice loss is only 0.1 dB per splice, comparable to connector pair loss. Most fiber reels hold 6 km (3.75 miles) of fiber. The table also shows that when all is said and done, there should be at least 3 dB left over as a safety margin. Anything above and beyond this is just *excess power*. If there is a lot of excess power, however, perhaps some money could be saved by selecting a less powerful transmitter and/or a less sensitive receiver. The safety margin is different from what is called the *operational margin* of the span. This requires some further explanation.

Consider the fiber span shown in the preceding figure. To compute the operational margin and the safety margin, two further steps are necessary.

First, it is necessary to allow for the additional span losses due to fiber attenuation, connector pairs, and splices. The fiber loss is multiplied by the distance of the span, and the others are added on a per-component basis. For example, assume that the two sites are separated by 20 km (12.5 miles) of fiber and have six connector pairs and four splices, as shown in the figure. The fiber link will operate at 1,310 nm (the standard SONET/SDH wavelength).

Optical power budget losses are always computed from largest loss contributor to smallest. That way, if the figure ever goes negative, which indicates that the transmitter and receiver selected will not work over the link, the entire power budget procedure can stop right there. So fiber losses come first, then connectors, then splices. This gives the operational margin of the span.

Optical Power Budget Factor	Typical Loss
Standard value for connector pair loss	0.75 dB
Typical multimode fiber attenuation loss at 850 nm	3.75 dB/km
Typical single-mode fiber attenuation loss at 1,310 nm	0.4 dB/km
Typical single-mode fiber attenuation loss at 1,550 nm	0.3 dB/km
Typical splice attenuation	0.1 dB/splice
Typical distance between splices (typical reel length)	6 km (3.75 miles)
Typical safety margin for power budget	3 dB

Table 4-2. Typical Optical Power Budget Values

It is easy enough to multiple 20 times 0.4 dB to get a contribution of 8 dB loss from the fiber. There are six connector pairs. Exact connector pair loss figures can be provided by the connector vendor or installer. The number of connectors is usually easy to determine from visual inspection at the building sites. Multiplying the six connectors in the example by 0.75 dB per connector pair gives an additional span loss of 4.5 dB.

Last come the losses due to splices. The hardest part here is for the user to determine just how many splices there are. Fortunately, the losses due to splices are so low that they are seldom the make or break factor in a power budget. Cable engineering and "as built" drawings should show splice number and location, but these might not be available for a user to consult. (Sometimes it seems as if they are not even available internally!) When in doubt, use this guideline: one splice per 6 km (3.75 miles) between sites, assuming a point-to-point fiber run. In this case, however, there are four splices. So 4 times 0.1 dB is a 0.4 dB loss contribution due to splices.

The available power less the losses due to attenuation, connectors, and splices gives the operational margin on the span. The process so far is shown on the optical power budget worksheet shown in Figure 4-14.

Optical power budget worksheet (incomplete)

		Minimum transmit power:	−3 dBm	
		Minimum receiver sensitivity:	−31 dBm	
		Available power:	**28 dB**	

Fiber loss:	20	km of cable ✕ <u>0.4</u> dB/km	<u>8.0 dB</u>
Connector pair loss: 6	pairs ✕	<u>0.75</u> dB/pair	<u>4.5 dB</u>
Splice loss:	4	splices ✕ <u>0.1</u> dB/splice	<u>0.4 dB</u>

	Total losses:	<u>12.9 dB</u>

Operational margin (available power − total losses): <u>15.1 dB</u>

Figure 4-14. Optical power budget worksheet

If the operational margin is negative, then there is not enough optical power to drive the link with the selected transmitter and receiver. But even if the operational margin number is positive, there is more to be done. There are two more factors to consider in a complete power budget calculation.

Fiber cuts happen more frequently than is often assumed. Buried fiber is vulnerable to floods and simple cable dig-ups ("backhoe fade") and more. Aerial fiber is at risk from tornadoes and hurricanes and earthquakes and fires and even backhoe fade (the backhoe was being transported on a tall truck and snagged the cable). It is easy enough to splice the cable back together again, or use two splices to drop in a new fiber cable section, but this affects the attenuation on the span. So an estimate of the number of repairs made to the fiber over its lifetime is needed.

Longer spans are more at risk than shorter spans. There are just more places for bad things to happen. Most optical power budgets use five as the number of repair splices expected during the lifetime of the fiber span, which is usually considered to be 25 years and sometimes more. This is just another way of saying that a fiber cut should be expected every 5 years or so on a given span. The number can be adjusted up or down, of course.

There are even a few other factors that affect fiber span performance. Temperature extremes are one important factor. Thermal expansion and contraction of connectors and some types of splices will cause additional losses over the lifetime of the fiber. There are other, more arcane, factors, but temperature and other considerations are usually lumped together in the *safety margin* entry in the power budget. The most common safety margin value used in optical power budget computations is 3 dB. (In no case should a value of less than 1.7 dB be used for a safety margin.) Conservative organizations often use higher values, and organizations willing to assume greater risk can use smaller values for the safety margin.

When the final figures for the repair splices and safety margins are subtracted from the operational margin, the result should still be positive. The greater the value, the more confidence in continued link operation over time. If the value of the power budget is negative at this point, this means that the link might still work, but perhaps not after a few repair splices or during a heat wave. The failure might not be catastrophic, but even a modest rise in the bit error rate on the span at the wrong time can be disastrous.

Figure 4-15 shows the completed optical power budget worksheet for the example span configuration.

In this example, the excess power is 11.7 dB. This is quite high (a figure of about 5 dB is more typical). Most likely, a designer performing this power budget calculation would seek to try to save money by selecting lower-powered transmitters and less sensitive receivers. But without a detailed power budget calculation, this possibility would not even have been considered.

Optical power budget worksheet (incomplete)

Minimum transmit power:	−3 dBm
Minimum receiver sensitivity:	−31 dBm
Available power:	**28 dB**

Fiber loss: <u>20</u>   km of cable × <u>0.4</u>   dB/km   <u>8.0</u> dB

Connector pair loss: <u>6</u>   pairs ×   <u>0.75</u> dB/pair   <u>4.5</u> dB

Splice loss: <u>4</u>   splices ×   <u>0.1</u> dB/splice   <u>0.4</u> dB

                                                                Total losses: <u>**12.9 dB**</u>

Operational margin (available power − total losses):   <u>**15.1 dB**</u>

Repair splice loss: <u>4</u>   splices   <u>0.1</u>   dB/splice   <u>0.4</u> dB

                                                                Safety margin: 3.0 dB

Excess Power (operational margin − (repair + safety)):   <u>**11.7 dB**</u>

Figure 4-15. Completed optical power budget worksheet

PART II

Optical Fiber and Electrical Nodes

This book opened with an overview of many of the key technologies needed to create optical networking. This section will greatly expand on the early applications of optical networking and introduce the key concepts behind wavelength division multiplexing (WDM) and dense wavelength division multiplexing (DWDM). Along the way, some of the details of SONET/SDH, an electrical network with optical links, will be presented to better appreciate the breakthroughs that WDM and DWDM represent. At the end of this three-chapter section, the impact of true optical networking will be apparent.

The main emphasis in Chapter 5 will be SONET/SDH. Today, SONET/SDH represents the most common WAN application of optical networking techniques. Certainly, SONET/SDH remains the method by which most users and service providers are first exposed to optical networking. But in spite of its many advantages, SONET/SDH remains flawed when it comes to direct application of optical networking. Chapter 5 will investigate exactly why SONET/SDH is not particularly well-suited for the new world of optical networking.

Chapter 6 leads into a discussion WDM and DWDM concepts. The whole idea is to examine the limitations of single-wavelength systems and introduce early efforts to combine wavelengths over a single strand of fiber. The differences between WDM and DWDM, and they are not trivial by any means, are fully discussed as well. There is much more to DWDM than just jamming WDM wavelengths closer and closer together. A whole host of issues that more or less remain hidden in simple WDM now come to the forefront with DWDM. All major types of DWDM impairments are detailed. However, DWDM remains the foundation of optical networking.

Chapter 7 closes this section with a look at the fundamentals of wide-area optical networking nodes. There are details on all of the key components of a truly optical network. The whole point is that with true optical networking, some of the tasks previously done electrically as in SONET/SDH network nodes are now done optically. So all of the details on optical cross connects, optical "switches" (such as they are today), and optical rings are presented in Chapter 7. Although the emphasis here is on optical networks in the wide-area networking arena, many of the concepts will apply equally as well once optical LANs become common.

All of the information in the first part of this book, with a few notable exceptions such as GBE, essentially ended with the state of optical networking around 1988. The time has come to see what has been happening in the optical networking arena since then. It is safe to say that the optical networking revolution has begun.

CHAPTER 5

SONET/SDH

T he development of the North American standard Synchronous Optical Network (SONET or often just Sonet) and the international standard version of the same technology, Synchronous Digital Hierarchy (SDH), marks a watershed in optical networking. Before SONET/SDH, optical solutions for the WAN were intensely vendor specific, haphazard in data support, and without clear migration paths as higher-speed optical transmission became possible and even common. SONET/SDH changed all of this, and very rapidly, because SONET/SDH was very quickly perceived to be a better way of deploying optical networks. With SONET/SDH, there were firm standards for vendor interoperability (although full interoperability with SONET/SDH remains as elusive as ever), several ways to support data more effectively, and scalable speeds to take a link from 155 Mbps or less to more than 10 Gbps, as needed.

So SONET/SDH was embraced by the large national and international telecommunications carriers, although mainly for voice. But once these carriers became Internet service providers (ISPs), or newer IP-only ISPs entered the arena, SONET/SDH played a crucial role in the fast and efficient deployment of high-speed backbones connecting routers.

There was only one issue with SONET/SDH: it was a single-wavelength technology. The bandwidth available on the fiber might be terabits, but once SONET/SDH was provisioned on the fiber at (for example) 622 Mbps, that was all the fiber could handle. The 1,310 nm laser operated at 622 Mbps, period. A second laser could not be added at a different wavelength because SONET/SDH was defined to operate at 1,310 nm only. If more bandwidth were needed, the laser (and receiver) had to be upgraded to operate (for instance) at 5 Gbps. There is always one wavelength and one serial stream of bits in SONET/SDH. In 1988, when SONET/SDH was standardized, this single-wavelength operation was not seen as a limitation in any way, but just as way of preventing too many variables for equipment vendors to deal with and service providers to configure. After all, who could ever fill up a SONET/SDH pipe running at more than 10 Gbps?

As a result of this single-wavelength operation, SONET/SDH is in somewhat of a state of crisis today. If SONET/SDH must run at a single wavelength, and new optical networks routinely use multiple wavelengths with wavelength division multiplexing (WDM) and dense wavelength division multiplexing (DWDM), what is the future of SONET/SDH?

There appears to be no easy answer, but this chapter will at least make it easier to understand the precise nature of the question. It should be mentioned that this chapter forms a very abbreviated introduction to SONET/SDH with the emphasis on architecture, equipment, and deployment. Of course, enough could be said about SONET/SDH to fill a very large book.

SONET/SDH ARCHITECTURE AND PROTOCOLS

SDH is an international standard from the International Telecommunication Union (ITU) that defines a hierarchical set of transmission rates and transmission formats for optical fiber systems. SONET is a set of coordinated ITU, ANSI, and Bellcore standards that defines the same thing for use in North America. SONET standards are to be used by equipment vendors and service providers at or above a basic optical transmission rate of 51.840 Mbps, which forms the base transmission rate for SONET. In contrast, SDH starts its hierarchy at 155.52 Mbps. SONET/SDH transmission speeds top out at 10 Gbps and are being extended to 40 Gbps.

This chapter describes the fundamental architectural layers of SONET and the protocols used by SONET/SDH at the various layers. The basic SONET frame formats are introduced, as well as virtual tributaries and super-rate frame structures. Virtual tributaries are used for backward compatibility with existing digital hierarchies, especially for voice. Super-rate frame structures are used for the transport of the cells that are used in Asynchronous Transfer Mode (ATM) networks and IP packets.

A more complete discussion of key SONET concepts such as clocking and timing, jitter, and network synchronization is impossible in a short chapter and will not be attempted. Interested readers are referred to a more complete book on SONET/SDH.

SONET/SDH Speeds

Table 5-1 shows the basic speeds of the SONET/SDH hierarchy. The SONET/SDH hierarchy is made up a few basic building blocks of terms and speeds. Each level has an Optical Carrier level (OC) and an electrical level transmission frame structure to go with it, called the Synchronous Transport Signal (STS). An STS-3 frame is sent on a OC-3 fiber optic link, and so on. Note that the SDH terminology differs from SONET levels. The table shows the SDH Synchronous Transport Module (STM) level.

Table 5-1 also shows the physical line bit rate, the *payload* bit rate, and the overhead bit rate for each SONET/SDH level shown. The payload is just the bit rate left over after the overhead bits, which cannot be used for customer data, are subtracted. The payload is carried inside an *envelope* in SONET. The "counting units" in SONET are the basic 51.84 Mbps bit rate of OC-1. STM levels increase by units of 155.52 Mbps (3×51.84 Mbps). The last entry shows the bit rate for a new SONET/SDH level, OC-768 or STM-256 (?), that operates at about 40 Gbps. This is the current top speed for serial bit transmission, but many of the details about SONET/SDH at 40 Gbps are still to be determined (TBD).

It is an absolutely crucial point about SONET/SDH that unless other arrangements are made in equipment configurations, *SONET remains as channelized as a link used for voice transmission.* That is, an STS-3 contains three STS-1s. An STS-12 contains twelve STS-1s, and so on. All of the STS-1s inside an STS-48 on a OC-48 fiber optic link run at 51.84 Mbps, the STS-1 rate. So ironically SONET, despite the incredible speeds available at the higher ends of the hierarchy, would seem to limit networking to a bunch of channels running at 51.84 Mbps from any individual user.

However, just as voice links can be used for IP packets and other data streams as an *unchannelized* transport without any structure (save overhead bits), so can SONET/SDH. An unchannelized STS-12 would offer a user a raw bit rate (no more 12 STS-1s) at about 622 Mbps. A few more details on "unchannelized" SONET levels (they are not called unchannelized STS frames in SONET, but *concatenated*) are given later on in this chapter.

Optical Level	Electrical Level	Line Rate (Mbps)	Payload Rate (Mbps)	Overhead Rate (Mbps)	SDH Equivalent
OC-1	STS-1	51.840	50.112	1.728	—
OC-3	STS-3	155.520	150.336	5.184	STM-1
OC-9	STS-9	466.560	451.008	15.552	STM-3
OC-12	STS-12	622.080	601.344	20.736	STM-4
OC-18	STS-18	933.120	902.016	31.104	STM-6
OC-24	STS-24	1244.160	1202.688	41.472	STM-8
OC-36	STS-36	1866.240	1804.032	62.208	STM-13
OC-48	STS-48	2488.320	2405.376	82.944	STM-16
OC-96	STS-96	4976.640	4810.752	165.888	STM-32
OC-192	STS-192	9953.280	9621.504	331.776	STM-64
(OC-768)	(STS-768)	39813.12	TBD	TBD	(STM-256?)

Table 5-1. The SONET Digital Hierarchy

The terms used in the table come up over and over again in the rest of this book. Since these terms form the basis of the SONET/SDH architecture, a review of these terms will aid in the understanding of the architecture.

▼ **OC-N** This notation refers to the SONET transmission characteristics of an Nth level transmission link. N can take on values between 1 and 255, but not all values are supported. Within SONET, an OC-3 transmission link is assumed to have an STS-3 frame structure.

■ **STS-N** In this notation, STS (Synchronous Transport Signal) refers to the SONET frame structure of an Nth level transmission link. N can take on values between 1 and 255, as before. Although most SONET links should properly be referred to as STS-N, it is much more common (and just as correct) to speak of OC-N.

■ **Payload** The term used to indicate user data within a SONET/SDH frame. Payloads can be sub-STS-1 level (e.g., T-1s, E-1s, DS-2s), STS-1 level (e.g., DS-3s) and super rate (e.g., SMDS cells, ATM cells).

■ **Envelope** The portion of the frame used to carry payload and end-system overhead.

■ **Overhead** The portion of the frame used to carry management data for operations, administration, maintenance, and provisioning (OAM&P).

▲ **Concatenation** A term that refers to the linking together of multiple frames to form an envelope capable of carrying super-rate payloads, such as when SONET/SDH is used to carry ATM cells. A concatenated SONET link is referred to as an STS-3c or an OC-3c, or whatever the SONET level happens to be. In SONET, the "c" really stands for "unchannelized," as will become obvious later on in this chapter.

Note that not all possible values of N are represented in SONET. For example, there is no OC-5 or OC-7. Although the value of N can technically take on any value from 1 to 255, SONET standards define only a few of the levels. Otherwise, there would potentially be 255 different types of SONET equipment, making a joke of interoperability in spite of the presence of standards.

The table does not even show that some speeds of SONET are more popular than others. It seems strange to think of standards in this light, but the fact is that customers and equipment vendors have favored some speeds of SONET over others. Sometimes there is a technical reason behind this preference. For instance, early equipment vendors found that it was actually more cost effective to build SONET gear operating at the OC-12 level than the OC-9 level (because of laser chip costs and economies of scale). Thus little OC-9 SONET equipment has been built, while the OC-12 equipment market has flourished.

A few final words are in order about the distinction between the STS and OC designations. The remainder of the SONET frame after the overhead bits is the SONET payload itself, more commonly (but less accurately) known as *user data*. This payload might be information from the existing T-carrier hierarchy (e.g., DS-1, and so on), ATM cells (if the SONET link is concatenated, like an OC-3c), or even something entirely different (e.g., MPEG video).

The designation STS, or Synchronous Transport Signal, refers to the *electrical* side of SONET. This seems odd at first, since SONET is supposed to be an optical network. But while modern network equipment excels at the sending and receiving of serial bits over a fiber optic link, the same is not true of any other network task. All SONET frames must be created, multiplexed, managed, and so on. This must be done in the electrical world, rather than the optical world. The outbound user-to-network side of this process is illustrated in Figure 5-1.

The whole point is that SONET (and SDH) devices do a lot in the electrical realm. In the figure, user data is sent to a SONET STS-1 multiplexing device. This device adds Path Overhead (POH) to yield an SPE, and then adds the Transmission Overhead (TOH) to yield the entire SONET STS-1 frame. Multiple user data streams can be combined as well. The SONET multiplexing device next sends the STS-1 frame to a SONET transmitting device, which can be up to 450 feet away (an STSX-1 interface for an STS-1) or 225 feet away (an STSX-3 interface for STS-3). Multiple SONET multiplexers cannot share a transmitter,

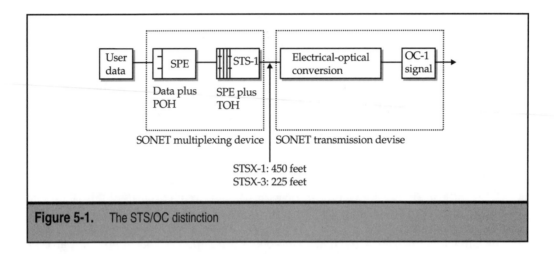

Figure 5-1. The STS/OC distinction

although several user data streams can share a SONET multiplexer, as in the case when three STS-1s are using a single STS-3.

For transmission on fiber optic links, the STS-1 structure is converted and an Optical Carrier Level 1 (OC-1) signal results. The OC signal is sent after scrambling to avoid long strings of zeros and ones, and to enable receiver clock recovery from transitions in the incoming data stream. The OC terminology is typically used when the carrier system itself is discussed, but even customers seldom use the STS terminology. Most users and service providers alike typically refer to SONET speeds by OC references.

SDH uses the same concepts, but with slightly different terminology. The details of these differences are not important here. What remains important are both the single wavelength nature of SONET/SDH and the degree of electrical operation. Both SONET and SDH are essentially *electrical networks with optical links.*

STS-1 Frame Structure

The basic building block of the SONET digital transmission hierarchy is called the Synchronous Transport Signal, Level One, or STS-1, frame. The SONET frame is a transmission frame used to transport a package of bits over a physical link. So a SONET frame exists at a lower level on a network than the more well-known Ethernet LAN frames. The SONET frame is the exact analogy of the T-carrier level transmission frames in function, such as the T-1 frame structure.

The basic STS-1 SONET frame consists of 810 bytes, transmitted 8,000 times per second (or once every 125 microseconds) to form a 51.840 Mbps signal rate. This basic rate, the fundamental building block of SONET, is derived as follows:

810 bytes/frame × 8,000 frames/sec × 8 bits/byte = 51.840 Mbps line rate

In other words, the 810 bytes of the basic SONET frame structure are sent 8,000 times per second, and since each byte consists of 8 bits, the signaling rate on the link is 51.84 Mbps.

Figure 5-2 shows the basic structure of the SONET frame in visual format. The STS-1 frame is 9 rows of 90 columns. It is always shown in this format, 9 rows of 90 columns, so that the overhead bytes will line up properly at the beginning of the frame. The STS-1 frame is transmitted one row at a time, from top to bottom, and from left to right within each row. So the byte in row 1, column 1 is sent first, and the byte in row 9, column 90 is sent last. After the ninetieth byte is sent at the end of row 1, the next byte sent is the first byte in row 2, the byte in column one.

One frame is sent every 125 microseconds, so SONET can maintain time slot synchronization that is required for delivery of normal, uncompressed PCM voice (8 bits 8,000 times per second, or 64 Kbps). SONET also adheres to frame synchronization timing with existing asynchronous network standards (for example, DS-1, E-1, and DS-3). This adherence can be an annoyance when it comes to transmitting data such as IP packets, but the roots of SONET/SDH are firmly in the voice world.

An STS frame is composed of two main sections, each with its own structure. The first three columns of the STS-1 frame form the Transport Overhead (TOH) for the entire frame. All of the SONET overhead information (divided into Section, Line, and Path) that is used to manage defined parts of the SONET network and transported data (called a Payload) is in the first three columns of the frame. This overhead section therefore

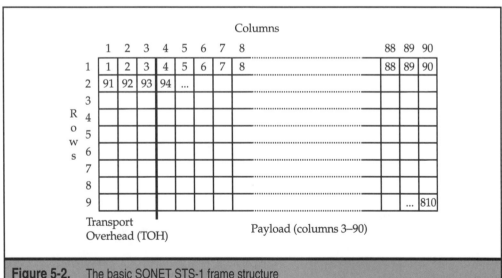

Figure 5-2. The basic SONET STS-1 frame structure

consists of 27 bytes (9 rows × 3 bytes/row) sent as part of each and every SONET frame. This overhead cannot be eliminated, nor used for user data at all.

The SONET payload is carried in the Synchronous Payload Envelope (SPE). The capacity of the SPE is 9 rows of 87 columns (since the first 3 columns in each row are for overhead). This comes to 783 bytes (9 rows × 87 columns) of payload in each frame, giving a total user data rate of:

783 bytes/frame × 8,000 frames/sec × 8 bits/byte = 50.112 Mbps payload rate

Actually, there is a little more to the SPE structure than just this. For example, more overhead, called Path Overhead, is contained in the SPE but considered part of the user data nonetheless. However, this is enough for understanding the basic SONET STS-1 frame structure. The SPE, like the entire SONET frame that contains it, is sent 8,000 times per second. The basic structures of the STS-1 frame are as follows.

Section and Line Overhead

The SONET STS-1 frame transports a considerable amount of overhead associated with operational functions within the SONET network. The first three columns of the STS-1 frame are dedicated to Section Overhead (SOH) and Line Overhead (LOH), performing a wide variety of functions. Together they are considered the Transport Overhead (TOH). The first three rows of the TOH make up the Section Overhead (9 bytes). The last six rows of the TOH make up the Line Overhead (18 bytes).

The Synchronous Payload Envelope Capacity (Information Payload)

The remaining 87 columns of the Synchronous Transport Signal constitute the Synchronous Payload Envelope capacity. It is more precise to say that these 87 columns represent the SONET *Information Payload* and that the SPE is the structure of the Information Payload, but usually only the SPE itself is singled out for attention.

The Synchronous Payload Envelope

The SPE contains both overhead information, called Path Overhead, and the actual user data. The Path Overhead, combined with the user data (Payload) that follows, constitutes a Synchronous Payload Envelope, or SPE. The Path Overhead can begin in any byte position within the SPE capacity. This typically results in the SPE overlapping into the next frame.

It should be noted that the extensive amount of overhead in SONET is useful for surveillance and network management, as well as other activities. And while SONET overhead is criticized as excessive, the fact that it does not increase proportionately at higher levels of the hierarchy is a real plus.

It is easy to construct the structure of a frame at any level of the SONET hierarchy once the basic STS-1 format is understood. All SONET frames are sent 8,000 times per second. All SONET frames are exactly 9 rows. The only variable is the number of columns. This may sound simplistic, but it is true and results in all of the different speeds that SONET runs at.

For example, an STS-3 frames consists of 9 rows and is sent 8,000 times per second. But an STS-3 frame is not 90 columns wide. The STS-3 frame is three times wider (N = 3, after all). So the STS-3 frame is 270 columns wide, and that is not all. The STS-3 overhead columns are multiplied by 3 as well, as are the SPE capacity columns. So an STS-3 frame is 270 columns wide (3 × 90), of which the first 9 columns are Transport Overhead (3 × 3) and the remaining 261 (3 × 87) are payload capacity. The whole STS-3 frame is 2,430 bytes (9 rows × 270 columns). The line rate for an OC-3 therefore must be:

2,430 bytes/frame × 8,000 frames/sec × 8 bits/byte = 155.52 Mbps line rate

The same result can be derived simply by multiplying the OC-1 line rate of 51.84 Mbps by 3. So the "trick" of SONET frame structures and line rates is that each frame is N times larger in terms of columns than an STS-1 frame, so the line must signal N times faster than the OC-1 line rate to get rid of the same 8,000 frames per second. After all, the STS-3 frames may be carrying hundreds of digital voice samples, all of which must arrive every 8,000th of a second.

The SDH Frame Structure

SDH frames differ in structure from their SONET counterparts. This section will detail the SDH frame concept and briefly describe the SDH STM-1 frame.

The same two-dimensional frame format and overhead structure used in SONET is followed in SDH also. Actually, the ITU-T sort of invented this format for E-1 and ISDN's primary rate (23B+D) format. In current networking standards, the two-dimensional frame structure used in SONET and SDH standards is all but universal. So just as in SONET, an SDH frame will have a given number of rows and columns and have a portion of overhead bytes that will occupy the "left-hand" side of an SDH frame. This basic structure of an STM-1 frame is shown in Figure 5-3.

As can be seen from the figure, an SDH STM-1 frame is very much the same as the SONET STS-3c frame. Both have a 9 row by 270 column structure. The octet in the first

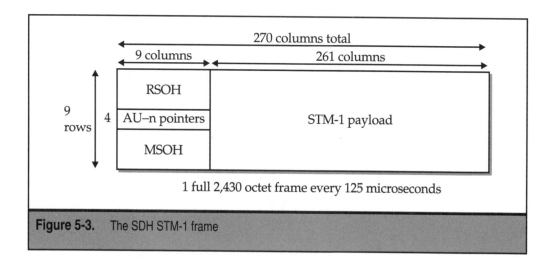

Figure 5-3. The SDH STM-1 frame

column of the first row will be transmitted first. This is followed by the octet to its right, also in the row 1. The transmission sequence is exactly the same as in SONET. That is from left to right, and from top to bottom. A frame is sent 8,000 times a second, or once every 125 microseconds.

All overhead in an SDH frame is just section overhead (SOH) or pointer octets. But the SOH in SDH is divided into two main parts by the pointer octets. In fact, an STM-1 frame consists of three main parts:

▼ **The regenerator section overhead (RSOH) octets**   This occupies an area of 3 octets (rows) by 9 octets (columns), has a capacity of 27 octets, and runs at 1.728 Mbps.

■ **The multiplex section overhead (MSOH) octets**   This can be seen as the multiplex section overhead octets plus the *AU-n pointer* octets. There are good reasons for including the AU-n pointers in the MSOH, and these will be discussed shortly. The AU-n pointers (where n = 3 or 4) serve the same purpose as in SONET: to locate the synchronous payload in within the frame. These pointers occupy the fourth row of the frame and so consist of 9 octets. The MSOH proper occupies rows 5 through 9 and thus has a capacity of 45 octets.

▲ **The STM-1 payload (envelope) itself**   This frame structure is 9 octets (rows) by 261 octets (columns) and so has a capacity of 2,349 octets and runs at 150.336 Mbps. The payload is designed to carry an ITU-T E-4 signal running at 139.264 Mbps or its equivalent.

SONET Architecture Layers

SONET, like just about everything else in computers and telecommunications today, has a layered architecture. Layers help to break up the communications tasks into more manageable pieces, and they can furnish a certain amount of flexibility for implementers making the software used as the layers more or less functionally independent of each other. So vendors have a fair amount of latitude when implementing layer functions internally, as long as the interface between layers is maintained in a standard fashion.

Although SONET has a layered architecture, it is important to note that SONET mainly deals with the Physical Layer of the OSI Reference Model (OSI-RM), similar to T-1 or T-3 services. In other words, there is really no relationship at all between the SONET layers and the layers of the OSI Reference Model, except for the existence of SONET at the Physical Layer of the OSI-RM. A "Layer 3" in SONET has no relation at all to Layer 3 (the Network Layer) of the OSI-RM.

There are four layers in the SONET architecture. These are shown in Figure 5-4. The Photonic Layer refers to the optical properties of the transmission path, which involves the sending of the serial 0's and 1's from a sender to a receiver. SONET lasers and LEDs and their paired receiving devices operate at this layer. The Photonic Layer deals with the transport of bits across this physical medium. No overhead is associated with this layer—it is only a stream of 0's and 1's. The main function of this layer is the conversion of STS electrical frames into optical OC bit signals.

Associated with each of the remaining levels is a set of overhead bytes that govern the function of each layer. These overhead bytes are logically known as Section Overhead

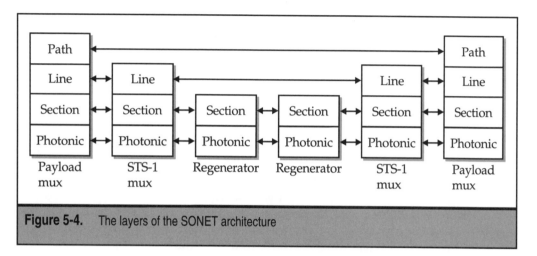

Figure 5-4. The layers of the SONET architecture

(SOH), Line Overhead (LOH), and Path Overhead (POH). The SOH and LOH make up the Transport Overhead (TOH) located in the first columns of the STS frame. The POH bits are included as the first column of the SPE.

The Section Layer refers to the *regenerator section* of the transmission link. In fiber optic transmission, as in most digital schemes, 0's and 1's must be *regenerated* at fairly regular intervals. Of course, the signal does not necessarily terminate at this point, and so a single SONET link may have regenerators spaced every 10 kilometers or so on a 100- kilometer link. In this case, every segment between regenerators forms a Section in SONET.

The Section Layer manages the transport of STS frames across the physical path, using the Photonic Layer. The functions of this layer include section error monitoring, framing, signal scrambling, and transport of Section Layer overhead. Section Layer overhead consists of nine bytes of information that contain the information required for the Section Layer to perform its functions. This overhead is created or used by what is known in SONET as *section terminating equipment (STE)*.

The Line Layer refers to the *maintenance span*. A maintenance span forms a segment between two SONET devices, excluding the lower-layer regenerators. A single SONET link from one user site to another may consist of many such spans. In fact, much of the terminology in this chapter that refers to SONET "links" would be more properly expressed in terms of SONET "spans." However, outside of SONET network maintenance and engineering groups, the term "span" is rarely heard. To most, a SONET network consists of many SONET "links" between sites. However, the term *mid-span meet* is a constant reminder of the importance of the span concept in SONET networks, especially when maintenance and protection are considered.

The Line Layer manages the transport of entire SONET payloads, which are embedded in a sequence of STS frames, across the physical medium. The Line Layer functions include multiplexing and synchronization, both required for creating and maintaining SONET payloads. Line Layer overhead consists of 18 bytes of information (twice as much as the SOH) that provide the Line Layer with its capability to perform its functions, its capability to communicate with the layers that surround it, and certain protection and maintenance features. This overhead is used and created by SONET *line terminating equipment (LTE)*.

Finally, the Path Layer covers end-to-end transmission. In this case, "end-to-end" refers to customer-to-customer transmission. One end of the Path is always where the bits in the SONET payload (the SPE) originate, and the other end of the Path is always where the bits in the SPE are terminated. This may not be the actual end-user device (such as a desktop workstation or server) but usually refers to some kind of premises SONET multiplexing device. Such devices can combine the bit streams from many end-user devices and send them on to another site, where the data streams are broken out in the companion SONET CPE. The Path Overhead (POH) associated with the SONET Path is considered to be part of the SPE for this very reason. The POH is passed unchanged through the SONET Line, Section, and Photonic Layers and is thus indistinguishable from user data to the SONET equipment (but the POH still must be there).

The Path Layer transports actual network services between SONET multiplexing equipment. These services would include the transport of customer DS-1s, DS-3s, ATM cells, and so on. The Path Layer maps these service components into a format that the Line Layer requires and then communicates end-to-end, using functions made possible by the Path Overhead bytes (nine in all) to ensure overall transmission integrity. The Path Overhead remains with the data (payload) until it reaches the other end of the SONET link.

SDH Layers

Just as in SONET, SDH has framing octets, parity-check octets, and so on. These network OAM&P (which is just OAM in SDH) functions are performed by three sets of SDH overhead octets. These are the *path, multiplex section, and regenerator section* overhead octets. Figure 5-5 shows a typical SDH end-to-end connection. This link connects two plesiochronous digital hierarchy (PDH) network links such as E-3s, but of course other configurations are possible. Some of the terms defined in this SDH figure have the same meaning as in SONET. When the definitions used in SDH are identical to those used in SONET, it will be so stated and left at that. Detailed information is beyond the scope of this book.

In SDH, a *path* is a logical connection between the point at which the standard SDH frame format for the signal at a given rate is assembled and the point at which the standard SDH frame format for the signal is disassembled. The Path Terminating Equipment (PTE) is the SDH network element (NE) that multiplexes and demultiplexes the payloads. The PTE can originate, access, modify, or terminate the VC-n path overhead, in any combination of these actions. The definitions of path and PTE in SDH are exactly the same as for SONET.

In SDH, the following NEs are typical PTE devices, using their proper SDH names:

▼ Low-order (or low-speed) multiplexer

■ Wideband cross-connect system

▲ Subscriber loop access system

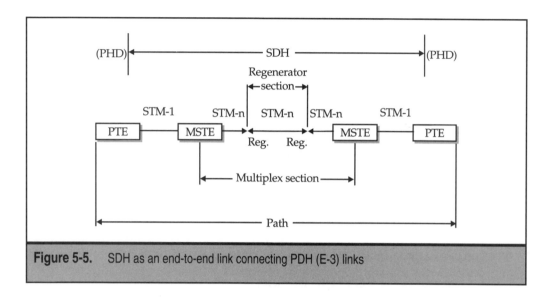

Figure 5-5.   SDH as an end-to-end link connecting PDH (E-3) links

As is also shown in Figure 5-5, an SDH *multiplex section* is defined as the transmission medium, together with the associated equipment, required to provide the means of transporting information between two consecutive network elements (NEs). One of the NEs originates the line signal and the other NE terminates the line signal in each direction. An SDH Multiplex Section Terminating Equipment (MSTE) is a network element that originates and/or terminates STM-n signals. The MSTE can originate, access, modify, or terminate the multiplex section overhead, or it can perform any combination of these actions on the STM-n signal. The definition of an SDH multiplex section is the same as the "line" definition used in SONET. Examples of SDH MSTEs are called:

▼   Optical line terminal

■   Radio terminal

■   High-order multiplexer

▲   Broadband cross-connect system

Finally, a regenerator section in SDH ("Reg" in the figure) is the portion of a transmission facility, including terminating points, between either a terminating network element and a regenerator, or two regenerators. The Regenerator Section Terminating Equipment (RSTE) is the network element that regenerates an STM-n signal for long-haul transport. The RSTE can originate, access, modify, or terminate the regenerator section overhead, or it can perform any combination of these actions. The SDH definition of regenerator section is the same as the "section" definition used in SONET.

SONET calls it...	Section	Line	Path
SDH calls it...	Regenerator Section	Multiplex Section	Path

Table 5-2.   SONET and SDH Architecture Terminology

Since these terms are so fundamental to SDH, and yet differ slightly from SONET, these differences are shown in Table 5-2.

Ironically, the three terms used in SDH actually have better names to represent the meanings of the terms than those used in SONET. For instance, a regenerator section in SDH simply and clearly means an SDH link connecting two adjacent network elements, at least one of which must be a regenerator. The SONET "section" term is less distinctive. Likewise, a multiplex section obviously means an SDH link connecting two adjacent SDH multiplexers (or two adjacent terminating equipment devices performing some sort of multiplexing functions). So a digital cross-connect system or a radio terminal device that acts as a multiplexer may serve as an MSTE, too.

"Super-Rate Payloads" in SONET

This chapter has pointed out that a SONET STS-N frame contains N STS-1s. This is just another way of saying that SONET is as channelized as T-carrier. So a DS-1 contains 24 DS-0s running at 64 Kbps each. But since a single DS-1 runs at 1.5 Mbps, perhaps the user's needs would be better addressed with a single channel running at the full 1.5 Mbps instead of 24 channels running at only 64 Kbps each. This is, of course, what an unchannelized T-1 does for a user. Technically, the term should be *nonchannelized*, but everyone just says unchannelized.

This need may be even more critical in SONET. The unprecedented speeds available in SONET are capable of solving a number of user problems. But not if an OC-3 running at 155.52 Mbps is only useful as three STS-1s running at 51.84 Mbps. What is needed is an unchannelized SONET link. This would give a user just a raw bit rate at 155.52 Mbps (minus the STS-3 overhead). Of course, such an unchannelized frame structure exists in SONET. But in SONET, these are known as *concatenated* SONET frames that give a user a *super-rate payload* (since the rate is now higher than the usual 51.84 Mbps).

Since channelized STS-3s are the rule rather than the exception, the STS-3 frame structure contains a number of features designed expressly for combining three STS-1s into a single payload. First of all, the overall frame structure is derived by simply multiplexing the three input STS-1s a single byte at a time into the outgoing STS-3 frame structure. This overall STS-3 frame is still 9 rows, but it has 270 columns (90 × 3), of course. The effect of this byte-wise multiplexing is to effectively interleave the *columns* of the input STS-1s.

So the 1st, 4th, 7th columns, and so on up to the 268th column of the STS-3 frame are totally derived from the bytes of the first STS-1. Likewise, the 2nd, 5th, 8th columns, and so on up to the 269th column of the STS-3 frame are totally derived from the bytes of the second STS-1. And naturally, the 3rd, 6th, 9th columns, and so on up to the 270th column of the STS-3 frame are totally derived from the bytes of the third STS-1. Since the first three columns of the STS-1s are the SONET overhead bytes, the net result is that the first 9 columns of the STS-3 frame are SONET overhead, while the remaining 261 columns are Information Payloads derived from the three input STS-1s.

But there are three sets of overhead bytes, and three column-interleaved payloads in each STS-3 frame. Since the error checking and many other SONET overhead functions only have to be done once for the whole STS-3 frame, overhead bytes as the B1 Section BIP-8 K1/K2 APS bytes are undefined and ignored except in the "first" STS-1 of the STS-3 (columns 1, 4, and 7). Also, what should be done if two error fields did not agree, such as the B1 Section BIP-8s? Their undefined status neatly solves the problem. The general structure of the STS-3 frame is shown in Figure 5-6.

The figure shows a normal, completely channelized, STS-3 with three STS-1s inside. Note that there is one frame (the STS-3 frame), but three sets of overhead and payloads

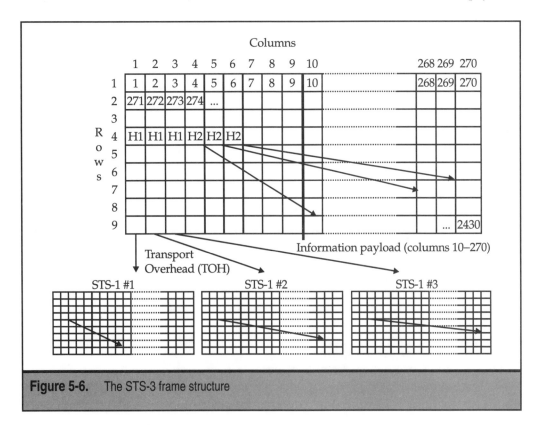

Figure 5-6.   The STS-3 frame structure

(the three STS-1 payloads). Therefore, there are three sets of live payload pointers and three completely separate Path Overheads. Of course, it is still possible to lock the three STS-1 payloads to the beginning of the STS-3 frame, as with the individual STS-1s. But this introduces a delay and need for buffering.

Multiplexing is by columns, as usual, and the three sets of columns are "exploded out" of the STS-3 frame in the figure.

STS-3c (OC-3c) SPE

As nice as it is to combine three STS-1s into a single STS-3 for transmission over a single OC-3 fiber link (which is the whole point), this does little to directly benefit the customer. Many customers need not three STS-1s, but a single STS-3 running at 155.52 Mbps (including overhead). Fortunately, SONET allows for the creation of *unchannelized* STS-3s, as well as other speeds. In SONET, this process is known as *concatenation*. So a concatenated STS-3 is an unchannelized STS-3 and contains not three, but one, SPE. The designation for a concatenated STS-N is to add a lowercase "c" to it. So a concatenated STS-3 becomes and STS-3c.

It is important to distinguish the lowercase SONET "c" in terms of function from the uppercase "C" used in T-carrier. Both mean "concatenation," but each uses the term in a different sense (surprise!). An FT-3C, for instance, is two T3s running on fiber "pasted" together. But there are still two distinct T3s, each running at 45 Mbps. And it makes no difference whether either is channelized into 28 T1s or not. There are still two bit streams running at 45 Mbps. But in SONET, an STS-3c is strictly unchannelized into a single bit stream running at 155.52 Mbps. There are 9 columns of overhead (the overhead stays in SONET, no matter what), but only one SPE that spans the entire 261 columns of the Information Payload.

It is technically proper only to speak of an OC-3 as the physical transport for either an STS-3 or an STS-3c. After all, senders and receivers care nothing about presence or lack of structure in the frames they transfer between them. However, just as T-carrier people came to speak of a "T1" as an overall term for "DS-1" and much else, SONET people have come to speak of "OC-3c" to indicate that the STS-3 frames sent on the link are STS-3c frames. But whether spoken of as STS-3c or OC-3c, the structure is the same, as Figure 5-7 shows.

Figure 5-7 shows the structure of an STS-3c or OC-3c frame. The concatenation is indicated by putting special values in the H1/H2 pointers (called the *concatenation indicator*) where the pointers for the other STS-1s would ordinarily be. The concatenation value is not a valid pointer value (that is, not between 0 and 782), so there is no confusion for receivers. Obviously, there is only one SPE, so one set of H1/H2 pointers will do. Since there is only one SPE, there is only one set of Path Overhead bytes. But they still form the first column of the SPE. As with STS-1 SPEs, the H1/H2 pointers must point to the first byte of the Path Overhead.

As has been pointed out, an STS-3c has only a single SPE, which cannot be used for transporting existing DS-1s or DS-3s (that is what STS-1s are for). So what good is an STS-3c? The fact is that the STS-3c was created expressly for the transport of ATM cells as part of a

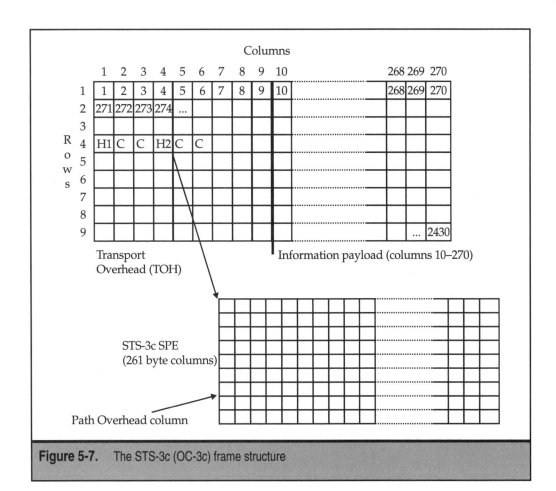

Figure 5-7.   The STS-3c (OC-3c) frame structure

B-ISDN network. This corresponds to the STM-1 in the SDH hierarchy. In fact, the initial ITU B-ISDN specification for ATM cell transport had no provisions for the transport of ATM cells at speeds lower than 155.52 Mbps. Broadband needs speed above all.

The SDH hierarchy basically starts with the STS-3 or STS-3c structure. But in the case of SDH, this is an STM-1 in both cases. More information is needed in SDH to determine if the payload is concatenated or not.

Lower-Rate SONET Payloads (VTs)

One of the advantages of the SONET standard is that it will transport all of the existing higher-rate signals such as DS-3. An STS-3c can carry a stream of ATM cells or IP packets. But what about the lower-rate signals of the T-carrier hierarchy? It seems wasteful to use a 51.84 Mbps pipe just to transport a 1.5 Mbps payload.

The SONET standard handles this situation by mapping the "lower" rate (which means rates lower than DS-3 in SONET talk) signals into "sections" of an STS-1 frame. These sections are each called a *virtual tributary*. Apparently the thought was that each lower-rate payload forms a "tributary" flowing into the main data "stream." Whatever river-based image was intended, the terminology has stuck. Each of the virtual tributary sections is independent of the others and of the virtual tributaries can carry a different type of *subrate payload*. A subrate payload is just a payload that does not map directly to an STS-1, but requires virtual tributaries.

For subrate payloads, each SONET frame can be divided into sections, or virtual tributaries. Each virtual tributary can contain any type of lower-rate data. So they need not all contain the same type of information. DS-1s, CEPT-1s, and DS-2s could all be in one basic STS-1 frame.

In SONET, each STS-1 frame is divided into exactly seven VT groups (VTGs). And it is all or nothing. A single STS-1 frame cannot have, for example, only four VT groups and use the rest of the payload for something else (like ATM cell transport). Either the STS-1 frame is chopped up and sectioned off into exactly seven VT groups, or it is not.

Four virtual tributary sizes are currently defined in SONET, listed as follows:

▼ **VT1.5**   Carries enough bandwidth to transport a DS-1 signal of 24 DS-0s at 64 Kbps. Each is contained in three nine-byte columns (27 bytes). A single VTG can carry 4 VT1.5s.

■ **VT2**   Carries enough bandwidth to transport an E-1, 2.048 Mbps signal. It is contained in four nine-byte columns (36 bytes).

■ **VT3**   Carries enough bandwidth to transport a DS-1C signal. It is contained in six nine-byte columns (54 bytes).

▲ **VT6**   Carries enough bandwidth to transport a DS-2 signal. It is contained in twelve nine-byte columns (108 bytes).

The seven Virtual Tributary Groups (VTGs) in an STS-1 frame consist of 108 bytes each. Since each column in a SONET frame has 9 rows, each VTG occupies 12 columns. Now, 7 VTGs of 12 columns each take up 84 of the 87 columns of the SPE. One of these "leftover" 3 columns is the Path Overhead, which must be present. The other two columns are reserved and have no currently defined function in a VTG-divided SPE.

The mapping of various lower-rate payloads into a Virtual Tributary is shown in Figure 5-8.

There is a pleasing symmetry in the VT mappings. Since each and every VTG consists of 12 columns of SPE payload, all of the four VT mappings can be carried in a VTG simply by varying the number of columns used. For instance, a DS-1 signal at 1.544 Mbps can be carried by three columns of an SPE. Three columns have 3 × 9 or 27 bytes. Since each DS-1 frame has 24 bytes plus a framing bit, there is plenty of room in three columns for the VT1.5, with a few bytes left over. Not surprisingly, these bytes are used for VT overhead. So four

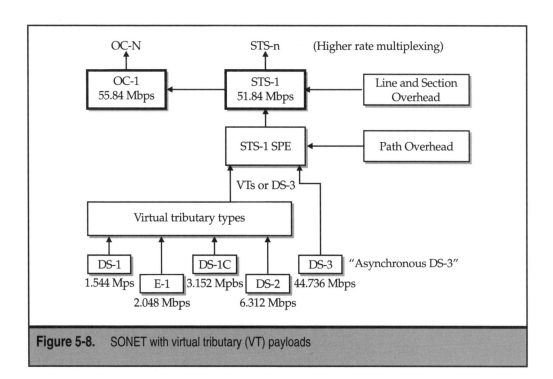

OC-N   STS-n   (Higher rate multiplexing)

| OC-1 | | STS-1 | | Line and Section |
| 55.84 Mbps | | 51.84 Mbps | | Overhead |

STS-1 SPE ← Path Overhead

VTs or DS-3

Virtual tributary types

DS-1		DS-1C		DS-3	"Asynchronous DS-3"
1.544 Mps	E-1	3.152 Mpbs	DS-2	44.736 Mbps
	2.048 Mbps		6.312 Mbps	

Figure 5-8. SONET with virtual tributary (VT) payloads

VT1.5s can be carried in a single VTG (12 divided by 3). This is especially good because the 28 DS-1s inside a single DS-3 can be carried by an STS-1 efficiently (four DS-1s per VTG, and 7 VTGs per STS-1). It should be noted that a DS-3 can be mapped directly into an STS-1 without concern for any DS-1s that happen to be inside. This is called an *asynchronous DS-3*, and is not covered by a VT mapping, as shown in the preceding figure.

Similar logic to that employed in the preceding discussion reveals that an E-1 running at 2.048 Mbps can be carried by 4 columns of a VTG, a DS-1C can be carried by 6 columns, and a DS-2 can be carried by all 12 columns. Happily, 3, 4, 6, and 12 are all factors of the column VTG, so things fit quite nicely.

The maximum capacity of an STS-1 used for Virtual Tributaries will depend on the structure of the VTGs. Any VTG can technically carry any of the subrate VT types (VT1.5, VT2, and so on), but it will be much more common to see STS-1s used for one type of Virtual Tributary alone, especially DS-1, of course.

SDH forms lower-rate payload "containers" differently than SONET. The entire process is quite complex in SDH and need not be discussed further. The same major points made about SONET VTs are equally valid for SDH container contents.

Other SONET/SDH Payloads

SONET and SDH can carry payloads other than channelized voice or data inside of basic 64 kbps voice channels. Quite a few payload types are defined for SONET/SDH, but only four of them are actually widely used. These are:

▼ Virtual Tributaries (the various "voice" channel structures previously described)

■ Asynchronous DS-3 (unstructured DS-3 frames in SONET: SDH has other methods)

■ ATM cells (SONET/SDH as a B-ISDN transport)

▲ Packet over SONET (and SDH) (POS: any Layer 3 packet will do, but the packet is almost always IP)

It is safe to say that the vast majority of SONET/SDH links carry one of these four types of traffic in the SPE. There are more than just these four possibilities for SPE content, but most service providers have never seen anything else inside an SPE but one of the four types of traffic listed. And many service providers have seen nothing but 64 Kbps voice inside their SPEs.

All of these payload types apply to the basic levels of the SONET/SDH hierarchy. In addition, the ATM cell and Packets over SONET payloads can be carried inside the larger SPEs such as STS-3c (OC-3c) and STS-12c (OC-12c) frames and links.

SONET/SDH NETWORK ELEMENTS (NEs)

SONET and SDH networks consist of more than just single links connected by fiber optic cable. The devices in a SONET/SDH network must perform different tasks. Yet all of these components must work together to transport bits around the network.

The major components of a SONET or SDH network are shown in Figure 5-9. Note that some of the components form rings in the diagram. Rings have become a distinguishing feature of SONET/SDH, much as touch-tones have become a distinguishing feature of telephones. Both are simply a much better way of doing things. A full discussion of SONET rings is beyond the scope of this book, but this is the place to introduce rings, since they also play a key role in WDM and DWDM optical networking.

As shown in Figure 5-9, there are six main elements to any major SONET network today. These are labeled in the figure as the

1. **Terminal multiplexer (TM)**   An "end point" device on the SONET network. This device will gather bytes to be sent on the SONET network link and deliver bytes on the other end of the network. The TM is intended to be the most common type of SONET CPE, but ironically it is still rare at the customer site. In most current SONET configurations, the TM is still found in the serving office.

2. **Add/drop multiplexer (ADM)**   Really just a "full-featured" TM. However, it is more accurate to refer to the TM as an ADM operating in what is known as

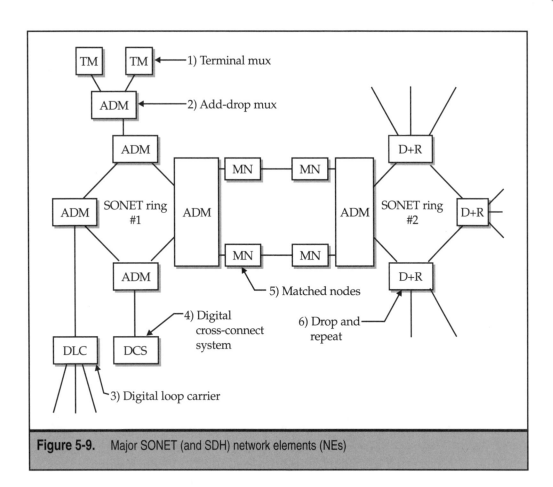

Figure 5-9.   Major SONET (and SDH) network elements (NEs)

"terminal mode." This device usually connects to several TMs and aggregates or splits (grooms) SONET traffic at various speeds.

3. **Digital loop carrier (DLC)**   This SONET device is used to link serving offices with ordinary analog copper twisted pair local loops in order to support large numbers of residential users in what is known as a Carrier Serving Area (CSA). The devices are also commonly referred to as Remote Fiber Terminals (RFTs) by many vendors.

4. **Digital cross-connect system (DCS)**   This SONET device can add or drop individual SONET channels (or their components in a VT environment) at a given location. It is basically an even more sophisticated version of the SONET ADM.

5. **Matched nodes (MN)**   These SONET devices interconnect SONET rings. They provide an alternate path for the SONET signals in case of equipment failure. This feature is commonly known as signal protection, but it has a variety of other names.

6. **Drop and repeat nodes (D+R)**   These devices are capable of "splitting" the SONET signals and sending copies of the bytes onto two or more output links. The devices will be used to connect DLC devices for residential video (or even voice) services.

Figure 5-9 shows a more complete network deployment that exploits the features of SONET. It should be noted that there are no "official" designations for the various pieces of SONET equipment shown in the figure. Although the figure generally follows terminology of such SONET equipment vendors as Nortel, it is not uncommon for vendors to have their own names for the devices shown in the figure. However, the emphasis here is on function, not so much on the names themselves. The equipment found in this network is separated into central office terminal equipment and outside plant equipment.

The central office equipment consists of a SONET-compatible switch (i.e., digital cross-connect switch) and digital loop carrier equipment. This equipment forms the network terminating point for the SONET transport services.

The outside plant equipment consists of a fiber ring topology for survivability, and access multiplexers to this ring. The multiplexers are functionally separated into two categories, terminal multiplexers (TM) and add/drop multiplexers (ADM). The terminal multiplexers are functionally similar to channel banks in the T-1 networks. They provide conversion from the non-SONET transmission media to the SONET format. The add/drop multiplexers are an integral component of the SONET architecture; they allow access to the SONET transmission network without fully demultiplexing the SONET signal. Non-SONET signals may be added to or taken from the SONET transmission signal via this equipment. The difference between SONET TMs and ADMs is often just a difference in network location and in function, rather than a fundamental or intrinsic difference. A SONET or SDH TM is really just an ADM operating in "terminal mode."

The major difference between a SONET ADM and TM is that the SONET ADM has two network connections, which must be of equal speed. Channels or even other OCs (as long as they are less than the network OC, naturally) can be dropped or added as desired. These channels or OCs can lead to SONET TMs, or course. On the other hand, a SONET TM has only one network connection. The "terminal mode" just multiplexes various bit rates onto and off of this single network link. The whole point is to allow "low speed" (relatively) user access to the SONET network. The user access may be at DS-3 or even OC-3 rates, as long as the network link can handle the speed.

Physically, the SONET equipment described is of quite modest size. The solid-state electronics that characterizes SONET results in components that are typically rack-mounted in standard communications cabinets. The power requirements are correspondingly modest as well.

Figure 5-9 can be misleading in one sense. The figure implies that the six major SONET NE functions might be all found in all large SONET networks and with equal likelihood. In fact, the truth is just the opposite. The SONET NE packages offered by most equipment vendors are ADMs, first and foremost. Many of the other NE functions are added (or subtracted) from the basic ADM package to yield the required NE for the application. So an ADM with non-SONET interfaces and a single SONET interface becomes a

SONET TM package. DSC functions can be added to the ADM in the same way, but this is seldom done. Most DCSs remain separate gear attached to the ADM. If the ADMs have multiple links used to connect their rings, this makes that pair of ADMs into Matched Nodes. In other words, the MN function is just another feature of the ADM.

The ADM/TM/MN package is normally what a SONET equipment vendor emphasizes, and that might be it. If a SONET DCS NE is offered, it is usually not an ADM feature, but separate equipment altogether. The D+R and DLC functions can be addressed in the same way. The D+R NE is a special ADM. The DLC is a special TM optimized for residential voice applications.

SDH NEs and configurations are basically the same as those used in SONET.

SONET/SDH RINGS

Perhaps the most distinctive feature of SONET/SDH networks in general and the most obvious way that SONET and SDH differ from earlier transmission systems is the area of rings. The capability of SONET/SDH to be deployed in ring architectures, rather than strictly point-to-point or multipoint architectures, has become the defining feature of SONET/SDH to date. This section explores some of the basic aspects of SONET/SDH rings. These rings are built for their advanced protection features, which allow service to continue even after a fiber cut or ring node failure.

SONET/SDH rings can be a few miles long and span a few city blocks or stretch to literally thousands of miles and span continents. Rings of international proportions are not uncommon, and small rings in downtown metropolitan areas are no longer newsworthy. SONET/SDH rings are built by local exchange carriers (LECs), competitive access providers (CAPs) or competitive local exchange carriers (CLECs), and interexchange carriers (IECs). There are even a few totally private SONET/SDH rings intended for the exclusive use of a single organization. All share the distinctive protection characteristics of all SONET/SDH rings.

But SONET/SDH rings come in many shapes and sizes, and all have differing characteristics that make them suitable for one application or another. This section will deal with some of these characteristics.

SONET/SDH Ring Basics

This section discusses SONET/SDH ring basic configurations and automatic protection switching (APS). APS is the process whereby a SONET/SDH ring recovers from a fiber or node failure. The discussion starts with a definition of a SONET/SDH ring, explores various types of ring architecture, and defines *unidirectional* rings and *bidirectional* SONET/SDH rings. It then examines the differences between a *2-fiber* and a *4-fiber* SONET/SDH ring, *line* (or *ring*) switching versus *span* switching, and the characteristics of these forms of SONET/SDH ring APS. Finally, application examples of Bidirectional Line Switched Ring (BLSR) and Unidirectional Path Switched Ring (UPSR), along with the definitions and functions of the SONET K1 and K2 overhead bytes for ring APS applications, will be discussed in some detail.

The rest of this section uses SONET ring terminology. The SDH ring terminology differences will be discussed at the end of this section. A *SONET ring* is defined as a collection of more than two SONET network elements (or just *nodes*) forming a *closed loop*. That is, the SONET NEs are linked together with the last forming a closed loop by being connected to the first. Each NE is thus connected to two *adjacent* nodes by means of at least one set of fibers operating in a duplex (one transmit, one receive) fashion.

A SONET ring provides many benefits to the service providers who deploy them and the customers that are served by them. They can provide redundant bandwidth for protection, redundant network equipment and electronics, and in most cases, both. A SONET ring is commonly known as a *self-healing* ring, but the term is a little misleading since it implies that the rings somehow repair themselves. However, since the rings' services can be automatically restored following a failure or degradation in the network signals, the term has stuck for better or worse.

SONET rings are intended to bring *loop diversity* to networks. While ring architectures are distinctive to SONET, recall that loop architectures and the protection they provide have been around since T-carrier days. The preferred way of meeting SONET ring loop diversity is by deploying the fiber links according to three major guiding principles. These are by employing a separate fiber sheath, including separate conduits, and taking different physical routes from source to destination. This provides what is known as *route diversity* in SONET rings. Loop diversity can also be achieved by employing a separate fiber sheath, including separate conduits, and taking the *same* physical route from source to destination, but this is not the preferred method. This is known as *structural* diversity, which is not quite as good as true route diversity but may be the only form of SONET ring diversity feasible in many metropolitan areas.

It should be noted that both forms of loop diversity, route or structural, often must still use the same entrance facilities in either a central office (the cable entrance facility, or CEF) or a business location (just known as the entrance facility). Route diversity is defined as separate routes from the first diverse point achievable at the source to the last diverse point achievable at the destination. Usually, this is within a few hundred feet of the endpoints, but not universally. In any case, route diversity is much better than no diversity at all.

It has already been mentioned that there are several SONET ring architectures to choose from. There are in fact three main features that characterize all SONET rings, each with two alternatives. These three basic distinguishing attributes are listed in Table 5-3.

Naturally, when there are three attributes with two choices, the total number of combinations is eight (2 × 2 × 2). So there are eight different SONET ring configurations that differ in at least one of the major attributes. Usually, the ring architecture designation just strings together the three major attributes to name the SONET ring type. So one hears of "2-fiber, unidirectional, line-switched rings" and "4-fiber, bidirectional, path-switched rings" with regularity.

Attribute	Choices
Number of fibers per link	Two Fiber Four Fiber
Direction of the signal	Unidirectional Bidirectional
Level of protection switching	Line Switching Path Switching

Table 5-3.   Types of SONET Rings

All of this is a mouthful, so naturally various abbreviations are used for the ring types. Unfortunately, the abbreviations do not mention the fiber numbers, information about which must be provided separately. But it still helps. The abbreviations are:

▼   Unidirectional Line-Switched Ring (ULSR)

■   Bidirectional Line-Switched Ring (BLSR)

■   Unidirectional Path-Switched Ring (UPSR)

▲   Bidirectional Path-Switched Ring (BPSR)

Any type of these rings can be either a 2-fiber ring or a 4-fiber ring. So there are *2-fiber UPSRs* and *4-fiber BPSRs* and so on. SONET (and SDH) standards provide for all eight types of ring for network element node interconnection.

Now, with eight architectures to choose from, things might seem confusing for service providers ready to build SONET rings and customers ready to buy the services delivered on them. In actual practice, however, with very few exceptions, only three of these eight types of rings have been built on a large scale outside the field trial stage. These are:

▼   Two-fiber, unidirectional, path-switched (2-fiber UPSR, or just *2-fiber path-switched ring*)

■   Two-fiber, unidirectional, line-switched (2-fiber ULSR)

▲   Four-fiber, bidirectional, path-switched (4-fiber BPSR)

Most local exchange carriers have tended to favor 2-fiber rings of the unidirectional sort with either line or path switching. Interexchange carriers and competitive carriers, on the other hand, have favored 4-fiber BPSR for reasons that will become clear later.

Unidirectional versus Bidirectional

Although SONET rings typical mention the number of fibers first, it may be wiser to start with the directionality of the rings while exploring the major ring attributes. Since the terms *unidirectional* and *bidirectional* are exclusively SONET ring terms and not exactly obvious in meaning, a few words of explanation may be helpful. In a unidirectional SONET ring, normal routing of the working traffic is done so that both directions of a connection between two endpoints travel around the ring in the same direction. So each connection uses bandwidth capacity along the entire circumference of the ring. For example, the SONET ring network in Figure 5-10 is a 2-fiber, unidirectional ring.

In the figure, each fiber link or span has been labeled for clarity. If a user attached to the NE1 SONET node is to connect to a user attached to the NE2 node, then Span #1 in the figure is used for that traffic. In order for the user at NE2 to talk back to the user at the NE1 node, Spans #2, #3, and #4 are used, not Span #5, as might seem more logical. Note that both transmission paths (from NE1 to NE2, and from NE2 back to NE1) are clockwise. That is, for two-way communication, only one direction, or a *unidirectional* transmission path, has been established. By the way, all working traffic in SONET unidirectional ring drawings travels clockwise by convention.

Also note that Spans #5, #6, #7, and #8 are not used for working traffic. What then are they there for? These are the protection spans or links. They are used when one of the working (clockwise) spans fails. Unidirectional rings are quite common when the geographical area to be spanned is not too large, for example a metropolitan area.

On the other hand, in a bidirectional ring, normal routing of the working traffic is done so that *both* directions of a connection travel along the ring through the same ring nodes, but in two *opposite* directions. A 2-fiber, bidirectional SONET ring is shown in Figure 5-11. Note that physically, it is indistinguishable from its unidirectional cousin. The difference is the traffic flow.

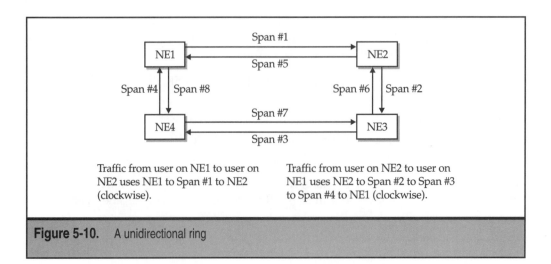

Traffic from user on NE1 to user on NE2 uses NE1 to Span #1 to NE2 (clockwise).

Traffic from user on NE2 to user on NE1 uses NE2 to Span #2 to Span #3 to Span #4 to NE1 (clockwise).

Figure 5-10. A unidirectional ring

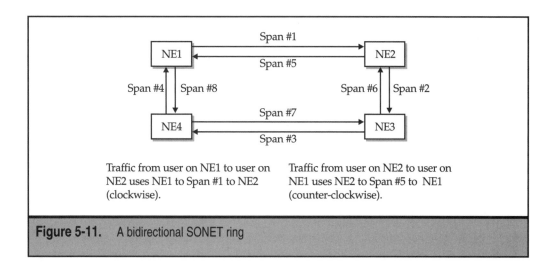

Figure 5-11.   A bidirectional SONET ring

In this case, if a user on the NE1 node is to talk to the NE2 node, Span #1 is used, as before. But when the user on the NE2 node wants to reply to the user on the NE1 node, Span #5 is now used for working traffic, instead of Spans #2, #3, and #4 as before. Note that in a bidirectional ring, one transmission path is clockwise, while the other path is counterclockwise. That is, for a complete two-way connection, both directions, or *bidirectional* paths, have been used between the NEs. Protection switching now uses the spans between NE2 and NE3, NE3 and NE4, and NE4 and NE1, if the links between NE1 and NE2 fail.

Two-Fiber versus Four-Fiber Rings

Both types of ring, either a unidirectional ring or a bidirectional ring, may have two or four fibers run between each SONET NE node. However, it is quite uncommon to have four fibers run for unidirectional rings (two fibers are robust enough for smaller geographical areas), so the discussion here will be limited to a bidirectional ring.

Not surprisingly, a 2-fiber bidirectional ring requires two fibers for each span of the ring. Each fiber span carries both the working traffic channel and the protection channel. This means that on each fiber, only a maximum of *half the channels* are defined as working channels; the other half must be defined as protection channels. This is shown in Figure 5-12.

In the figure, one fiber, which may carry an STS-n signal and has only one set of SONET transport and path overhead bytes, is assigned for the working channel as well as for the protection channel. Each channel takes half of the bandwidth (capacity) of the STS-n signal.

Contrast this operation with a 4-fiber bidirectional ring. This type of ring requires four fibers for each span of the ring. Working and protection pairs are carried over different fibers. That is, two fibers, each with some form of loop diversity and transmitting in opposite directions, carry the working channels; and two other fibers, also diverse and in opposite directions, carry the protection channels. A set of overhead is dedicated either to working or protection channels for the 4-fiber ring. This can be seen in Figure 5-13.

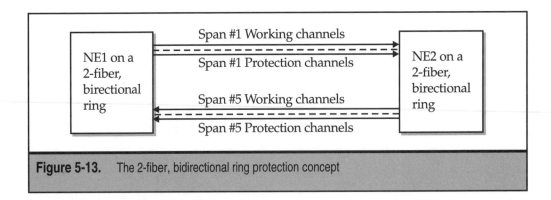

Figure 5-13. The 2-fiber, bidirectional ring protection concept

An advantage of using four fibers over two can be immediately seen. Each fiber carrying an STS-n can be used to its full capacity, instead of half as before. Of course, twice as many fibers are required, and cables must be diversely routed or the whole scheme makes no sense at all. But the real advantage of 4-fiber rings is that they may suffer multiple failures and still function. Usually, two failures will effectively segment a 2-fiber ring of any type. This makes 4-fiber rings very attractive for interexchange carriers, with wide geographical areas to cover, and competitive carriers, who typically have fewer maintenance personnel and staff than do incumbent LECs.

Line Switching versus Path Switching

The last major ring characteristic to be explored is the type of protection switching employed. SONET rings are either line switched or path switched. Much ink has been spilled debating the relative merits of each. SONET documents mention "ring" and "span" switching as well, but only the line switching and path switching are useful concepts in SONET networks. After all is said and done, both the line and path forms will work. And switching type is only one characteristic of SONET rings.

Line switching works by restoring all working channels of the entire OC-n capacity as a single protection operation. Note that the protection capacity is idle, and must be idle,

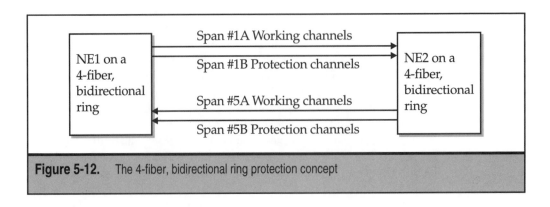

Figure 5-12. The 4-fiber, bidirectional ring protection concept

when the ring is operating normally. In the event of a failure, the live traffic is switched to the protection fiber at both ends of the span. All channels within the "line" are switched this way, which is why it is called line switching in the first place.

On the other hand, path switching can restore all working channels at a level below the entire OC-n capacity. That is, the protection switching can be done on an STS-1 in an OC-3, or even at the VT level if that is what the STS payload is carrying. In order to do this, the signals are sent onto both the working and protection channels across a span. The receiving end constantly monitors both and selects the best signal to be used.

Path switching is almost universally done on 2-fiber unidirectional rings. Line switching is characteristic of 2- or 4-fiber bidirectional rings. Because line switching can operate at the line level of SONET, line switching has a reputation for being more efficient than path switching. Switch the line, and all paths on the line are protection switched as well. Path-switched rings have a reputation for having less fundamental capacity than line-switched rings, but the fact is that the aggregate bandwidth on both types of rings is pretty much the same.

It is nonetheless true that *unprotected* traffic can and often is carried on any type of SONET ring. Such unprotected traffic, a service usually offered at a deep discount, rides the spare capacity reserved for protection purposes on the SONET ring. In case of a fiber or node failure, the protected traffic is shunted onto the protection links, while the unprotected traffic is simply dropped for the duration of the outage. Today such outages rarely exceed four to eight hours, depending on location.

Line switching systems must be able to restore service and reverse direction in 50 milliseconds or less (for comparison, the human eye blinks in 100 milliseconds) for rings with circumferences of 1,200 kilometers (750 miles) or less.

SONET rings are also limited to 16 "nodes" (in reality, SONET ADMs) on a ring. There used to be no limit at all, and huge rings with up to 32 or more nodes were constructed in the early 1990s. Then the structure of the K1/K2 APS bytes was defined, and only four bits were reserved for ring node "address" or position of the node in the ring. The four-bit field limited the number of nodes on the ring to 16 or less. Now rings have grown large enough, and ring nodes are common enough, that there are proposals for working around this apparent ring circumference and node count limit.

A summary of the major differences between the three major SONET ring types is shown in Table 5-4. There are also OC-192 (STM-64) rings, but these are not yet very common and suffer from a lot of issues related to the far higher speeds used.

Although this chapter has dealt with SONET rings almost exclusively, ETSI has recently defined everything needed to construct similar rings for SDH. Due to the terminology used in SDH, BLSRs are usually called *MSP rings* in SDH.

This chapter has presented a brief overview of SONET/SDH. The most important points of the chapter are the fact that SONET and SDH are really electrical networks with optical links, use only a single wavelength with strict serial bit transmission, are optimized for voice channel support and are used with IP packets awkwardly, and are deployed

Characteristic	Unidirectional Path Switched	Bidirectional Line Switched
Number of fibers between NEs	2	2 or 4
Typical speeds	OC-3 (STM-1) or OC-12 (STM-4)	OC-12 (STM-4) or OC-48 (STM-16)
Common application	Metro area, loop plant	Long distance, regional
Specifications of interest	Telcordia TR-496 ANSI T1.105.01-1998	Telcordia GR-1230 ANSI T1.105.01-1998

Table 5-4. Major SONET Ring Types

normally in rings. When WDM and DWDM are considered, only ring deployment still applies to optical networks.

So the time has come to look at WDM and DWDM in more detail, which you can do in Chapter 6.

CHAPTER 6

WDM and DWDM

This excursion into optical networking now leads into the realm of wavelength division multiplexing (WDM) and dense wavelength division multiplexing (DWDM). In a very real sense, there is no difference at all between what WDM does and what DWDM does in optical networking. Both methods involve placing multiple wavelengths over a single strand of fiber optic cable. The only difference, which, however, turns out to be crucial enough to force engineers to consider WDM and DWDM as almost totally different technologies, is the *density* of placement of the separate optical wavelengths. In ordinary WDM, the wavelengths used are generally spaced about 4 to 10 nanometers (nm) apart. In DWDM, the wavelength spacing is usually 1 nm or less, down to about 0.1 of a nanometer in some cases.

So if an optical spectrum of 1,540–1,570 nm is available for use on an optical link, this means that transmitters, receivers, and other components will exhibit acceptable error and speed performance within this range. Even with early WDM systems, at least four channels could be used over this optical link, with wavelengths 1,540, 1,550, 1,560, and 1,570 in use at the same time. Once DWDM became practical, the same fiber could be used to carry at least 32 channels. Since each channel runs serially as fast as the other channels (for the most part), all WDM systems represent a huge increase in bandwidth on the same fiber (in many cases) previously used for a single wavelength.

However, as more and more wavelengths in DWDM are jammed into less and less of a portion of the total optical spectrum available for use, various impairments that are merely annoying in WDM become critical in DWDM. Indeed, some of these impairments were only discovered in prototype WDM and DWDM systems. This chapter investigates all of the implications and issues that arise when multiple wavelengths are used on a single strand of fiber.

SINGLE-WAVELENGTH LIMITATIONS

But if there are tricky issues involved with using multiple wavelengths on one fiber, why bother with multiple wavelengths of light at all? Why not just use a single wavelength and make it go faster and faster? Because there are real limitations that apply to single-wavelength optical networks. These were the incentives behind a lot of research into WDM and DWDM in the first place.

The major limitations of single-wavelength optical network systems are:

▼ Small serial bit intervals

■ Speed-specific repeaters

▲ Expensive new fiber runs

Each of these is important enough to deserve a few words of explanation, although some of these ideas were first mentioned in Chapter 2 and Chapter 3.

Small Serial Bit Intervals

As the speed in terms of bits per second of a serial link increases, the characteristic bit time or bit interval of each individual bit shrinks proportionately. As pointed out in Chapter 3, at 1 Kbps each individual bit is represented by 1 thousandth of a second (1 millisecond, or 1 ms). Now, it is true that on the link itself, bits might be grouped and represented by a single *baud*, or change in line signaling condition. But this makes little difference to the serial devices connected to the link. The serial ports must still deal with one thousand bits per second. And when it comes to optical networking, most modulation is done by simply varying the intensity of the light above or below the threshold of the receiver, as discussed in Chapter 2 and Chapter 3. This is *intensity modulation* and is essentially a one-bit-per-baud method of serial transmission.

At a line speed of 1 Mbps, the characteristic serial bit time is down to 1 millionth of second, or 1 microsecond (1 microsecond), which is of course only one-thousandth as long as a millisecond. Transmitters and receivers must be able to generate and detect bits one after the other within 1 microsecond or the link will drop bits and produce bit errors.

Gigabit Ethernet (GBE) has to produce and react to bits sent in only 1 billionth of second, or 1 nanosecond (1 ns). 10 GBE bits last only 1/10th of a nanosecond (0.1 ns). The top current optical devices operate at 40 Gbps and produce bit streams with characteristic bit intervals of only 0.025 ns. The time slots available for transmitters and receivers to function are too small at present beyond 40 Gbps. But even if this limit were overcome tomorrow, that is not the end of the story.

Why is this bit interval so important? Because of an impairment to all transmission systems known as *jitter*. Jitter is just a term for timing variations on a transmission link. Jitter is a time-domain impairment just as noise is a frequency-domain impairment in transmission systems, and in fact the two are related. Neither can be ignored in communications system design, and both limit the effective speed of any communications link, electrical or optical. And jitter, like noise, cannot be eliminated. At some point, jitter will cause bits to arrive in the wrong time slot at the receiver. As the speed of the link increases, and the bit interval and time slot in which transmitters and receivers must operate shrinks, at some point jitter makes it impossible for the link to function with an acceptable error rate.

The effect of jitter on small bit intervals is shown in Figure 6-1. Both portions of the figure show bits sent by a transmitter and the effect of jitter on the bits received in each time slot. In the upper portion of the figure, a bit is lost because jitter causes the bit interval to wander beyond the "edges" of the bit time at the receiver. In the lower portion of the figure, a bit is duplicated because jitter causes the bit interval to shrink smaller than anticipated by the receiver. Both cases will cause a bit error.

Jitter effects are *not* limited to high-speed links, of course. Enough jitter will cause errors at low speeds. The point is that jitter effects always limit the effective line rate, and time slot available, for serial bit transmission. The only way to increase bit rates on links limited by jitter effects is to add WDM or DWDM.

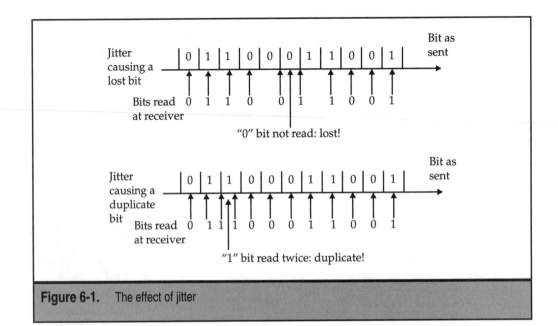

Figure 6-1. The effect of jitter

Speed-Specific Repeaters

Digital repeaters on an optical link restore, reshape, and retime a light pulse that represents a bit by operating in the electrical domain (where all bits reside by definition). As pointed out in Chapter 3, the biggest problem with digital repeaters is that they are *code* and *timing sensitive*. This means that in order to reshape a pulse, a repeater must know what line code is being used to represent the 0's and 1's. To retime the signal, the repeater must know the bit rate of the link to derive the bit interval. A repeater that works with a 10 Mbps link will not work if the transmitter and receiver are changed to operate at 100 Mbps.

These aspects of digital repeater operation pose an interesting problem when WDM and DWDM are in use. A digital repeater will not function on a fiber optic link if there are multiple wavelengths present. The transmitter in a simple repeater, designed to be an inexpensive and yet robust device for field operation, can operate only at one wavelength. Multiwavelength repeaters could be made, but these are experimental devices at present and anything but inexpensive and robust.

When WDM and DWDM are in use, it is therefore necessary to use optical amplifiers instead of digital repeaters. All aspects of optical amplifiers will be discussed later in this chapter.

Expensive New Fiber Runs

Once a single-wavelength fiber optic link has reached its upper limit for serial bit transmission, whether 1 Gbps or 10 Gbps or 40 Gbps, there is no other way to increase the capacity on the link except to add new fiber. In some cases, this is relatively easy. Fiber is often run in conduits with plenty of spare room for new fibers. But all this does is postpone the inevitable problem of what to do when capacity of the conduit is exhausted and no more fibers can be jammed into the conduit.

So just how expensive is it to run new facilities to handle new fiber installs? New facilities such as conduit are normally necessary because fiber optic cables are not usually just buried in the ground and covered up with dirt (although this is not unknown).

A wide range of figures have been published in a wide variety of sources. Figures released a few years ago by a major U.S. interexchange telephony service provider put the cost at about $70,000 per mile. This was a few years ago, so current costs would probably be even higher. This cost covers more than just the cost of the new fiber. Trenches must be dug, conduit installed, the trench closed, fiber placed in the conduit, and new transmitters and receivers purchased and deployed. Undersea fiber is a special case that has its own concerns: these figures are for land-based fiber installations. All of this costs money—a lot of money.

If it is possible to use WDM or DWDM, a lot of the expense of running new fiber goes away. In many cases, it is possible to use the same fibers that are in use with single- wavelength systems, but it is not always possible to use the older fiber. But even if new fiber is required for WDM and DWDM, in many cases the new fiber can replace or add to the older fiber without the need to run new conduit. Even with the cost of new fiber factored in, WDM or DWDM will require the purchase and installation of new multiwavelength transmitters, receivers, and optical amplifiers.

The need to replace digital repeaters with optical amplifiers on a fiber span is a serious cost consideration with WDM and DWDM. This is not only an equipment cost, but a labor-intensive process as well. In many cases, there will be fewer optical amplifiers than digital repeaters on a WDM or DWDM link, but the need to remove the digital repeaters is still there.

But the major cost factor in DWM and DWDM installation is the cost of the transmitters and receivers. Obviously, there is one set of transmitters and receivers for each wavelength channel in use on the link. These are quite costly not only in terms of quantity, but also in terms of intrinsic cost, since the transmitters must be appropriate for the WDM and DWDM window at around 1,550 nm. The current cost of DWDM components has lead some observers to believe that DWDM will probably remain cost effective only for long WAN links for some time to come.

WDM AND DWDM MADE POSSIBLE

In order to overcome the limitations of single-wavelength systems and move forward to WDM and DWDM, four key optical components are needed. These are:

▼ Special fibers with nonzero dispersion and no water peak characteristics

■ The in-fiber amplifier, such as the Erbium-Doped Fiber Amplifier (EDFA)

■ The tunable laser diode operating in the "third window" at around 1,550 nm

▲ The in-fiber Bragg grating to isolate individual wavelengths at the receiver

Some of these have been discussed in general in earlier chapters in this work. Here the emphasis is on using special fibers and special lasers for WDM and DWDM in particular. But the fiber and laser discussions were very general. And two of the key concepts, the EDFA and the in-fiber Bragg grating, have not been discussed in any detail at all. So a little repetition is necessary here just to bring all of these ideas together.

Figure 6-2 shows each one of these concepts in use in a WDM system. The major challenges of optical networking were to make each one of these key components practical and at the same time cost effective.

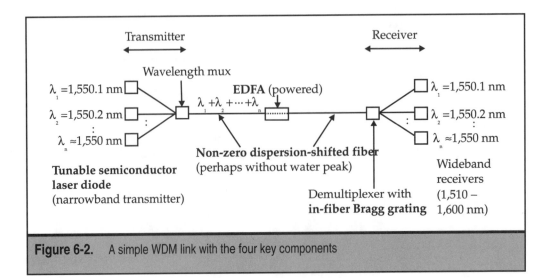

Figure 6-2.   A simple WDM link with the four key components

Each one of these key components is important enough to deserve considerable discussion. In this chapter, there will be no specific mention of the other pieces of optical networking, such as optical cross-connects. However, the optical networking concepts introduced here apply equally to all other networking architectures that include fiber optic links, such as Gigabit Ethernet.

Before jumping right into the details of optical networking, this is a good place to bring together a lot of the ideas first discussed in Chapter 2 and Chapter 3. These ideas are only summarized here to make the full picture of WDM and DWDM more understandable.

Special Fibers for Optical Networking

Chapter 2 outlined the main differences between multimode and single-mode fibers and the major forms of fiber optic cable. All of the advantages of fiber optic cable in general were detailed in that chapter, but mention should also be made of the fact that the connectivity in an optical network is optical, not electrical. This is not a trivial difference. Electrical communications systems always have the possibility of "ground loops." Grounding, an important part of electrical networks, basically prevents static electricity from building up on the cable and preventing signals from being sent. But if there is an appreciable voltage potential difference between two grounding points, the resulting current flowing over the cable can easily swamp the actual signal sent. Furthermore, electrical connections can be hazardous, as high voltages can harm people touching the wire, as anyone touching a "wet" T-1 can assert. Finally, electrical communications cables can be struck by lightning, even when buried underground. Lighting strikes form a hazard in many tropical areas around the world. Optical networks suffer from none of these drawbacks, except when electrical power is distributed to powered fiber components along the link. But this is just an argument for as many passive (nonpowered) optical network elements as possible.

The potential drawbacks of fiber optic networks include the difficulty in joining the fiber cables, the risk of bending or kinking the cable, and effects of gamma radiation and very high voltage electrical fields. The best way to join fiber is by fusion splicing: melting the glass so that the ends fuse to one another. This is a problem outdoors during harsh weather. Connectors can be used but add considerable loss to the link. The attenuation can be about 0.3 dB for multimode fiber, and higher for single mode due to the smaller core diameter.

If fiber is bent too much, the light is no longer reflected into the core properly, and it escapes. The allowable bend radius depends on the actual difference in the refractive index (RI) between the cladding (lower RI) and the core (higher RI). The RI is just the ratio of the speed of light in a vacuum to the speed of light in the material, which is glass in the case of fiber optic cable. Confusingly, in formulas relating the fiber optics, RI is represented by the letter n. If $n = 1.5$, for example, the speed of light in the fiber is about 200,000 km/sec, not 300,000 km/sec as in a vacuum (300,000/200,000 = 1.5).

The bigger the difference in core and cladding RIs, the tighter the allowable bend radius. Unfortunately, the special fibers discussed in this section all tend to minimize the RI difference between core and cladding, leading to a larger allowable bend radius. Kinks in fiber are tight, permanent bends caused by physical damage during installation, such as dropping a heavy weight on a fiber cable. A more correct term for a kink is a micro-bend.

The whole point is that, in many cases, special fibers must be used for WDM and DWDM. However, modern single-mode fibers, especially those installed since around 1996, might still be used on short spans.

Interference (Interaction) of Light

Mention has been made that the traditional picture of light traveling down a fiber optic cable as a little arrow is somewhat misleading. This arrow representation is know as the *ray theory of light* and was common right up until the last part of the nineteenth century. The current theory of light acknowledges that in accordance with quantum physics, light is both a wave and a particle. Light normally travels as an electromagnetic wave and is detected as a particle. Both forms can be called photons, although it is more common to call a particle of light a photon and refer to light in its wave form as just a light wave.

This point is important because light does not really travel down a fiber as a ray in the form of little arrows, bouncing off the walls between core and cladding from transmitter to receiver. The truth is much more complex. Light consists of fluctuating electric and magnetic waves, orthogonal (90 degrees apart) to each other (so they do not interfere). As light propagates, the whole structure normally rotates in a pattern known as *circular polarization*. When the light travels down a fiber, a whole array of effects takes over. Light may be elliptically polarized, as when the period of rotation of the fields is not the same as the wavelength of the light or when the propagation speeds of the electric and magnetic fields are not quite the same, as in a fiber. The main point is that this complex light structure actually snakes its way down a fiber in a spiral known as a helical pattern.

Interference is another important issue in understanding how modern fibers function. Light interferes even with itself in ways that can be described perfectly with mathematics but are hard to explain or understand in an everyday sense. Even a single photon of light can interfere in odd ways with itself when a system is set up just right. It is one thing to think of light as a wave phenomena and understand interference as similar to water waves on a pond interfering with each other. It is quite another to think that a single photon can somehow interfere with itself when obviously there is only one of them to begin with.

Interference is important in all aspects of optical networking, even in the simple act of injecting light from air onto the fiber itself. Typically, about 4 percent of the incident light is reflected, and the higher the angle, the greater the reflection. Differences in RI always cause reflection, 4 percent into the fiber at the sending end and 4 percent into the air at the other end of the fiber link. But when monochromatic light (one wavelength) shines on a thin sheet of glass, the total reflection is not 8 percent, as might be expected. There could be anywhere from 16 percent of the light reflected to no light reflected at all. The result depends on the thickness of the glass in terms of the wavelength of the light. The maximum of 16 percent reflection occurs when the "round trip" thickness of the glass is exactly a multiple of the wavelength of the light. This is constructive interference. The minimum of zero reflection occurs when the round trip thickness of the glass is an odd number of half wavelengths. This is an example of destructive interference in the fiber.

It is important to point out that interference does not destroy the energy present on the fiber. Energy is always conserved, so the lost energy just shows up in some other part of the optical system, probably as heat or noise. With this in mind, modern fibers include anti-reflection coatings of special thickness and RIs to keep reflections to a minimum. This is not to say that interference is always a negative. In optical networks, filters, gratings, and other related components all depend on light interference to function.

Wavelengths are always important in optical networking. Details on the wavelengths used for WDM and DWDM were discussed in Chapter 3 and need not be repeated here. It is enough to point out that the wavelength range used for modern optical networking is from 1,510 nm to 1,600 nm (the 1990s window or third window). This range is used because it is at the lowest point of attenuation on the fiber.

As pointed out in Chapter 3, opening the third window was not easy. Light sources are expensive in this range, and there is more to the fiber story than simple attenuation. Dispersion is always an issue. There are many reasons why a pulse of light sent along a fiber will degrade or be changed. The light will be weakened, a pulse lengthened in time, and so on. The five major reasons a pulse of light may degrade when sent along a fiber optic cable were discussed fully in Chapter 3 but should be repeated briefly here.

▼ **Attenuation**   A light pulse will always weaken because the fiber absorbs light. If the fiber is not manufactured with precise uniformity in diameter and so on, flaws cause scattering of the light. Scattering effects are the biggest cause of light loss in fibers today.

■ **Dispersion**   A pulse of light becomes spread out (disperses) in time as it travels down a fiber and all of the pulses eventually merge together. Three types of dispersion are of primary concern. *Material dispersion* (also called chromatic dispersion) results from the fact that fibers present slightly different RIs at different wavelengths and so cause light to travel at slightly different speeds. *Waveguide dispersion* is caused by the shape of the core and RI profile across the core. Waveguide dispersion causes the light pulse to "bunch up" in time, and so can counteract the effects of material dispersion. (Some references use the term "chromatic dispersion" as the sum of material dispersion and waveguide dispersion). *Modal dispersion* occurs in multimode fibers because the distance traveled, and hence the arrival time, is slightly different for each mode.

■ **Power limits**   The amount of optical power coupled onto a fiber is limited by nonlinear effects. Doubling the input does not double the output, and in fact it might decrease the output to lower levels than before. Optical networks must function in the linear power region to be useful. The nonlinear effects in a fiber are mostly caused by the intense electromagnetic fields in the core when light is present.

- **Polarization effects** The core of fiber optic cable should be perfectly round and symmetrical, but there are always imperfections. So light in the fiber is changed in polarization.

▲ **Noise** It is true that optical networks will not pick up noise from *outside* the network. In optical networks, *modal noise* has a complex effect present on multimode fibers, and *mode partition noise* is a problem with single-mode fibers as power "hops around" (is partitioned) among many close wavelengths.

As pointed out in Chapter 3, none of these optical impairments are fatal and most can be overcome by a variety of methods. It is desirable to make the fiber core as narrow as possible, which is the idea behind single-mode fiber. The wavelength(s) chosen for operation must be selected with consideration to the characteristics of the fiber itself. Dispersion effects that depend on wavelength can be minimized by narrowing the spectral width of the light source. Semiconductor lasers can have spectral widths of around 1 nm to 5 nm, while lasers can have spectral widths as low as 0.01 nm. LEDs will have spectral widths between 30 nm and 150 nm. And material and waveguide dispersion, which act to spread out and narrow waves respectively, can thus be used together to counteract each other to the benefit of the optical system.

When it comes to optical network design, the two most important considerations are signal strength and dispersion effects. Noise is rarely a problem in simpler optical systems.

As pointed out in Chapter 2, optical fibers are not made out of the same kind of glass used in windows. In scientific terms, a "glass" is any substance (mostly containing large portions of silicon) that does not undergo a transition to a solid state when cooled from a liquid. Glass acts as a solid but is still technically a liquid.

Pure fused silica, or silicon dioxide, is the primary ingredient in optical fiber cables. Doping is used to change the RI of the silica core, or cladding, or both. Doping is just the process of intentionally adding impurities to a substance during fabrication. In optical fiber, 4 percent to 10 percent of germanium oxide is added to the silica to increase the RI. Boron trioxide will decrease the RI.

However, dispersion effects still have to be taken into consideration. The fiber zero dispersion point, where material and waveguide dispersion exactly cancel each other out, was originally around 1,310 nm. By changing fiber fabrication methods (doping) and manipulating the RI profiles across the fiber core, manufacturers were able to produce a whole generation of dispersion-shifted fibers that moved the zero dispersion point into the 1990s window, centered around 1,550 nm.

A few other important points about the fibers used with WDM and DWDM systems should be summarized here. Wavelengths in the 1,310 nm range (used for SONET/SDH) can be used with dispersion-shifted fibers (sometimes called DSF). However, the dispersion (and attenuation) will be higher at 1,310. For WDM and DWDM, it would be better to have a range between 1,510 and 1,600 nm, where dispersion is low but still nonzero. Again, clever doping and RI profiles made the manufacture of such non–zero dispersion–shifted fiber (NZ-DSF) possible.

When dispersion is a problem, special efforts to control dispersion effects must be taken at the transmitter and/or receiver. These forms of *dispersion compensation* add expense and complexity to the optical network.

The Water Peak

Fiber attenuation usually varies dramatically with wavelength. There is usually a "water peak" at around 1,400 nm (it is actually closer to 1,385 nm) caused by water ions that are always present in the fiber. In 1998, however, Lucent announced a new fiber they called "AllWave" that had almost no water peak at all. Lucent claimed that the fiber would provide 50 percent more useful wavelengths and 100 nm more bandwidth than conventional single-mode fibers.

Fibers without a pronounced water peak allow operation of transmitters in the 1,400 nm range, where dispersion is lower than it is in the "normal" WDM and DWDM range around 1,550 nm. The 1,400 nm range is a very attractive range for metropolitan fiber networks, since this wavelength range is better for short-haul networks and allows short-haul networks to use fewer amplifiers or repeaters. Applications in metropolitan area DWDM rings, in "coarse" (as opposed to "dense") WDM for Internet access and the like, and in cable TV HFC systems seem assured.

Non–water peak fibers almost completely eliminate the water ions that were an unavoidable by-product of fiber optic cable fabrication. The materials are just very, very pure. These new fibers make it possible to use the entire range from 1,300 to 1,600 nm without skipping over the 1,400 nm water peak with its high attenuation. And the new fiber retains the attenuation and dispersion characteristics of most single-mode fibers at 1,310 and 1,550 nm, making backward compatibility a nonissue.

The attenuation and dispersion profile of the new non–water peak fiber is shown in Figure 6-3.

The "full spectrum" characteristics of non–water peak fiber are very attractive. If services are provisioned on a per-wavelength basis, then having the 1,400 nm range available for use makes a huge difference. No dispersion compensation is needed at 1,400 nm due to the inherently low dispersion in that region to begin with. So non–water peak fiber can function with the most attractive wavelengths, provide a lot more capacity per fiber, support high transmission rates (10 GBE) without the need for dispersion compensation, and of course allow for the provisioning of services based on wavelength.

The Erbium-Doped Fiber Amplifier (EDFA)

The second key piece of the optical networking puzzle, in addition to the creation of fibers with special characteristics in the 1,500 nm to 1,600 nm range suitable for WDM and DWDM, was the invention of the in-fiber optical amplifier. There are also semiconductor optical amplifiers (SOAs), but these are not considered here. The most well known and most widely deployed in-fiber optical amplifier is the optical amplifier based on the rare earth element erbium. When erbium is used, the amplifier is called an EDFA, or erbium-doped fiber

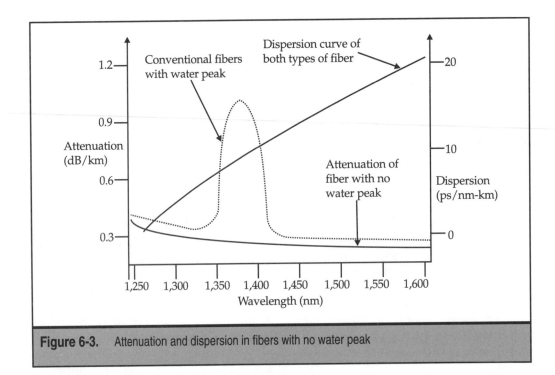

Figure 6-3.   Attenuation and dispersion in fibers with no water peak

amplifier. As might be imagined, an EDFA is created by doping an otherwise ordinary single-mode fiber with erbium during fabrication.

The use of EDFAs has become so common that the acronym EDFA is in danger of becoming a synonym for fiber amplifier in general. This is unfortunate, as other elements can be just as effective as erbium when it comes to amplifying optical signals. For example, praseodymium and ytterbium can be used, which are also rare earth elements. Sometimes, this whole class of optical devices are called rare earth doped optical amplifiers (REDFAs), but this terminology is far from uniform.

It is important to note that SONET/SDH does not, and right now cannot, use EDFAs or other forms of optical amplifiers. This is because SONET/SDH, as a firmly entrenched 1980s technology, uses the 1980s optical communications window, usually at 1,310 nm. EDFAs, on the other hand, were developed in the 1990s and so function in the 1990s optical communications window from 1,500 nm to 1,600 nm. The essential characteristic of any optical amplifier is that these devices operate across a wide range of wavelengths, independently and effectively. The electro-optical repeaters used in SONET/SDH typically function only at the SONET/SDH wavelength of 1,310 nm. If SONET/SDH links are to be mapped onto WDM or DWDM systems, it is necessary to use a transponder to shift the native SONET/SDH wavelength into the 1,500-1,600 nm range.

It is somewhat ironic that, due to the cumulative noise effects of cascading amplifiers, such amplifiers were used in analog transmission systems but abandoned in favor of digital regeneration and repeater techniques culminating in SONET/SDH electro-optical repeaters. In other words, in an analog system, noise is added by each amplifier, since analog noise is indistinguishable from the analog signal. In a digital system, noise can be removed at each regenerator as the signal is strengthened and "cleaned up." Noise also accumulates in optical amplifiers. What saves optical amplifiers and allows them to be used in WDM and DWDM systems is the fact that noise in optical systems is very low to begin with. However, given enough EDFAs, low optical noise or not, eventually a long enough cascade will result in more noise than signal (this is the case with 80 or more EDFAs on a link several thousand kilometers long).

The attractions of EDFAs and optical amplifiers in general are many. Electro-optical repeaters are complex and can fail. EDFAs are built into the fiber itself. Repeaters are tuned to specific wavelength and bit rate. Amplifiers just amplify anything on the link. With WDM, each wavelength would have to be demultiplexed, converted to electricity, regenerated, converted back to optical signals, and be remultiplexed. In-fiber amplifiers make WDM technically feasible. Finally, repeaters cost much more than optical amplifiers. The differences between electro-optical repeaters and in-fiber optical amplifiers is shown in Figure 6-4.

The operation of an EDFA is simple to describe in principle. A relatively high-powered (compared to the weakened input signal) laser source at 980 nm is mixed with the arriving input signal on a special section of erbium-doped fiber, typically about 10 meters of fiber coiled to reduce the form factor of the device. This is called the "pump" signal (there is also a second pump window at 1,490 nm). The erbium ions present in the EDFA fiber section are excited to a higher energy state by the pump laser. When photons from the input signal in the range of about 1,530 to 1,620 nm (the higher wavelengths are not as effectively amplified) strike the excited erbium ions, some of the erbium ions' energy is transferred to the signal and the erbium ions return to a lower energy state (until they are pumped again). The nice thing is that the photons have exactly the same phase and

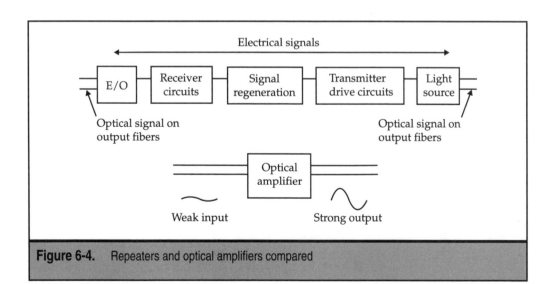

Figure 6-4.   Repeaters and optical amplifiers compared

direction as the incoming photons. There are just more of them, so the input signal is amplified. That is all there is to it.

Actually, EDFAs are a little more complex than that. For example, reflections at the end of the EDFA section can cause the EDFA fiber itself to become a laser, so an optical isolator is usually employed to prevent these reflections. (Whenever light is confined in a small enough space, it will "lase.") Also, noise in an EDFA manifests itself as amplified spontaneous emissions (ASEs), which can emerge from both ends of the fiber link and must be dealt with. ASEs in the same direction as the signal become a major source of cumulative noise, while backward-propagating ASEs can harm source lasers if not filtered out.

Oddly, the pump signal can actually be injected into the EDFA section in the opposite direction from the input signal! In fact, there are firm technical reasons for doing so. The erbium ions do not go anywhere, of course. EFDAs can also be pumped from *both* directions. EDFAs can even be pumped remotely, from the transmitter or receiver side. In some configurations, an EDFA can be pumped *through* another EDFA section, as long as the attenuation at 980 nm is not too bad. Usually the pump laser is right at the EDFA section it pumps, however, although the electrical power needed to operate the pump laser can be delivered locally or remotely (sometimes in the metallic fiber protection sheath). Obviously, EDFA design is a very active area of research. EDFAs have been used, not only as line amplifiers, but as preamplifiers directly in front of a receiver and as power amplifiers placed directly off a transmitter. A more detailed schematic picture of EDFA operation is shown in Figure 6-5.

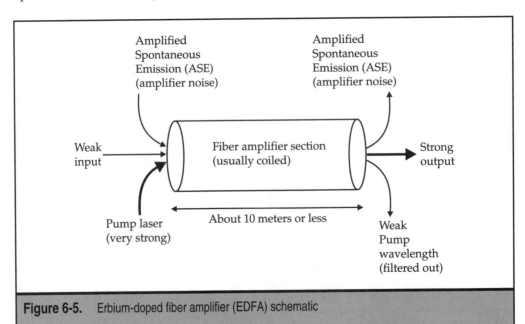

Figure 6-5. Erbium-doped fiber amplifier (EDFA) schematic

So optical amplifiers have numerous advantages over repeaters. Optical amplifiers are simpler, smaller, and much less expensive than repeaters and yet have much longer mean times to failure. They operate over a range of speeds and wavelengths at the same time and produce a gain in the weakened optical signal of anywhere from 25 to 50 dB. They amplify all forms of optical signals, whether digitally encoded or analog, and in fact the coding on the fiber can be changed without changing the in-fiber amplifier. They add no appreciable delay to the fiber run, and if designed properly they can allow the operation of diagnostic tools such as optical time domain reflectometers (OTDRs) right through the amplifier.

Naturally, there are drawbacks to optical amplifiers as well, especially since they are relatively new optical networking devices. The biggest problem is that they add noise to the system that can accumulate as in any cascade of amplifiers until the noise swamps the signal. They do not yet operate over the full WDM and DWDM wavelength window, and the gain produced is not the same at all wavelengths of interest (that is, the gain spectrum is not flat). They do not perform any "cleanup" of the optical signal as regenerators do, such as reshaping pulses. This means that, if an incoming optical signal is dispersed (distorted), the output of the optical amplifier will be stronger, but still distorted. Finally, in many cases the optical amplifiers must be placed more closely than electro-optical repeaters.

But when all is said and done, the in-fiber optical amplifier will be a key piece of any optical network.

The Tunable Laser Diode Operating at 1,550 nm

The third key piece of the optical networking puzzle, in addition to the creation of special fibers and in-fiber amplifiers for WDM and DWDM, was the creation of the tunable laser diode operating at around 1,550 nm, right in the middle of the 1990s optical communications window. There are actually three pieces to this invention. The first part is the *laser diode* itself. The second part is the *tunable* nature of the laser, which enables the same physical device to be used to generate different wavelengths of light. Finally, the third part is operation at around 1,550 nm. All are equally important for optical networking today, and each aspect will be discussed in some detail. Some of the fundamentals of laser diode operation were discussed in Chapter 3. This section summarizes the major points and adds to them.

In the early days of fiber optic systems, lasers were bulky affairs that were very expensive and hard to work with in spite of their numerous advantages over LED-driven fiber systems. Not only was the cost often prohibitive, but lasers often had to be locked in a closet with a warning sign on the door and anyone entering had to wear protective eyewear. Today, both LEDs and lasers are semiconductor devices mounted on a simple computer card assembly. LEDs remain simpler than lasers when implemented in a chipset, but they also share a lot of features in common. So it makes sense to start with a quick look at semiconductor LEDs.

An LED is just a *forward biased p-n junction*, but this is not the place of a full discussion of semiconductor electronics. All that is needed is an appreciation that when the proper electric potential (voltage) is applied to a wafer of p-semiconductor material and n-semiconductor material, light will be emitted at the junction between the materials. That's the easy part. The hard part is getting the LED light at the desired wavelength out of the junction and onto the fiber. In practical LEDs, sometimes the light emerges from the surface of the device (SLEDs) or from the edges of the device (ELEDs). In most SLEDs, the n-semiconductor has a "hole" in the middle to allow the light to emerge from the junction. A very simple diagram of a SLED (sometimes called a "Burrus LED") is shown in Figure 6-6. This figure also appeared in Chapter 3 but is important enough to show again.

The characteristics of LEDs as light sources (transmitters) in optical networks compared to lasers are important. LEDs have been lower in cost than lasers, although this price differential has all but disappeared in recent years. Much of the price differential was based on production volume. Certainly the use of lasers in simple CD-ROM data and audio devices has leveled the cost playing field drastically. LEDs have had lower power outputs than lasers (100 microwatts or so), and they produce light not at a single wavelength, but in a large band or "spectral width" of about 5 percent of the wavelength (50 to 100 nm). The spectral width can be reduced through the use of filters, but this filters out part of the signal power also, of course. LED light is not coherent as is a laser's light, and

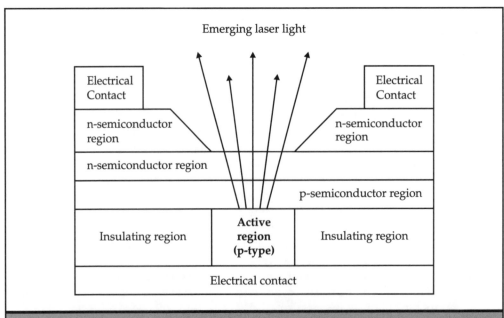

Figure 6-6.   Conceptual structure (components not to scale) of a surface emitting LED (SLED)

so the LED light must be focused with a lens onto the fiber (this is why LEDs are not used with single-mode fibers: it is too hard to get much light into the small core). LEDs cannot be turned on and off fast enough to drive gigabit-per-second systems as lasers can. Most LEDs top out at about 300 Mbps. On the other hand, LEDs can easily be used with analog modulation techniques. This can be done with lasers, but only with much more difficulty.

In contrast to LEDs, lasers are much better to use in optical networks as light sources. Laser light in theory is "coherent" and consists of a single wavelength. This is not strictly true in practice, as all lasers used for communications exhibit a characteristic *linewidth*. But lasers can be turned on and off very quickly, in some cases on the order of femtoseconds (one millionth of a nanosecond, or 10^{-15} seconds), and produce higher powers than LEDs (about 20 milliwatts for communications systems, and up to 250 milliwatts for optical amplifier pump lasers). Laser light, which emerges highly parallel from the device, can inject 50 percent to 80 percent of its power onto a fiber.

But however close lasers and LEDs come in price, the fact remains that temperature control and power control is needed in lasers. These controls require electronic circuits to function. And lasers are generally nonlinear with respect to input power.

In spite of some claims to the contrary, most semiconductor lasers used for communications do *not* produce a single wavelength of light. Just like LEDs, they produce a range of wavelengths, and the spectral width is the measurement of this. This turns out to be important if a *tunable* laser is desired. Usually, the spectral width of the laser is about 8 nm, scattered across eight peaks or "modes" (not the same as fiber modes). Not all of these wavelength peaks are of equal strength. The laser first outputs power at a "dominant mode," then another mode and another, switching back and forth unpredictably. Thus power is distributed in a bell-shaped curve, with ill-defined edges. So it is common to quote spectral width as between the points on the curve where the power drops to one half the maximum "mode" power. This is called the full width half maximum (FWHM) and is a valid measure for LED spectral widths as well.

Spectral width is a crucial laser parameter for a number of reasons. First, the wider the spectrum, the greater the dispersion on the fiber. When using WDM or DWDM, the closer the wavelengths can be packed (i.e., the smaller the spectral width), the more channels can be used on a single fiber. Also, sophisticated modulation techniques such as phase modulation cannot be used unless the spectral width is very narrow. On the other hand, very narrow spectral widths, usually desirable, can actually cause unwanted nonlinear effects in the fiber (such as stimulated Brillouin scattering (SBS), discussed more fully later in this chapter).

In addition to spectral width, semiconductor laser output has a characteristic *linewidth*. So not only are the "peaks" spread across the spectral width, but each individual peak is slightly different from the others in width. The relationship between spectral width and linewidth was fully discussed in Chapter 3.

There are many other characteristics of semiconductor lasers, such as coherence length and time, power, operational range in terms of wavelength, wavelength stability, and so on. However, these are beyond the scope of this discussion.

It is conceptually easy to make a semiconductor laser out of a semiconductor LED. Whenever light is confined in a proper material of the proper shape and size, the material will "lase." Sometimes this can even happen when not desired, as with reflections in a fiber amplifier section. So to make a semiconductor laser out of a semiconductor LED, just add mirrors at the ends! In practice it is a little more complicated, but not by much. The end result is a Fabry-Perot laser, or FP laser. Essentially, the FP laser is an edge-emitting LED (ELED) with mirrors on the ends. The only problem is that simple FP lasers have very wide spectral widths, so a lot of time and effort is spent repackaging FP lasers to get just the right operational characteristics needed.

Usually, some form of diffraction (Bragg) grating is placed within the lasing cavity between the mirrors to narrow the spectral width, typically to the 0.2 to 0.3 nm range. A diffraction grating is just a series of etched lines, discussed more fully in the next section, which refracts (bends) the light in the laser cavity in a certain way. There are two major classes of these types of lasers: the distributed feedback laser (DFB) and the distributed Bragg reflector laser (DBR). Of the two, the DFB laser is of the most interest in optical networking, because it leads directly to the tunable laser. A simple diagram of the active region of a DFB laser is shown in Figure 6-7.

The quarter-wavelength phase shift in the middle of the grating is optional, but it narrows the linewidth of the laser output significantly. The main purpose of the grating is to reduce the spectral width of the basic FP laser. The grating produces just a small variation in the RI of the material with a characteristic period (wavelength). In fact, since much of the action of the bending of the light below the lasing cavity (where the grating is usually located today) is reflection, the DFB laser does not even need end mirrors, but they are usually present. The form and position of the grating can be manipulated to yield a tunable laser as well.

The major drawbacks of DFB lasers, besides expense, are that they are sensitive to reflections (light entering the device from the attached fiber will eventually destroy the DFB laser) and that they are sensitive to temperature variations (not only environmentally, but also due to the *internal* heat generated by sending a long string of 1 bits). So DFB laser bit streams are typically "balanced" to guarantee an equal number of 0 and 1 bits over time.

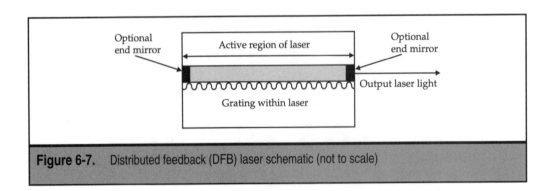

Figure 6-7. Distributed feedback (DFB) laser schematic (not to scale)

The inclusion of a Bragg grating within the structure of the DFB laser leads directly to consideration of tunable lasers. In WDM and DWDM systems, having physically identical light sources that can be tuned to slightly different wavelengths is highly desirable. In one sense, tunable lasers make a virtue of necessity. When attempting to construct simple WDM systems, it proved difficult to fix the wavelengths of the transmitters exactly. This was because slight variations in the semiconductor laser fabrication process meant that the lasers produced varied slightly in wavelength anyway. Whenever a precise wavelength was needed, many lasers had to be made, tested, and selected, and the rest discarded. This was wasteful and expensive. Also, the materials that make up the laser diode deteriorate over time and change the wavelength, leading to a risk of the wavelengths overlapping. If the laser can be tuned, this will correct both situations.

But the most important reason for using tunable lasers in optical networks is the interest in *optical wavelength routing* (more precisely, direct optical cross-connecting). Tuning lasers hooked into single strands of fiber allows optical network devices to quickly accept a signal arriving on one wavelength from an input fiber and "route" the signal onto an output fiber at another wavelength. The arriving and departing wavelengths can be changed quickly to establish new routes or paths through the optical network. It is even possible to set up wavelength paths just like telephone calls across such an optical network.

Typically, tunable lasers can be tuned directly over a range of about 10 nm, mostly centered around 1,550 nm, which is right in the middle of the third optical communications window. This remains a very active area of research, naturally. To extend the tunable range even further, a *sampled grating* can be added to the basic DFB laser design. A sampled grating is just a short, but otherwise normal, refraction grating with periodic *blanked sections* between the grating sections. In the tunable laser, two sampled gratings are made to interact with each other so that a small, induced change in the RI of the device causes a larger change in the wavelength output.

The gratings produce reflection peaks at slightly different wavelengths. Most of the peaks from the two gratings will be at slightly different wavelengths, but one set of peaks can be made to coincide, reinforcing that wavelength and effectively tuning the laser to that wavelength in particular. When a tuning current is applied to one grating, the "comb" of reflection peaks shifts slightly, reinforcing another set of peaks at another wavelength. The tuning range can be extended in this fashion to 100 nm. The basic principle of operation of the tunable laser diode is shown in Figure 6-8.

The two gratings essentially produce two sets of output wavelengths with slightly different spectral widths. By "tuning" these sets of wavelengths, one pair of peaks can be made to coincide and reinforce each other, while the other peaks are made to interfere. This is somewhat of an oversimplification of what occurs, but close enough for this level of detail.

There are other forms of tunable laser currently under research, such as the external cavity DFB laser that employs a tunable grating on a crystal structure. There is more than enough interest in optical "routing" to make tunable lasers a highly active field for some time to come.

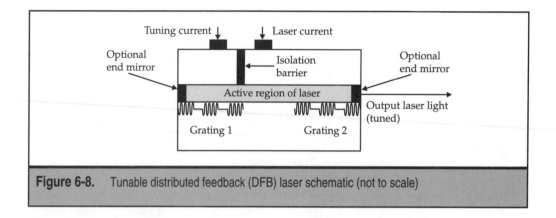

Figure 6-8. Tunable distributed feedback (DFB) laser schematic (not to scale)

In-Fiber Bragg Grating

The fourth and final key piece of the optical networking puzzle, in addition to the creation of special fibers, in-fiber amplifiers, and tunable lasers for WDM and DWDM, was the invention of the in-fiber Bragg grating. The in-fiber Bragg grating is used to isolate individual wavelengths at the receiver of the WDM or DWDM system. Some form of isolation is needed because the general-purpose receivers used in most optical systems today are wideband receivers, meaning they will register photons not only from one specific wavelength, but across a wide range of wavelengths. This problem is shown in Figure 6-9.

Of course, narrowband receivers tuned to a specific wavelength can be developed and used, but this approach is an expensive proposition, as well as an increasingly difficult task for DWDM systems where wavelengths are jammed closer and closer together all the time. There are other ways to filter out the unwanted wavelengths at the receiver besides the in-fiber grating, naturally. The attraction of the in-fiber Bragg grating, like the in-fiber optical amplifier, is that these optical devices are part of the actual fiber itself and allow the continued use of a standard, wideband optical receiver.

Some observers of the optical networking scene rank the invention of the in-fiber Bragg grating (sometimes called an FBG) as second only to the invention of the laser itself as a milestone in optical networking. The FBG is simple in the extreme to make, very inexpensive, and highly reliable. Not much more could be asked for. The FBG is just a short (only a few centimeters) of ordinary single-mode fiber into which variations in the core RI have been etched using beams of ultraviolet radiation. When etched with just the right spacing, the in-fiber Bragg grating will reflect (select) one precise wavelength back in the direction the light entered the fiber and pass the other (nonselected) wavelengths through

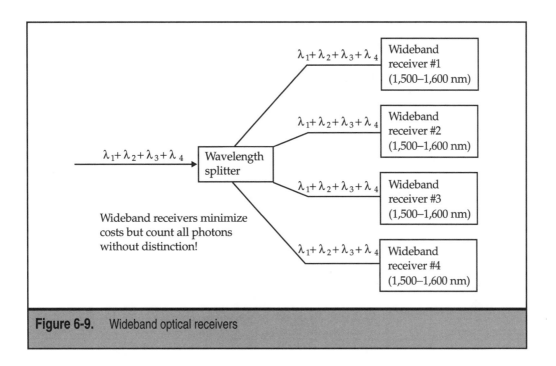

Figure 6-9.   Wideband optical receivers

the fiber unchanged (there is some small attenuation effect on the other wavelengths). The general idea behind the FBG is shown in Figure 6-10.

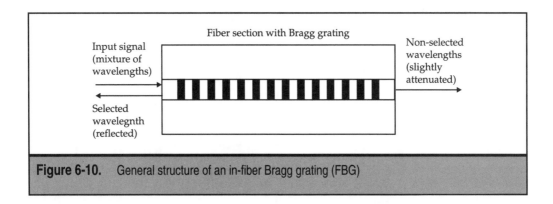

Figure 6-10.   General structure of an in-fiber Bragg grating (FBG)

The small changes in RI lengthwise in the fiber core of an FBG section will reflect a small amount of light at each discontinuity. If the wavelength of the light and the spacing (period) of the grating are the same, then there is positive reinforcement in the backward direction. In technical terms, the grating forms an *electromagnetic resonant circuit*.

There are many variations on this basic in-fiber Bragg grating design. Some of these can be used to deal with the issue that the wavelength selected is reflected *back* toward the source. It would be more useful to pass the desired wavelength *through* the fiber and eliminate the others, as a true filter should. And so it turns out it is possible to etch many different FBGs into the same section of fiber. Then a *blazed grating* can be added that is built at an angle to the fiber core and reflects light out of the core. So it is possible to combine many FBGs and blazed gratings to filter out all but the desired wavelength.

The use of FBGs as filters at the receiving end of a simple WDM link, allowing the use of identical wideband receivers, is shown in Figure 6-11.

The only problems with FBGs have been some sensitivity to temperature variations and stretching of the fiber section with the gratings. Both effects have been relatively easy to control in practice. FBGs can not only be used as filters in WDM and DWDM systems, but also to control dispersion in the 1,550 nm band and to stabilize the output of a laser, tunable or otherwise.

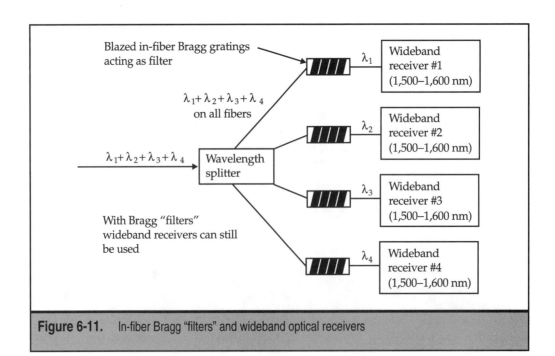

Figure 6-11.   In-fiber Bragg "filters" and wideband optical receivers

WDM

Early WDM devices and links involved two-channel multiplexing. These used the 1,310 and 1,550 nm laser operating regions in short-span applications, or two widely spaced channels in the 1,550 nm region (such as 1,538 and 1,558 nm) in long-span applications.

These WDM systems could be unidirectional or bidirectional. A bidirectional WDM device will pass a number of multiplexed frequencies in both directions on one fiber. For example, in a four-channel bidirectional WDM span, signals at 1,549 and 1,557 nm are multiplexed for transmission in one direction, and 1,533 and 1,541 nm signals are multiplexed for transmission in the other direction, all on the same working fiber. This is shown in Figure 6-12.

On the other hand, a unidirectional WDM device multiplexes a number of different frequencies for transmission in one direction on one fiber. For example, in a four-channel *unidirectional* WDM span, signals at 1,557, 1,549, 1,541, and 1,533 nm are multiplexed for transmission in one direction on the fiber. This type of WDM is shown in Figure 6-13.

Both of these techniques allow for such fiber configurations as combining four 2.4 Gbps serial bit streams on a single fiber span for an aggregate signal speed of 10 Gbps. This WDM technique can use a previously installed fiber base to support extremely narrowly spaced channels. This simple WDM system did have a number of nice features. For instance, using WDM to combine four channels makes maximum reuse of any existing 2.4 Gbps equipment, such as OC-48 (as long as the operational wavelengths are shifted to those of

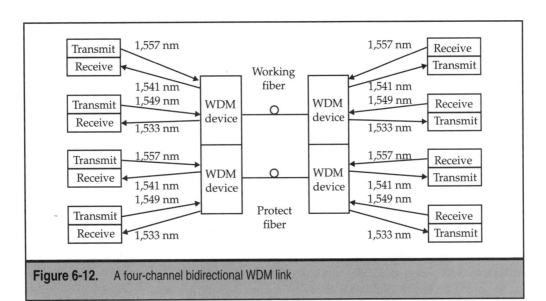

Figure 6-12. A four-channel bidirectional WDM link

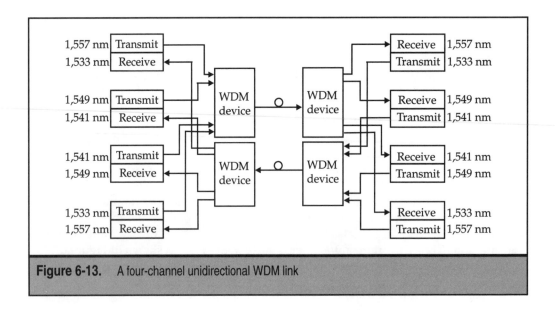

Figure 6-13.   A four-channel unidirectional WDM link

the WDM system). There is also not only a minimum of four-wavelength operation, but the capability to upgrade to eight or even 16 different wavelengths (or more) in the future. Finally, there is the capability to add or subtract wavelengths (provisioning services by wavelength), and unidirectional and/or bidirectional operation.

INTRODUCTION TO DWDM

Not only is WDM with wavelength spacings of about 10 nm an accepted optical technology, but WDM has moved along to the point where *dense* WDM (DWDM) systems are the major topics of discussion and consideration. The major difference between WDM systems and DWDM systems is that in DWDM, the wavelengths used are packed within 1 nm or even 0.1 nm of each other, or about ten to one hundred times denser than in WDM. The current record for wavelength packing on a single fiber is 1,022 channels, although the total throughput is only about 40 Gbps. However, since the current serial bit transmission record on fiber is 160 Gbps, potentially DWDM systems with throughputs in excess of 160 Tbps (160,000 Gbps or 160,000,000 Mbps) are possible.

So it would seem that there is very little real difference between WDM and DWDM. The latter offers closer spacing of wavelength channels, true enough, but essentially WDM and DWDM are the same technologies. But this is not really true. The close spacing of wavelengths in DWDM systems translates to a whole new set of concerns.

This is because in practice the closer spacing of wavelengths in DWDM makes a big difference. If the packing of wavelengths to within 0.1 nm (or less!) over a single fiber core were trivial, DWDM would merit no more than a footnote in WDM discussions. But there is a world of difference between 10 nm spacing and 0.1 nm spacing of light waves. For example, using 0.1 nm spacing allows 100 light sources to concentrate their energy over a 10 nm bandwidth into a single fiber core nine microns or so in diameter. This is a huge energy flux concentrated in a very small space, although in terms of raw power the total amounts are still relatively small. But confining this electromagnetic energy in the fiber core for long distances without debilitating attenuation, dispersion, and other fiber effects is a challenge that deserves a section all its own.

Everything is magnified in DWDM, so that even small effects in WDM become crippling in DWDM unless steps are taken to compensate for these impairments. Fortunately, there are methods available that make DWDM systems not only practical, but attractive.

To take a simple example, in DWDM systems the center wavelength of a channel must be tuned carefully to make sure it is exactly where it is supposed to be. This is because even simple modulation of the carrier signal will essentially double the bandwidth required (in precise accordance with Nyquist) and might otherwise overlap and interfere with wavelengths still within their band, but not properly centered. Naturally, as the wavelengths move to within a fraction of a nanometer of each other, these *band edges* become more and more important.

Before getting into a full discussion of DWDM system impairments and what must be done to control their effects, this is the place to say a few words about two often neglected components of optical networks. These are the *combiners* and the *splitters* that mix wavelengths onto a single fiber at the sender and split the mixed wavelengths onto separate fibers at the receivers. These components have come a long way as well. In many cases, the same device can serve both functions. Used in one direction, it is a splitter. Turn it around, and it is a combiner or coupler.

Simple couplers make use of a phenomenon called *resonant coupling*. This occurs when two fiber cores are brought into close proximity. If the lengths are just right, half of the input power ends up in each fiber core. This is called a *3 dB coupler*. The basic building block can be employed to make (for example) eight-way combiners or splitters. However, each stage causes some signal loss and attenuation. This is a physics phenomenon, and even two light streams directly coupled onto one fiber *must* lose half their power.

When operating at different wavelengths, DWDM couplers and splitters are more efficient. Modern systems can use optical technology such as *beamsplitter prisms* or *bifringent materials*. Many splitters make use of diffraction gratings of one form or another, including planar diffraction gratings.

At the source, some lasers are now *multiline lasers*. Instead of using the "comb" of spectral width wavelengths to tune the laser, multiline lasers attempt to even out linewidths and equalize the power of the several wavelengths produced. This is probably a good idea in the long run. If DWDM systems evolve to 1,000 inputs or more, 1,000 input lasers might be needed. But if multiline lasers can produce 10 usable wavelengths, only 100 input lasers would be needed. Of course, the multiline laser must be modulated externally at each wavelength to send bits, since there is only one "drive" device, but this is simple enough to do.

DWDM Impairments

It has been established that jamming many wavelengths onto a single fiber core is difficult. This is due to a number of fiber impairments that are annoying in any optical system, but absolutely crippling to a DWDM system. Oddly, some effects can be controlled by jamming the wavelengths even closer together, and other impairments require keeping the wavelengths well apart. Some effects can be dealt with by boosting the transmitter power, while others require power limits. But just what are they, and why do they interact this way?

All fiber impairments can be roughly characterized as dispersion effects. In this book, they are further divided into modal dispersion effects, scattering effects, and miscellaneous effects, but they all manifest themselves as dispersion. Table 6-1 lists the effect, the cause, and the workaround in DWDM systems. Keep in mind that this table is a distillation of dispersion effects that most impair DWDM systems. There are many other types of dispersion and fiber impairments.

It should be noted that many of these effects also vary with the serial bit speed used in the fiber channel of a DWDM system. As a general rule, doubling the speed from 5 to 10 Gbps will quadruple the dispersion on the same length of fiber. This only reinforces a truism in networking: to go farther, go slower. The following sections take a quick look at each effect in some detail.

Modal Effects

The first modal effect, *intermodal dispersion*, is really included here just for the sake of comparison. Intermodal dispersion is the result of the interplay between the various physical paths by which a beam of light can make its way down a fiber core. The workaround is to use single-mode fiber (SMF), and SMFs are used in all practical DWDM systems. Even older single-mode fibers designed for 1,300 nm use can be employed, but the attenuation is high and runs must be short.

But there is also *intramodal* dispersion that affects even a single propagated mode. The causes of this impairment are varied according to type, and the workarounds all concern fiber construction methods. *Material (chromatic) dispersion* is a result of the RI of the fiber varying with wavelength. *Waveguide dispersion,* which acts to counteract the effects of material dispersion, is the result of 20 percent of the light being carried not in the fiber core, but in the cladding. The difference in RI in the cladding causes dispersion. These effects usually cancel out at around 1,300 nm, but dispersion-shifted fibers move this point to around 1,550 nm. So most DWDM systems will use dispersion-shifted single-mode fibers. The final modal effect is *polarization mode dispersion*. This is the result of the RI of the fiber varying with the polarity of the signal. It is generally only of concern at serial bit speeds of 10 Gbps and above. The workaround is to use fiber with a perfectly round core.

Modal Effects	Cause	DWDM Workaround
Intermodal dispersion	Multiple modes interact	Single-mode fibers
Intramodal dispersion	Various:	Fiber construction methods:
Material (Chromatic)	RI varies with wavelength	Effects cancel at 1,300 nm
Waveguide	20% light travels in cladding	(Can be shifted)
Polarization mode	RI varies with polarity	Perfectly round core (difficult)
Scattering Effects		
Rayleigh	RI varies with fiber density	Perfectly uniform density
Raman (SRS)	Low wavelength pumps high, produces "power tilt"	Lower power, limit span, pack wavelengths close together
Brillouin (SBS)	RI varies with acoustic waves	Lower power, limit number of channels, avoid two-way systems
Miscellaneous Effects		
Linear cross talk	Mostly result of SRS	Use filters, separate wavelengths
Four-wave mixing (FWM)	Additive wave response	Watch channel wavelengths, avoid 10+ Gbps systems
Cross-phase modulation	RI varies with intensity (called optical Kerr effect)	Balance power in system
Self-phase modulation		Limit power in system

Table 6-1.   Fiber Impairments

(Surprisingly, there are times that perfectly round cores are not desired and polarization mode dispersion is actually encouraged!) Now, most fiber fabrication techniques do their best to produce round cores. However, it is easy for the fiber to be stepped on or otherwise deformed during installation.

Scattering Effects

The *scattering effects* are an interesting group. These are named, curiously enough, after physicists and investigators into optics. *Rayleigh scattering*, common to all fibers, is unavoidable and results when light is scattered, or deflected, by particles in the fiber. It is a concern in DWDM systems because variations in the density of the fiber make the dispersion worse. Again, the only way to compensate is to construct the fiber to very exacting standards.

Of more interest is *stimulated Raman scattering (SRS)*. SRS is caused when multiple wavelengths interact on one fiber through the crystal lattice of the fiber core, and so it became a concern only in WDM systems. With SRS, lower wavelengths will "pump" higher wavelengths, resulting in a power loss at lower wavelengths and a power gain at higher wavelengths. This causes a pronounced "power tilt" in the received power of the wavelengths, even though all might be sent with the same power. SRS is not always bad: optical amplifiers are nothing more than fibers in which the pumping from the lower wavelength (the pump laser's wavelength) to all the higher ones is encouraged to the extreme. SRS can be controlled by lowering the input power, limiting the span distance, and packing the wavelengths (generally, the further apart the wavelengths, the more pronounced the SRS between them).

Finally, there is *stimulated Brillouin scattering (SBS)*. SBS, which occurs even at low power levels, is the result of acoustic vibrations in the fiber. That is, the atoms of the fiber literally move back and forth, creating an acoustic (pressure) wave in the fiber. The pressure change affects the RI, which scatters the light, producing SBS. Both SRS and SBS are *stimulated* in the sense that the light that is itself scattered produces the effect. Oddly, SBS is less severe at higher data rates as the bit time shrinks. However, SBS is much worse when light signals propagate in both directions at the same time in bidirectional WDM systems (because of the complex acoustic waves generated). In fact, SBS is one of the main reasons that two fibers are used, one for each direction, in most optical systems. But if runs are short enough, even DWDM can be used as long as the number of wavelengths used is limited. So the workarounds are limited power (which limits runs), limited wavelengths, and avoiding two-way transmission on a single fiber.

Miscellaneous Effects

The final category of *miscellaneous effects* is also an interesting group. The first member, *linear cross talk*, is not really a separate impairment at all. Linear cross talk is caused by SRS, but many sources treat linear cross talk separately. Cross talk is just the interaction or interference between two wavelengths. The workaround is to keep the wavelengths well separated, and filter out extraneous wavelengths not used for information carrying.

The name *four-wave mixing (FWM)* is an unfortunate one. This impairment was first noticed in early WDM systems with four channels, hence the name. But it really should be *N-wave mixing*, since the interactions of any number of wavelengths can lead to this phenomenon. FWM is just an additive response to multiple wavelengths, the optical equivalent of *beats* in electrical systems. Three wavelengths will combine to yield a fourth wavelength at the receiver according to a formula (so FWM is appropriate in this sense). The formula, in frequencies used, is $f_{new} = f_1 + f_2 - f_3$. Naturally, if the new wavelength is also used in the DWDM system, cross talk results. The workaround is to watch the wavelengths used and space them out unequally (the ITU grid, discussed in the next section, allows for this). FWM is also worse at 10 Gbps and above.

Finally, there are two impairments known as *phase modulation* effects. Both are the result of the RI of the fiber varying with the intensity of the light, a phenomenon called the Kerr effect. Cross-phase modulation (CPM) occurs between the wavelengths, and self-phase modulation (SPM) occurs within even a single wavelength of light. CPM is compensated for by carefully balancing the power between the channels in a DWDM system, and SPM is dealt with by limiting the power used (which also limits distance, of course).

The ITU DWDM Grid

The ITU-T has been active in exploring international standards for DWDM. One specification is G.692, entitled "Optical interfaces for multichannel systems with optical amplifiers," which contains a number of interesting points about DWDM. Unfortunately, DWDM developments have been proceeding so fast that the ITU-T and other standards organizations have been hard pressed to keep up with system capabilities. These words might soon be obsolete, if they are not already.

The ITU-T specifications mainly address point-to-point DWDM systems using 4, 8, 16, or 32 channels. The maximum specified span without amplifiers for a DWDM link is 160 kilometers (100 miles) or up to 640 km (400 miles) when amplifiers are used.

A standard wavelength grid is established based on multiples of 50 GHz spacing between the wavelengths used. There is also a center reference frequency ("wavelength") at 193.1 terahertz (THz: one thousand GHz). The whole grid is sometimes called the "ITU 193 THz grid." In keeping with the tradition of citing frequency in ITU-T specifications, all of the wavelengths are specified in Hz (frequency), not nm! But conversion is easy enough to do. The reference frequency of 193.1 THz equates to 1,553.5 nm, right in the middle of the 1990s window, as might be expected. The 50 GHz spacing works out to about 0.4 nm between the wavelengths. The wavelengths used are in the range from 1,528.77 nm to 1,560.61 nm.

The G.692 draft proposal allows:

▼ 4 channels with 3.2 nm (400 GHz) spacing

■ 8 channels with 1.6 nm (200 GHz) spacing

■ 16 channels with 1.6 nm (200 GHz) spacing

▲ 32 channels with 0.8 nm (100 GHz) spacing

Supervisory wavelengths are allowed at 1,310, 1,480, 1,510, and 1,532 nm. In the draft, users are allowed to use any wavelength on the grid, but only those wavelengths on the grid. However, working DWDM systems have progressed to 128 channels or more, and spans in excess of 640 km are possible, challenging the ITU-T plans. Allowable speeds for each channel are 622.08 Mbps, 2.488 Gbps, and 9.9 Gbps. These correspond to SONET/SDH speeds of OC-12/STM-4, OC-48/STM-16, and OC-192/STM-64. Optical amplifiers of the EDFA type are expected to be used, operating in the 1,550 nm region. In addition to the basic 50 GHz spacing, systems are allowed to use spacings of 100 GHz (8 or more channels), 200 GHz (4 or more channels), 400 GHz (4 or 8 channels only), 500 GHz (8 channels only), 600 GHz (4 channels only), and even 1000 GHz (4 channels only).

Potentially, the ITU-T grid spacing could be firmly allocated on a national basis, with each country limited to using the wavelengths assigned to it. This is actually a good idea, since the day may come (and many feel this day is closer than anyone thinks) when a single strand of fiber will literally circle the globe carrying hundreds or even thousands of wavelengths with in-fiber optical amplifiers. Each country would just attach to this global fiber ring and never fear interfering with some other countries' wavelengths. To communicate with any other country, just send on a channel allocated by the receiving country. To receive, just listen on the channel assigned for the return path.

There are a number of ways to use SONET/SDH as a *management channel* in a DWDM system. One way is just to couple in a SONET/SDH channel at 1,310 nm with the DWDM channels at 1,550 nm. After all, there is only one fiber. If the SONET/SDH management channel indicates a problem, there is obviously a problem in the whole fiber. Alternatively, a transponder can be used to shift the 1,310 nm management channel into the band centered around 1,550 nm. Sometimes the SONET/SDH management channel is used to carry additional network management information about the other DWDM channels, such as error correcting codes. Additional issues concerning the future of SONET/SDH in a DWDM world will be considered in Chapter 8.

CHAPTER 7

Optical Networking Nodes

I t has been pointed out several times in this book that, in spite of definitions and claims, SONET/SDH is *not* an optical network. SONET and SDH are still firmly electrical networks with optical links. Anytime anything happens worth mentioning in SONET/ SDH, the electrical frame structures, not the light waves themselves, are needed.

In contrast, optical networking, by definition, has network elements (NEs) and components that perform at least some of the essential networking tasks at an optical level. Optical network components currently include "regenerators" (really optical amplifiers), optical ADMs (OADMs), and optical cross-connects (sometimes misleadingly called optical "switches"). Optical amplifiers were discussed in the previous chapter and will not be considered further. This chapter will deal with OADMs and optical cross-connects, however. True optical switching has been a dream for many years, and optical switches in DWDM systems are almost, but not quite, ready to go. The chapter will go on to consider possible optical routers, but right now such optical network nodes are only research topics and not products.

DWDM is still an essential part of an optical network. DWDM allows for the packing of many wavelengths onto the same physical path, the optical equivalent of FDM. But if all there were to optical networking was DWDM, then optical networking would be as exciting as watching 100 TV channels whiz by on a coaxial cable. The excitement of optical networking is what happens before, during, and after the wavelengths come together on the fiber.

One of the problems when mentioning DWDM and optical networking in the same chapter is that the constantly evolving capabilities of DWDM in terms of channels and aggregate speeds make written statements almost instantly obsolete. Fortunately, this chapter is not so much concerned with feeds and speeds as with how DWDM and optical NEs come together to make a true optical network. It is safe to assume that the DWDM is a 100-channel (current maximum is over 1,000) system, with each channel running at 10 Gbps, the OC-192/STM-16 rate (current serial maximum is 40 Gbps), for a total throughput of 1 terabit per second. Later in this chapter a "terabit router" will be introduced as the perfect companion for this configuration.

After the principles of operation of the OADM and optical cross-connect are outlined, an overall architecture for optical networking will be introduced. OADMs in particular make use of the in-fiber Bragg grating and optical coupler to selectively separate and merge wavelengths on a single strand of fiber.

THE OPTICAL ADM (OADM)

In electrical networking, a basic network building block is the cross-connect and add-drop multiplexer (ADM). Cross connecting simply rearranges the channels used on a multiplexing system. So a cross-connect in Memphis can take channel #1 inbound on a multiplexed link from Cincinnati and make it channel #5 on the multiplexed link outbound to Nashville and so on. Although the cross-connect function can be handled by a simple patch panel, this function is usually done with some sort of intelligent network element that can be reconfigured electronically. This configuration capability alleviates the need to have technicians on site with physical access to the equipment in Memphis to handle the cross-connecting.

Closely related to the cross-connection function is the ability to add and drop channels locally. So channel #1 from Cincinnati can be dropped locally in Memphis to service local users. The traffic on channel #5 outbound to Nashville can be picked up locally from another Memphis customer. Typically, channels are dropped, cross-connected, and then added back into the multiplexed traffic stream. These twin functions, cross-connecting and add-drop multiplexing, can be implemented in the same piece of equipment, but normally they are not. This increases flexibility of system arrangement and keeps the cost of each individual piece of equipment down.

In optical networking, these closely related functions are implemented in separate devices, the optical add-drop multiplexer (OADM) and the optical cross-connect (sometimes seen as OXC). There is no technical reason why these functions cannot be combined into the same device, of course. But the same reasons of flexibility and cost apply to optical networking.

So one of the first steps in optical networking is to create a device that does not add and drop electrical channels, but acts on the *wavelengths* arriving and departing on a DWDM optical link. Consider the adding and dropping of wavelengths locally. The fundamental principle behind the optical add-drop multiplexer (OADM) is simple enough. A filter is used to isolate, or drop, the desired wavelength from the multiple wavelengths arriving on a fiber. Multiple wavelengths can be dropped, of course, or allowed to pass, depending on the physical configuration of the filters. Once a wavelength is dropped, another channel employing the same wavelength can be added, or inserted, onto the fiber as it leaves the OADM. This is shown in Figure 7-1.

This simple OADM has only four input and output channels, each with four wavelengths (λs). Within the OADM, the absence of a dropped wavelength is indicated by a space in the wavelength profile on the fiber inside of the OADM itself. These are "empty lambdas" in the sense that there is no wavelength present on the internal connections within the OADM.

There is also no absolute relationship between dropped or inserted wavelengths and the ports on the OADM. This is a configuration issue. Naturally, two channels using the same wavelength cannot "share" an output fiber. On their way through the OADM, the wavelengths might be amplified, equalized, or further processed. The important point is that all of this takes place in optical form, with no electrical-optical (E/O) conversion required. The figure shows how the same wavelength could be inserted, as long as the wavelengths are used on different output fibers.

All an OADM does is basically isolate wavelengths for some purpose. It is useful to be able to access a wavelength selectively, but the real need is to be able to rearrange wavelengths from fiber to fiber. Since the basic transport link is still the physical fiber itself, the OADM must have some means of rearranging the wavelengths from input fiber to output fiber. This is where the optical cross-connect, or "switch," comes in.

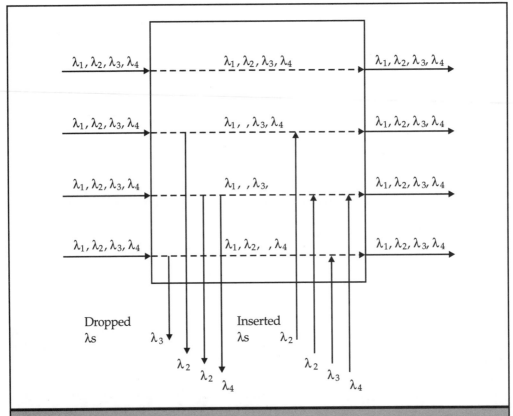

Figure 7-1.   An optical add-drop multiplexer (OADM)

OPTICAL CROSS-CONNECTS

Electricity easily can be made to follow paths through electronic components, a fact of nature that makes electrical cross-connecting a trivial exercise. It is not a problem to make electricity jump from wire to wire, or from path to path. Electricity always follows the easiest path when confronted with a choice. But the whole idea behind fiber optics in general is to make light *stay* in the fiber. How can light waves be induced to change fibers without changing the light into electricity? This is what the optical cross-connect does.

An optical cross-connect can take four input fibers (for example), each carrying four wavelengths (not a lot for DWDM, but used for ease of illustration), and rearrange the 16 wavelengths onto four output fibers. Naturally, if all four fiber channels use the same wavelengths, then there is a risk that two input channels of the same wavelength will want to share the same output fiber. In this case, a simple *transponder* inside the optical cross-connect will shuffle one of the wavelengths to an available channel.

Transponders have typically been electrical devices. That is, in order to make a traffic-bearing optical signal change wavelengths, it has been necessary to convert the optical signal to electricity, and then use the recovered bit stream to drive the modulation of a second wavelength. Recently, however, pure optical transponders have been developed, although their performance has been limited in several respects. Optical transponders will be discussed later in this chapter.

The optical cross-connect process is sometimes called "switching," but this is misleading. Switching (and routing, for that matter) takes place by examining some header or label inside the arriving traffic stream. Needless to say, cross-connecting does no such thing. It is a simple port-by-port rearranging process, much more related to multiplexing than switching and routing. The "proof" is that if an output fiber can carry only four channels, then only four channels can be cross-connected onto the output fiber. In the example, 16 input channels are rearranged, and wavelengths reassigned, but always become 16 output channels. True switches can switch all arriving traffic onto one output port, if need be, and sometimes it seems like this is always the case. The optical cross-connect principle applied to DWDM is shown in Figure 7-2.

Note that a wavelength can arrive on one fiber and leave on another, as long as that is the way the cross-connection table had been configured. The wavelength used can also change, and often does. This "wavelength shift" just requires a simple transponder to detect one wavelength and transmit another (wavelengths can even shuffle within a fiber).

The wavelength and port mappings from input to output in the figure can be represented by a simple table. These tables are similar to switching tables used in ATM switches and help to make sure no port is "double-booked" or left out. The table for this

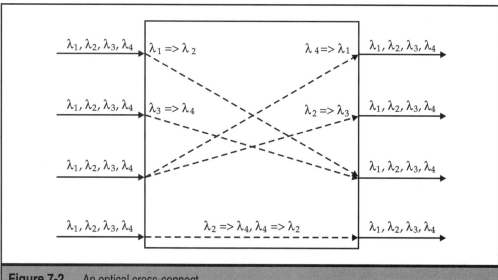

Figure 7-2. An optical cross-connect

cross-connect configuration is given in Table 7-1. Cross-connected ports that are not passed straight through the device are marked with an "*."

The optical cross-connect performs exactly the same port-by-port reconfiguration functions as the electrical cross-connect. The traffic streams are not STS-1s (or whatever) inside an STS-12 (for example) as in SONET. The traffic streams cross-connected are wavelengths on the fibers. The trick is to do all this entirely in the optical domain. The key component that allows this to happen is the *optical cross-connection element*, or *fiber switching junction* (which is one reason the entire device is sometimes called a "switch").

Ironically, there is plenty of electricity involved in the optical cross-connection element. Not so much in terms of electrical power, but just in terms of role.

The trick, if there is one, is simply to fabricate the "fibers" (in some cases they are actually optical channels etched onto the silicon substrate of a chip) in the optical cross-connection element so that the fiber cores actually come in contact with each other for a length. Usually, the length is no more than 3–4 millimeters, but there is a lot of variation here. Since light is optical, but still within the realm of electromagnetic phenomena, the

Input Port	Input Wavelength	Output Port	Output Wavelength
1*	λ_1	3	λ_2
1	λ_2	1	λ_2
1	λ_3	1	λ_3
1	λ_4	1	λ_4
2	λ_1	2	λ_1
2	λ_2	2	λ_2
2*	λ_3	3	λ_4
2	λ_4	2	λ_4
3	λ_1	3	λ_1
3*	λ_2	2	λ_3
3	λ_3	3	λ_3
3*	λ_4	1	λ_1
4	λ_1	4	λ_1
4	λ_2	4	λ_2
4	λ_3	4	λ_3
4	λ_4	4	λ_4

Table 7-1. Cross-Connections Performed by the Optical Cross-Connect in Figure 7-2

proper application of a modest voltage (or absence of voltage) across this junction area can make or prevent light from changing cores from one fiber to the other. The principle behind the optical cross-connection element is shown in Figure 7-3.

The wavelengths input into the optical cross-connect element are shown as λ_A and λ_B because optical cross-connects function at the same wavelengths. But the two input streams do "flip-flop" onto the output fibers.

The basic optical cross-connect element can be combined into a simple optical cross-connect device. This is shown in Figure 7-4. No DWDM is shown, just four wavelengths being shuffled around among input and output fibers according to how the device is configured. Only the path for λ_A is shown as a dashed line, but the others are "routed" in a similar fashion.

The output pattern in the figure (D, B, C, A) is not an arbitrary one. This pattern is produced when all wavelengths shift output fibers in the device-switching fabric. But this is only a simple configuration of cross-connect elements for illustration purposes. The wavelengths do not share a fiber within the device, so only simple "flip-flops" of wavelengths are possible. In other words, there cannot be "collisions" of wavelengths within a given switching element of the architecture. The output pattern is accurate, but not very exciting, since all combinations of output are not possible with this simple collection of cross-connect elements. Many more elements would be needed to produce a full matrix of all input and output ports in a working device.

DWDM capabilities can be added to the basic optical cross-connecting, perhaps by the creative use of transponders to shift wavelengths into and out of the optical cross-connect elements. This is a giant step closer to the optical "switch," since what is being "switched" are multiple traffic streams on each input fiber. Care must be taken, of course, to prevent two wavelengths from trying to share one output port.

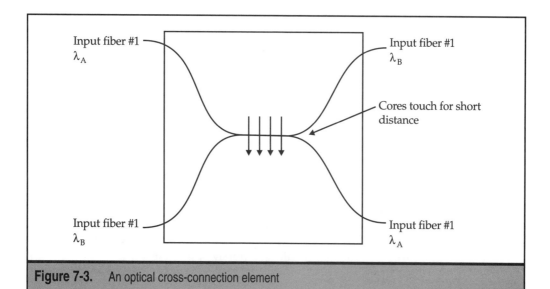

Figure 7-3.   An optical cross-connection element

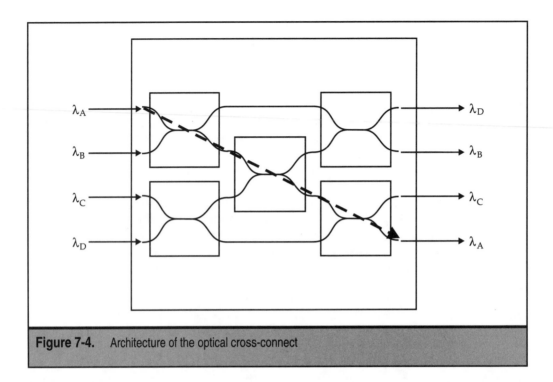

Figure 7-4. Architecture of the optical cross-connect

Whenever SONET/SDH is wanted as a wavelength in DWDM systems featuring OADMs and even optical cross-connects, transponders are almost always used. This is because SONET/SDH signals all want to use the standard wavelength established for the SONET/SDH, 1,310 nm. But DWDM operates in the 1,550 nm range. So transponders not only shift the native SONET/SDH wavelength into the DWDM range but also prevent "collisions" when two or more OC-12/STM-3 (for example) inputs are being carried.

OPTICAL TRANSPONDERS

Naturally, if it is necessary to convert optical traffic streams to electricity in order to change the modulated wavelength in an optical cross-connect, this is a considerable limitation when it comes to optical networking. The goal of optical networking is to do as much as possible in the optical realm and to rely on electricity as little as possible except for necessary control functions until truly optical computers become available.

Until recently, all wavelength shifting required in optical network systems had to be done by recovering the bit stream represented by the optical signal and then remodulating the new wavelength with the recovered bit stream. (The term "optical transponder" described what the device did, not how the device did it.) Not only was this slow, but it introduced an undesirable but necessary electrical stage into what should be a purely optical device.

But truly optical transponders have been at least demonstrated and announced as products. However, most are not yet shipping to customers or are available only for testing purposes.

The general idea behind the optical transponder is similar to the optical cross-connect element just discussed. An intensity-modulated optical signal on wavelength λ_1 enters the optical transponder on one fiber. At the same time, an unmodulated carrier wavelength λ_2 enters the device on a second fiber. Keep in mind that these "fibers" might be waveguides etched in silicon.

Next, as in a simple optical cross-connect element, the cores of the fibers or waveguide areas are brought very close together. The active section of the optical transponder has special optical characteristics that vary too much in prototypes to be easily characterized. However, the point is that modulated wavelength λ_1 and unmodulated carrier wavelength λ_2 will interact in such a way that the intensity modulation representing bits on wavelength λ_1 are picked up by the carrier wavelength λ_2. In this way, the output of the optical transponder is a modulated wavelength λ_2. The output wavelength λ_2 now carries the same bit sequence as the input wavelength λ_1 that can safely be ignored. After some optical amplification, wavelength λ_2 is now ready to be further cross-connected. The principle behind pure optical transponders is shown in Figure 7.5.

In practice, optical transponders have proved to be expensive and inefficient. It is currently much more cost-effective and reliable to convert the optical input on wavelength λ_1

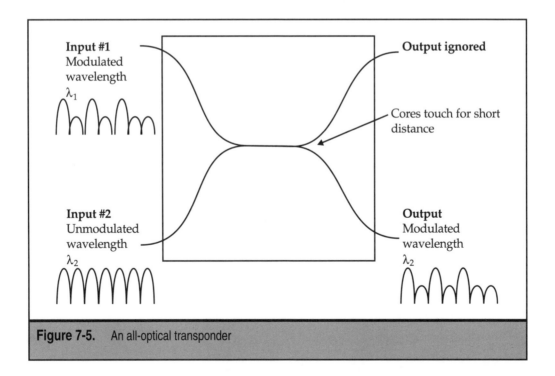

Figure 7-5. An all-optical transponder

into electricity and then use the recovered bit stream to drive electrical modulation of the output wavelength λ_2. Such electro-optical conversion is not the main philosophy of optical network nodes, however. So little by little all-optical transponders will find their way into all major optical network elements from OADMs to OXCs to switches.

PUTTING THE PIECES TOGETHER

The basic components of optical networks—DWDM, OADMs, transponders, and optical cross-connects (sometimes called "ODCSs," although there is nothing really "digital" about them, or OXCs)—can be combined to create an optical node on a fiber ring. This *optical fiber ring* is no longer a SONET/SDH ring, of course, although SONET/SDH ADMs and the like might still be present on the ring. The fiber is now carrying DWDM wavelengths, and might even be new fiber installed just for DWDM (although if the spans are short enough, SONET/SDH single mode fiber works just fine). This basic node structure is shown in Figure 7-6.

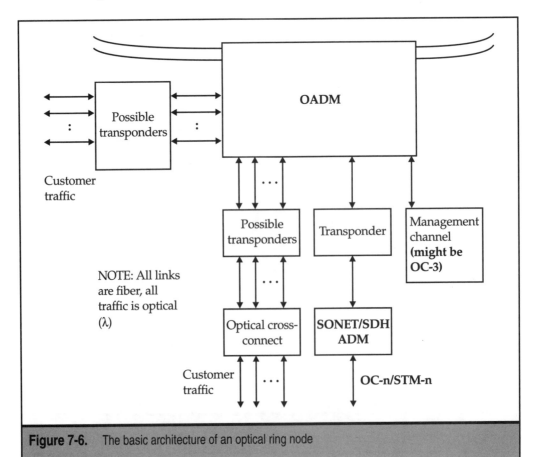

Figure 7-6.   The basic architecture of an optical ring node

One important part of the figure for the purposes of this book is that SONET/SDH is no longer the reason for the ring to exist. The role that SONET/SDH plays in the overall architecture is reduced to that of network management, or perhaps even less. The ring architecture can still be a four-fiber BLSR, or other ring type, and often is.

Protection switching is still a key element of any ring architecture, DWDM optical rings included. The nice things about the continued presence of SONET/SDH on the new DWDM ring are twofold. First, transponders can be used to allow one or more of the SONET/SDH ADMs to use the DWDM ring. In fact, ring capacity for SONET/SDH is greatly enhanced, since there are many wavelengths available. This also allows "legacy" SONET/SDH equipment to share the ring, of course. Second, since there is still only a single physical fiber ring to protect, once the SONET/SDH line level overhead makes the protection switching decision to use another span or even ring-wrap, then all of the other DWDM channels go along for the ride. In other words, a SONET/SDH ring can use the status of the line overhead bits to decide whether to perform some ring protection function, even wrapping signals back around the ring. This protection switching is only done by SONET/SDH, but if the SONET/SDH signal is one wavelength on a DWDM ring, then all of the other wavelengths will enjoy the protection that SONET/SDH provides.

But what about all of the other wavelength channels? There is no need to fill these with SONET/SDH streams, although that is always possible. Perhaps other *direct-to-fiber* interfaces could be used, and this has always been a goal of DWDM and optical networking systems. Many frame structures other than SONET/SDH use fiber as a transport medium. For instance, the Fiber Distributed Data Interface (FDDI) is a 100 Mbps fiber ring architecture, but FDDI frames are not SONET frames and fit very poorly into SONET STS-3c payloads (about 55 Mbps is "wasted"). Fibre Channel is another architecture not aligned with SONET, and even 10 Mbps Ethernet has run on fiber for a number of years (10Base-F). None of these are SONET, or benefit from the presence of SONET directly (SONET protection switching is really an indirect benefit, since FDDI and some others have their own forms of protection switching). The somewhat unique position of Gigabit Ethernet (GBE) and 10 GBE will be considered in Chapter 12. The point here is that to allow direct-to-fiber traffic to most efficiently use a fiber ring, the requirement to use SONET/SDH framing should be dropped. If SONET/SDH nodes are present, just plug them into an OADM, assign a wavelength, and stand back.

Optical networking offers a way for service providers just to *sell a wavelength* to a customer for almost any form of traffic. The customer could use the wavelength for anything at all, SONET/SDH or not. The OADMs, transponders, optical cross-connects, and so on take care of making sure the traffic stream carried by the chosen wavelength finds its way through the ring nodes, across rings, between rings, and so forth, properly. The customer need not know (and need not care) what wavelengths are used for the traffic internally, although this knowledge might be provided routinely for troubleshooting and network management purposes.

This flexibility is the real promise of optical networking to customers and service providers. SONET/SDH runs at fixed speeds, but many things are not aligned with SONET/SDH speeds. SONET/SDH frames can carry many things, but in a very real sense both

SONET and SDH are optimized for VT voice channels. Other things fit, but some only poorly with much wasted bandwidth and redundant overhead functions.

If all channels of a DWDM optical network become non-SONET/SDH, how then do the customer and service provider gain any benefit from SONET/SDH protection switching and network management functions? If there is no SONET/SDH present on the optical network, then standard protection switching and network management must be added to each and every direct-to-fiber technology, right?

Probably not, although there have been proposals just along these lines. The DWDM standards allow for a *network management channel* at 1,310 nm, outside the normal DWDM traffic-bearing range. This channel coincides with the standard SONET/SDH wavelength, and use of SONET/SDH for this network management channel makes a lot of sense but is not required. The presence of SONET/SDH on this channel protects and manages all of the other services on the single physical fiber.

This SONET/SDH management channel need not carry any live user traffic at all. This might seem quite wasteful, since only the overhead is needed for protection and management purposes. However, some DWDM vendors have made creative use of this channel by placing forward error correcting (FEC) code information derived from blocks of bits on the other channels. So there is the added benefit of cutting down on the need for resends, which is the whole point of FECs in the first place. Some vendors even carry this FEC on "in-band" SONET/SDH channels, simply replacing idle stretches in the SONET/SDH payloads with the FEC.

OPTICAL SWITCHING ARCHITECTURES

It has already been pointed out that optical cross-connects are often called "optical switches." This is more than a little misleading, however. Cross-connects rearrange traffic port by port and/or wavelength by wavelength based on configuration parameters. Switching, as the term is usually applied in communications, implies that traffic output port determination is made on the basis of *data stream content* such as packet headers, not physical port or wavelength configurations.

The difference in definition of switching is important. Suppose an optical cross-connect can be reconfigured fast enough that its performance is not that much different—if at all—from a traditional switch. If the reconfiguration is directed by the content of the input data stream, is this device any different from a conventional switch or router? After all, it is what the device does that is important, not so much how it does it. Nevertheless, many would insist that switches and routers operate in a fundamentally different fashion than a quickly reconfigurable optical "switch" or rapid optical cross-connect.

There is no need to take sides in this debate in this section. Perhaps all that is needed is to note that an "optical switch" is not quite the same as an ATM switch or IP router and let it go at that. The "optical switch" terminology seems to have stuck, and this section will explore some real-world and practical optical switch architectures.

Electrical switches used as network nodes require optical links to slow down and their content to be converted to electrical bits for switching. Reconversion to light on the output link makes this architecture costly and adds delay to the network. Waiting for faster electronic components is not the answer, because optical rates are increasing even faster than their electrical counterparts. More than a handful of DWDM links will always be overwhelming to an electronic switch. Optical switches are required for optical networks.

Optical switches today are built with mirrors and bubbles and more exotic technologies. A better term for what these extremely fast and flexible cross-connects do would be "wavelength switching." Each wavelength on a DWDM link can be thought of as a multiplexed channel carrying thousands of voice calls or a stream of packets. So manipulation of the wavelength is enough for an optical switch on a large enough backbone. Only at the edge of the optical network will it be necessary to convert light to electricity to access the individual streams of bits.

It is important to realize the scale at which DWDM optical switches operate. An optical switch with 100 input and output fibers (not very large compared to a 10,000-line telephone central office) each carrying 100 channels (many more are possible) has to deal with 10,000 input wavelengths. Each wavelength could be carrying information at 10 Gbps. So the switch would have to deal with 100 terabits or 100,000 Gigabits of data per second if it were forced to operate in the electrical realm. Optically, there are "only" 10,000 wavelengths to be dealt with.

In 1992, state of the art optical switches could handle 16 inputs and 16 outputs (16×16), with each fiber carrying only a single wavelength. A few years later, this was upped to a 48×48 matrix, but once DWDM is thrown into the mix, it is obvious that such approaches to optical switching are just inadequate.

Newer optical switches employ radically different approaches to the problem. One of the most promising approaches uses what is called MEMS (MicroElectroMechanical Systems) technology. Others use thermo-optical switching, bubble switching, and liquid-crystal switching. All are discussed in this section.

MEMS (MicroElectroMechanical Systems)

MEMS technology is not limited to optical networking applications. MEMS is actually a term for a new type of device fabrication technique and is not limited to optical networking applications. MEMS technology builds many types of "micromachines" used in the automotive, aerospace, and electronics industries, among others, as well as telecommunications. MEMS are a type of mechanical integrated circuit. It is the same large-scale integrated circuit technology using etching processes the produce circuits and components on silicon wafers. But with MEMS, layers are etched away to produce not only electrical and optical pathways, but also microscopic structural elements. So instead of producing transistors, the MEMS technique produces mechanical devices tens of microns across that can be moved as directed by tiny electrical currents.

Several companies, among them Lucent and Nortel Networks, have chosen MEMS to create optical cross-connects. The MEMS process yields components that are small and inexpensive and that can be combined with integrated circuits to produce very efficient optical devices. MEMS micromachines are robust, scalable, and long-lived enough to place many such devices on a single wafer. Such scalability is important in optical networking when it comes to DWDM of course.

A MEMS optical switch uses mirrors to direct light from input to output. The mirrors are very small, and the MEMS mirrors are positioned so that the input wavelengths from the light inside the fibers fall on one or more of the MEMS mirrors. The MEMS mirrors can be tilted under electronic control from side to side or up and down to direct the wavelengths inside the fiber to an output fiber.

There are many types of MEMS optical switch architectures. In one of the most promising switches, there are 256 input fibers and 256 output fibers (256 × 256). The operation of this MEMS switch is very simple to understand. Imagine a person sitting in a dark room with several windows. Outside one window, an automobile's two headlights shine on the person. The twin headlights represent two wavelengths and the window represents an input fiber. The person can hold a mirror and tilt the mirror to direct the light from the headlights out any one of the other windows in the room. The mirror can be tilted quite rapidly to redirect the light to another window. The other windows flashing with reflected light represent the output fibers of the device. In the MEMS switch, it is the stream of photons of all carried wavelengths that strike a series of small mirrors in the MEMS mirror array. The mirrors can be controlled very rapidly and very precisely to reflect all of the wavelengths onto the proper output fiber.

Control software in the MEMS switch processor makes the output fiber decision. The control software sends a signal to an electrode on the MEMS chip. The electrode creates a tiny electromagnetic field that causes the mirrors to tilt properly. Arriving light from as many as 256 input fibers is filtered and focused so that light from each fiber will fall on one of the 256 MEMS input mirrors in the total mirror array on the chip. Another mirror reflects all 256 light streams back into 256 output mirrors. The output mirrors direct the light to the proper output fibers. The basic principle of the MEMS optical switch is shown in Figure 7-7.

In the figure, only 16 fibers and mirrors are shown for simplicity. Two light paths are detailed with different arrow structures. In many optical MEMS architectures, there are twice as many mirrors as fibers. Each fiber can carry multiple wavelengths as well. In this case, the light path is a little more complicated. When multiple wavelengths are present, each must be sent through a filter and imaging lens arrangement. But the end result is that any wavelength arriving on any fiber can be directed onto any output fiber.

MEMS switching elements are very small. One MEMS architecture employs mirrors only one-half millimeter across, about the size of the head of a pin. These mirrors are spaced one millimeter apart, and the whole array of 256 mirrors fits on a silicon wafer 2.5 centimeters on a side. This set of mirrors is about 32 times more dense than the switch matrix required if the light had to be converted to electricity. The entire optical MEMS switch is about the size of a grapefruit. And since there is no need to power 256 electro-optical converters, the switch consumes about one-hundredth the power required by traditional optical cross-connects.

The mirrors are fabricated out of silicon, not metal. This has two advantages. First, regular silicon-based fabrication techniques can be used. Second, silicon mirrors are more

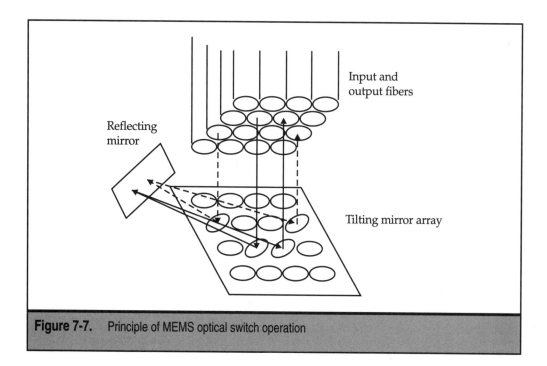

Figure 7-7.   Principle of MEMS optical switch operation

stable than metallic ones in terms of temperature and electrical field performance. In one manufacturing process, the mirrors are made in large part by *self-assembly*. In the final fabrication stages, tiny springs behind the mirrors on the surface of the silicon wafer are released. A frame lifts each mirror just far enough beyond the wafer surface and locks them in place with just enough leeway to allow for the proper ranges of motion required.

The mirror array in the 256 × 256 switch uses one mirror for input and one for output. The mirrors have to be quite stiff to avoid drifting out of alignment. The silicon mirrors can be easily scaled to handle more and more ports and/or wavelengths. One architecture supports a total of 10 terabits per second switching capacity, and each fiber can operate at up to 320 Gbps aggregate speed. This is about ten times the capacity of the fastest electro-optical routers.

Thermo-Optical Waveguides

Optical switches based on MEMS technology are very promising. There are over ten MEMS startup companies in California's Silicon Valley, each hoping that their approach will catch on in a big way. But MEMS is not the only way to produce an optical switch architecture that uses many small and inexpensive components to control the flow of light from input to output. One interesting approach is to use what are known as *thermo-optical waveguides*.

Waveguides are paths of integrated circuit chips that have all of the properties of optical fibers. Waveguides can be built by the same standard processes used to make integrated circuits and so look like "fibers on a chip." Waveguides have a core and cladding made of glass with differing indices of refraction, just like normal fiber optic cables.

There are several types of waveguide optical switches. One current design being explored by several companies for optical switching products uses what is known as the *thermo-optical effect*. The thermo-optical effect takes advantage of the fact that a slight temperature change applied to the waveguide will change the phase of the light as it travels as a wave. Changing the phase in turn can change the waveguide pathway taken by the light.

So an optical switching element based on the thermo-optical effect resembles the basic switching elements based on electrical fields discussed earlier in this chapter. But instead of relying on electrical currents to do the job, the thermo-optical switch employs heat with the same effect. In is easy to apply heat in just the right place and in just the right amount along the waveguide, easier than applying electricity. After all, the whole switching element has many electrical circuits for control, and adding more electrical fields in small places can cause unwanted side effects. Thermal effects are more controllable and efficient.

The basic thermo-optical switching element has an input waveguide and two possible output waveguides. In between, there are two other short, internal waveguides that first split the input light and then couple the two internal waveguides together again. Splitting and coupling a single light wave is easy to do, and not much power is lost if the pathway is short enough. Ordinarily, this recombining of the light would be unexceptional. The re combined light would proceed down the "default" output waveguide. But the thermo-optical effect makes it possible to use this coupling of the light as a switching element.

The temperature is changed slightly on one of the internal waveguides, usually by adding a small amount of current and so heating a small electrical resistor slightly. The slight rise in temperature in one of the internal waveguides makes this path grow slightly longer than the second, unheated internal pathway. This lengthening changes the phase of the light passing through the internal waveguides. When the light is brought back together again, the difference in phase "bumps" the light wave onto the second output waveguide. There is nothing odd or strange about this self-interaction of the light, at least when quantum physics is concerned. This is just an application of light wave interference or interaction in practice.

The general principle of thermo-optical switching elements is shown in Figure 7-8. In the figure, an input light wave is split onto two separate waveguides. If no heat is applied to the lower branch in the figure, the coupler will output the waveform onto the waveguide labeled Output #1 in the figure. The figure shows the heating element activated, and a slightly different phase induced into the waveform on the lower branch. So the output light wave does not take the default output waveguide but ends up on the waveguide labeled Output #2 instead.

Only one of the internal waveguides needs to be heated, saving energy and making the whole element simpler in design. Sometimes the default output waveguide pathway is the actual desired output path, and this saves energy and effort as well. Waveguides, since they share many of the characteristics of integrated circuits, can be made in large

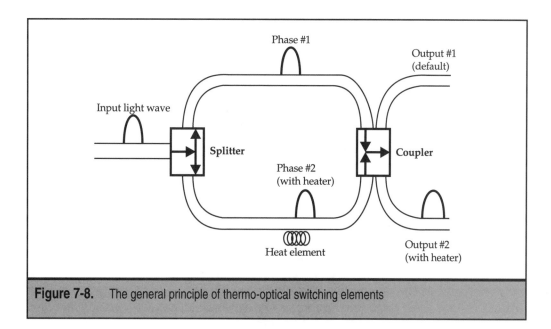

Figure 7-8.   The general principle of thermo-optical switching elements

batches and are small and inexpensive. The common wafer substrate can also serve as a platform for integrated lasers and detectors. Many components can be integrated into the basic chipset design, building up more elaborate switching engines for optical networks.

The Bubble Switch

Most people are familiar with the common inkjet printers that are often included in the price of a new PC. Inkjet printing technology is simple and inexpensive, and employs bubbles in several ways. For example, the inkjet head has tiny holes to blow ink onto a sheet of paper. Letters are formed by selectively blowing bubbles of ink out of the holes of the inkjet printhead. Surprisingly, the same basic bubble jet technology used in these printers can be used to build an optical switch.

A bubble optical switch employs waveguides on a wafer substrate, the same fundamental packaging used in all optical switches today. But in the bubble optical switch, these waveguides are etched to form a mesh. At the junction of each waveguide intersection, there is a tiny hole. The hole can be the source of a tiny bubble of fluid, not ink. This bubble optical switch fluid has the same refractive index properties as the waveguides. So if no bubble is present at the junction, the light proceeds down the default waveguide path. If a bubble of fluid is present at the junction, the light is shifted onto the second output waveguide. The holes at the junctions connect to an inkjet printer head below the surface of the wafer substrate. This modified printhead can both excrete and absorb the fluid bubble at the junction. It is the coordinated presence and absence of bubbles at the junctions of the waveguide intersections that make the bubble optical switch work.

The general principle of the bubble optical switch is shown in Figure 7-9. The actual structure of the bubble switch is a little more complicated than shown in the figure. The bubbles typically are in *channels* connecting *fill holes*. A channel might have several fill holes and span several junctions in the switch. Neither the channels nor the fill holes are shown in the figure.

In Figure 7-9, the upper portion of the figure shows the path that a light wave will take through the bubble switch when no bubble is present at any junction. The holes are shown in the upper portion of the figure, but the printhead has not inserted a bubble into any of the junctions. However, the lower portion of the figure shows what happened to the path of the light wave when a bubble is injected and present at one of the waveguide junctions. The bubble acts as a mirror that reflects the light wave to another branch of the

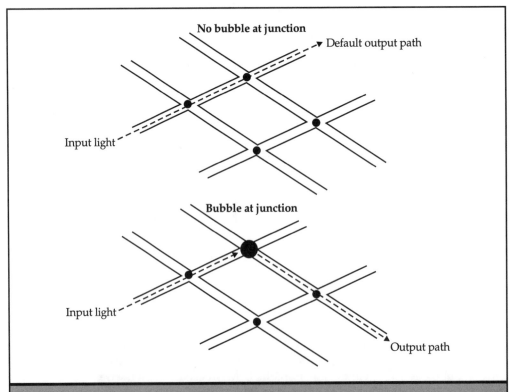

Figure 7-9.   The general principle of the bubble optical switch

switching element. The printhead beneath the wafer substrate can remove the bubble very rapidly as well. Quite complex patterns of output can be built up in this fashion.

Oddly, the complexity of the bubble switch is in the printhead, not the wafer. The bubble switch is small, fast, and efficient. However, bubble switches might not scale well. This is because of the dependence on a printhead hole for each junction. A bubble switch with 10 inputs and 10 output paths has 100 junctions, each requiring a printhead hole to insert and remove the bubble. But a bubble switch with 1,000 inputs and 1,000 output paths, only really 10 times larger than a 100 port switch, requires 1 million printhead holes. There is a hole at each junction, or 1,000 times more than in a 100 port switch.

Liquid Crystal Switches

Even more people are familiar with the liquid crystal displays found in digital watches and some forms of computer output devices than are familiar with inkjet printers. Liquid crystals can also be used as a basis for optical switches as well. This is because liquid crystals are molecular chains with a long, thin, almost one-dimensional shape. When an external electrical field is applied to the liquid crystal, the molecules line up and so can become opaque. The same optical properties can be used to build an optical switch element.

The thermo-optical effect relies on a change in the phase of optical signals with the application of heat to make a switching element. Liquid crystal switches rely on a change in the polarization of optical signals with the application of electrical voltage to make a switching element. Because the liquid crystal molecules are so long and thin, they will let only light of a particular orientation pass through the liquid crystal.

Liquid crystal switching elements are built with two active components: the *cell* and the *displacer*. The liquid crystal cell is formed by placing the liquid crystals between two plates of glass. The glass is coated with an oxide material that conducts electricity and is also transparent. The glass plates form the electrodes of the cell portion of the switching element. The main function of the cell is to reorient the polarized light entering the cell as required. The displacer is a composite crystal that directs the polarized light leaving the cell. Light polarized in one direction is directed to one output waveguide by the displacer, while light polarized at a 90-degree angle is directed to a second output waveguide.

The general structure of the liquid crystal switching element is shown in Figure 7-10. As with other optical switches, the liquid crystal switch has a default output path when no voltage is applied to the cell, and a second output path when voltage is applied.

The upper portion of the figure shows the path of a light wave when no voltage is applied to the cell. Input light of arbitrary polarization lines up with the default polarization orientation of the liquid crystals inside the cell. The displacer also has a default orientation, and the light emerges as shown in the figure. The lower portion of the figure shows the path of a light wave when voltage is applied to the cell. Note that the liquid crystals in the cell and those in the displacer both change their orientation under the influence of the voltage. The polarized light now takes the second output path, as shown in the lower portion of the figure.

Liquid crystal optical switches have been around for a while. In the past, these types of switches have been quite slow and have suffered from a lot of internal optical signal

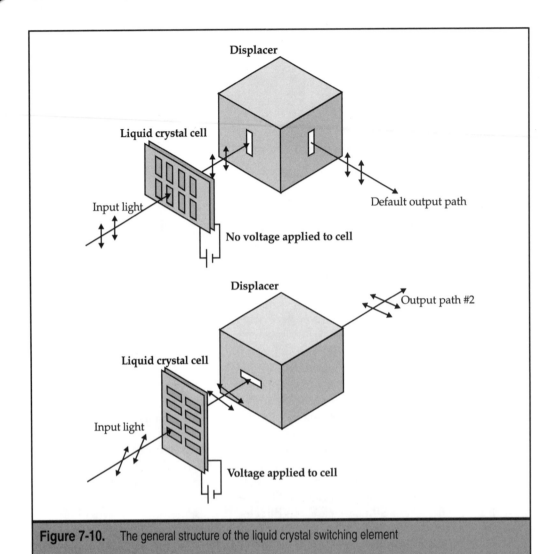

Figure 7-10. The general structure of the liquid crystal switching element

loss. A lot of this had to do with the polarization of the light. If alignment of the components was off only a small bit, the losses were immense. Recent research has found ways to minimize these effects. Scaling still remains an issue, however, for the same reasons as in the bubble optical switches.

NonLinear Kerr Effect Switches

The preceding chapter mentioned the optical Kerr effect in describing optical network impairments. This effect occurs when the intensity of the light entering a fiber optic cable or waveguide varies. The intensity variation causes a variation in index of refraction of the fiber cable or waveguide. The can be used to change the path that the light would ordinarily take. The effect is also nonlinear in nature, meaning input power doubling does not double the output of a nonlinear system. Nonlinear effects change the nature of the material linking input to output and so the simple linear relationship between input and output.

All optical systems exhibit nonlinear properties. In fact, this nonlinear behavior strictly limits the amount of power that can be used in an optical system. But nonlinear fiber optic effects can also be used to build very rapid optical switching elements. Such elements can operate in femtoseconds, or millionths of a nanosecond.

One switch architecture that employs nonlinear effects has switching elements that consist of a *nonlinear optical loop mirrors.* An optical loop mirror is formed when two light beams interact in an *interferometer* arrangement. An interferometer is formed whenever two light waves are made to interact intentionally in a certain fashion.

In the nonlinear optical switch architecture, the incoming light wave is divided in two by the splitter portion of a splitter/combiner device. Each segment of light will then travel through the loop mirror in opposite directions. The light then recombines with itself and exits the loop. There are two ways that the light can exit the loop. The light can exit the loop in the same direction it entered, or the light can exit the loop in the opposite direction from which it entered the loop.

The choice of exit mode is determined by the presence or absence of a pulse of control light which can enter on a control waveguide on one side of the mirror loop. If the control pulse is absent, the light wave exits in the same direction it entered the loop. If the control pulse is present, then the light wave exits in the opposite direction. The control pulse changes the index of refraction in one half of the loop and so also changes the phase of the recombined light wave. The change in phase is just enough to cause the recombined light wave to take a different exit path.

The main idea behind nonlinear optical switching is shown in Figure 7-11. Generally, nonlinear effects occur only at relatively high power levels in optical systems. However, the control pulse on a nonlinear switch does not need to be very high powered because an optical amplifier can be used internally to boost the power and lower the threshold.

In Figure 7-11, the upper portion shows how output path determination is made by a nonlinear optical switch. There are five steps to the process.

1. An input light wave enters the nonlinear switching element.

2. The splitter/combiner splits the wave into two identical (in terms of wavelength, polarization, phase, etc.) waveforms and sends them in opposite directions around the mirror loop.

3. The control pulse, shown as a dotted waveform, enters one branch of the loop.

4. The control pulse alters the refractive index of that branch of the loop in a nonlinear fashion, changing the phase of the waveform in that branch. The original and control waveforms are shown overlapped to emphasize this.

5. The output wave enters the splitter/combiner again and is "merged" in the interferometer onto the output waveguide. Again the waveforms are superimposed for clarity.

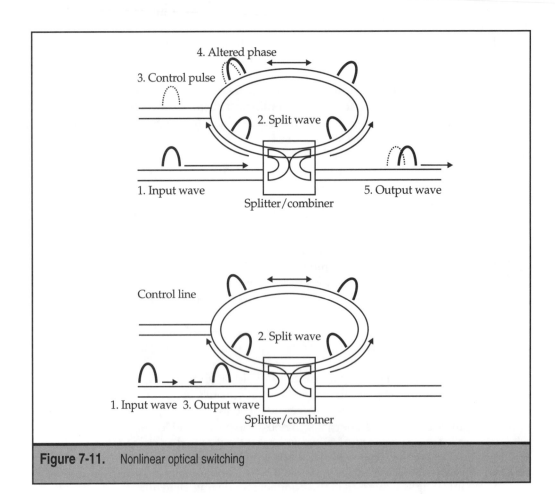

Figure 7-11.   Nonlinear optical switching

The lower portion of the figure shows the "default" path of an input light wave when no control pulse is present. The path of the light wave is simpler, but it exits the loop in the same direction that it entered.

1. An input light wave enters the nonlinear switching element.

2. The splitter/combiner splits the wave into two identical (in terms of wavelength, polarization, phase, etc.) waveforms and sends them in opposite directions around the mirror loop.

3. The output waveform emerges the way it entered from the splitter/combiner. The bidirectional path is very short, so SBS is not a problem here. Naturally, overall switch element configuration is more complex in a nonlinear switch.

Nonlinear optical switches are currently in the research phase, but they show promise. This section can only hint at even newer optical switching element technologies such as holograms or acoustic switching. To judge from the recent past, evolution of new optical switching techniques will be rapid and alternatives will be plentiful. The trick will be to find the best architecture among the many choices.

OPTICAL ROUTERS

It is all well and good to explore a world where service providers sell wavelengths and optical switches carrying GBE frames with larger and more numerous IP packets than ever before. But all of these huge IP packets must come and go from an IP router. An optical switch is not an IP router. How can IP routers possibly keep up with as many as 2,000 maximum-sized IP packets per second arriving and departing on a single GBE DWDM *channel?* Even 10 channels used for this purpose would require the router to process and route 20,000 maximum-sized IP packets per second per link!

Fortunately, today terabit (1,000 Gb) IP routers are available. These new routers are sometimes called "carrier-class" routers to distinguish between these fiber backbone nodes and the simple IP routers used as CPE for the enterprise. These routers are claimed to support link speeds as high as 20 to 60 Gbps, and extreme claims as high as 150 Gbps have been made. This is on a per-port basis, but all of these units support multiple links, of course. Most terabit router boxes are deployed in "clusters" and can be made to look like one unit operating at up to 19 terabits per second in aggregate. The distance between units in a cluster can vary from "right here in this room" to 15 km (about 10 miles) or even more away, and so units can be spaced much like ring nodes.

Ironically, most terabit routers still support SONET/SDH rates, from OC-3/STM-1 to OC-192/STM-16, and are not particularly aimed at GBE or DWDM. However, it is still impressive to see a router handle 15 or 16 OC-192/STM-16 ports running at 10 Gbps each.

Eventually, the terabit router, the GBE switch, DWDM channels, and OADMs will come together to create a new form of optical networking. Whether this is called photonic networking or IP over λ or something else is not important. In spite of the interface speeds and formats supported, there is little room for SONET/SDH in many of the new equations. SONET/SDH might remain useful as a network management channel for the new optical networks, but the IP world typically uses other OAM&P techniques such as SNMP and Java.

Of course, none of this may happen if the cost of optical networking components such as GBE direct-to-fiber interfaces remains higher than paying for the equivalent SONET/SDH capacity. Oddly, this is not the case. Newer line cards (boards that support more than one input or output link) with fiber optic interfaces such as for GBE cost significantly *less* than their SONET/SDH counterparts, and they offer higher speeds.

Consider a customer or service provider given a choice between configuring a router/switch with an OC-12 port (622 Mbps) or a GBE port (1,000 Mbps). Both will work, since either one can be carried on a separate wavelength over a DWDM system. In other words, the unit purchased and offered for sale is the λ, not the speed. In that case, the choice will more than likely boil down to the cost per bit and the cost per port (usually several ports will be desired, of course).

An OC-12 interface for a SONET ADM can cost up to $12,000. A GBE interface can go for about $2,500. And once the overhead is stripped off the OC-12, the throughput available is about 600 Mbps, a little more than half what is available on GBE.

Why the difference? Because to comply with specifications, the OC-12 board must be able to handle all SONET OAM&P, timing distribution, VTs, and so on, just in case anyone would want to put channelized voice onto the SONET link. GBE suffers from no such requirement to support channelized voice. Want to do voice over GBE? Use voice over IP (VoIP) and forget about the need to cross-connect!

Strangely, this GBE versus SONET price differential closely mimics the early price differential between 10 Mbps Ethernet and 4 Mbps Token Ring LANs. Token Ring was and is arguably the better technology: there were no collisions, delays were predictable, and so on. But Ethernet won the war. Ethernet boards cost about half (or less) what Token Ring boards cost, and the speed difference more than made up for the limitations of Ethernet. Why did not Token Ring vendors just lower prices to be competitive? They really could not, because the extra circuitry needed to perform token distribution, beaconing, and so on was required to be compliant with the Token Ring specification. Wherever chip prices fell for Token Ring, the same economies applied to the much simpler Ethernet chips, so the differential held until Ethernet dominated the LAN marketplace. This same phenomenon might repeat itself with GBE and SONET/SDH.

This might be a good place to examine the issue of the relationship between OSI-RM frames like GBE and PPP frames and transmission frames like SONET/SDH and DS-1 frames. Essentially, SONET/SDH frames exist because the type of traffic that both SONET and SDH are optimized for, namely 64 Kbps DS-0 voice, has no frame structure in and of itself. And the OSI-RM protocol concepts of the frame and packet apply to *data* networks, not voice networks. So in order to get all of the benefits of frames in the voice world, such as error control and the like, it was necessary to invent a voice frame structure.

This is hardly a SONET/SDH innovation. The T-1 and E-1 frames are just voice application frame structures to allow for voice access and cross-connections. Naturally, the issue is now whether such a frame structure for voice is really needed if VoIP becomes the rule, as appears to be the case.

The whole point is that since GBE has a perfectly adequate frame structure for VoIP and even video over IP and anything else over IP, there is no longer any need to provide SONET/SDH framing on a fiber link. A more detailed examination of the role of SONET/SDH in the DWDM world of optical networking world will be given in a later chapter.

The rising sun of DWDM, optical networking, GBE, and the like will probably be mirrored by the sunset of SONET/SDH. All technologies give way to newer methods eventually. This section looks at the possible deployment of the "missing piece" of optical networking, the optical switch/IP router. But even in the photonic networking world, SONET/SDH might still have a place as a legacy, or even consumer, networking method.

Optical switches in the optical networking world of DWDM are really sophisticated optical cross-connects. The shuffling of wavelengths and fiber links that occur within optical cross-connects are done by configuration, not packet by packet or cell by cell on the basis of header content as true routing and switching does. But there are architectures that actually route or switch packets through the device without ever changing the input photons to electricity, at least the payloads. One such architecture for a terabit router is shown in Figure 7-12.

The trick of photonic switching and routing is to literally do it with mirrors. The figure shows how. Note that the routed traffic enters and emerges as a stream of light, and yet individual packets present in the stream can be routed to different output ports. This fulfills the definition of photonic routing, but that is not the end of the story. True photonic switching and routing can only occur when a true *optical computer* and optical circuitry control the entire process. This development is still a ways off. In the meantime, the best that can be done is to control the whole process with electrical circuits and table lookups.

Here is how the optical router works:

1. Light of a given wavelength enters an input port. The input fiber could have many wavelengths. If so, each is first isolated for routing purposes. When a packet header is detected in the input stream of photons, the optical signal is split through a simple process so that there are two copies of the header portion of the optical packet. The header can be detected by several means, including fixed-length data units (e.g., cells) or a special optical pattern used as a delimiter.

2. The light representing the packet header is converted to electricity by using a standard optical receiver. Only the header need be converted in this fashion. The process is quite efficient since the header is presumably much smaller than the packet payload. This is sometimes called the electrical "wrapper" for the optical packet.

3. Meanwhile, the optical signal representing the packet is fed into a simple delay loop. Routing takes time, and this time is gained by delaying the optical packet until the path through the device is set up.

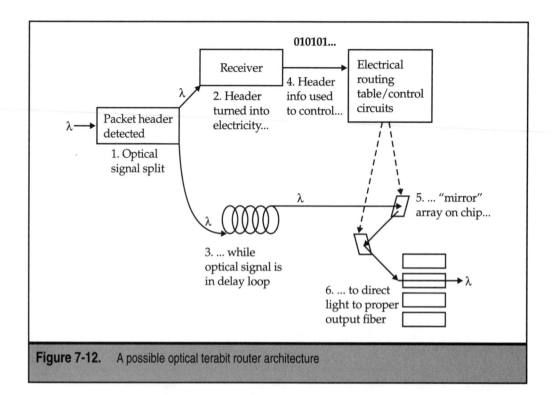

Figure 7-12. A possible optical terabit router architecture

4. The electrical header information, now a string of 0's and 1's, is used to perform a normal routing table lookup. Routing table maintenance can be performed by existing routing protocols, either electrically or optically, since most routing information flows between adjacent systems and is not routed itself.

5. Once the proper output port has been determined for the optical packet, an array of mirrors etched onto a silicon substrate are adjusted—electrically—so that the optical packet emerging from the delay loop is "aimed" at the proper output fiber.

6. Finally, the optical packet is directed to the proper output. Naturally, a stream of packets is conducted through the optical router very quickly, and capacities are in the terabit ranges.

A more realistic architecture for an optical router will be investigated in a later chapter. This is just an architectural possibility.

OPTICAL RINGS

Once the optical terabit router is added to the optical networking mix, many things change. Transport networks now carry IP by wavelength and "legacy" traffic such as SONET/SDH by normal time division multiplexing (TDM). The basic access node is the DWDM/OADM device. The services provisioned and sold are either "IP over λ," good for almost everything from VoIP to video to data, or "legacy" voice, frame relay, and ATM services, good for backward compatibility. GBE is easily added to the IP side of the ledger as a framed transport.

Once placed onto the DWDM fiber(s), the traffic is passed through an optical cross-connect node and onto a fiber ring (or mesh). This photonic core ring differs from SONET/SDH rings in the sense that it is the optical cross-connects that are the ring nodes, not the DWDM/OADMs themselves. This makes sense because the basic transport unit of the photonic core is the wavelength, not the frames that ride the wavelength. This type of fiber network of the future is shown in Figure 7-13.

It is highly debatable how near or how far in the future this architecture might be. Some see it around every corner, some see such a bandwidth-rich infrastructure as total overkill except in a handful of countries around the world (the United States is one of the handful). One skeptic noted that plans to place a gigabit fiber link to the Fiji Islands was

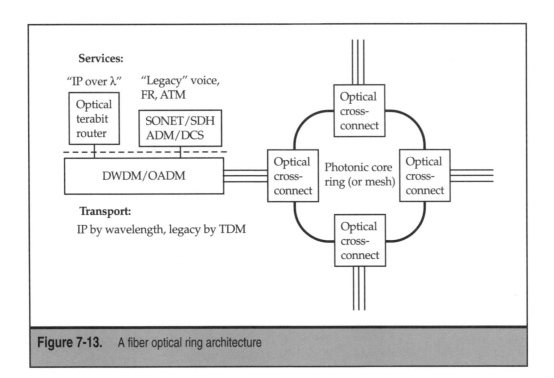

Figure 7-13.   A fiber optical ring architecture

like running a gigibit link into a 30,000 person community in the United States—the economies are about the same size. Even with Internet demand what it is, it would be hard to cost-justify the link in the United States, let alone to the Fiji Islands.

So SONET/SDH will provide enough bandwidth for present needs for some time to come. As for the future, many technologies make their way from service provider backbone to access network to consumer product. Modems have certainly made the trip, and DS-1 might follow (most plans for new residential networks feature speeds of 1.5 Mbps or better at least downstream). Why not SONET/SDH?

Stripping off the VT and timing support in SONET/SDH could reduce the price of an OC-12/STM-4 interface to about $3,300 for 622 Mbps. Perhaps the backbone SONET/SDH ADM once used as the ring node, and now replaced by the OADM, could make its way to the end of the access fiber link serving someone's house. This might be the best long-term hope for SONET/SDH, but there will be competition at every step along the way.

WAVELENGTHS AND SONET/SDH

So is there a place for SONET/SDH in the optical networking world? If services are provisioned and sold by wavelength, who needs SONET/SDH? For now, the answers are "yes" and "anyone concerned with standard telecommunications network management" given the role that "legacy" SONET/SDH equipment and services can play in the optical networking world. But this might not always be the case. There are other considerations that need to be examined in the next part of this book.

There are many transport frame structures besides SONET/SDH and GBE: FDDI, Ethernet, and Fibre Channel are but a few. But given the domination of the Internet and the Web in the world of networking today, most frames exist for one purpose only: to carry IP packets between client (e.g., Web browser) and server (e.g., Web page). The IP packets themselves can carry voice, video, or data.

It is time to consider the impact of optical networking on traditional WAN architectures such as SONET/SDH, ATM, and IP itself. This is the subject of the three chapters in the next part of this book.

PART III

Wide-Area Optical Networking

arts 1 and 2 introduced fiber optic technology in general and then used these principles to highlight the development of DWDM systems and optical network elements. The emphasis in these two sections was on the fundamentals of the underlying technologies that enable the deployment of optical networks. Both advantages and limitations were discussed.

Part 3 takes the principles outlined in the first two parts of this book and applies them to existing wide area networks. The impact of DWDM and optical networking on three major WAN architectures is explored as well.

Change is often abrupt and jarring, and the changes that optical networks bring to SONET/SDH, ATM, and IP are no exception. It might seem odd at first to consider IP as a WAN protocol in the same breath as SONET/SDH and ATM, since many IP packets ride over SONET/SDH and ATM links in the first place. IP exists at a higher level than either SONET/SDH or ATM. However, the impact of the Internet and the IP protocol on WAN networking, coupled with the rise of the IP router as the network node of choice for a lot of WANs today, warrant consideration of the impact of DWDM and optical networks on IP as a whole.

Chapter 8 examines the relationship of single-wavelength SONET/SDH and optical networks based on DWDM in more detail. The emphasis here is not on the basics of SONET/SDH, but on how and why—and even if—SONET/SDH has a place in a multi-wavelength world. The role of SONET/SDH might be reduced, but SONET/SDH will be present in optical networking, especially for some services, for some time to come. The values that SONET/SDH adds to DWDM is first explored, then the role of SONET/SDH as an optical networking management channel or protocol is investigated in some detail. Finally, this chapter closes with a look at how SONET/SDH remains essential for some forms of voice transport.

Chapter 9 explores the relationship between ATM networks and optical networks. Once a promising new technology on its own, ATM has fallen on hard times as faster hardware, increased bandwidth, and the prevalence of IP-based networking has transformed the world. ATM is briefly outlined, and then a simple ATM network architecture is given. Finally, the position of this ATM architecture in the DWDM world is assessed. If optical network nodes, such as those outlined in Chapter 8, take over many of the functions of current ATM switches, then the future of ATM is bleak indeed.

Chapter 10 investigates the relationship between optical networking and IP routing. As the protocol of the Internet and Web, the position of IP is important even though IP is a higher-layer protocol than either SONET/SDH or ATM. IP is introduced in brief, and then the current structure of the Internet and Web using IP routers is detailed. Finally, an "optical Internet" is introduced, and the role of IP examined anew. It turns out that even if optical network nodes take over many of the functions of current IP routers, this will only further establish IP as the protocol of choice.

CHAPTER 8

SONET/SDH Migration

Chapter 5 of this book introduced the basics of SONET/SDH as a transport protocol using fiber optic links. This chapter builds on those concepts and considers how SONET/SDH might migrate to a world of DWDM and full optical networking. In order for SONET/SDH to occupy more than a niche in the new optical networking arena, and continue to be of value in this new world, SONET/SDH must add value to optical networking. At present, there is considerable value that SONET/SDH can add to an optical network. However, there are ways of more or less replacing the functions that SONET/SDH currently provides in optical networks. So the future of SONET/SDH might provide more and more of a transitional role in optical networks. This chapter explores the reasons behind this reduced role for SONET/SDH.

ADDING VALUE TO DWDM

Chapter 6 in this book ended by stating that it is possible to use SONET/SDH as a *management channel* in a DWDM system. Current international standards allow for such a management channel to ride a DWDM ring at 1,310 nm, the standard SONET/SDH wavelength. Operational channels on the DWDM system would still run in the 1,550 nm neighborhood, as might be expected. There is of course only one 1,310 nm management channel. Now, performance of fibers can be optimized for 1,310 nm operation, or 1,550 nm operation, but seldom for both. So future fibers might justifiably favor multiple-wavelength DWDM performance over single-wavelength SONET/SDH performance. In order for SONET/SDH to add enough value to an optical network using DWDM, there must be something that SONET/SDH can do better than DWDM can do on its own. Right now, this appears to be true. That is, there are several things that SONET/SDH handles better than DWDM. But this might not be the case for long.

Right now, there is a good reason that SONET/SDH is often used as a management channel in DWDM systems. Optical networks at the DWDM level are, by definition, defined only as wavelengths of light. There are no bits to speak of, and since the whole goal of optical networking is to make light remain light, the less conversion of modulated wavelengths into electrical bits, the better. As pointed out in the previous chapter, optical network nodes such as optical cross-connects and even transponders can be made to operate on wavelengths without any need to convert anything to electrical bits.

But SONET/SDH has extensive overhead dedicated to network management. The whole scheme of the SONET/SDH network management architecture is defined in a series of international standards that fall under the ITU-T's Telecommunications Network Management (TMN) series of recommendations. The details of TMN are beyond the scope of this book, but the main point is that in order to manage an optical network at the WAN level according to TMN standards, it is easiest to retain SONET/SDH as a network management channel.

The extent and versatility of this SONET/SDH overhead should never be underestimated. SONET/SDH has been around since 1988, and many transmission systems have had stable hardware and software implementations to handle SONET/SDH overhead for operations, administration, maintenance, and provisioning (OAM&P) for many years. All of the basic OAM&P functions are important for overall network management, which includes things like failure recovery, performance monitoring, and error compensation.

Without going too much into detail, it will be enlightening to examine the SONET/SDH overhead bits and bytes in a little more depth. After all, if optical networks are to perform these network management tasks in a standard fashion *without* SONET/SDH, then there must be some means of putting these basic network management tasks into an optical network. Perhaps it is best not to have to reinvent the wheel.

SONET/SDH OVERHEAD

This section will examine the SONET overhead bits and bytes first and then discuss their implementation in SDH. This is a good approach because SONET had a considerable head start over SDH when it came to deployment and standardization. Also, SDH frame structures are more complex than SONET, so it is easier to explain basics in SONET and then apply them to SDH than the other way around.

SONET Overhead

The overhead used in SONET for operations, administration, maintenance, and provisioning (OAM&P) is a distinctive feature of SONET. Recall that the overall architecture of SONET is a three-layered architecture of section, line, and path from bottom to top. This section will outline the functions of the Section, Line, and Path Overhead bytes in SONET. Many of these functions overlap, since there are separate Section, Line, and Path components on a SONET link that all need overhead benefits like error monitoring and failure protection.

Mention is made during this discussion of *composite* SONET signals. When several SONET data streams are multiplexing into one higher-rate data stream (such as from three STS-1s to a single STS-3), the resulting SONET frame structure is called a *composite.* Every frame except a basic STS-1 frame is a composite frame of one form or another. Composites will still retain *all* of the overhead bytes from each signal source. But obviously an STS-3 frame that has nine columns of Transport Overhead instead of just the three from each of the STS-1s still forms a unit and not just three frames traveling together. So many of the "repeated" overhead bytes are essentially ignored ("undefined" in SONET talk) in composite frames. The full set of 27 Transport Overhead bytes is retained only in the first STS-1 of any composite. In the other STS-1s, all of the Transport Overhead bytes may be present, but many of them are neither examined nor processed at all.

Section Overhead

The Section layer contains overhead information used by all SONET equipment along a network path, including signal regenerators. The Section Overhead bytes are shown in Figure 8-1.

A1	A2	J0/Z0 (STS-ID)
Framing	Framing	Trace/growth
B1/undefined	E1/undefined	F1/undefined
BIP-8	Orderwire	User
D1/undefined	D2/undefined	D3/undefined
Data Com	Data Com	Data Com

Figure 8-1. SONET Section Overhead

The Section Overhead is contained in the top three bytes of the first three columns in the basic STS-1 frame structure. The Section Overhead's nine bytes function as follows:

Framing Bytes (A1, A2)   This two-byte code (1111-0110-0010-1000, or F628 in hexadecimal notation) is used for frame alignment. These bytes uniquely identify the start of each STS-1 frame. When multiple STS-Ns are sent in a higher rate STS-N+, framing bytes must still appear in each and every STS-1 of the composite signal.

Section Trace (J0)   This byte is used to trace the origin of an STS-1 frame as it travels across the SONET network. It is obviously a plus to be able to trace the origin of any particular STS-1 frame to a particular piece of equipment from any point on the network. The source sends the SONET network address of the originating device in this field one byte at a time, over and over. In the case of multiple STS-1s in an STS-N, the J0 byte is defined only in the first STS-1 (since all of the frames must come from the same device). In other STS-1s in an STS-N, the use of this byte is just for Growth (Z0).

BIP-8 (B1)   A single interleaved parity byte is used to provide STS-N error monitoring. BIP is an abbreviation for Bit Interleaved Parity. It performs a routine even parity check on the previous STS-1 frame, after scrambling. The parity is then inserted in the BIP-8 field of the current frame before scrambling. SONET relies on even parity for its error checking calculations. During parity checking, the first bit of the BIP-8 field is set so that the total number of ones in the first positions of all octets in the previous scrambled STS-1 frame is always an even number. The second bit of the BIP-8 field is used in exactly the same way, except it performs its check on the second bits of each octet, and so on. This octet is only defined in the first STS-1 frame of a composite STS-N signal.

Orderwire (E1)   A leftover from the days of copper wire carrier systems, this channel was historically available to craft personnel for maintenance communications. In SONET, the channel is a 64 Kbps voice path used for communication between remote terminals and regenerators. A telephone can be "plugged in" at the repeater location into a more or less standard voice jack, allowing craft personnel instant voice access on the fiber span. This octet is only defined in the first STS-1 of an STS-N frame.

User (F1)   The Section User Channel is reserved for use by the network services provider for its own application management activities. Its design is unspecified because it can be tailored by the network provider. F1 is only defined in the first STS-1 of an STS-N frame. Many vendors of SONET equipment use the F1 byte, which runs at 64 Kbps, for equipment-specific OAM&P information.

Data Communications Channel (DCC) (D1-D3)   These three bytes are used as a 192 Kbps data channel for operations functions such as OAM&P. These capabilities constitute a (if not *the only*) huge advantage offered by SONET technology over DWDM and optical networking today. The Data Communications Channels ship such things as alarms, administration data, signal control information, and maintenance messages. D1 through D3 are only defined in the first STS-1 of an STS-N frame. The TMN standard defines the use and protocols employed on these bytes.

Line Overhead

Line Overhead is contained in the bottom six bytes of the first three columns of the SONET STS-1 frame. This overhead is processed by all pieces of SONET network equipment, except for the regenerators. The Line Overhead bytes are shown in Figure 8-2.

One of the most important functions of the Line Overhead is to allow for the shifting of the Synchronous Payload Envelope (SPE) inside the SONET frame. This shifting is usually caused by timing differences between different parts of the overall link, which might link different service providers, each with its own voice clocking mechanisms. The H1 and H2 pointers always allow a receiver to find the start of the SPE, and the H3 byte allows the receiver to detect when the starting SPE location shifts to the left or right inside the SONET frame.

Pointer (H1, H2)   The H1 and H2 Pointer is composed of two bytes used to indicate the offset between the pointer bytes themselves and the beginning of the STS Synchronous Payload Envelope (SPE). It allows dynamic alignment of the SPE within the allowable capacity of the envelope itself. In other words, since the SPE can start anywhere within the Information Payload of a SONET frame (that means anywhere but the overhead fields), it would be nice to give the receiver a hint as to exactly where this might be.

Pointer Action (H3)   The Pointer Action byte is allocated to compensate for the SPE timing variations. In the course of otherwise normal operation, it may happen that the arriving SPE data rate will exceed the frame capacity. This means that within a 125-microsecond period, more than 783 bytes may be ready to be sent out in an SPE. If this excess is less

H1 Pointer	H2 Pointer	H3 Pointer action
B2/undefined BIP-8	K1/undefined APS	F1/undefined APS
D4/undefined Data Com	D5/undefined Data Com	D6/undefined Data Com
D7/undefined Data Com	D8/undefined Data Com	D9/undefined Data Com
D10/undefined Data Com	D11/undefined Data Com	D12/undefined Data Com
S1/Z1 Sync status/ growth	M0 or M1/Z2 REI-l/ growth	E2/undefined Orderwire

Figure 8-2.   SONET Line Overhead

than 8 bits, the extra bits are just buffered sequentially and sent out as the first bits of the next frame. But when a full byte (8 bits) has accumulated in the buffer, the pointer action byte is used to carry the "extra" 784th byte. This is called a *negative timing justification* and is governed by strict rules of SONET equipment operation. Naturally, a companion *positive timing justification* is used when there are not 783 full bytes in the CPE buffer to fill the SPE exactly.

The H3 byte must be provided in all STS-1 signals within an STS-N signal. The value contained in this byte, when it is not used to carry the SPE data from a positive justification, is not defined and ignored. The H1 and H2 pointer bytes tell the receiver when the H3 byte is used for useful information. This byte is necessary due to the variations in timing across different service provider's networks or when CPE clocking feeds a SONET link.

BIP-8 (B2)   Similar to the B1 byte, this bit interleaved parity byte is used for STS-1 error monitoring. It calculates its parity check from all bits of the Line Overhead and the STS-1 frame capacity of the previous frame. Note that the Section Overhead is specifically excluded, reflecting the hierarchical nature of the SONET overhead bytes. This field must be present and used in all STS-1s of an STS-N signal.

APS (K1, K2)   These two bytes communicate Automatic Protection Switching commands and error conditions between pieces of line terminating equipment. They are specifically used for link recovery following a network failure. These two bytes are only used in the first STS-1 of an STS-N signal. The full purpose and use of the K1 and K2 bytes are not completely standardized, but they are used in many pieces of SONET equipment to switch lines on SONET rings.

The K2 byte is also used to detect other types of line level alarms on SONET links. APS use is another key area where DWDM has no standard way to implement these functions which are a key part of SONET.

Data Communications Channel (D4–D12)   These bytes represent a 576 Kbps message-based channel used for shipping OAM&P messages between SONET line-level network equipment. Typical messages might be for maintenance, administration, or alarms, and can be either internally generated or from an outside source. In some cases they may be manufacturer-specific. The Data Communications Channel bytes are only defined in the first STS-1 of an STS-N signal. Another advantage over DWDM, of course.

Synchronization Status (S1)   This byte is used to carry the *synchronization status* of the SONET device. The use of this byte is defined for the first STS-1 in an STS-N. Only bits 5–8 are currently defined, leaving bits 1–4 for future functions. The purpose is to allow SONET equipment to actually choose the best clocking source from among several potential timing sources. This helps to avoid the creation of disastrous *timing loops* within the network. In other STS-1s present in an STS-N, this byte is defined as a Growth (Z1) byte. Growth bytes are used for future functions as yet undefined by SONET standards.

STS-1 REI-L (M0) or STS-N REI-L (M1)   The use of this byte is complex, but a long discussion here is unnecessary. If the SONET link is an OC-1, which can only contain an STS-1, or an electrical STS-1, then this byte is the M0 byte. The M0 byte is used for a Line level Remote Error function (REI-L), formerly called the line Far End Block Error. This conveys the Line BIP-8 (B2) error count back to the source. Bits 5–8 of the M0 byte are used for this purpose, and bits 1–4 are undefined.

On the other hand, for an STS-N frame structure, this becomes the M1 byte. Not all STS-1s in an STS-N have the M1 byte. Only the "third" (as defined by SONET documentation) STS-1 in the STS-N will carry the M1 byte. Obviously, the value of N must be 3 or greater. The M1 byte has the same purpose and function as the M0 byte, with minor differences above the STS-48 level.

Growth (Z1, Z2)   When the first two bytes of the last row of the SONET frame are not used for the S1 and M0 or M1 functions, these overhead bytes are defined as Growth bytes. Growth bytes are reserved for future functions as yet undefined by SONET standards.

Orderwire (E2)   Similar to the E1 byte, this is a 64 Kbps voice channel. It is only defined in the first STS-1 of an STS-N signal.

Path Overhead

In addition to user data, the SPE contains Path Overhead bytes. These are processed at SONET STS-1 terminating equipment (usually CPE, but not always) because they travel as part of the payload envelope and are processed everywhere it is processed. The Synchronous Payload Envelope contains nine bytes of Path Overhead. These bytes form a "column" in the SPE portion of the SONET frame, meaning the Path Overhead bytes are always in a column. However, since the position of the SPE can "float" within the STS-1 frame Information Payload area, the position of the Path Overhead bytes can float as well.

The Path Overhead bytes are shown in Figure 8-3.

STS Path Trace (J1)   This field transmits a repeated, 64-byte, fixed-length string that enables the receiving Path Terminating Equipment to verify its connection to the device sending the SPE. This is a user-programmable field. If no message has been loaded by the user, then a string of 64 null characters is sent. This field can be something as simple as the IP address or E.164 address (telephone number) of the CPE device. Since few customers have SONET CPE, retaining T-carrier equipment that just feeds SONET links, this field is typically used by the service provider.

When used by the service provider, in most cases the local telephone company, the J1 byte contains the CLLI code (pronounced "silly code"), which stands for Common Language Location Identifier. The CLLI code is 8 to 11 bytes long and identifies a particular End Office (central office) within the national telephone system. Since the CLLI code is much less than 64 bytes long, the rest of the J1 bytes are padded with zeros.

Path BIP-8 (B3)   This field's function is analogous to that of Line and Section BIP-8 fields. It uses even parity and is derived from the parity of the previous Synchronous Payload Envelope. Note that the Line and Section Overheads are specifically excluded, again reflecting the hierarchical nature of SONET overhead.

STS Path Signal Label (C2)   The Signal Label tells network equipment what is contained in the SPE (i.e., how it is constructed). This allows for transport of multiple services simultaneously. This means that a single SONET device can actually interleave SONET frames (technically the SPEs) containing a DS-3 with frames containing ATM cells, or Fiber Distributed Data Interface (FDDI) data frames and so on.

Path Status (G1)   This field notifies the originating end of the Path performance and status of the entire duplex path. It carries two maintenance signals, the B3 error count known as the Path Remote Error Indicator (REI-P), formerly called Path FEBE, in bits 1–4,

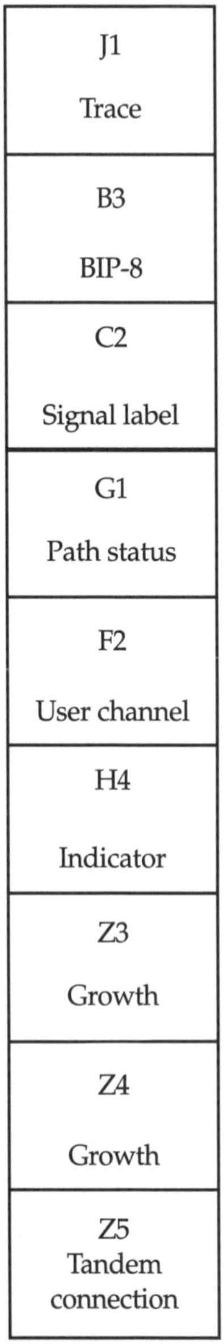

J1
Trace

B3
BIP-8

C2
Signal label

G1
Path status

F2
User channel

H4
Indicator

Z3
Growth

Z4
Growth

Z5 Tandem connection

Figure 8-3.   SONET Path Overhead

and a Path Remote Defect Indicator (RDI-P) in bits 5–7. This allows the entire Path to be monitored from either end. The remaining bit is undefined.

Path User Channel (F2)   This field, like its line/section cousin, can be used by the network provider for internal network communications. The F2 byte is also used in one odd case to carry Layer Management information generated by Distributed Queue Dual Bus (DQDB) networks.

Indicator (H4)   This field is used when a frame is organized into various *mappings*. Mappings determine the actual structure of the portion of the SPE carrying user data. One of the most common mappings is *virtual tributaries*. With virtual tributaries, it is possible for users to have a T-1 at each end of a private line, but SONET in the middle. SONET transports virtual tributaries easily as ATM cells or anything else, using different mappings. In ATM networks, the Indicator byte was used to denote cell boundaries, but this use is now officially "not recommended" (ITU talk for "totally obsolete"). This feature allowed SONET to transport such cell relay services as B-ISDN (ATM). In DQDB networks, the Indicator byte is used to carry Link Status information.

Growth (Z3–Z5)   These bytes are reserved for future use. In DQDB, however, the Z3 byte is used to carry more DQDB Layer Management information.

Tandem Connection (Z5)   ANSI has defined the Z5 byte for use as a Tandem Connection Maintenance Channel and a Path Data Communications Channel (DCC). A *tandem* is a switching office that switches between trunks and so has no CPE devices at all on the end of the links. This can be a problem with SPE handling, since there is no direct contact with the originator of the SPE. Finally, the Path DCC performs the same functions as the Line and Section DCCs. The TMN standard defines the use and protocols employed on the line, section, and path DCCs.

SDH (and OC-3c) Overhead

As has been pointed out in Chapter 5, an STS-3c frame is the equivalent of, and virtually identical to, an STM-1 frame. The apparent differences are easily explained by the more international scope and intention of SDH compared to SONET.

Nevertheless, it is appropriate here to outline the use of overhead in SDH. Optical networks with DWDM today are in many cases using SDH frames rather than purely SONET frames. So this section is a chance to see what the overhead functions are like for SDH STM-1 and SONET STS-3c (OC-3c) links. The structure of the overhead for a SONET STS-3c and SDH STM-1 frame is shown in Figure 8-4.

Just as in SONET, an SDH STM-1 frame has two parts of STM-1 transport overhead. But these are now the regenerator section overhead (RSOH), which is located at rows 1–3 and columns 1–9, and the multiplex section overhead (MSOH), which is located at rows 5–9 and columns 1–9, as shown in the figure. This seems to leave out row 4 altogether. Row 4 in the STM-1 frame consists of the AU-n pointers from column 1–9. More precisely, these are either AU-4 or AU-3 pointers. Theoretically, these pointer octets can be considered to be

9 columns — POH

A1	A1	A1	A2	A2	A2	J0	R	R		J1
B1	R	R	E1	R	R	F1	R	R		B3
D1	R	R	D2	R	R	D3	R	R		C2
H1	H1	H1	H2	H2	H2	H3	H3	H3		G1
B2	B2	B2	K1	R	R	K2	R	R		F2
D4	R	R	D5	R	R	D6	R	R		H4
D7	R	R	D8	R	R	D9	R	R		F3
D10	R	R	D11	R	R	D12	R	R		K3
S1	Z1	Z1	Z2	Z2	M1	E2	R	R		N1

(9 rows) **SONET STS-3c**

9 columns — VC-3/VC-4 POH

A1	A1	A1	A2	A2	A2	J0	*	*	Regenerator SOH	J1
B1	**	**	E1	**	R	F1	*	*		B3
D1	**	**	D2	**	R	D3	R	R		C2
AU-n Pointer										G1
B2	B2	B2	K1	R	R	K2	R	R	Multiplex SOH	F2
D4	R	R	D5	R	R	D6	R	R		H4
D7	R	R	D8	R	R	D9	R	R		F3
D10	R	R	D11	R	R	D12	R	R		K3
S1	Z1	Z1	Z2	Z2	M1	E2	*	*		N1

* = Reserved for National Use

** = Media Dependent Octet

SDH STM-1

Figure 8-4. STM-1 frame with overhead

part of the multiplex section overhead (MSOH), and there are in fact good reasons for doing so. Most importantly, the SONET pointer bytes are part of the Line Overhead (rows 4 through 9).

And just as in SONET, these pointers and the transport section overhead have fixed positions within a 125-microsecond frame. In other words, the pointer octets have a fixed phase with respect to the STM-1 frame. On the other hand, the VC-3 or VC-4 path overhead may occupy any column from columns 10 to 270 of the STM-1 frame. So the path overhead is "floating" with respect to the whole STM-1 frame. Therefore the VC-3 or VC-4 path overhead has a variable phase with respect to the STM-1 frame (actually, the A1 and A2 framing octets). The whole concept of "floating" path overhead is the same as in SONET. Instead of the Synchronous Payload Envelope (SPE) of SONET, SDH had either a VC-3 or VC-4 payload. Naturally, when the format of the payload is a VC-3, the pointers are AU-3 pointers, and when the format of the payload is a VC-4, the pointers are AU-4 pointers. A more detailed explanation of the differences is not needed here.

Since the names and functions of each overhead octet have already been detailed for SONET, they will be examined here by comparing them with those of an SDH STM-1 frame. The functions are almost identical to the overhead structure of an STS-3c SONET frame. By cross-referencing the two frames in the figure, it will be possible to observe the following similarities and differences between the SDH and SONET overhead standards, even though only STM-1 overhead octets are discussed here.

SDH and SONET Overheads Compared

The first row of the regenerator section overhead of an STS-3c and an STM-1 are virtually identical (Al, A1, A1, A2, A2, A2, J0, ...), with the exception of the last two bytes or octets (indicated by "*"). For SONET, these two bytes are not yet assigned and so are designated as "R" bytes, while for SDH they are assigned for national use. This means that each country gets to assign their meaning. In the United States, in SONET, they are "R" bytes, which is perfectly valid usage.

In SDH, the second row has three octets (second, third, and fifth columns) assigned for media-dependent usage (indicated by "**"). This means that the usage varies depending on whether the STM-1 is carried on fiber optic links (normal) or on electrical media such as coaxial cable (for short runs between equipment). The last two octets of this row are also assigned for national use. On the other hand, in SONET standards, these bytes are unassigned or are reserved for future use.

In the third row, the only difference is the two media-dependent octets in columns 2 and 3 for SDH. These two bytes are reserved in SONET. All of the other octets have the exact same functions for both SDH and SONET.

The pointer row is exactly the same in both SONET and SDH, but this is hard to see. The only difference is the application of the pointers. For SONET, an STS-3 signal can be used to carry three DS-3 signals. For this SONET application, these nine pointer bytes would be split up into three groups. Each group would have a set of H1, H2, and H3 pointers, and each set is used to locate the payload of one DS-3 signal. This arrangement is also used in SDH. That is, an STM-1 can carry three DS-3 signals also. This helps in international situations. In SONET, an STS-3c can also be used to transport a B-ISDN signal, such as stream of ATM cells, or even an FDDI 100 Mbps signal. In such cases, only one set of H1, H2 pointers is required. In the same fashion with SDH, an STM-1 can be used for the same purpose and so has the same arrangement of pointers. And since the original purpose of STM-1 is to carry an E-4 signal at 139.264 Mbps, only one set of H1, H2, H3 octets is required.

Rows 5 through 8 are identical for both SDH and SONET. But in SDH the last row has two octets assigned for national use. In SONET STS-3c, these two bytes are reserved.

As far as the path overhead is concerned, there is only one byte or octet that is different for SDH and SONET. N1 is assigned as Network Operator Byte (NOB) for SDH. This is to be used for network operators to communicate with SDH equipment. In SONET, this is the Tandem Connection Maintenance (TCM) byte.

REINVENTING THE WHEEL

As long as there is no easy way to add the essentials of TMN and OAM&P directly to optical networks, it makes sense to retain SONET/SDH as a management channel for optical systems. However, there are several vendor-specific and consortium-based proposals for adding the same types of management capabilities that SONET/SDH provides directly to optical networks.

One possible way that SONET/SDH functions in terms of network management might migrate to a DWDM optical networking environment is by "wrapping" the information carried on an optical link inside a *digital wrapper* of one form or another. The digital wrapper would carry many of the management information channels provided by SONET/SDH but would add key features such as forward error correction (FEC). The idea of adding functions to a digital wrapper for optical networking is a good one. SONET/SDH OAM&P methods are good, but optimal for the heavily mixed electro-optical environment of SONET/SDH. Features such as FEC are optimal when it is desired to keep electro-optical conversions to a minimum. With FEC, no mechanism for requesting resends from the other end of the link or end-to-end was needed. If the errors were few, the FEC provided all of the information a receiver needed to correct the bits in error and send the content on its way. Also, links with high delays, such as undersea fibers, were not performing at their best if the end users had to ask for retransmissions.

At first sight, the idea of adding a digital wrapper to an optical network seems self-defeating. But it makes perfect sense when the issue of migration from SONET/SDH to an optical networking environment is considered. It is practical and attractive to deploy optical networks. But it neither practical nor attractive to expect all of the management systems that operate the new optical networks to also migrate to something optical. For the present, network operations and management must remain in the digital and electrical realm.

So the idea of a digital wrapper around light-wave content is not as startling as it first might seem. After all, at some point network management equipment must look at digital 0's and 1's to make sure that the network is running as it should.

G.975 and FEC

Most of the current digital wrapper schemes add FEC to the optical links content. This is even done when SONET/SDH is the content of the optical link. This approach began in 1996 when the ITU-T released Recommendation G.975 titled "Forward Error Correction

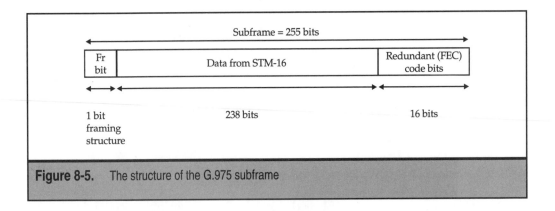

Figure 8-5. The structure of the G.975 subframe

for Submarine Systems." FEC was an important part of this wrapper, and in fact just about the only thing G.975 added to SDH management (SONET was not covered in the ITU-T recommendation). The idea was that an optical receiver could have some way to correct a few bit errors here and there, and so the end users did not have to request re-sends. Also, digital wrappers gave service providers a way to monitor link performance without relying on the SONET/SDH BIP checks. This freed up optical links from their heavy reliance on SONET/SDH.

G.975 is mainly concerned with STM-16 (similar to SONET STS-48) data running at about 2.5 Gbps. The process organizes the STM-16 data into subframes 256 bits in length. The first bit forms a framing structure across many subframes, the next 238 bits are from the STM-16 data itself, and the last 16 bits are the redundant codes that make up the FEC itself. This G.975 subframe structure is shown in Figure 8-5.

Now, the are a lot more than 238 bits in a full STM-16 frame. In fact, there are 9 rows, 4,320 columns, and therefore 8×9×4,320 = 311,040 bits in an STM-16 frame. So G.975 just assembles as many subframes as there are FEC encoders (and decoders) and makes them a unit. Each FEC encoder generates 8 subframes (one for each bit the FEC encoder acts on). If there are four FEC encoders, for example, G.975 will assemble 4×8 = 32 subframes into a unit called a *FEC frame*. It is the FEC frame that is really the digital wrapper.

Figure 8-6 shows the structure of a digital wrapper with four FEC encoders and decoders. Note that the FEC frames are sent column-wise, not row-wise, as in SONET/S5DH.

When it comes to digital wrappers, neat frame structures are not a concern. The subframes do not contain an even number of bytes (238 bits / 8 = 29.75 bytes). So FEC frames do not respect SDH frames boundaries at all (311,040 / 238 = 1306.8907 subframes per STM-16 frame). So in spite of the name, digital wrappers are optimized for optical needs, not electrical SONET/SDH needs. There is no need to respect SONET/SDH frame boundaries, only FEC frame boundaries.

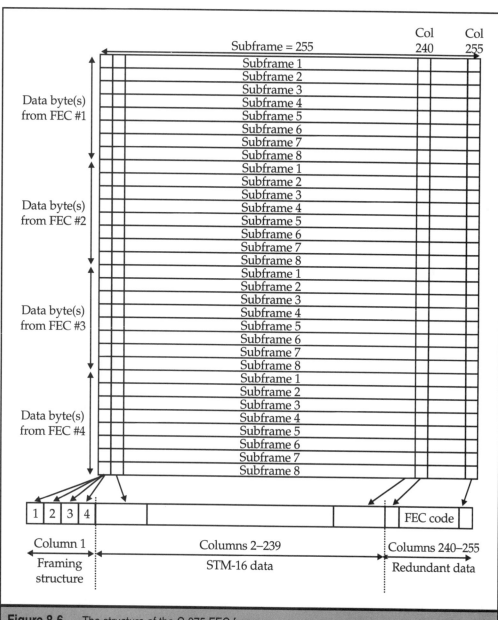

Figure 8-6.   The structure of the G.975 FEC frame

Digital Wrappers and Optical Networking

So digital wrappers have been around for a while. What many vendors have done is to extend the G.975 concept to DWDM systems. When it comes to optical networking, there are so many light waves to manage that adding a FEC to data is not just a luxury, it is a requirement.

SONET/SDH has good mechanisms for per-link OAM&P. But optical networks require per-wavelength OAM&P methods. New methods are needed to perform OAM&P at the *optical channel* level. The new methods would not require that each optical channel carry SONET/SDH-formatted information and therefore are a key step on the way to SONET/SDH migration to the world of optical networking.

New digital wrappers give each optical channel the functionality and reliability of a simple SONET/SDH link. They apply TMN concepts to allow for optical layer performance monitoring, FEC, and ring protection switching on a per-wavelength basis, and all without any reliance on the format of the input signal.

Most digital wrappers add information such as restoration signals, traffic type, destination, and so on to the basic FEC capability. Optical network nodes can then use this information in the wrapper to determine signal status, rerouting needs, if the receiver is properly configured and provisioned, and so on. All of this happens on a per-wavelength (per–optical channel) basis without regard to information format. This is exactly what is needed to free optical networks from dependence on SONET/SDH OAM&P.

The general idea behind digital wrappers for optical networks is shown in Figure 8-7. This is just a general model, since details vary widely among equipment vendors. There are proposals from vendors before ANSI and the ITU-T to standardize one or another of these specific formats.

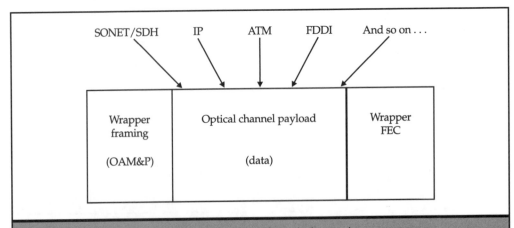

SONET/SDH   IP   ATM   FDDI   And so on . . .

| Wrapper framing (OAM&P) | Optical channel payload (data) | Wrapper FEC |

Figure 8-7.   The general idea of a digital wrapper for optical networks

The most appealing aspect of digital wrappers is that they are totally independent of optical channel speed, format, or information content. Naturally, this is exactly what is needed to manage optical networks without SONET/SDH. As more and more network traffic becomes IP-oriented, this content independence is a real plus. This is also shown in Figure 8-7.

G.975 is an international standard. Many of the current digital wrapper methods used by optical networking equipment vendors are based on G.975, but they go far beyond G.975 in form and function. Interoperability at the digital wrapper level will be a challenge but is not impossible. Nevertheless, such convergence to a type of digital wrapper standard for DWDM and optical networking seems many years away as vendors strive to differentiate themselves in the marketplace.

WHY BANISH SONET/SDH?

New technologies such as wrapping optical frames inside of error detecting overhead packages are promising. But it takes years in many cases for such technologies to mature to the point where they are stable in implementation, to say nothing of the new technology becoming widespread enough to be considered in any way standard.

If SONET/SDH is to be truly banished from the world of optical networking, there must be more to the story than just the need to replace SONET/SDH OAM&P methods with newer ones for optical networking. As it turns out, SONET/SDH is limited when it comes to provisioning and delivering services in the rigid hierarchy that SONET/SDH represents.

Mention has already been made of wavelength provisioning, where carriers have the luxury of defining real-time, on-demand bandwidth outside of the TDM hierarchy that SONET/SDH represents. Traffic priorities in optical networks can be defined and altered on a circuit-by-circuit, route-by-route, or even user-by-user basis. With optical networking, carriers can serve more customers and offer more service levels and types than ever before.

This flexibility is just not possible with SONET/SDH. Optical networks require the following services, which SONET/SDH struggles to provide if at all.

Fractional Optical Transports

The current SONET/SDH TDM hierarchy defines bit rates such as 155 Mbps, 622 Mbps, 2.5 Gbps, or 10 Gbps. The problem is that these rates were established for reasons totally independent of any consideration of the type and speed of networks feeding the SONET/SDH links. Multiple voice streams at 64 Kbps are easy enough to fit into high-speed SONET/SDH backbones and rings. But fast Ethernet runs at 100 Mbps, and Gigabit Ethernet (GBE) generates frames at a line rate of 1.25 Gbps. Putting a fast Ethernet stream onto a 155 Mbps SONET/SDH link wastes 33 percent of the bandwidth on the fiber. Putting GBE onto a 2.5 Gbps link wastes 50 percent of the bandwidth, and so on. On the other hand, optical network nodes are "bit rate agnostic" and care little to nothing about the code form and line rate of the wavelength at the electrical bit level. Banishing

SONET/SDH is a way to gain back the 33 percent or 50 percent of wasted bandwidth. And having flexible access to the bandwidth represented by a wavelength in an optical network means that bandwidth is not stranded if a customer requires a different line rate or service. Manual network grooming is not required in an optical network, since all of this is more or less automatic in an optical network node. Customers would not have to pay for an increase from 2.5 to 10 Gbps if all they needed was a way to carry three GBE traffic streams instead of two.

Dynamic Bandwidth Delivery

If a customer needs a coast-to-coast OC-12 SONET link in the United States today, this often requires months of planning and lead time to install equipment and provision the circuit. Many SONET network nodes are involved, all of which must be carefully groomed and loaded with the extra traffic stream. In an optical network environment, a wavelength can be often provisioned and supplied literally within seconds or minutes of the request. Now, there are always cases where adding bandwidth requires new equipment, but with optical networks, this should be much rarer than in single-wavelength SONET/SDH systems. Optical networks can reconfigure bandwidth rapidly on the basis of one-time events, changing time-of-day routing requirements, or trade-offs in bandwidth applications. Differentiated services are easy to provide in optical networks featuring real-time "bandwidth on demand."

Bandwidth Bursting

SONET/SDH systems are firmly based on TDM principles. This means that a fixed amount of bandwidth is tied up in a SONET/SDH link regardless of how much bandwidth is actually used. A 155 Mbps SONET/SDH link carrying fast Ethernet frames carries 155 Mbps worth of bits whether the fast Ethernet LAN feeding the SONET/SDH link is generating live frames and traffic or not. It would be nice for other traffic flows to be able to use this "excess" bandwidth for bursts when a SONET/SDH link is otherwise idle. However, due to the way that SONET/SDH works, other devices on the same route cannot use this bandwidth at all. But optical network devices can easily pool such bandwidth on a wavelength-by-wavelength basis, and optical cross-connects or switches can rapidly shift gears between traffic flow mappings much more rapidly than SONET/SDH equipment can be reconfigured. Optical network links have no maximum or minimum bandwidth requirements.

Optical Virtual Private Networks

Virtual private networks (VPNs) have been around for a few years. When deployed on a LAN, VPNs rely on two main features: making a public facility look as secure as a private SONET/SDH link and making the public facility deliver the same quality of service (QoS) parameters that a private SONET/SDH link can. SONET/SDH easily supports VPNs because SONET/SDH links are usually marketed and sold as private links. But optical networks can offer a new breed of *transparent wavelength services* in which the form and content of the link never need be visible to the service provider. Such transparent wavelength services are not unknown in the SONET/SDH world but are extremely limited by the single-wavelength limitation of SONET/SDH and the need for constant electrical-to-optical conversion and back again. With optical WANs, the customer can manage their own secure bandwidth without the need for the service provider to set up and tear down connections or adjust bandwidths.

Variable Bandwidth User Connections

Bandwidth bursting applies to network backbone connections. Variable-bandwidth user connections extend this concept directly to the end-user devices. For example, SONET/SDH would allow for the provisioning of a 622 Mbps connection end to end across the SONET/SDH network from user site to user site. Provisioning additional bandwidth in a SONET/SDH environment can be time consuming and costly. Bandwidth bursting in an optical network would allow this 622 Mbps to be an allocated maximum and not have to carry many megabits of idle or fill patterns when used below 622 Mbps. However, variable-bandwidth user connections apply this idea to bandwidth *above* the 622 Mbps. The user would have premises equipment capable of many 155 Mbps or 622 Mbps bit streams on multiple wavelengths. Initially, only one or a few wavelengths are needed to deliver the required 622 Mbps. If the user requirements exceed 622 Mbps, more and more optical wavelengths can be turned on rapidly from site to site to provide bandwidth up to 2.5 Gbps or beyond. This can be done with no incremental cost to the service provider.

Market Differentiation

One SONET/SDH box works pretty much like another. Service providers seeking to differentiate their offerings and market according to time-honored "mine's better than theirs" principles have been forced to exploit minor SONET/SDH features for this purpose. Mesh versus ring protection, faster backbone aggregate speeds, and so on have become the ways that market share is gained and lost in SONET/SDH. Standardization to a large degree often becomes stagnation. But the wide-open field of optical networking offers a green field of opportunity when it comes to differentiating service provider offerings. New advances in optics, silicon, and the software to bring it all together in a manageable network environment will drive optical networking for some time to come.

When all of this is put together, it makes sense *not* to rely on SONET/SDH for features not currently found in optical networking. There may well be a period of transition, perhaps an extended one. Early automobiles looked exactly like carriages without a horse and were even called "horseless carriages" for this very reason. But cars quickly evolved to have their own distinctive shape and size once unfettered from the idea that there must be a horse around somewhere. So optical networks might go through a "SONET/SDH-less optical" phase where the network has to look like a SONET/SDH network just to be manageable. But this is not an ideal situation.

THE IMPORTANCE OF LEGACY VOICE

So there are firm reasons to "banish" SONET/SDH from optical networks. However, there is one more often compelling reason to retain SONET/SDH in an otherwise all-optical network. This is due to the presence of what is known as *legacy voice applications*.

Legacy voice is the traditional way that voice conversations are carried across a WAN. Legacy voice is characterized by several features. Legacy voice is

▼ Carried on time division multiplexed facilities

■ Switched at voice optimized circuit-switches known as central offices or local exchanges

▲ Digitized as a constant, full-duplex bit stream at 64 Kbps

Legacy voice has recently been challenged by voice over IP (VoIP) and related packetized voice methods that break all of the "rules" of legacy voice. Newer voice technology such as VoIP is

▼ Carried as packets on statistical multiplexed links such as between IP routers

■ Routed by IP routers along with other flows of IP packets

▲ Digitized as a fluctuating (depending on coding method), half-duplex (depending on silences) bit stream much lower than 64 Kbps (as low as 4–5 Kbps)

Packetized voice is much easier to combine with data packet flows in a router-based environment such as the Internet. However, the vast majority of all voice is still carried by legacy voice facilities and will be for some time to come.

Why is this important for optical networks? Because until more voice makes the transition from legacy voice to packetized voice in terms of users, facilities, and equipment, optical networks must be able to carry legacy voice along with other forms of traffic. SONET/SDH is optimized for legacy voice traffic. So SONET/SDH will more than likely be around for a long time to come even in pure optical networks.

The statement that "SONET/SDH is optimized for legacy voice traffic" is a powerful one. SONET/SDH can carry ATM cells, IP packets, and other types of traffic. Yet even a quick look at how SONET/SDH carries legacy voice traffic will show that legacy voice support is more than just something that SONET/SDH can do. Legacy voice traffic support is basically what SONET/SDH was designed for. That is, SONET/SDH constituted a way to carry more and more legacy voice traffic on networks with fiber optic links. If SONET/SDH could also support ATM and/or IP packets, so much the better. But no one would have implemented SONET/SDH in 1988 unless there was solid legacy voice support.

Virtual Tributaries

This section will discuss virtual tributaries (VTs) from a SONET perspective. As important as voice is to SDH as well, it is enough to outline the major aspects of carrying legacy voice on SONET. This section will be somewhat limited in the amount of detail offered. All the essentials are here, but the operational details of things like VT alarms are not given exhaustive treatment. There are also slight differences as well from vendor to vendor.

The intention of SONET VTs is to allow existing voice cross-connects and SONET-aware cross-connects to treat the voice channels in a SONET SPE exactly the same way they would a voice channel riding a T-1. This same statement would only change slightly to apply to SDH and E-1, of course.

The figures in this section will offer slightly different representations of the SONET frame seen previously. Since the topic in this section is the form and function of the SONET SPE, the figures will always represent only the SPE itself. So the total size of the SPE is 783 bytes (9×87), but the area available for information is 9 bytes less (due to the presence of the POH column), or 774 bytes. This is shown in Figure 8-8.

When the Signal Label POH byte (C2) is set to a value of 02x (02 in hex, or 0000 0010), then the SONET receiver knows that the arriving SPE must be carrying VTs. The first thing to realize is that the number of VTs present is not arbitrary. Each SPE must be divided into exactly seven Virtual Tributary Groups (VTGs). This is the first major division of the SPE for VTs.

The columns assigned to the seven VTGs in the SPE mimic the pattern established for multiplexing three STS-1s into a single STS-3 frame. That is, multiplexing proceeds byte by byte from each source until an entire row of the frame is filled. The result is that the multiplexing process creates a column-by-column structure where each column belongs

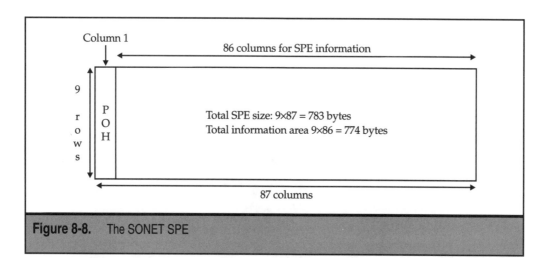

Figure 8-8.   The SONET SPE

to a separate multiplexing source. (This frame creation was detailed in the previous chapter.) An SPE consisting of seven multiplexed VTGs follows the same pattern. The intent is the same: to minimize the impact of delay on voice by interleaving sources instead of employing buffers.

The only complication is that seven does not divide 86 columns equally: there are two columns left over. So 86 columns / 7 VTGs = 12 columns for each VTG with a remainder of 2 columns. The solution here is to "skip over" two columns when multiplexing the seven VTGs into an SPE. The two columns "reserved" (really "ignored": their content is defined as "fixed stuff" bytes that must be ignored by the receiver) are columns 30 and 59 of the SPE, counting the POH column as column 1. Once this interleaving process is defined, the structure of an STS-1 SPE containing seven VTGs can be represented, as in Figure 8-9.

In fact, now that this structure for the VTGs has been defined, it is a simple matter to generate a table showing the columns occupied by each of the VTGs, which are always numbered VTG 1 through VTG 7. This table is shown in Table 8-1.

Several features of the table are noteworthy. First, the table starts with column 2 of the SPE. This just reflects the position of the POH column always as column 1 of the SPE.

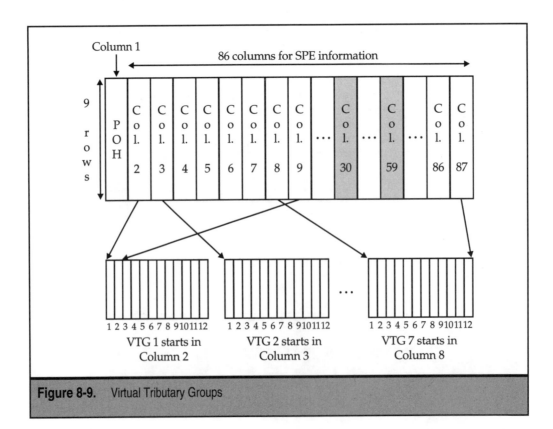

Figure 8-9. Virtual Tributary Groups

Virtual Tributary Group	SPE Columns Used by the VTG
1	2 9 16 23 31 38 45 52 60 67 74 81
2	3 10 17 24 32 39 46 53 61 68 75 82
3	4 11 18 25 33 40 47 54 62 69 76 83
4	5 12 19 26 34 41 48 55 63 70 77 84
5	6 13 20 27 35 42 49 56 64 71 78 85
6	7 14 21 28 36 43 50 57 65 72 79 86
7	8 15 22 29 37 44 51 58 66 73 80 87

Table 8-1.   SPE Columns Occupied by the VTGs

Next, the VTG column assignments skip over columns 30 and 59, columns that would normally be between VTG 7 and VTG 1.

It is important to realize that when an SPE is divided into seven VTGs, the SPE cannot carry information organized into anything *but* VTGs. Once the VTG columns are established inside the SPE, that is all that the SPE can be used for. This does *not* mean that the Signal Label cannot change with the arrival of the *next* SPE: Signal Labels can actually change SPE by SPE. However, this feature of SONET was controversial when introduced and has never been used on a large scale. Certainly it makes no sense to flip-flop Signal Labels on an STS-1 link when the intention is to carry DS-0 voice. There is just no room for anything else on that link. The 8,000 SONET frames (SPEs) generated each second will carry 8-bit voice samples in each column and row byte, yielding 64 Kbps, but no more.

Although this discussion of VTGs is limited to SONET in this section, it is worth mentioning that the SONET VTG structure is exactly the same in SDH. Of course, the name and acronym must be different. So in SDH, a VTG is called a Tributary Unit Group-Level 2, or TUG-2.

So far, the process of dividing the SPE into VTGs has produced seven identical structures consisting of 12 columns each, spread across 84 columns of the SPE. So each VTG has exactly 108 bytes (9 bytes in a column, and 12 columns). The next step is to look inside each of the VTGs and see the structure of this basic 108-byte unit.

Inside a VTG

Just because all VTGs form identical 108-byte units at different places in the SPE does not mean that the content of all VTGs is the same. There are four different forms that an individual VTG can take on. Although VTGs are optimized for the transport of DS-0 channels, not all DS-0s arrive at a SONET NE inside the same level of the T-carrier hierarchy. So the four possible VTG formats are

▼ VT1.5, used for DS-1 transport (1.544 Mbps)

■ VT2, used for E-1 transport (2.048 Mbps)

■ VT3, used for DS-1C transport (3.152 Mbps)

▲ VT6, used for DS-2 transport (6.312 Mbps)

The only real surprise on the list is the definition of the VT2, used for E-1 transport. E-1 support is provided in SONET because whenever links cross a border beyond the United States, international standards apply. With a VT2, a service provider in another country can easily hand off an E-1 to a SONET NE and all the DS-0 channels inside will be carried to the end user equipment. As an aside, a real problem before SONET was that 30 voice channels on an E-1 arriving in the United States had to be handed off to a DS-1 with only 24 voice channels! Either 6 channels could not be used (but were paid for) or a second DS-1 had to be purchased (but only 6 channels were used). SONET VT2s are a much better solution.

Of the four possible VTG structures, the most important is the VT1.5. Since there are many more DS-1s in service in North America than DS-1Cs or DS-2s, SONET links will most likely encounter DS-0s inside a DS-1 than in any other form. Even when T-carrier links arrive at a SONET NE is the form of a DS-3, chances are that the DS-3 is carrying exactly 28 DS-1s, each with 24 DS-0s. As it turns out, when an STS-1 is divided into seven VTGs and each VTG is structured as a VT1.5, the carrying capacity of the STS-1 is exactly the same as the capacity of a DS-3: 672 DS-0 channels running at 64 Kbps.

So a SONET SPE consisting of VT1.5s can accept and carry and cross-connect 28 separate DS-1s or a single DS-3 containing 28 DS-1s. This is no accident. As has been pointed out, the whole reason that the STS-1 level exists in SONET is to be backward compatible with existing DS-3 links (and the DS-1s most likely inside them). This allows an easy, gradual migration to an all-SONET network with flash cutovers and the like.

VT Structures

Each VTG is 108 bytes long (9 rows, 12 columns). However, the structure and carrying capacity of the VTG varies with VT type. In other words, a VTG containing DS-1s is fundamentally different from a VTG containing E-1s, and so on for all four VT types. The carrying capacity of each VT type is determined by the frame structure of the T-carrier or E-carrier link feeding the SONET NE. It is probably easier to look at each VT type in turn by example rather than attempting to describe the structure in the abstract.

Consider VT1.5. It is designed to carry DS-1s. DS-1s consist of frames generated 8,000 times per second with a structure of 193 bits. The 24 DS-0s, each 8 bits, make 192 bits. The 193rd bit is the framing bit. The framing bit is important, but look at the 24 DS-0s by themselves. The 24 DS-0s do not divide the 108-byte VTG evenly. There would be 12 bytes left over (108 / 24 = 4 with 12 bytes remainder). So four DS-1s would fit nicely inside a single VTG. But what about the "left over" 12 bytes? Well, there are four DS-1s. The 12 bytes are distributed between the four DS-1s inside the VT1.5 so that each DS-1 gets three bytes of

VT overhead (VTOH). What is this overhead for? The same types of things as SONET overhead: alarms, performance information, and other OAM&P functions.

So each DS-1 now requires 24 bytes of the VT1.5 for the DS-0 and gets another 3 bytes for overhead. That makes 27 bytes total. And since SPEs (and SONET frames) are 9 rows, the DS-1 fits nicely in three columns of the 12 column VTG! Further, 4 DS-1s of 3 columns each precisely fill the 12-column VTG. So the structure of a VTG according to VT1.5 looks like Figure 8-10.

In the figure, each column of the VTG is labeled A, B, C, D. These represent the four DS-1s inside the VTG. The multiplexing of the DS-1s inside the VTG is done column by column, as expected and in keeping with the general SONET/SDH columnwise multiplexing approach. Above the VTG, the columns that form each VT1.5 are given under their respective representations. The first VT1.5 (labeled "VT1.5 #1 (A)") has arrows pointing to the "A" columns of the VTG. The other three VT1.5s do not have arrows, but only to keep the clutter in the figure to a minimum.

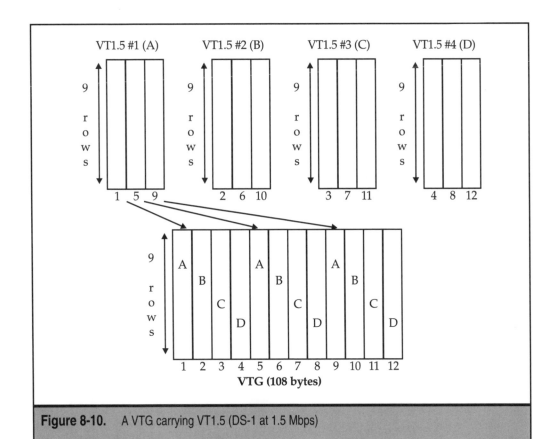

Figure 8-10.   A VTG carrying VT1.5 (DS-1 at 1.5 Mbps)

The key point is that each VTG structured according to VT1.5 carries four interleaved DS-1s. If each VTG is structured according to VT1.5, then the total DS-1 capacity of the SONET STS-1 SPE is 28 DS-1s (4×7). It is no accident that this is also the number of DS-1s inside a DS-3. Notice that the DS-3 *frame structure* is not carried directly inside the SONET SPE. (Recall that DS-3 uses bitwise multiplexing, not byte-multiplexing, as does SONET/SDH.) Rather, the 28 DS-1s inside the DS-3 frame are extracted by the SONET NE and then repackaged as 28 DS-1s inside seven VTGs.

The other three VT types' carrying capacities can be determined in the same way by considering the size of the frame each VTG must carry. For VT2, the E-1 frame has 30 DS-0s, plus one byte for synchronization and one byte for signaling. These 32 bytes do not divide the 108-byte VTG evenly. There can be three E-1s inside the VTG and once again 12 bytes left over (3×32 = 96). These 12 bytes form VT overhead as before, but in the case of VT2s, the 12 VTOH bytes are distributed among three E-1s, not four DS-1s. So the number of overhead bytes allotted to each E-1 is four, not three. When all seven VTGs are carrying VT2s, the carrying capacity of the SONET SPE is 21 E-1s (3×7).

A VT3 is designed to carry a DS-1C. A DS-1C is basically two DS-1s pasted together with some additional overhead bits. So the DS-1C frame structure is twice as large as the DS-1 frame and carries 48 DS-0s. Thus a VTG can carry only two DS-1Cs, again with 12 bytes left over for VTOH (48×2 = 96). The number of overhead bytes allotted to each DS-1C is now six. When all seven VTGs are carrying VT3s, the carrying capacity of the SONET SPE is 14 DS-1Cs (2×7). Note that this is half the DS-1 capacity, but each DS-1C carries twice as many DS-0s, so the overall channel capacity is still the same (672 DS-0 channels).

A VT6 is for DS-2. The frame structure of a DS-2 is rather complex, but the bit rate of a DS-2 is 6.312 Mbps and a DS-2 frame carries 96 DS-0s. These 96 bytes fit inside the 108-byte VTG nicely, with the same 12 bytes left over for VTOH. This time, all the VTOH goes for the single DS-2 inside the VTG. When all seven VTGs are carrying VT6s, the carrying capacity of the SONET SPE is 7 DS-2s (1×7).

All of this is most easily expressed in a table. Table 8-2 lists all of the important parameters of the four VT types.

VT Type	Tributary	Columns per VT	Number in VTG	Max. Number in SPE	Max. DS-0s
VT1.5	DS-1	3	4	28	672
VT2	E-1	4	3	21	630
VT3	DS-1C	6	2	14	672
VT6	DS-2	12	1	7	672

Table 8-2.   SONET Virtual Tributaries

In the table, the numbers in the *Columns per VT* and *Number in VTG* must multiply out to 12, which is the number of columns in each VTG. The *Maximum Number in SPE* column applies when all seven VTGs carry that type of VT, as does the *Maximum DS-0s* entry.

Before looking at each VT type in a little more detail, there are a few rules that apply to VTGs and the VT types they contain that should be examined first. The first rule has already been stated: when a SONET SPE in divided into VTGs, there are exactly seven VTGs, as detailed in Table 8-1. The two rules that apply to VTs themselves follow:

▼   **Each VTG can have a different VT type**   That is, VTG 1 can be structured according to VT1.5, VTG 2 structured according to VT6, and so on through all seven VTGs. Although this mixing of VT types is allowed, in practice this is seldom done.

▲   **Each individual VTG can contain only one VT type (VT1.5, VT2, VT3, or VT6)**   That is, if a given VTG is structured according to VT1.5, then no other type of VT can be mixed into that particular VTG, either in that frame or in subsequent SPEs. VT types are an all-or-nothing proposition. VT types are configured at each end of the SONET link to be one type. There is no real justification for this limitation, besides a wish not to make life too hard for SONET NEs. There *is* a VT Signal Label in each VTG, but this Signal Label is not used for the same purpose as the POH Signal Label (C2). A further problem is that the basic unit of the VTG is *not* a single SPE. The basic unit of the VTG is a sequence of four SONET SPEs called a VT superframe. Allowing changes of VT types would require careful handling of VT superframes. So each individual VTG must contain only one VT type, as identified by the configuration of the SONET NEs.

VT1.5 Superframes

In this section, the basic unit under discussion is the VT itself. In other words, each of the VT1.5s (for example) in Figure 8-10 will be presented as if it were a unit in and of itself. There is nothing wrong with thinking of the VT1.5 like this, but it should be kept in mind that things are really a lot more complicated. For instance, consider the VT1.5 in Figure 8-10 labeled "VT1.5 #3 (C)." Note that the three columns of this VT1.5 actually occupy columns 3, 7, and 11 of the VTG that the VT1.5 is configured in. And suppose that this is the fifth VTG of the SPE. A look at Table 8-1 will show that this VTG occupies columns 6, 13, 20, and so on of the SPE itself. Not only are the VT1.5 columns spread out, but so are the VTG columns! Where in the overall 87-column SPE is VT1.5 #3 of VTG #5?

As it turns out, once all of this "reverse mapping" is done, VT1.5 #3 of VTG #5 occupies columns 20, 49, and 78 of the overall SPE. This is a good place to present a table for VT1.5 detailing the location of all 28 DS-1s inside the SPE. Note that this table is completely valid only when all seven VTGs are VT1.5s. The only complexity in constructing such a table is that the DS-1s are organized and numbered by VTG order, not VT order. So *all* of the first VT1.5s (the As) in all seven VTGs load first into the SPE. The VT1.5 #2s (B) follow, and so forth. In other words, the DS-1 numbering is consecutive by SPE position, not VT position. To find the 15th DS-1 inside a VT1.5-structured SPE (perhaps in order to cross-connect the whole DS-1 or a DS-0 inside), all a SONET NE need do is extract the 16th, 45th, and 74th columns of the SPE. Naturally, other VTG structures locate other VTs in the SPE using other rules.

The whole point of the exercise is to point out that SONET can carry low-bit-rate, channelized, legacy voice as easily (if not more so) than SONET can carry masses of IP packets or ATM cells.

Mappings for Other VT Types

Just as VT1.5 has a distinctive format for the VT superframe, the other VT types also have distinctive VT superframe formats. There are also mappings established for the other three VT types (VT2, VT3, and VT6). It is not necessary here to detail all three to the degree afforded VT1.5, if only because this section would then be very long. More details will be explored for VT2s, but only because E-1 transport is the way in which international circuits must be supported.

Only the VT2 mappings, intended for E-1 links running at 2.048 Mbps and carrying 30 DS-0 channels, require some additional comment. First of all, a SONET SPE carrying all VT2s in the seven VTGs transports 21 E-1s (3 VT2s×7 VTGs) and thus 630 DS-0 channels. The positions of the E-1 columns in the SPE columns is shown in Table 8-4, which might be compared to Table 8-3 for the DS-1 equivalents.

DS-1 No.	VTG No.	VT No. (Letter)	VTG Columns	SPE Columns
1	1	1 (A)	1, 5, 9	2, 31, 60
2	2	1 (A)	1, 5, 9	3, 32, 61
3	3	1 (A)	1, 5, 9	4, 33, 62
4	4	1 (A)	1, 5, 9	5, 34, 63
5	5	1 (A)	1, 5, 9	6, 35, 64
6	6	1 (A)	1, 5, 9	7, 36, 65
7	7	1 (A)	1, 5, 9	8, 37, 66
8	1	2 (B)	2, 6, 10	9, 38, 67
9	2	2 (B)	2, 6, 10	10, 39, 68
10	3	2 (B)	2, 6, 10	11, 40, 69
11	4	2 (B)	2, 6, 10	12, 41, 70
12	5	2 (B)	2, 6, 10	13, 42, 71

Table 8-3. DS-1s Inside a VT1.5 SPE

DS-1 No.	VTG No.	VT No. (Letter)	VTG Columns	SPE Columns
13	6	2 (B)	2, 6, 10	14, 43, 72
14	7	2 (B)	2, 6, 10	15, 44, 73
15	1	3 (C)	3, 7, 11	16, 45, 74
16	2	3 (C)	3, 7, 11	17, 46, 75
17	3	3 (C)	3, 7, 11	18, 47, 76
18	4	3 (C)	3, 7, 11	19, 48, 77
19	5	3 (C)	3, 7, 11	20, 49, 78
20	6	3 (C)	3, 7, 11	21, 50, 79
21	7	3 (C)	3, 7, 11	22, 51, 80
22	1	4 (D)	4, 8, 12	23, 52, 81
23	2	4 (D)	4, 8, 12	24, 53, 82
24	3	4 (D)	4, 8, 12	25, 54, 83
25	4	4 (D)	4, 8, 12	26, 55, 84
26	5	4 (D)	4, 8, 12	27, 56, 85
27	6	4 (D)	4, 8, 12	28, 57, 86
28	7	4 (D)	4, 8, 12	29, 58, 87

Table 8-3. DS-1s Inside a VT1.5 SPE *(continued)*

E-1 No.	VTG No.	VT No. (Letter)	VTG Columns	SPE Columns
1	1	1 (A)	1, 4, 7, 10	2, 23, 45, 67
2	2	1 (A)	1, 4, 7, 10	3, 24, 46, 68
3	3	1 (A)	1, 4, 7, 10	4, 25, 47, 69
4	4	1 (A)	1, 4, 7, 10	5, 26, 48, 70
5	5	1 (A)	1, 4, 7, 10	6, 27, 49, 71
6	6	1 (A)	1, 4, 7, 10	7, 28, 50, 72
7	7	1 (A)	1, 4, 7, 10	8, 29, 51, 73
8	1	2 (B)	2, 5, 8, 11	9, 31, 52, 74

Table 8-4. E-1s Inside a VT2 SPE

E-1 No.	VTG No.	VT No. (Letter)	VTG Columns	SPE Columns
9	2	2 (B)	2, 5, 8, 11	10, 32, 53, 75
10	3	2 (B)	2, 5, 8, 11	11, 33, 54, 76
11	4	2 (B)	2, 5, 8, 11	12, 34, 55, 77
12	5	2 (B)	2, 5, 8, 11	13, 35, 56, 78
13	6	2 (B)	2, 5, 8, 11	14, 36, 57, 79
14	7	2 (B)	2, 5, 8, 11	15, 37, 58, 80
15	1	3 (C)	3, 6, 9, 12	16, 38, 60, 81
16	2	3 (C)	3, 6, 9, 12	17, 39, 61, 82
17	3	3 (C)	3, 6, 9, 12	18, 40, 62, 83
18	4	3 (C)	3, 6, 9, 12	19, 41, 63, 84
19	5	3 (C)	3, 6, 9, 12	20, 42, 64, 85
20	6	3 (C)	3, 6, 9, 12	21, 43, 65, 86
21	7	3 (C)	3, 6, 9, 12	22, 44, 66, 87

Table 8-4.   E-1s Inside a VT2 SPE *(continued)*

Note the absence of SPE columns 30 and 59 in Table 8-4. These are always fixed-stuff columns and cannot be used in any VT type.

The VT payload capacity of the VT2 is not 108 bytes (27 bytes in 4 SPE frames), as are VT1.5s. Because a VTG structured according to VT2 contains only three E-1s instead of four DS-1s, the columns unit is four columns, not three as in VT1.5. So the basic unit for an E-1 is 36 bytes (nine rows, four columns), and the VT2 superframe is 144 bytes long.

SONET/SDH TO OPTICAL NETWORK MIGRATION

When it comes to long-distance links today, SONET/SDH is no longer the only option. In fact, DWDM channels have many advantages over SONET/SDH configurations. Some of the most important of these advantages are shown in Figure 8-11.

Most long-haul systems today based on SONET/SDH use multiple fibers with electrical regenerators spaced 40 kilometers apart (some "intermediate" systems have regenerators spaced only 15 or so kilometers apart). Optical fiber amplifiers can be spaced 120 km apart and use only a single fiber pair. Consider eight SONET-SDH links that are 360 km in length. There are 7 regenerator pairs on each link, or 56 in total. But only 3 optical amplifiers are needed to cover the same span.

So one optical amplifier can replace more than 18 regenerators! And none of the electro-optical conversion associated with digital regenerators is needed with optical amplifiers. And this is just a simple DWDM link with eight wavelengths.

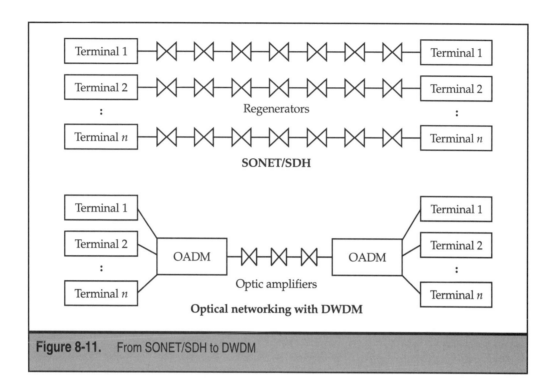

Figure 8-11. From SONET/SDH to DWDM

Upgrading a link is more efficient with optical networking as well. There is no need to add regenerators or fiber. All that is needed is an unoccupied wavelength. The amplifiers and fiber are already in place.

SONET/SDH is not the only technology whose future is threatened by optical networking. ATM has been under assault for a number of years, especially in the LAN arena. But optical networking might just be the catalyst that dooms ATM in the WAN as well.

CHAPTER 9

Optical Networking and ATM

Not too long ago, Asynchronous Transfer Mode (ATM) was in a position similar to that of optical networking. That is, careers were based on ATM expertise, products were announced almost daily, trade shows were organized, meetings were held, courses were taught, books were written, capital was raised, and companies were founded, all in the name of ATM. The ATM bandwagon was one everyone wanted to jump on, and the ATM parade was one no one wanted to miss.

But things have changed, and quickly too. ATM, once promised as "the last network we will ever need" and "the only network equally at home on LAN, MAN, and WAN," now struggles to hold on in the WAN backbone arena. Anyone who rises in a meeting to suggest a new ATM network or solution to solve a business problem is greeted with quizzical looks or stunned silence, and sometimes both. This erosion of ATM started long before optical networking came to town, but optical networking just might be the straw that causes the back of the ATM camel to buckle.

This chapter explores in a little more detail what ATM was intended for and how ATM works. This will do a lot to explain why ATM is under assault today, and how optical networking might just be fatal for ATM as a mainstream networking technology.

WHAT'S WRONG WITH ATM?

The theme of this chapter is not so much what is *wrong* with ATM as what is *right* with optical networking. In a nutshell, ATM is a high-overhead networking technology not well suited to work with either the Internet or the IP protocol stack in general. So any use of ATM requires the use of expensive and complex interworking equipment. ATM remains firmly tied to the fixed-bandwidth and fixed-services world of traditional time division multiplexing (TDM) and circuit switching. Now optical networking still time division–multiplexes and "switches," but at the wavelength level rather than at the ATM layers. And that, as it turns out, makes all the difference.

None of this should be interpreted as claiming that ATM is dead or useless. What is implied is that ATM will survive as a niche technology, squeezed from the desktop and LAN by faster versions of Ethernet, and now squeezed from the backbone by DWDM and more advanced optical networking techniques. ATM's future seems more and more to be squeezed into a corner of the access network, and ATM cannot even handle all of the access network technologies well at all.

Even now, as optical networking is emerging, ATM is most often seen handling virtual private network (VPN) access for private organizations. ATM remains popular with VPNs due to the security that ATM virtual circuits add the creation of the *closed user groups* that VPNs rely on to provide security on a public network. And often ATM must share this VPN access role with frame relay, a more cost-effective "fast packet switching" relative of ATM. A typical service provider network employing ATM today looks much like the general architecture shown in Figure 9-1.

In the figure, customer premises equipment (CPE) at a site is of three major types: a PBX circuit switch for legacy voice, an IP router for Internet access, and an ATM switch

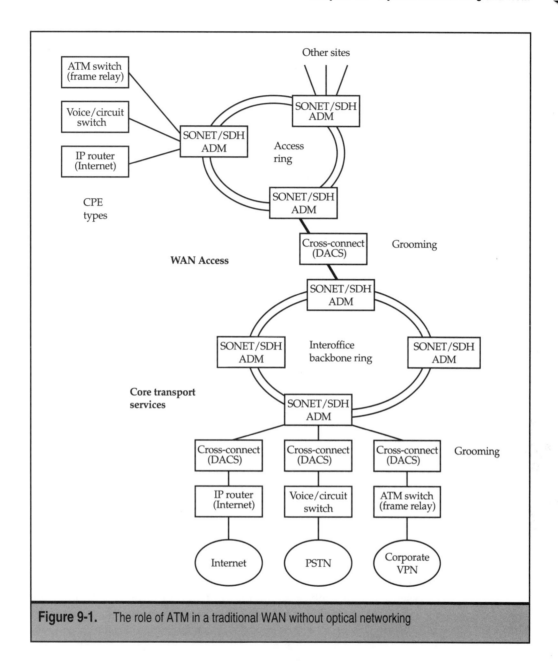

Figure 9-1. The role of ATM in a traditional WAN without optical networking

for specialized data applications and (increasingly) VPN access. In many cases, frame relay is used to ride on top of the ATM architecture for reasons of cost. Customer traffic is gathered on a SONET/SDH access ring formed from SONET/SDH ADMs and then rearranged at the service provider's network node, a process known as *grooming*. Grooming is performed by special cross-connects most commonly known as digital access cross-connect systems (DACS). Once groomed, the traffic is sent onto a larger and faster interoffice backbone SONET/SDH ring. These rings link the core transport and services sites together. At the service site, the traffic is split up again and individually groomed for the proper network destination through DACS equipment. IP router traffic makes its way onto the Internet; legacy voice finds its way onto the public switched telephone network (PSTN). ATM (and frame relay) traffic makes its way through an ATM switch to form the corporate VPN, with virtual circuits providing a large measure of the security needed in a VPN. The tragedy of ATM is that almost every "box" in the figure could have been labeled "ATM" and these ATM devices could have carried VPN data, legacy voice, and IP packets with equal ease. But ATM devices could not perform this task with equal efficiency or cost effectiveness.

Note the role of SONET/SDH in the figure. ATM cells ride the SONET/SDH links and rings along with other forms of traffic such as IP packets and traditional circuit-switched (legacy) voice. Ironically, since "ATM is for everything," IP packets and legacy voice can share a stream of ATM cells with whatever other types of traffic there are on the service provider's network. But again, the crushing overhead of the ATM "cell tax" makes other means of transport for IP and voice traffic much more attractive than ATM. And since 1990, many other technologies such as Ethernet have evolved even faster than ATM. ATM links and switches, once blazingly fast compared to other serial links and routers, have lagged behind the curve when it comes to aggregate speeds and throughputs.

At the end of this chapter, some plans to "rescue" ATM and bring ATM into the world of optical networking will be investigated. Then a service provider network using optical networking techniques will be presented and the role of ATM reassessed. The architecture presented at the end of this chapter might then be compared to the one that started the chapter.

First, however, a whirlwind tour of some of the more important characteristics of ATM is in order. Only by detailing ATM's current limitations can the assault on ATM by optical networking be fully appreciated.

THE BASICS OF ATM

Asynchronous Transfer Mode networks began as an idea in the minds of a group of carrier technicians (essentially telephone company personnel) around 1986. These far-sighted individuals basically looked around at the computer industry and saw the writing on the wall. The demand for faster networks to accommodate higher-speed end systems and the need for higher-bandwidth networks to accommodate the virtual explosion of file sizes and applications (DOS itself went from 360 Kbytes to over 7 Mbytes for MS-DOS 6.22 in less than 10 years) had left these carriers in an uncomfortable position. The carriers—the traditional source for networking solutions for all organizations—faced the prospect of a rapidly approaching future that would leave them with a customer demand that the carriers could not supply. They would have no high-band-width, low-delay solutions to sell unless they acted to invent one.

The need was for both elements: not only high-bandwidth technology, but the fast switching that would have to take bits from one high-bandwidth link and put them on another link inside the network very rapidly. So, working inside the ITU (then the CCITT) and its own Study Groups, these technicians very quickly came up with a very high-bandwidth family of fiber-based networks of digital trunks and a multiplexing and switching strategy to connect these trunks as a full network architecture. The whole system was specified in raw form in a series of standards published in 1988 as a part of the CCITT "Blue Books."

The high-bandwidth digital trunks were a whole family of standard fiber links known as the Synchronous Digital Hierarchy (SDH). In the United States, Bellcore had developed SONET as a basis for the SDH development effort. The differences between SDH and SONET were mainly in terminology, not function.

The multiplexing and switching technology that was specified especially for SDH and SONET was Asynchronous Transfer Mode. ATM was a method of building and operating networks that was fundamentally different than anything that had ever been done before. And there was a very good reason for that.

ATM was an attempt to "leapfrog" the trend of end system and application size expansion by getting ahead of the pack. The developers also attempted to ensure that the technology would be agile enough to stay ahead of the network bandwidth and delay demand no matter how fast it grew in the future. They did this by changing not the technology that formed the network, but the network that implemented the technology. This needs some explanation.

In the past, the kind of network an organization built heavily depended on the distance spanned. If the network had to link end systems located a mile or so apart, a LAN would do nicely. For larger and longer distances, a WAN was usually needed, although various groups addressed the need for linking systems less than 60 miles or so apart with a MAN. The whole point is that the hardware, software, and protocols used to build any one of these networks was unique to that network.

This means that the hardware, software, and protocols developed to run on a LAN could not be made to run over WAN distances without extensive changes or a whole new kind of network technology installed in the middle. For example, linking two LANs required a WAN between them, with a router to essentially perform as a device to be both a LAN node and WAN node at the same time.

ATM was designed to be fundamentally different. By aligning the network links with SONET/SDH, ATM networks were based on a family of standard, interoperable hardware, software, and protocol implementations that could supply more and more bandwidth well into the gigabit ranges if need be. And by basing the multiplexing and switching technology on ATM, the result was to make the end-to-end network nodal processing delay through the network an insignificant portion of the propagation delay, which cannot be addressed due to speed of light constraints.

ATM networks are intended to be the ultimate converging technology: one basic under-lying network infrastructure that is the same for everything. The philosophy of ATM was simple: once the most efficient way of sending 0's and 1's from Point A to Point B has been built, there can exist no end system or application that can ever need more bandwidth or less delay. Building networks that can do this is both the promise and the challenge of ATM.

How ATM Builds a Network

ATM gets its "magic" from its potential capability to mix the different kinds of networks (voice, video, data) into one big *unchannelized* physical network. This method of multiplexing the ATM cells defines the concept of an *asynchronous* "transfer mode." In this context, "asynchronous" refers to the capability of an ATM network to only send the data associated with a connection when there is actual live data to send. This is in contrast again to channelized TDM networks where even if a channel is idle, a special bit pattern (called, naturally, the "idle" or "interframe fill" bit pattern) must be sent in every time slot representing the channel. Otherwise, the receiver would not be able to recover the infor-mation present in the other time slots. This is the essence of *synchronous* transfer mode networks, which most people would just refer to as time division multiplexing.

A "transfer mode" has been defined as the main techniques of transmitting, multiplexing, switching, and receiving information in a network. Many observers use it in a more restrictive fashion. In this usage, every communications technology from tele-phone to broadcast TV has had its own transfer mode, although people seldom talk about it in those terms because it is so obvious. *Circuit mode* is for voice: pick up the phone (ter-minal device) and dial the number (network address). *Packet mode* is for data: take the user's data from a LAN client (terminal device) and add a header with destination infor-mation (network address) for connectionless services or a circuit identifier for connec-tion-oriented services. Until now the network transfer mode was specialized for network functions. This is the key to understanding just what ATM is for. A packet network could not be built using voice packets, since this packet "transfer mode" was developed and op-timized for data. Of course, this tends to be a very inefficient resource (e.g., bandwidth, etc.) usage. And putting the voice and data on the same *channelized* TDM link with fixed bandwidth assignments does not really help; it just shifts the problem.

All these methods involve *synchronous* use of the network. ATM, by contrast, is struc-tured to work in an *asynchronous* manner. The terms "synchronous" and "asynchronous" as applied to transfer modes refer to the scheme of multiplexing: mixing traffic from many sources together on the same physical network path. In a "synchronous transfer mode," each source is assigned a fixed bandwidth based on *position* in the data stream: a frequency band in FDM or a time slot in TDM. In contrast, ATM is not based on position in the data stream at all: a header identifies whose traffic it is and where it goes. All traffic is sent based on demand: no traffic, no bandwidth drain. Therefore, an ATM network is not service dependent: it works well for voice and video as well as data. It is not inflexible: as bandwidth requirements for video decrease (the blue VCR "idle" screen, for example), ATM networks can easily adjust. It is not inefficient: resources assigned for now to a voice connection can be used later for data traffic. Everything in ATM is done on the basis of connections or *virtual circuits,* not channels or time slots, as is done in traditional time division multiplexing.

All ATM networking is based on the ATM *cell* as the unit of data exchange. (The structure of an ATM cell is shown in Figure 9-2.) A cell is defined as a fixed-length block of information. Previous networks all used a simple stream of 0's and 1's that were organized into different structures depending on the service and network. This organization into different structures is still done with ATM networks, but at the end points of the network. At the physical (bit) level, everything is sent and received as cells: fixed-sized blocks of bits that replace the time slots in TDM.

The cell starts with a 4-bit generic flow control (GFC) field that has never been used in a consistent fashion to control the flow of cells across an interface. This is followed by 8-bit Virtual Path Identifier (VPI) and 16-bit Virtual Channel Identifier (VCI) fields. These bits identify the ATM connection that the cells flow on. Then comes a 3-bit Payload Type Identifier (PTI) field so that ATM switches can pick out information bearing cells from other types of cells. Then there is a 1-bit Cell Loss Priority (CLP) field for choosing cells to discard under conditions of network congestion. Finally there is an 8-bit Header Error Control (HEC) field that corrects errors in the cell header, but not the cell payload. The cell payload section is 48 bytes, not bits, long. Network ATM switches should never have to examine the contents of the cell payload to make destination decisions.

So ATM is easy to understand: it is simply a method of transferring information as it is generated by a source using fixed-length cells. The "asynchronous" part refers to the "as it arrives" phrase in the definition. The use of cells in ATM is related to the concept of *cell relay.* Cell relay is the general term for networking with cells. ATM is the international standard for cell relay technology. Cell relay just says "there shall be cells." ATM sets the size of the cells and their structure and use, and so on. Much of the cell technology in ATM is closely related to packet switching systems such as X.25 and frame relay. That is, it is a connection-oriented network method based on switches as network nodes, not routers.

In an ATM network, the network nodes (switches) switch ATM cells. The ATM cell structure consists of 53 bytes. Bytes, or groups of 8 bits, are usually called *octets* in ATM literature, but the two terms mean the same thing today, for all intents and purposes. The 53 bytes are divided into a 5-byte header and a 48-byte information section, known as the *payload.* Since "ownership" of cell contents is not determined by position in the data stream, determination of this ownership is one function of the cell header.

The structure of the ATM cell header is also shown in the figure. Officially, this is the *B-ISDN User-Network Interface (UNI) cell header.* Most of the bits in the header, 24 out of 40, are used for a hierarchical network connection identifier (the VPI/VCI field).

As previously mentioned, the *asynchronous* in ATM refers to the sending of data on the network as it arrives. Data is packaged into the fixed-sized cells and sent out onto the network. It is this use of cells by data "as it arrives" that has lead to the label *bandwidth on demand* being applied to ATM. Of course, ATM cannot make bandwidth out of nothing, but ATM does make the most flexible use of the available and presumably scarce bandwidth when the bandwidth must be shared by a number of users. The term *flexible bandwidth*

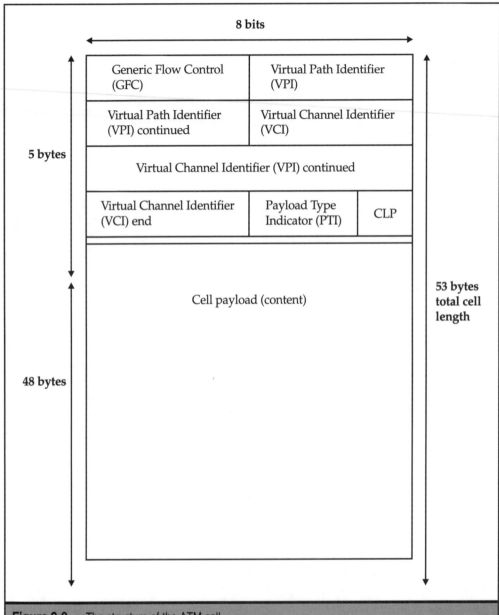

Figure 9-2. The structure of the ATM cell

allocation is technically more accurate, but "bandwidth on demand" remains entrenched even in ATM circles.

The cell stream leaving a network node may be filled by the same user's data, or a different user's data. Most likely, it will be a combination of many users taking up varying amounts of cells at different times: bandwidth on demand.

ATM is sometimes referred to as "label switching" (a term that shows up again at the end of this chapter and plays a large role in the next), "label multiplexing," or even "asynchronous time division multiplexing" in older documentation. All of these terms were used to refer to ATM when the standards were still under development. The "label" is the connection identifier in the cell header that tells the switch what to do with the cell and tells the receiver which connection the cell is to be associated with.

How ATM Multiplexes

One of the major attractions of ATM is how ATM multiplexes cells in a restricted bandwidth environment. ATM multiplexing works by establishing a fixed number of cells per second available for use, depending on link speed. Each cell has the 5-byte header whose primary purpose is to identify cells belonging to the same virtual circuit, or connection. Cells are transmitted according to what is called in ATM the user's "instantaneous real need." The idle patterns and fixed bandwidth channels have been eliminated where applicable, to the benefit of both the user (who gets to send data faster) and the network service provider (who gets more efficient use of the scarce network bandwidth).

The cell length in ATM is only 53 bytes. By contrast, the smallest allowable frame size on an Ethernet LAN is 64 bytes. The cell length in ATM is set so small for a number of reasons. It is basically a compromise between the needs of voice and the needs of data applications like file transfer. The whole idea is to avoid the long and unpredictable delays waiting for long packets to finish transmission. This scheme gives acceptable voice delay—but it has been shown to be very inefficient for data transfer at lower bit rates (meaning below Ethernet LAN speeds).

It may not be obvious why long, variable-length data packets may be detrimental to short, fixed-length legacy voice samples when they are mixed on an un-channelized link. Why bother to make cells at all? Here is a simple analogy.

It the United States today there are still railroads in use for moving large, bulky loads of freight. Railroad engines pull variable-length trains of boxcars, up to 200 or so, on special "networks" of railroad tracks. Highways were built as another special kind of network, but this time for fixed-length cars that are optimized for carrying people. As long as the networks are separate, and the trains stay on the tracks and the cars stay on the roads, the networks both work fine.

The problem arises when the railroad tracks must cross the highway. At a railroad crossing, once the train starts across the highway, the cars all must stop. They have to wait until the entire train is across the intersection. This causes a delay, which may be a long one, considering the number of boxcars that may be in the entire train. The cars must be "buffered" until the train is past. And even when the train is clear of the crossing, the backup of automobile traffic still takes a while to make its way through the crossing.

In this analogy the train is the large, variable-length data frame, and the cars are the smaller, but more numerous, fixed-length, delay-sensitive legacy voice traffic. The entirely separate railroad tracks and highways are the channelized time division multiplexed networks still commonly built today. The crossing situation represents an attempt to mix the data and delay-sensitive voice samples on the same un-channelized network, but without ATM cells as a possibility.

Clearly, this causes a problem. The backup of voice traffic will cause the receiver to think that something is wrong if the absence of "cars" arriving persists too long. And even if they arrive, the delay is variable and can distort the voice. But exactly how long this situation will persist depends on the length of the train.

There are actually two potential solutions to this problem. It is important to keep this in mind because ATM solves the "crossing" problem of mixing delay-sensitive traffic such as voice with bulk data transfers one way, and optical networks solve the problem the other way. If an entire industry embraces the optical networking solution, then ATM's approach is essentially rendered obsolete. And this is precisely what appears to have happened in the world of networking. ATM makes optimal use of available but scarce bandwidth. But what if bandwidth is no longer scarce, as envisioned with optical networking?

In the first possible solution, if the engine pulling the train runs very much faster than usual, the delay for the cars waiting for the train to pass will be correspondingly shorter. That is, if the train runs twice as fast, the wait is half as long. If it runs ten times as fast, the wait is one tenth as long, and so on. The point is that if the train runs fast enough, the time that the cars have to wait, and the number built up at the crossing, can be made almost insignificant. This solution is then to make an un-channelized network that runs much faster than ever before, and trains to run on them.

There is a potential drawback to this "fast train" approach, however. Application bandwidth requirements have grown in leaps and bounds in the past, and the trend is upward. The trains keep getting longer and longer. Rebuilding the railroad to stay ahead of the trend may be a self-defeating proposition in the long run. Faster railroads may actually encourage longer trains. People only have to look at the evolution of Ethernet from 10 Mbps (in fact, early Ethernet ran at only 3 Mbps) to 100 Mbps to 1,000 Mbps (Gigabit Ethernet, or GBE) and beyond to suggest that there will never be enough bandwidth to keep everyone happy.

So if bandwidth remains scarce, maybe there is another solution. ATM tries to make each boxcar a self-propelled unit about the same size as a car. Now if a "train" starts across the intersection and a car arrives, there is only a very short delay before the boxcar is through the crossing and the car can pass. This very different than the previous "throw bandwidth at it" solution and requires no re-engineering of the railroad or the highway. This is the ATM solution, with both the boxcars and cars becoming cells and no distinction between train tracks and highways.

Broadband ISDN (B-ISDN) and ATM

In the majority of organizations today, the voice, video conferencing, and data networks are implemented completely separately. If they have any parts in common at all, it is generally only the fact that these networks all use channelized links. However, the channels are not shared in any real sense. It is true that some organizations have begun to merge voice, video, and data onto IP router networks, and this trend will increase. But most organizations are still building and maintaining separate voice, video, and data networks.

Not only does ATM offer the hope of merging of these separate networks in an organization's private networks, it offers these same capabilities to the voice, video, and data public network service providers. ATM was envisioned as a public network offering, but many private ATM networks have been built.

Voice network service providers today are not only interested in providing voice communications between people on the telephone, but they seek to support add-on voice applications like messaging, call forwarding, and the like. ATM was intended to supply these service providers an opportunity to offer these services in a standard fashion from provider to provider.

Before everybody started to want to do everything on the Internet and the Web, and ATM was still a cutting-edge technology, all telephone companies were very interested in offering "video dial-tone" services as well. Video dial-tone was intended to give customers the ability to watch digital movies on demand and access other services like "living life online" that are now seen as possible applications for the Internet and the Web.

Advanced video services are offered today by the cable TV companies. But the cable TV companies are all considering offering voice services as well. In fact, studies claim that as many as 33 percent of cable TV subscribers would actually prefer to get phone service through the cable TV company.

Data services are offered through ISPs (often voice companies as well). But often these services are usually accessed at low speeds through a dial-up arrangement through the local telephone companies. However, in many places around the world, Internet access is available at rates of about 500 Kbps or 1 Mbps and even above through the set-top box provided by the cable TV company or a DSL modem from another service provider.

Until recently, these businesses all built very different networks: telephone companies built voice networks, cable TV companies build coaxial cable networks, and so on. But in many cases, any company is able to offer any service it is able to provide. So there is a *convergence* of the voice, video, and data service businesses. ATM was seen as the only technology that would be efficient for these new "anything goes" companies to build into their new networks. ATM can deliver adequate voice, video, and data services over the same physical network, as long as the bandwidth remained scarce enough for the "fast train" solution to be too expensive to consider.

How ATM fits into the converged services vision of B-ISDN is shown in Figure 9-3. The layers of the ATM protocol stack deliver the converged services defined by B-ISDN. Notice the use of various "convergence" components in several ATM layers.

Convergence is an important B-ISDN and ATM concept. It means the following: There are multiple options that may be employed above or below some layers in the

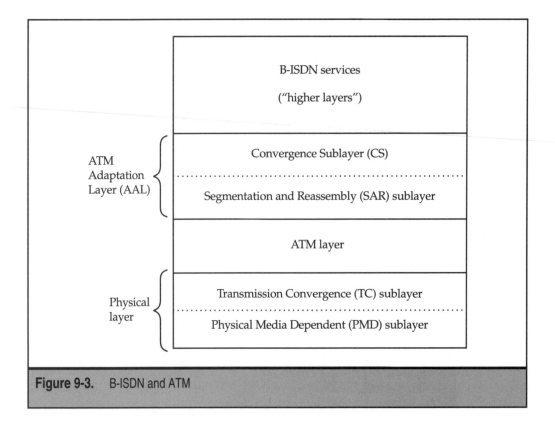

Figure 9-3. B-ISDN and ATM

model. Bits may be framed or sent "raw." They may be sent on fiber or coax. They may come from a constant bit rate (CBR) service like voice, or from a bursty variable bit rate (VBR) service like LAN interconnection routers. But whatever the diverse options at one layer or another, the convergence layers help to present a uniform interface to other adjacent layers, just to make it easier for implementers.

The lowest layer of the ATM model is the Physical layer, divided into two sublayers: the Transmission Convergence (TC) sublayer and the Physical Media Dependent (PMD) sublayer. The Physical layer is concerned only with functions that are completely dependent on the physical medium itself. These physical media dependent (PMD) functions include all bit transmission and bit alignment functions for transmitting 0's and 1's across the link. The line coding is therefore done here, and if the electrical signal from the device is being sent over an optical link, this conversion is provided here as well.

The ATM layer performs all cell switching and multiplexing functions. The ATM layer only looks at the cell header to do this; it never looks inside the cell payload. All of the ATM layer switching and multiplexing is done according to the VPI and VCI fields in the cell header.

The ATM Adaptation layer (AAL) is not required on end systems and is not needed on internal network nodes, such as the ATM switches themselves. It is divided into two sublayers: the Segmentation and Reassembly (SAR) sublayer and the Convergence sublayer (CS).

The SAR sublayer supplies a "bookend" function so that the receiver can associate a sequence of cells into the original data unit the sender broke down into cells. The CS provides the mechanism for mixing the different requirements of voice, video, and data by defining a number of *classes of service,* each with the appropriate parameters for the service. These are used to provide the proper Quality of Service (QoS) parameters on that connection.

How these layers are implemented on an ATM network is shown in Figure 9-4. Note the presence of the AAL only on the ATM end systems. Between ATM switches, the interface is not the UNI, but the network node interface (NNI).

The four classes of service currently defined in B-ISDN are mapped onto "types" of AAL for ATM implementation. Typical services using these classes might be

▼ **Class A**   Circuit Emulation, constant bit rate voice and video

■ **Class B**   Variable bit rate audio and video (compressed)

■ **Class C**   Connection-oriented data transfer

▲ **Class D**   Connectionless data transfer (sometimes called Switched Multimegabit Data Services (SMDS), a connectionless form of ATM)

The ATM Forum has defined additional service types at the ATM layer. The most important are constant bit rate (CBR), variable bit rate (VBR), and available bit rate (ABR).

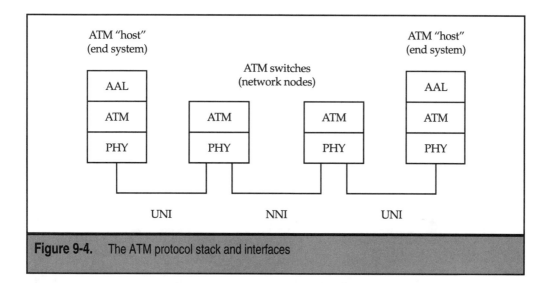

Figure 9-4.   The ATM protocol stack and interfaces

ATM VIRTUAL CHANNELS AND PATHS

As has already been pointed out, ATM is a network architecture that uses un-channelized network transports. But of course traffic still must be identified as voice, video, or data, if for no other reason than to preserve the required Quality of Service parameters for each separate service. Since there are no physical channels to distinguish the traffic in an ATM network, their place is taken by logical connections. Instead of voice or video channels, ATM networks have voice or video connections. In ATM networks, these logical connections are established and maintained by means of a two-part identifier structure: the Virtual Path Identifier (VPI) and the Virtual Channel Identifier (VCI).

VPIs and VCIs are in cell headers, and they are hierarchical: Many Virtual Channels may make up a Virtual Path. Basically, the VCI field identifies device-to-device connectivity, and the VPI field identifies site-to-site connections.

This is how the ATM network uses the VPI/VCI values in a cell header to deliver traffic through the ATM network. The ATM network service provider has previously agreed with users at Site A, either at the time the ATM service was first provided, or by means of a signaling protocol, that when Site A sends cells to Site B, Site A generates cells with the VPI field in the cell header set to 69. Also, the ATM service provider has set up the network (again, either by prearrangement or a signaling protocol) so that cells generated by users at Site A with VPI = 76 will be delivered to Site C. Therefore, the connections indicated in the VPI fields take the place of physical channel destinations. This is the equivalent of a permanent virtual circuit (PVC) in a packet-switched network.

The channels themselves are represented by the VCI field values. VCI values may be set up by signaling or prearrangement, as are the VPIs. However, since VCIs are supposed to represent "dynamic" or device-to-device resources, VPIs are supposed to be set up at service provision time (PVCs: permanent virtual circuits) and VCIs are supposed to set up by a signaling protocol (SVCs: switched virtual circuits). Note that the terms "PVC" and "SVC" are not ATM terms per se, but they are useful analogies to the actual ATM connection arrangements.

When it comes to discussing ATM switches, it is always a good idea to keep in mind that these devices are switches, not routers. In routing, all hop-by-hop output link decisions are made on the basis of information included in the packet header, and each packet is sent repeatedly through the routing process at each node. In switching, all hop-by-hop link decisions are made *ahead of time,* and network resources are allocated for the connection at this point. Thereafter, *no* other routing decisions in a switch are necessary, other than a quick table lookup. Header information may be kept to a minimum.

The other advantage of using switches instead of routers as network nodes is that switch-based networks assure sequential delivery. If delivery of data is not guaranteed to be sequential, as is true in connectionless, router-based networks such as the Internet, then if a destination receives packets 1 and 3, but not 2, the destination must wait to see if packet 2 shows up. This means an additional delay must be built in to the destination to allow for this. On the other hand, packet 2 may be "lost" in the network (a link or router fails, etc.) and never arrive. So the timeout interval must be balanced between the need to wait for an out-of-sequence packet and the need to request the sender to resend the missing packet.

Connection-oriented, switch-based networks have no such limitation. With sequential delivery guaranteed, if the destination receives packet 1 and then packet 3, packet 2 must be missing in action in the network itself. Immediate steps may be taken by the destination to correct this error condition, which may involve notifying the sender that a retransmission is needed, but not necessarily.

ATM SIGNALING PRINCIPLES

So ATM switches make decisions based on the information in a VPI/VCI table. But how does the ATM switch VPI/VCI table information get there in the first place? There are actually several ways.

Connectionless networks employ a sophisticated routing protocol to update tables stored in their network nodes, such as OSPF (Open Shortest Path First). These protocols may add a large amount of traffic to the network, and they essentially run independently of the users of the network itself.

In ATM switch networks, things are different. The information on Virtual Channels stored in the ATM network nodes gets there from the users themselves. There are several ways a user can inform the ATM network what connections, either Virtual Paths or Virtual Channels, need to be set up on the ATM network. This whole process is known as *call control*, where a *call* is any connection on the ATM network.

The simplest way is to do it by hand. That is, the user of the ATM network service simply writes down all the connections needed on a piece of paper and hands it to the service representative of the ATM network service provider. The ATM network service provider may then configure the tables in the network and inform the user: "When you wish to send cells to the Boston video-conference site, use VPI 45 and VCI 186." (It may be up to the individual user to configure the site software to generate cell headers with these VPI/VCI field values, or the service provider may actually do it.) This whole process, known as setting up these PVCs at "service provision time," has the advantage of being a straightforward process. These connections may be changed in the future, but this tends to be a slow process, relying as it does on the manual coordination of several parties.

Alternatively, there may be a signaling protocol that runs between the customer premises equipment (CPE) at the user's site and the ATM network node. This protocol must be a standard way for the devices themselves to set up, maintain, change, and terminate connections across the ATM network. In this scenario, the CPE sends a message to the local network node (the one that it is directly attached to) requesting a connection with Boston. The ATM network nodes set up the path and, if the connection is acceptable to the network (e.g., the destination will accept the connection), another message is sent back to the originator along the lines of "Connection OK, use VPI = 45, VCI = 186."

FIRST CRACKS IN ATM: WHERE'S B-ISDN?

The vision of ATM advocates around 1990 tied ATM popularity closely to the availability and desirability of converged services through B-ISDN. Who would not want one network device that could handle voice and video and data through one network connection? Convergence was already underway, but not toward some special (and expensive) B-ISDN device available from the local service provider. Convergence happened with the simple, but versatile, PC as the device of choice. There were attempts to reorient B-ISDN to make it more PC and LAN friendly, notably with a connectionless ATM data service known as Switched Multimegabit Data Services, or SMDS. In spite of some initial interest, SMDS was just too radical and expensive an approach to LAN interconnections to be either accepted by customers or championed by equipment vendors.

By the mid-1990s, it became obvious that the converged services that everyone wanted would be delivered not by a new B-ISDN network, but by the Internet and its new cousin, the Web. And the Internet philosophy was not so closely tied to ATM as B-ISDN was. If bandwidth was scarce, the Web could still be used for voice, video, and data, just with voice that sounded rather bad, video that lurched and froze, and data that took 30 seconds or more to load a busy graphical Web page. But in most cases the Internet and Web people did not want ATM to solve these scarce bandwidth problems obliquely. They were more direct: just give me the bandwidth! This was a huge incentive for the development of DWDM and optical networking, much to the detriment of B-ISDN and ATM.

So ATM in the mid-1990s suddenly found itself as the vehicle of choice to deliver B-ISDN services that would never exist. ATM could only prosper if a way could be found to adapt ATM to the world of LANs, routers, and the Internet and the Web. ATM advocates gave it a shot. The results were called LANE and MPOA.

ATM LAN EMULATION (LANE)

Many organizations have invested heavily in both physical networking LAN infrastructures, mainly based on Ethernet, and protocols and applications to run on these infrastructures. Any networking technology that wants to catch on must address the support requirements for this massive investment in software and ease the transition for the hardware.

ATM LAN Emulation (LANE) was designed to allow existing networked applications and network protocols such as IP to function over ATM networks. LANE uses ATM as a backbone for networks connecting "legacy" LAN networks. LANE was also designed to support ATM end systems directly attached to the ATM portion of the network and end systems attached through Layer 2 (bridging) devices. Also, LANE allows multiple emulated LANs to exist on the same physically interconnected ATM network.

The general idea behind LANE is shown in Figure 9-5. In the figure, an ATM network interconnects multiple Ethernet segments and ATM-attached end systems. End systems on the Ethernet segments communicate with other Ethernet-attached end systems through

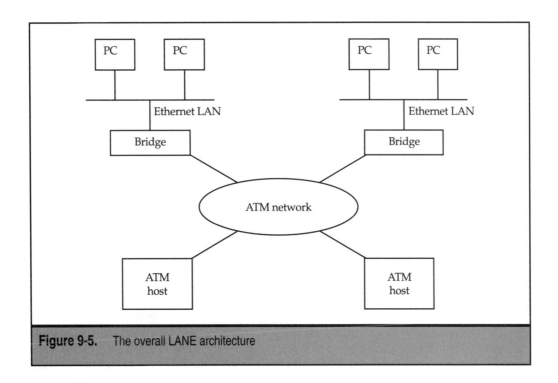

Figure 9-5.   The overall LANE architecture

bridges across the ATM network in a backbone configuration. ATM-attached end systems can communicate with ATM-attached servers, and both types of system can communicate with Ethernet-attached end systems through the bridges. A lot of the LANE protocol stack was designed specifically to address the bridging-related issues of hooking up LANs over ATM.

LANE development began in the fall of 1993, and LANE was released in the beginning of 1995. In LANE, the interface to a host is through a predefined MAC-layer interface such as the Network Driver Interface Specification (NDIS) from Microsoft, the Open Datalink Interface (ODI) from Novell, or the Data Link Provider Interface (DLPI) from various UNIX implementations. The whole reason for the LAN Emulation layer is to make the upper-layer MAC interfaces believe that they are still running on a standard Ethernet or Token Ring network (LANE only emulates Ethernets or Token Ring LANs).

LANE uses the AAL5 adaptation layer to fragment data frames into 53-byte ATM cells. The ATM switches transport the cells just as they would for any other cell stream on an ATM connection. The ATM-LAN Bridge device implements not only the MAC layer emulation features of LANE, but also the portions of the LANE protocol designed for edge devices. These portions include portions of the bridge forwarding table and forwarding unknown frames to the legacy LANs. On the LAN side of the bridge, the LAN Host shown emphasizes the fact that no changes are required in order to support connectivity with ATM-attached hosts providing client/server functions. This allows network migration by implementing ATM in the bandwidth-critical sections of the network and still allowing existing legacy LAN devices to operate essentially as before.

So the whole point of LANE is to create an "emulated LAN." The emulated LAN can be either Ethernet or Token Ring, but they cannot be mixed.

The biggest advantage of LANE is that existing applications and protocols can be adapted to ATM simply by installing new ATM interfaces and drivers. The disadvantage is that these new driver interfaces are not intended to take advantage of any ATM features, but rather to *hide* the ATM network from the end LAN systems. It is somewhat ironic that when successful, LANE hides the ATM completely.

So LANE allowed LANs to transparently use ATM networks. But what about the IP packets that flowed through the routers at the "end" of these LANs? Since LANE was essentially a bridging technology, LANE had little or nothing to do with IP routers. (IP routers could be converted to perform bridging, but this was a little like buying a limousine and using it instead of a pickup truck to haul gravel.) This proved to be a real issue when people considered deploying LANE to connect two sites over almost any WAN distance.

LANE makes two or more LANs act as one big LAN. From the IP router perspective, this means that the two sites must share a common IP address space. The details of precisely what this means are not important. What is important about this address space sharing is that most organizations do not assign and use IP addresses in this fashion. Usually, sites with LANs separated enough to require a WAN to link them had their own individual IP address spaces. These addresses were administered by local personnel, as might be imagined. LANE meant that the IP addresses had to be coordinated across as many emulated LANs as there were. Most organizations just did not operate in this fashion. Not only were shared IP address spaces an issue, but most sites used IP address spaces that allowed for only some 250 devices on a LAN (Class C IP addresses). So two sites could deploy 500 devices. With LANE, the single 250-device address space had to be stretched to cover both sites. LANE required many changes, sometimes too many changes, to become as common as ATM advocates wished.

MULTIPROTOCOL OVER ATM (MPOA)

When it came to linking LANs running IP with routers over an ATM network on the WAN scale, LANE suffered from a number of limitations. Most of these limitations were related to the relationship between LANE and the IP routers that were still all over the LAN connectivity scene. LANE just did not fit particularly well into the world of routers. LANE was addressing the LAN issue, not the router issue, as a challenger to ATM deployment. But LANs were never the issue that caused an organization to find reasons not to buy ATM switches and build or use ATM networks. The real issue was that organizations were buying and using IP routers instead of ATM to link LANs.

So no sooner was the ink dry on the LANE specification than work began on what was called Multiprotocol over ATM, or MPOA. The goal of MPOA was to make an ATM switch (or IP router) into a more effective ATM device by allowing the device to route IP packets as well as switch ATM cells. The use of the term "multiprotocol" was a little odd in this context, since everyone knew that the one protocol that ATM needed to accommodate was IP. Technically, however, MPOA could route and switch packet types other than IP packets.

MPOA devices performed an odd mixture of routing and switching. All packets went into ATM cells, of course. But sometimes these packets were routed according to the IP address inside the cell, which meant that MPOA devices had to look inside the cell payload, a big no-no up to that time in the ATM world. Once the destination of the packet was found on the ATM network, the MPOA device could switch the rest of the packets in the "flow."

Nobody at the ATM Forum wanted to rehash all the work done on LANE, so LANE became a prerequisite for MPOA. In other words, if an ATM switch vendor wanted to incorporate and offer MPOA inside their product, they first had to support and implement LANE whether the customer wanted to emulate LANs or not. MPOA added several key components to LANE to extend the capabilities of LANE into the realm of routers.

The three major components that MPOA added to LANE were a basic specification for carrying IP packets inside ATM cells that predated even LANE called IP over ATM, a way to find paths between IP routers on the ATM network called the Next Hop Resolution Protocol (NHRP), and a way to handle IP multicasts called Multicast Address Resolution Servers (MARS). IP over ATM added a way to encapsulate IP packets directly inside ATM cells without the need for a LAN frame as in pure LANE. NHRP added a way for IP routers connected to an ATM network to find the "best route" between themselves, a key feature of IP totally absent in virtual circuit networks such as ATM. MARS added the capability to send a single input stream of packets to multiple destinations, another thing that was easy to do in IP but that was difficult to translate to the world of point-to-point ATM connections.

MPOA made an ATM switch "as good as an IP router" in terms of connecting LANs. At least that was the hope of the ATM switch vendors and ATM service providers pushing for MPOA. What MPOA does for LANs normally connected with routers is shown in Figure 9-6.

Once MPOA was added to LANE, there was no longer any restriction on IP address space assignment. IP addresses could be deployed and used as usual, since the MPOA-enabled ATM switch was essentially an IP router as well. It was hoped by ATM advocates that the widespread use of MPOA in ATM switches would result in a world with fewer IP routers, since the ATM switch could be both ATM switch and IP router at the same time. The way that the MPOA protocol stack adds routing to basic LANE bridging is shown in Figure 9-7.

The idea behind MPOA of making an ATM switch as good as a router at handling IP (and other) packets seemed very attractive at first. But MPOA implementations were complex, processor intensive, and expensive. It was not just a case of deciding whether to buy a new router or put MPOA on an ATM switch. Too often it was a case of deciding whether to buy *many* new routers or put MPOA on an ATM switch and spend weeks or months getting the whole agglomeration of ATM switches to work properly with the existing IP routers. Most organizations chose to go with what they knew, and that meant IP routers.

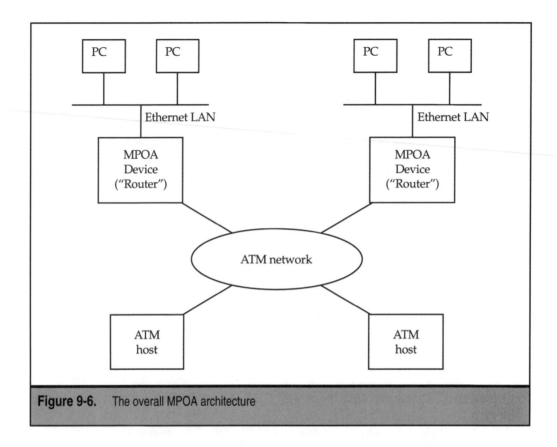

Figure 9-6.   The overall MPOA architecture

MPOA promoters pointed out that it would be just as easy to add MPOA to an IP router as it was to add MPOA to an ATM switch. But in the real world where IP companies did IP and ATM companies did ATM, there was little realistic hope that MPOA would somehow lead to a grand coalition of routing and switching. If no one made an MPOA router, then no one could buy one. Tied so closely to ATM, MPOA never seriously threatened the IP router market.

And once again, MPOA succeeded because it hid the ATM. This struck many as an odd strategy for a technology that was supposedly so much better than anything else.

WHITHER ATM?

So ATM started out as the switching and multiplexing methods for B-ISDN services. The ATM cell was a more flexible application of old TDM time slots, with ATM cells assignable according to "instantaneous bandwidth demand." ATM switches shuttled cells with voice, video, and data inside around very quickly over the same links, a feat impossible with fixed TDM circuits.

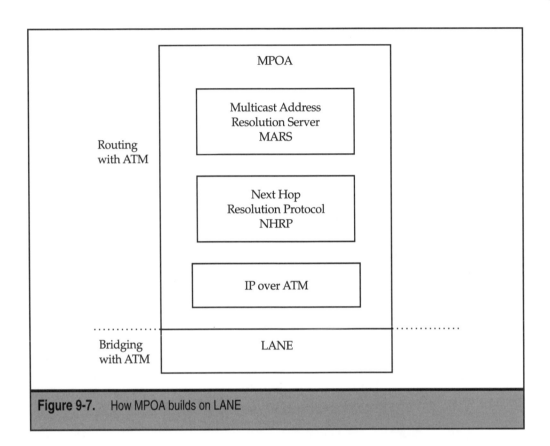

Figure 9-7. How MPOA builds on LANE

B-ISDN services were planned by all the major carriers, but once the Web hit town in the mid-1990s, the Internet started providing the types of services like home shopping and music videos in the here and now that B-ISDN could provide only in the future. And for the Web, all that was needed was adequate bandwidth, not ATM. ATM was intended to make a limited-bandwidth environment useful for many things. With adequate bandwidth (and fast enough routers), ATM became superfluous. All ATM adds to a high-bandwidth environment is expensive equipment and crushing cell overhead in terms of both bits and processing power. Studies claim that at gigabit speeds, cells are more of a hindrance than a help.

So when B-ISDN services failed to materialize, the scramble was on, mostly led by the ATM switch vendors, to adapt ATM for use in the world of IP and routers. These efforts eventually gave birth to LANE and MPOA. Once these were unveiled by the ATM switch vendors, often with much fanfare, the Internet people tended to shrug and ask "is that the best you can do?" By that time, it was apparent to everyone that ATM and IP did not mix particularly well anyway.

Consider LANE. Few organizations embraced LANE, because it required all hosts to use the same IP address space (bridging) in creating the virtual LAN. But most organizations assigned different IP address spaces to different sites so that the routers would function properly. So LANE on a WAN became a technology that was odd to implement with IP routers (IP routers are "invisible" to LANE, since LANE runs at a lower network layer). LANE had some success only at the campus network level, since groups of buildings tended to draw on the same IP address space. And LANE remained expensive due to this lack of wide interest. Years of effort on the part of ATM advocates had ultimately yielded a niche product with limited marketability.

MPOA attempted to apply LANE to the way IP was used in the real world. In fact, MPOA advocates claimed that MPOA would run in an IP router as easily as in an ATM switch. This was true enough, but the industry reaction to MPOA availability on both IP routers and ATM switches was not "if I get an ATM switch, I won't need an IP router" but "if I have an IP router, I don't need an ATM switch." And even with ATM interfaces and MPOA, an IP router was still very much an IP router. Moreover, the requirement to build MPOA upon LANE, even as LANE suffered from its own implementation woes, doomed MPOA to a few ISP backbone applications where MPOA was used to link IP routers over an ATM backbone.

As time goes on, the story gets even worse for ATM as a technology of choice even for large backbone networks. The best features of MPOA (which meant out went the LANE and even ATM prerequisites) were incorporated by the Internet people into IP routers as Multiprotocol Label Switching (MPLS). It made sense, given the bandwidth and traffic loads on the Internet, to make a switch out of an IP router for some forms of traffic. But it made little sense to make the IP router into an ATM switch. ATM was required to support mixed traffic types with cells. However, cells did not carry mixed traffic anymore: IP packets carried the mixed voice (audio) and video and data traffic. ATM cells only carried IP packets (which carried the important stuff) inside the SONET/SDH frames (which were needed on the SONET/SDH links) in many cases. Why not just put the IP packets directly inside the frames and switch the IP packets in the routers? This is what MPLS is for in a nutshell (more on MPLS is in the next chapter). So what was that ATM for again?

ATM IN THE OPTICAL WORLD

The answer to the last question, by the way, is "ATM is for delivering B-ISDN services in a restricted-bandwidth environment where it is necessary to have flexible bandwidth allocation because there is not enough bandwidth to give every form of traffic a fixed-bandwidth virtual circuit." Now, the Internet has replaced B-ISDN services as the first reason for ATM to exist. Optical networking with DWDM replaces the second reason. So in the world of optical networking, ATM has become a solution without a problem.

The current "squeeze" on ATM is shown in Figure 9-8. In the 1990s, the service provider vision for mixed-media services was B-ISDN for content delivery, ATM for switching and multiplexing, and SONET/SDH for data unit (cell) transport. In the 2000s, this vision has shifted to the Internet and Web for content delivery, IP packets for "switching" (routing) and the multiplexing of all these mixed-media streams together, consideration of non-SONET/SDH framing ("digital wrappers") for data unit (IP packet) transport,

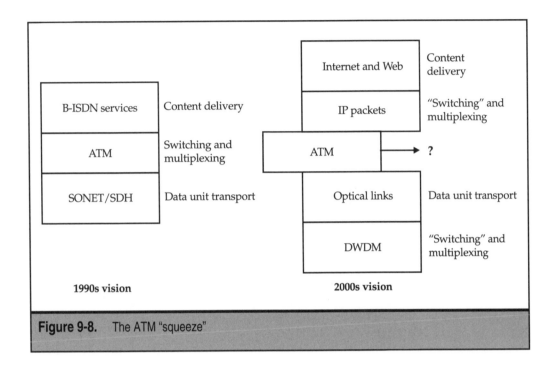

Figure 9-8. The ATM "squeeze"

and DWDM for even more wavelength "switching" (cross-connecting) and multiplexing. With all this switching and multiplexing going on already, is more switching and multiplexing with ATM really wanted or needed? Without B-ISDN above and SONET/SDH below, ATM in the middle makes little sense.

None of this should be interpreted to imply that ATM will go away anytime soon or even ever. Not only does ATM have a considerable installed base, although nowhere near the installed base of IP equipment, but ATM does enjoy some popularity as a vehicle for VPNs, as mentioned at the start of this chapter.

Even expensive ATM equipment could have been cost-justified by corporations in the early to mid 1990s if they saw that adding ATM to their corporate network was more cost-effective than trying to add bandwidth. Adding bandwidth required time to provision, and a long private line could be even more expensive than buying an ATM switch at each end of a link. In many cases, the carrier did not even have more bandwidth available to sell to some sites. In fact, many *carriers* used ATM to make more effective use of their own restricted backbone bandwidth internally.

Once invested in ATM, the corporate world felt a certain inertia to keep ATM going. If ATM could be used for anything that was perceived as more difficult to do than in IP, then ATM was the vehicle of choice for that purpose. As the Internet became a more and more essential part of the corporate world for all forms of information needed by employees in the course of performing their jobs (or even when taking a break), the utter lack of any form of security at all on the global public Internet became a common corporate concern. Open the corporate network to the Internet, and the hackers followed quickly.

But ATM (and frame relay) featured virtual circuits that formed closed user groups (sometimes seen abbreviated CUGs). One end of an ATM connection only went to the other end of the virtual circuit and nowhere else. When coupled with various forms of authorization and encryption, ATM (and frame relay) connections became an important part of a corporation's VPN effort. For the first time in a long while, ATM could do something easily that was much more costly to do in a pure IP environment with open Internet access. Nobody hacks into a virtual circuit. (A lot of this was just a matter of efficiency: why should a hacker spend a month learning how to hack into thousands of ATM switches when they can spend a month learning how to hack into millions of IP routers?)

This situation is slowly changing with IP initiatives such IPSec, but for the time being it is still true that it is relatively easier to turn an ATM cloud into a VPN than it is to turn a cloud of IP routers into a secure VPN.

So in the world of optical networking, IP routers will coexist with ATM (and frame relay) switches as CPE devices at a site. All voice, video, and data will be delivered either in ATM cells or in IP packets, but in no other way. Voice can go to ATM as VTOA or IP as VoIP. There will most likely still be a SONET/SDH collector ring for WAN access, and to link close sites. At the local service provider office, there will still be DACS equipment grooming the traffic. But instead of feeding another ring, the local office will separate IP and ATM traffic streams. Both can now have their own DWDM channel(s) and be sent onto a DWDM ring instead of a larger and faster SONET/SDH ring. As all forms of traffic migrate to IP packets for site-to-site connectivity, more and more DWDM channels can migrate from ATM to IP. The security for VPNs formerly provided by ATM virtual circuits is now provided by DWDM channels. Nobody in one DWDM wavelength can hack into another DWDM wavelength. So ATM is no longer required even for virtual circuits. ATM is present only for backward compatibility.

This is all shown in Figure 9-9. This figure can be compared with Figure 9-1 in this chapter. The trend is toward a simpler, leaner architecture. And if ATM can be incorporated into IP packets, a concept sometimes called "cells in packets," then IP is the only game in town.

The rise of IP, fueled by the popularity of the Internet and the Web, has been phenomenal. However, ATM and especially SONET/SDH once enjoyed their own popularity. Yet both ATM and SONET/SDH now find themselves threatened by new forms of optical networking. But is IP invulnerable to similar threats from optical networking? At the moment, the answer seems to be "Yes, but IP might be changed...."

To understand why IP might have to change in order to adapt itself better to the world of all-media traffic on an optical network, it is first necessary to take a closer look at IP itself.

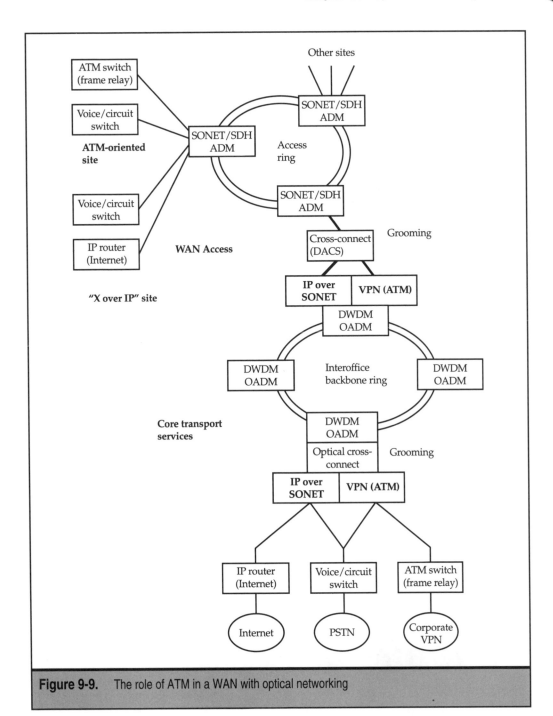

Figure 9-9.   The role of ATM in a WAN with optical networking

CHAPTER 10

Optical Networking and IP

The Internet Protocol (IP) is the Layer 3 (OSI-RM Network Layer) protocol used by the Internet and the Web. Needless to say, this aspect of IP alone makes consideration of the relationship between optical networking and IP a worthy topic for a chapter by itself. The fact that IP is a Layer 3 protocol is itself of some interest. This is because the other WAN networking methods examined in this part of the book are not Layer 3 protocols in any shape, manner, or form. SONET/SDH is firmly a Layer 1 (Physical Layer) transmission layer according to the OSI-RM, and although ATM is officially undefined in terms of layers, ATM is often positioned as a Layer 1 or Layer 2 (Data Link Layer) protocol. This is mainly because of ATM cell switching, of course. The key role of the Network Layer is routing, which ATM does not do (without the addition of MPOA).

This chapter assumes a basic familiarity with the Open System Interconnect Reference Model (OSI-RM) layers for networking. The OSI-RM remains, as it has been since 1979, the major conceptual architecture (model) for computer-to-computer communications whenever interoperability among different vendors is a concern. The OSI-RM has been mentioned before, in Chapter 5 and Chapter 7, but only in passing. This chapter mentions the OSI-RM frequently, especially the bottom four layers of the complete seven-layer OSI-RM. It might be a good idea to introduces these layers more completely and systematically. The names and major functions of these lower layers of the ISO-RM are:

▼ **Layer 1 (Physical Layer)**   This layer sends raw bits across the link between two directly connected systems. There is no structure to the 0's and 1's here. The Physical Layer also defines the voltages, physical connectors, wavelengths, bit rates, and so on used on the interface.

■ **Layer 2 (Data Link Layer)**   This layer organizes the bits into a *data unit* that can be sent on the wire. It shuttles *frames* between two directly connected systems. Frames are the last things that bits are before they leave a system, and the first things that bits become when they arrive at a system. Bridges link LANs at Layer 2. Frames usually delimit the bits, provide error detection, and allow the receiver to determine the structure of the frame contents. Frames have Layer 2 source and destination addresses on LANs, and usually connection identifiers on WANs (e.g., frame relay).

■ **Layer 3 (Network Layer)**   This layer shuttles *packets* (sometimes called *datagrams*, especially in IP) across many *intermediate systems* or network nodes. Packets can be sent between two directly connected systems as with frames, but usually the packet must be processed (and some packet header fields changed) by these intermediate systems or network nodes. Intermediate systems or network nodes can be loosely defined as systems that are neither the sources nor the destinations of packets. Routers or switches are examples of intermediate systems. Packets form the *data content* of frames. It is important to note that if two Ethernet LANs (for example) are linked by a router, the packet content and the basic header are the same on both LANs, but the frame is technically a different frame on each LAN and will have different Layer 2 addresses.

▲ **Layer 4 (Transport Layer)** Once called the end-to-end layer, this layer's information forms the *content* of the packets between applications running on the *end systems* (*hosts* in IP) of the network. In IP, the packet content is often called a *segment* when a protocol known as TCP is used, but the OSI-RM just calls this layer's information *messages* or *Layer 4 protocol data units (PDUs)*. There are source and destination *ports* in the segment header. The combination of IP packet address and segment port number completely identifies a particular client or server application anywhere on the network.

The bottom four layers of the OSI-RM are shown in Figure 10-1. For the sake of completeness, the OSI-RM also defines Layer 5 (the Session Layer) for keeping track of the history of a connection, Layer 6 (the Presentation Layer) for taking care of differences in internal representations of bits on different computer platforms, and Layer 7 (the Application Layer) for bundling functions common to specific applications (e.g., e-mail, remote login) into packages to make implementation easier.

Anyway, if IP is to be used with optical networking, what becomes of Layer 1 and Layer 2? If these layers remain SONET/SDH and/or ATM, then the issues covered in the previous chapters would suffice. But the interesting thing is that there are proposals and methods for making the relationship between IP and optical networking, especially the DWDM aspects of optical networking, much simpler and more streamlined than ever before. What if an efficient way to carry IP packets directly on optical networks were found and standardized? If it is not possible to directly carry IP packets on an optical network, what is the *minimum* that needs to be added to IP and/or optical networking to make this marriage easiest? And what additional network technology fits this minimum requirement the closest? These are the main issues to be examined in this chapter.

Layer 4 Transport layer	Major function: transfer of messages end-to-end across network. Data unit: Segment in TCP/IP.
Layer 3 Network layer	Major function: transfer of packets end-to-end across network. Data unit: packet (content of frame).
Layer 2 Data link layer	Major function: transfer of frames between directly connected systems. Data unit: Frame.
Layer 1 Physical layer	Major function: bit transfer. Data unit: unstructured stream of bits.

Figure 10-1. The bottom four layers of the OSI-RM

Before jumping into these issues, however, it is first necessary to at least outline what it is that makes IP so important today and how IP's relationship to optical networking differs fundamentally from how SONET/SDH and ATM fit into the same world. One of the key aspects of these features of IP is the fact that the IP network nodes are not multiplexers and cross-connects (as in SONET/ATM) nor switches (as in ATM) but *routers*. However, the network nodes in optical networks *are* multiplexers and cross-connects and switches. So the relationship between just what a router is and just what a switch is becomes very important in this chapter. The easiest way for IP to ride an optical network is if the network nodes are switches.

Fortunately, there are standard and interoperable ways to carry IP packet flows through a switch without requiring full IP routing. So this chapter introduces Multiprotocol Label Switching (MPLS), the alternative to MPOA for IP over ATM networks and the generic way to carry IP packets in a switched network. Naturally, the relationship between IP over MPLS is examined not only with regard to ATM, but with regard to a switched optical network.

At the end of this chapter, one possible *optical router* architecture is introduced, an architecture that requires no packet conversion to electricity at all. This architecture depends, not surprisingly, on MPLS or a related IP switching technology. Then the relationship between MPLS and optical networking is explored more fully, with the idea that MPLS can become the control channel for an optical network, reducing the dependence on SONET/SDH to fill this role.

THE INTERNET PROTOCOL SUITE

Until recently, the Internet protocol suite went by the simple acronym TCP/IP. The forerunner of TCP/IP had been the official protocol of "the Internet" since 1974 or thereabouts, although all current documentation is dated to 1981. This predates even the name "Internet" by a few years (back then it was the ARPAnet). Using the term "TCP/IP" to describe the Internet protocol suite had several drawbacks, however. First, TCP was not the only Layer 4 protocol that rode inside the Layer 3 IP packets (there were UDP and many others besides). Second, TCP and IP were only two major protocols running at two layers of the complete four-layer Internet protocol suite. Finally, even though 90 percent to 95 percent of all Internet traffic still uses TCP, a lot of newer Internet applications, such as voice over IP and video streams, do not use TCP at all. The term "Internet protocol suite" is more accurate and less limiting in this regard.

What the Internet protocol suite, or TCP/IP, actually does is to enable an organization to create an *internetwork*. Rather than tying all the computers at all sites into one big network, TCP/IP makes it possible to have individual networks (LANs) at various locations (and under local control and administration) and yet tie all of them together with a higher-level structure known as an internetwork, or just internet.

Such an internetwork would enable all users at all sites to interoperate as if they were all on the same network, when in fact they are not. There are several critical factors that have led the computer networking industry to create internetworks in the first place and to use TCP/IP as the protocol of choice for this task in the second place. TCP/IP has been so successful as a network interconnection protocol that recently many organizations have begun using it to interconnect desktop devices such as PCs and printers as well, not just when access to the Internet is required.

The network node used in TCP/IP is the IP router. A router may not only link two LANs, but many. This will keep down the number of routers needed, and thus the cost involved, by the organization. However, the router must be able to find the correct outbound link to send the data to the proper network. In TCP/IP, the networks are identified by a unique network address known as the IP address, since it is used by the IP protocol running in the router.

The use of the IP address for building internets is much more efficient than using physical port or LAN addresses because the IP address has a structure to it lacking in the other types of address. The details are not important here, but the IP address is divided into a network identifier portion (NET_ID) and a host (computer device) identifier (HOST_ID) portion. This enables the IP address to be used to unite the constituent networks into an internet. Each network will have a unique IP NET_ID portion and yet still be reachable by all other connected networks, since the IP address is understood by all network devices (routers) on the network.

The concept of network identifier and host identifier portions of the IP address is shown in Figure 10-2. Once there was a concept of IP address classes, still officially defined. But all IP routing today is done based on something called a *subnet mask* that is associated with each and every IP address. Officially, the subnet mask determines the IP address *prefix* to be used for routing purposes. The subnet mask determines where the break between NET-ID and HOST_ID (device) falls in the IP address. The prefix is the NET_ID portion of that address.

This hierarchical structure of IP addresses is similar to the structure of a telephone number in the United States, but with significant differences. A complete telephone number in the United States has a structure as follows: one-digit country code (omitted for domestic calls), three-digit area code, three-digit central office switch code, and four-digit line number code (for example +1 914-422-4191). When someone in area code 813 in Florida makes a long distance phone call to New York, the long distance telephone switching equipment does not need to know exactly which telephone in New York to ring. It is enough to know that the destination telephone is in New York (914), so the switch just routes the call north. Routers work the same way. But instead of using the area code portion of a telephone number, IP routers use the network portion of the full IP address to route data. Only the router directly attached to the destination network needs to know more information about the destination computer, just as in the voice telephone network. The major difference is that IP network addresses have never been consistently assigned on a strict geographical basis. Then again, "overlays" to numbers for highly mobile cell phones complicate the telephone network today also.

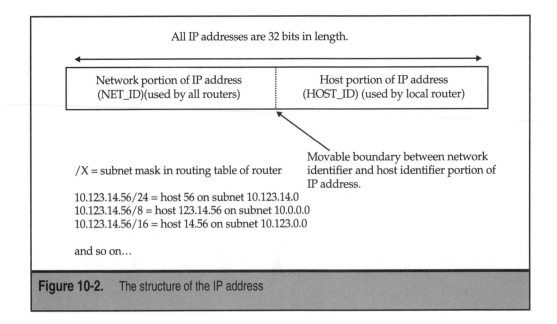

Figure 10-2. The structure of the IP address

The Internet and TCP/IP

Many organizations around the world are linking their LANs to a preexisting router-based network already connecting LANs in most of the world's countries. This is the global public Internet, the original home of the TCP/IP protocol stack. In fact, many organizations' first exposure to TCP/IP is when the organization links to the Internet through a router. Especially in a commercial environment today, with the emphasis on global, not only national, business contacts and opportunities for sales, more and more commercial users are using the Internet to exchange electronic mail, files, and other information with other organizations throughout the world. Instead of having to buy leased lines thousands of miles long to link two sites, companies can just have short access links to the Internet at each location.

The rise of the Web has only made the Internet more indispensable for business today. The Web added graphical point-and-click capabilities to the command- and prompt-driven Internet. This made the Internet much more accessible for common people who were not interested in typing commands, but just finding the information they needed. The massive "dot com" industry, for all its ups and downs, was spawned by the Web.

The TCP/IP protocol does not require the Internet. Any organization may purchase routers and link them with some WAN connections. In many cases, these routers are connected to each other and to the Internet with private, leased, point-to-point links from various service providers. The TCP/IP software usually comes preinstalled on the router. However, it is the customer's responsibility to install the proper TCP/IP software on the other devices on the customer's network so that these devices may communicate with the routers.

Origins of TCP/IP

TCP/IP grew out of an effort in the late 1960s in the United States to create a network that would link together the various organizations receiving funds from the Department of Defense. Most of these organizations were universities, but not exclusively. The organizations received funds from the Department of Defense for research into defense-related activities (better tanks, rockets, armor, and so forth). The funds were dispensed by the Advanced Research Projects Agency (ARPA), and by the late 1960s, ARPA had grown to such an extent that keeping track of the disbursement and spending of these funds became a nightmare. Not only that, the government found that many of the research organizations needed to exchange information both to make rapid progress and to avoid duplication of effort. In many cases one group was stymied by a problem that had actually been previously solved by another group across the country but whose solution had not been communicated due to the sensitive nature of the research and the lack of adequate contact between the groups.

By the 1960s, more and more of the information used by these research groups was on computers, although most universities had exactly one. Many people used the "host" (as most of these mainframe computers were called) by linking terminals to the central computer, but usually only on the campus, or even in the same building. These terminals had no real processing power of their own and were little more than keyboards and output devices (usually scrolls of paper, not even screens).

So a proposal was made to link all the computers used by the ARPA-funded research groups with a network. A new networking scheme was sought, mainly due to the fact that the overwhelming majority of network protocols available for the implementers were proprietary. That is, organizations linked IBM computers with IBM's Systems Network Architecture (SNA), DEC computers with DEC's network scheme, and so on. Although these two examples actually date from the 1970s, they are more familiar and better examples than obscure protocols. The point is the same. In the 1960s, there was only one real "network" around: the airline industry's reservation system called Sabre.

Linking computers from different vendors was in most cases difficult, if not impossible. Since the various organizations had many different kinds of computers, a proprietary scheme was not really feasible. The fact that these proprietary network protocols were expensive to license from the vendors was a drawback also.

When the initial proposals were accepted and funding provided, the government built the first working packet-switched network in the world: the ARPAnet. In a packet network, all data is organized into variable-length units of data known as *packets*, which are routed to their destinations through various network nodes. *Network nodes* are specialized computers with no users on them that merely move the packets on their way through the network. In the original ARPAnet, the network nodes were known as "gateways," but they are now universally referred to as "routers." The initial design proposal was made in 1967, and the contract was awarded in 1968.

In 1969, the ARPAnet consisted of a grand total of four network nodes or gateways. From these humble beginnings, the global public Internet was born. In the original network, the network functions were much too complex to be added to each and every participating computer. Other special computers known as "IMPs" (for Interface Message Processors) handled all the network tasks. The capabilities needed from the new network were straightforward:

1. Any terminal may access any host on the network, not just the one to which it is locally attached (known as "remote login").

2. Any file may be copied across the network to any other computer (known as "file transfer").

3. Any pair of users may exchange messages across the network (known as "electronic mail" or just "e-mail").

These three requirements formed the basis for the applications supported by the ARPAnet. But the protocols used on the network were not stabilized until almost five years later with the invention of what became TCP/IP by two researchers named Vinton G. Cerf and Robert E. Kahn. In 1974, it was a mature enough network technology to implement on the ARPAnet, although it was not split into TCP and IP layers until 1978.

In 1983, the ARPAnet had grown to encompass 300 "hosts," mainly mainframes and minicomputers. Newer personal computers (PCs) and workstations (similar to PCs, but usually running the UNIX operating system) linked by local area networks (LANs) came a few years later. A "host" to the ARPAnet was defined as any computer capable of running the TCP/IP programs, which were a considerable processing load for the simpler computers available in those days. That year, TCP/IP became the official Department of Defense networking protocol, meaning that the DOD would buy no computer systems that did not run TCP/IP. This was significant because the DOD was the largest buyer of computer equipment in the world.

The network continued to grow, and many military research installations were linked into the network until the government became concerned that the lack of security on the network would expose military bases to unauthorized access on the part of college students and enterprising individuals. Horror stories of missing data and other abuses were rampant and eventually formed the basis for many books and at least one movie (*War Games*). In 1984, the DOD acted by splitting off military bases into a "Milnet" with very restrictive links to the original network. In 1986, the National Science Foundation (NSF) started building a new backbone to replace ARPAnet. Most ARPAnet sites got NSFnet connections during the next few years. By 1990, the ARPAnet officially became the "Internet," and the router network backbone became the "NSFnet," administered and funded by the National Science Foundation. Although this technically severed the military connection, many people (especially in other countries) still associate the Internet with the United States Department of Defense.

One other development in the early 1980s gave an enormous boost to TCP/IP outside the relatively restricted environment of the ARPAnet. Many of the computers on the ARPAnet were running some version of UNIX as the operating system. There were a couple of good reasons for this, not the least of which was the same idea of "openness" that communications protocols benefited from. UNIX was flexible, meaning it ran on a variety of vendors' platforms and was not tied to any particular vendor's equipment. UNIX was also extensible, meaning that since the operating system itself was written in a high-level language (called C language), the UNIX operating system functions themselves could be enhanced, changed, or otherwise modified merely by writing more code and linking this new code to the UNIX core functions, known as the "kernel."

Many colleges and universities obtained UNIX and modified it in exactly this way. UNIX was unique in that it was the product of AT&T, not one of the traditional mainframe or minicomputer vendors. Part of the restrictions placed on AT&T as part of the regulated telephone environment of the 1950s included a prohibition from the federal government that did not allow AT&T to sell any computer hardware or software. (The philosophy was that AT&T's revenues from their monopoly telephone business would allow AT&T to undercut the prices of all other vendors of computers—even powerful IBM.) Of course, AT&T's Bell Telephone Laboratories was one of the premier developers of computer hardware and software in the world. But because of this restriction, Bell Labs was forced to license the entire UNIX operating system to anyone who asked for it for the princely sum of $1.00. Universities and colleges, ever conscious of their budgets, and looking for one operating system to run on many different computer platforms, found the offer irresistible.

One of the universities that got UNIX and modified it was the University of California at Berkeley. In 1983, Berkeley themselves began distributing their modifications to UNIX to other organizations. This was known as the Berkeley Systems Distribution (BSD) of UNIX. The key is that this new version of UNIX included TCP/IP as the communications protocol software bundled with the operating system. Naturally, because it was included, few of the cost-conscious organizations that obtained BSD 4.2 saw the need to go out and buy anything else, especially when the other organizations they wished to link to were running their own BSD TCP/IP as well. Today, TCP/IP is bundled with almost every version of UNIX from Linux to FreeBSD, and with Windows as well.

TCP/IP Architecture

Figure 10-3 shows the relationship between the four layers of the Internet protocol suite and the full seven-layer OSI-RM. Layer 3 was the "routing" layer in the OSI-RM, charged with taking connectionless Layer 3 PDUs (packets) from an input port, looking up the network (or rather internetwork) address in a table of some kind, and then forwarding the packet onto the correct output port to the destination network. In this fashion, the routers move packets "hop-by-hop" to the *next hop router* through the router network. In this context, a "hop" is by definition a single point-to-point link between any two adjacent routers.

Layer 4 was the "end-to-end" layer in the OSI-RM, charged with assuring the correct reception of information from one end system (source) to the other end system (destination). The Transport Layer (Layer 4) must also break up large files and messages into smaller units for transfer across the network (called "segments" in TCP) inside the packets themselves. The destination Transport Layer must of course recover these data unit segments, sequence them if necessary (connectionless networks often deliver out-of-sequence information), and even

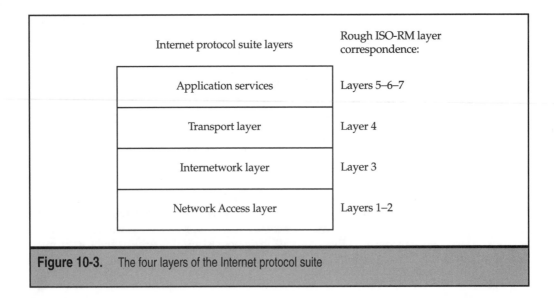

Figure 10-3.   The four layers of the Internet protocol suite

ask for a retransmission of data received in error (based on some error detection scheme). All in all, a considerable amount of work. Obviously, since the Layer 4 software on the end systems does this, there is no need at all for Layer 4 software in the intermediate systems (routers). (In practice, all routers today implement all four layers of the Internet protocol suite.)

The Internet layer holds the network (IP) address and performs the routing function in the TCP/IP architecture, exactly as does the Network Layer in the OSI-RM. The Transport layer, when TCP (Transmission Control Protocol) is used, ensures the orderly and correct delivery of data from end-to-end across the network (internetwork) exactly as the Transport Layer in the OSI-RM. The TCP protocol data unit is called the *segment*. But these two core protocols of the TCP/IP architecture only occupy two layers of the entire seven layer OSI-RM.

Layers 1 and 2, the Network Access layer of the Internet protocol suite, are not defined by the TCP/IP protocol itself. Instead of defining its own Layer 1 and 2 protocols, the TCP/IP architecture defines an *interface* with many existing international standard Layer 1 and 2 protocols. For instance, documentation has been established defining how IP packets are to be sent over Ethernet-type LANs. Other documents define how IP packets are to be transported by Token Ring LANs, X.25 networks, and many other international and de facto standards for Layers 1 and 2 as well. This means that the TCP/IP architecture is very flexible: it is capable of being implemented on many different kinds of networks (which TCP/IP will make into a large internetwork).

The top of the Internet protocol suite above the Transport Layer is a bit more complicated. In the full OSI-RM, the top three layers (the Session, Presentation, and Application Layers) are usually grouped together and implemented as a whole. Taken together, these layers will establish sessions (Layer 5), perform any data conversion needed between source and destination system (Layer 6), and supply specific functions for application programs running in the local memory of any system attached to the network (Layer 7). In the full Internet architecture, the functions of these layers are also grouped together.

In addition, the top of the TCP/IP protocol stack includes tools for interfacing with these basic applications. So developers may still write TCP/IP-enabled front-end applications that still do file transfer, for example, but add to the functionality of the very simple FTP application included as part of the TCP/IP architecture.

More details of the layers of the Internet protocol suite are shown in Figure 10-4. There is a lot going on here, much of which has helped to make TCP/IP the protocol of choice for almost any form of network traffic today. This also impacts how IP fits in with optical networking, so a few more words are in order here. (Thanks to Gary Kessler for suggesting this representation instead of the typical "tiny boxes" form that many books use.)

The Lower Layers of the Internet Protocol Architecture

The lower layers of Internet protocol include the Network Access layer and the IP layer itself. It is only these lower layers that need deal with the routing of packets through the routers that form the network devices in a TCP/IP network. The entire four-layer TCP/IP architecture need only be present on the end systems (sources and destinations of files, mail, etc.) on the network.

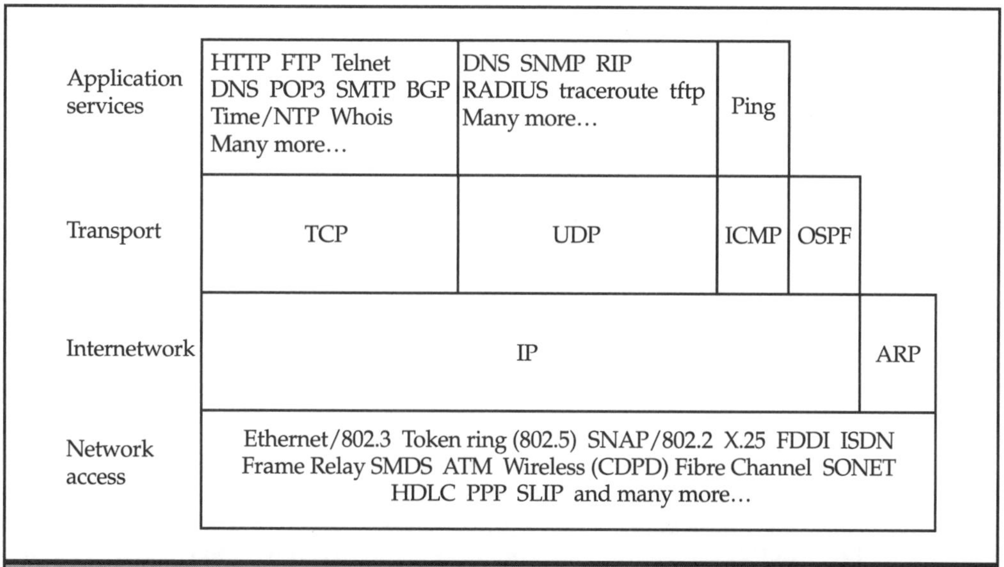

Application services	HTTP FTP Telnet DNS POP3 SMTP BGP Time/NTP Whois Many more…	DNS SNMP RIP RADIUS traceroute tftp Many more…	Ping		
Transport	TCP	UDP	ICMP	OSPF	
Internetwork	IP				ARP
Network access	Ethernet/802.3 Token ring (802.5) SNAP/802.2 X.25 FDDI ISDN Frame Relay SMDS ATM Wireless (CDPD) Fibre Channel SONET HDLC PPP SLIP and many more…				

Figure 10-4. The Internet protocol suite in more detail

The Network Access layer of TCP/IP deals with things like physical connectors, bit signals, and media access control—in short, everything that the OSI-RM covers in Layers 1 and 2. The Network Access layer need not be a LAN. In fact, IP packet interfaces have been defined for Frame Relay and X.25 wide-area networks also. All IP implementations must be able to support at least one of these interfaces.

Interfaces for IP packets have been defined for all of the following:

1. Ethernet from DEC, Intel, and Xerox (known as DIX Ethernet)

2. IEEE (Institute of Electrical and Electronics Engineers) 802.3 Ethernet-based LANs

3. Token Ring LANs (IEEE 802.5)

4. Point-to-Point Protocol (PPP from the IETF developers themselves, the closest thing TCP/IP has to an official Network Access layer protocol, a very popular choice)

5. X.25 (an international standard public-switched connection-oriented network protocol)

6. Frame Relay (the international standard public-switched connection-oriented network protocol)

7. FDDI (Fiber Distributed Data Interface, a LAN-like network running at 100 Mbps)

8. SMDS (Switched Multimegabit Data Services, a high-speed LAN-like public network service)

9. ISDN (Integrated Services Digital Network, a public-switched network similar to X.25)

10. SLIP (Serial Line Interface Protocol, an older but common way of sending IP packets over a dial-up, asynchronous modem arrangement)

11. HDLC (High-Level Data Link Control, similar to PPP)

12. CDPD (Cellular Digital Packetized Data, a way of sending IP packets over the cellular mobile radio network. CDPD is sometimes loosely called "wireless TCP/IP.")

There are more obscure interfaces as well (e.g., ARCnet, IEEE 802.4 LANs, HYPERchannel), but these need not be detailed due to their relatively uncommon implementation.

The Network Access layer of TCP/IP basically takes IP packets at the source and puts them inside whichever frame structure is used at Layer 2 (e.g., Token Ring frame, Ethernet frame). The Network Access layer will then send the frame as a series of 0's and 1's (bits) on the link itself, exactly as expected of Layer 1. At the receiver, the Network Access layer will recover the frame from the arriving sequence of 0's and 1's and extract the IP packet. The IP packet is then sent to the receiving Internet layer.

The advantage of not tying the IP layer to any specific network at the lower layers is flexibility. A new type of network interface (as for sending IP packets on an ATM [Asynchronous Transfer Mode] network) can be added without great effort. Also, it makes the linking of these various network types into an internetwork that much easier.

The IP layer functions in a connectionless manner. This means that IP packets are routed completely independently through the internetwork. This also means that packets may arrive out of sequence, or even with gaps in sequence (due to lost packets), at the destination. IP does not even care which application a packet belongs to: it delivers all packets without a sense of priority or sensitivity to loss. IP itself will not be concerned with this aspect of its operation, which is characteristic of all connectionless networks. Sequenced traffic, priorities, and guaranteed delivery (i.e., acknowledgments) must be provided by the upper layers of the TCP/IP protocol stack. These are not functions of the IP layer.

There are two other major protocols that run at the IP layer besides IP itself. The routers that form the network devices in a TCP/IP network must be able to send the end systems (hosts) error messages in case a router must discard a packet (due to lack of buffer space because of congestion, for example). This protocol (which must run at Layer 3 because the other layers are technically lacking in routers) is known as the Internet Control Message Protocol (ICMP). ICMP messages themselves are sent inside IP packets. All IP implementations must include ICMP.

The other major protocol actually has many different functions depending on the exact network type that IP is running on. These are known as Address Resolution Protocols (ARPs), and their main function is to provide a method for the IP layer protocol itself to determine the proper network address to place in the frame header in order for the network beneath the IP layer to deliver the frame containing the IP packet to the proper destination. For instance, if IP is running on a LAN, frames are addressed by MAC (Media Access Control) address, coded into every LAN adapter card. ARP offers a method for the IP layer to send a message onto the LAN saying, in effect, "Who has IP address 192.168.40.69?" The IP software on each system on the LAN will examine the ARP message, and the system that has the corresponding IP address (192.168.40.69) will reply to the sending IP address with the correct MAC layer address in the frame.

The Upper Layers of the Internet Protocol Architecture

The upper layers of the Internet protocol suites include the Transport layer itself and the Application Services layer. Again, this reflects OSI-RM terminology. If the Transport layer uses TCP, then this layer, unlike the Internetwork layer, is connection-oriented. That is, before an application running on an end system can send any information to another end system running TCP/IP, the TCP protocol running at the TCP layers on the two systems must establish a connection first. This connection establishment request may be rejected by the destination system for a number of reasons. The destination system may already be overwhelmed by traffic and now have enough resources free to process the information arriving on yet another connection. Or the intended destination may screen out connection requests from various source systems as a matter of organizational policy (e.g., security).

The TCP layer contains all the functions and mechanisms needed to make up for the "best effort" connectionless delivery provided by the IP layer. Packets may arrive with errors, out of sequence, or even with gaps in sequence due to lost or discarded packets. TCP must guarantee that the segment (the TCP layer protocol data unit) is delivered to the destination application error-free, complete, and in sequence. In keeping with the general philosophy of connection-oriented networks, TCP provides acknowledgments that periodically flow from the destination to the source to assure the sender that all is well with the data sent up to that point in time.

On the sending side, TCP passes segments to the IP layer for placement in packets, which IP will route connectionlessly to the destination. On the receiving side, TCP accepts the incoming segments from the IP layer and delivers the data to the proper application running at the layer above TCP in the exact order in which the data was sent.

Like the IP layer, the TCP layer has another major protocol that runs at the TCP layer besides TCP. This is the User Datagram Protocol (UDP). UDP is a protocol that is as connectionless as IP. When applications run with UDP at the TCP layer instead of TCP, there is no need to establish (or maintain) a connection between a source and a destination before sending data. Connection management, of course, adds overhead and delay to the network. UDP offers a way to send data quickly and simply. However, it offers none of the services that TCP supplies, such as resending lost segments. Applications cannot count on TCP to ensure error-free, guaranteed (via acknowledgments), in-sequence delivery of data at the destination.

For some simple applications, purely connectionless data delivery may be sufficient. Single messages between applications may be sent more efficiently with UDP because there is no need to exchange TCP segments to establish a connection in the first place. Most applications will not be satisfied with this mode of operation, however. So all of the primary TCP/IP applications use TCP, and most are expressly forbidden to use UDP. Similarly, several applications in TCP/IP must use UDP only and will not function correctly with TCP (such as streaming voice and video).

At the top of the Internet protocol stack, at the Application Services layer, are the basic applications and services of the TCP/IP architecture. Application Services offers several basic applications themselves. These are typically bundled with the TCP/IP software distributed by various vendors and, fortunately, are generally interoperable.

The standard application services include a file transfer method (File Transfer Protocol: FTP), a remote terminal access method (Telnet), an electronic mail system (Simple Mail Transfer Protocol: SMTP), and a Domain Name System (DNS) directory service for domain name–to–IP address translation (and vice versa). In addition, many TCP/IP implementations include a way of accessing records within files remotely (rather than transferring the whole file) known as Network File System (NFS). There is also the Simple Network Management Protocol (SNMP). Note that some applications are defined to run on TCP and others are defined to run on UDP. Some applications, such as DNS, use UDP for some operations and TCP for others.

IP Packet Header

The structure of the IP packet is shown in Figure 10-5. The minimum (no options: very common) IP header length is 20 bytes (always shown as 4 bytes per line), and the maximum length (rarely seen) is 60 bytes. Some of the fields are fairly self-explanatory, such as the fields for the 4-byte (32-bit) IP source and destination addresses, but others have specialized purposes.

These fields of the IP packet header are as follows:

▼ **Version**   Currently set to "4" for IPv4, the most common form of IP used. The new version of IP is IPv6.

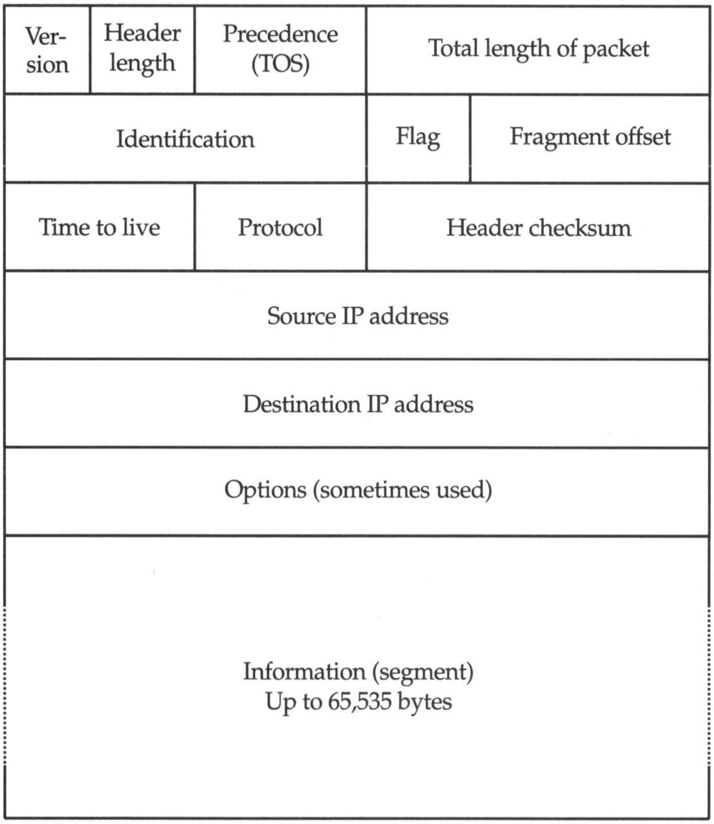

Ver-sion	Header length	Precedence (TOS)	Total length of packet	
Identification			Flag	Fragment offset
Time to live		Protocol	Header checksum	
Source IP address				
Destination IP address				
Options (sometimes used)				
Information (segment) Up to 65,535 bytes				

Figure 10-5.   The IP packet header

- **Header Length**   The length of the IP header in four-byte (32-bit) units known as "words."

- **Type of Service**   Parameters that affect how a packet is handled by routers and other equipment.

- ▲ **Total Length of Packet**   The length of the packet in bytes. The maximum is 65,535—a length that is approached by no TCP/IP implementation.

The next three fields directly figure in the *fragmentation* process. Fragmentation occurs when a larger IP packet (e.g., 9,000 bytes) must be broken up to fit inside a smaller frame (e.g. the 1,500-byte frame used in most types of Ethernet). These are:

- ▼ **Identification**   A number set for each packet fragment that helps the destination host to reassemble fragments that share the same identification number.

- **Flag**   Only the first three bits are defined. Bit 1 must be 0. Bit 2 is the DF ("don't fragment") bit. It is 0 if fragmentation is allowed (UDP should not allow fragmentation of the packet containing a UDP message) or 1 if fragmentation is *not* allowed (e.g., for UDP). Bit 3 is the MF ("more fragments") bit; it is 0 if the packet is the last fragment or 1 if there are more fragments to come.

- ▲ **Fragment Offset**   When packets are fragmented, the fragments must fall on an 8-byte boundary. That is, an 800-byte-long packet may be fragmented as two packets of 400 bytes, but not as eight packets of 100 bytes, since 100 is not evenly divisible by 8. This field contains the number of 8-byte "units," known as "blocks," in the packet fragment.

The rest of the IP header fields are as follows:

- ▼ **Time to Live**   This value is the number of *seconds,* up to 255 maximum, that a packet is supposed to have to reach the destination. Each router is to decrement this field by a preconfigured amount. If the packet has the field set to 0, it is discarded. Unfortunately, there is no easy way to track this time, so most routers interpret this field as a simple "hop count" between routers and decrement this field by 1.

- **Protocol**   This field contains the number of the Transport layer protocol that is to receive and process the data content of the packet. The protocol number for TCP is 6 and UDP is 17.

- **Header Checksum**   An error check on the IP header fields only, not the data fields.

- **Source and Destination Address**   The IP addresses of the source and destination hosts.

- ■ **Options**   The options are seldom used today and will not be described further.

- ▲ **Padding**   When options are used, the padding field is used to make certain the packet header ends on a 32-bit boundary. That is, the packet header must be an integer number of four-byte "words."

THE ROUTER AND THE SWITCH

Another important concept for understanding just how TCP/IP functions is the idea that all networks and network protocols are inherently connection-oriented or connectionless. In a connection-oriented network, such as frame relay or ATM, a connection must be established between a sender and a receiver before data transfer can take place. The connection may be established by a special signaling protocol (similar to dialing a telephone number to establish a connection on the public voice network) to set up a "circuit" across the network, or the circuit may be set up on a permanent (really semipermanent, since it may be changed with advance notice) basis and be available for use all the time. In modern connection-oriented networks, connections set up with a signaling protocol are known as "switched virtual circuits" (SVCs), and those that exist for long periods of time are known as "permanent virtual circuits" (PVCs). The term "virtual" refers to the fact that such connections are usually sharing a single communications link and therefore are not really circuits in the sense of a single wire from point A to point B. This setup takes time and effort on the part of the network, delays the start of data transfer for SVCs, and adds overhead to the network. For this reason, SVCs have been slow to come to frame relay and ATM in general. The signaling protocol for frame relay is called Q.933, and the signaling protocol for ATM is called Q.2931.

In a connectionless network, data may be transferred without a connection between the source and the destination. Such data is merely presented on the link to the network, and the network makes a "best effort" to deliver the data. However, there is usually no guarantee that the receiver is actually able to receive the data or is even on the network at all. On the other hand, on a connection-oriented network, the acceptance of a connection indicates both the presence of the receiver and its willingness to accept data. All connection-oriented networks will periodically send a special acknowledgment message from the receiver to the sender saying in effect "Everything I got up to this point is okay." These acknowledgments may be present on a connectionless network (an acknowledged connectionless network), but usually they are not. Any feedback from the receiver to the sender on a pure connectionless network must be sent by the application itself on the destination system.

TCP/IP can be both connection-oriented and connectionless. That is, one major component is connectionless (IP) and the other major component is connection-oriented (TCP). UDP is connectionless end-to-end, but studies show that 90 percent to 95 percent of all traffic on the Internet is still connection-oriented TCP. There is a very good reason for this heavy use of connections end-to-end even with connectionless IP routers along the way. Connections offer a network the benefit of knowing the destination system is reachable, and the acknowledgments provide needed feedback on the status of the end-points. Connections also guarantee that data arrives in order, an important consideration (otherwise the receiving program must order the data units). And the ISO-RM Session Layer 5 will only keep track of the state of the session if it is represented by a connection. There is no such thing as a connectionless session.

But connectionless networks are just faster, *if* the network consists of unreliable nodes and links. The way all this is usually expressed when referring to TCP/IP is to say that TCP/IP is connectionless on the network nodes (IP) and connection-oriented on the end systems (TCP). However, there are exceptions related to MPLS detailed later. (In fact, if the network nodes and links are considered to be reliable and hardly ever fail, then connection-oriented switching can actually be faster than connectionless routing!)

It might be of interest to point out that the concept of connectionless routers was first developed for the ARPAnet. There was a need to limit the complexity of the network and yet make the network as robust as possible for potential military purposes. With SVCs, every connection-oriented network must still route call setups connectionlessly with routers. Only the *user data* flows on connections. The leap of imagination that created IP was to realize that if *everything* on the network were routed as a call setup, the resulting network would be much simpler than building both a connectionless fabric for signaling and a connection-oriented fabric for user data.

So in a very real sense, a router is a connectionless switch, and a switch is a connection-oriented router. It was hard until recently to make a device perform both tasks well, so normally a network node was either a switch or a router but seldom attempted to function as both. This has changed as faster hardware and software have made connectionless switches (ATM switches with MPOA) and connection-oriented routers (IP routers with MPLS) possible.

Even in the telephony world, voice connections were made through the central office switch by "signaling routers," part of the Signaling System 7 (SS7) architecture. SS7 is the international standard for telephony signaling that makes it possible to directly dial almost any telephone in the world.

Routers determine output ports from full, globally unique destination addresses. Each packet is routed independently, so routing around trouble spots is easy. The "best" route is determined by a table lookup that holds all possible reachable destinations. The table is maintained by a routing protocol, so no manual configuration is necessary and the network reacts quickly to congestion and failures. The biggest advantage is flexibility. The biggest drawback is unstable paths that can lead to highly variable delays through the network and unpredictable bandwidth availability.

Switches determine output ports from smaller, locally unique connection identifiers. Each packet is routed along the same established path, so trouble spots will fail all connections mapped through that point on the network. The "best" path is determined by a table lookup that holds only those connections currently active. The table is maintained by hand in PVC networks, so manual configuration is necessary and the network reacts sluggishly (often in 24 hours) to congestion and failures. The biggest advantage is stable delays and predictable bandwidth availability (depending on the skill of the people setting up the virtual circuits). The biggest drawback is that the network cannot easily shunt traffic quickly around trouble spots.

Figure 10-6 compares the characteristics of IP routers, PVC networks such as most frame relay and ATM networks, and SVC signaling network nodes such as the telephone network with SS7. Note that IP routers and SS7 signaling nodes behave much the same.

As mentioned, the biggest drawback of a pure connectionless network such as IP is that unstable paths can lead to highly variable delays through the network and unpredictable bandwidth availability. This is a fact of life on the Internet today. So voice and video, which require stable delays and bandwidth allocations to function properly, struggle to perform well on the Internet. But as the Internet becomes the network of choice for all forms of traffic, some changes must be made to IP to accommodate voice and data.

One way to do this in an IP network is to simply add a switching fabric to the IP router. Happily, this change to IP makes IP much more accommodating for a world of optical switching. One of the leading contenders for adding switching to IP routers is Multiprotocol Label Switching (MPLS).

	IP routing	Switching and PVCs	Signaling call setups for SVCs
Traffic flows are…	Flexible	On paths set up ahead of time	Flexible
Nodes look at a…	Full destination address	Short connection identifier	Full destination address
Header information is…	Globally unique	Locally unique	Globally unique
A node…	Routes packets individually	Switches traffic onto stable paths	Routes setups individually
The table is a…	Routing table	Switching table	Routing table
The table has…	All reachable destinations	Only connected destinations	All reachable destinations
The table is therefore…	Very large	As small as it can be	Very large
Table is maintained…	With a routing protocol	By hand	With a routing protocol

Figure 10-6.   IP routers, PVC switches, and SVC signaling nodes

MPLS: SWITCHING COMES TO IP

A full discussion of MPLS is out of the question in this chapter. MPLS is a huge topic in and of itself. Only the features of MPLS that relate to the relationship between IP routing and optical network switching need be explored here.

With MPLS, a router becomes a switch for what are called *sustained flows* from a source to a destination. A *flow* is just a stream of related packets from one source application to a destination application. So simple request and response UDP messages inside IP packets would not constitute a flow, nor would packets with ICMP messages. But almost anything using TCP would make a flow across a series of IP routers.

When MPLS is added to an IP router, think of the IP routing as handling the initial TCP connection setup messages, UDP traffic, ICMP messages, and so on. Once the TCP connection is set up, the IP router can go into "switching mode" on these connections or flows. In fact, an MPLS router will set up the path using a signaling protocol just as SVCs are set up in a frame relay switch with Q.933 or in an ATM switch with Q.2931. An MPLS device even has a choice currently of two signaling protocols: RSVP and LDP.

RSVP is the Reservation Protocol used in some portions of the Internet and now adapted for MPLS usage. LDP is the Label Distribution Protocol invented expressly for MPLS. This is not the place to debate the relative merits of RSVP and LDP when it comes to MPLS. Suffice to say that both work, that RSVP has the attraction of being stable but the drawback of not being expressly for MPLS, and that LDP has the attraction of being invented just for MPLS but the drawback of being very new and therefore evolving.

MPLS is multiprotocol both "above and below." That is, not only can any type of packet be used with MPLS, not just IP, but multiple protocols can be used in the underlying switches that make up the MPLS *domain*, or portion of the whole network where MPLS is deployed in routers. MPLS can be used on frame relay or ATM networks, for example.

MPLS is label switching in the sense that each IP packet has a *label* attached to it. A label in this context is just a short header. The MPLS label has the connection identifier associated with the flow and whatever other information in the IP packet header that would be useful to the MPLS engine in the IP router.

The structure of the MPLS label is shown in Figure 10-7. Because the MPLS label is neither a Layer 3 packet header nor a Layer 2 frame header, it is common to represent MPLS as a sort of "Layer 2 ½" in terms of the OSI-RM. But this is just for convenience.

The figure shows that an MPLS "shim label" (header) is attached to a Layer 3 IP packet, but there are other possible packet structures that can be used with MPLS. At Layer 2, the MPLS label can be coordinated with an ATM network's VPI/VCI virtual circuit switching or the Data Link Connection Identifier (DLCI) used in frame relay. Of course, MPLS could also be used with network frames that do not have connections to map MPLS packet flows onto. For example, MPLS devices could be connected by Ethernet LANs or point-to-point links using the Internet protocol suite's Point to Point Protocol (PPP). In these cases, the "shim label" becomes a "shim header" for these types of networks.

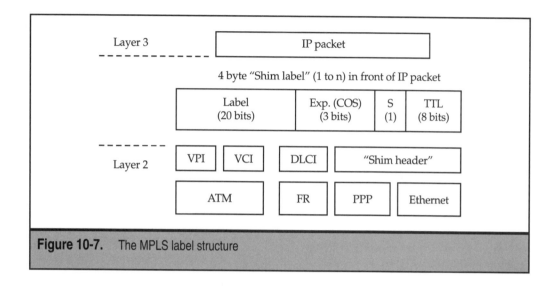

Figure 10-7.   The MPLS label structure

It is important to note that the shim label and shim header have exactly the same structure, the only difference being whether or not there are existing virtual circuits to map the MPLS labels to. If there are virtual circuits, it's a shim *label*. If there are no virtual circuits, it's a shim *header*.

The shim label or header is four bytes (32 bits) long. There can be multiple shim labels in front of any particular packet. So there can be MPLS domains inside other MPLS domains. For example, a series of routers could use the closed user groups provided by MPLS "virtual circuits" to form secure "tunnels" making up a VPN. Inside this VPN, other MPLS routers could map the packets onto a backbone of routers also using MPLS linked by ATM switches. So one MPLS label is pasted on as the packet enters the VPN MPLS domain, and a second MPLS label is pasted on as the packet enters the ATM "cloud." The respective labels are "popped" as the packet exits the ATM backbone and VPN tunnel.

Most of the bits in the shim label are dedicated to the label itself. This is the flow or connection identifier, a simple, locally significant number used for switching table lookups in the MPLS devices. This field is 20 bits long. The next field is a three-bit Experimental field, although some older MPLS documents call this the Class of Service (COS) field. In some cases, the encapsulated IP packet's Type of Service (TOS) header field is mapped into the Experimental field. Officially, however, the use of this field is "experimental." The next field is only one bit long. The S or "bottom of Stack" bit is used to tell MPLS devices that the label they are popping is the last one and all that is now left is the IP packet itself. Finally, there is an 8-bit Time to Live (TTL) field serving exactly the same purpose as in an IP router. That is, MPLS devices decrement this field, and if it ever hits zero, the packet is discarded instead of being forwarded. Various rules exist in MPLS outlining the relationship between the MPLS label TTL field and the IP packet header TTL field.

In the MPLS architecture, labels are put on by a device called a Label Edge Router (LER) or Edge Label Switch Router (Edge LSR). These devices form the "edges" of the MPLS domain. They perform full IP routing on at least one interface and MPLS switching on at least one interface as well. Inside the MPLS domain, labels are processed and swapped as the packets are switched by the core LSRs. No IP routing is needed on these devices, and LSRs could be frame relay or ATM switches with the DLCI or VPI/VCI used as the flow identifier for MPLS purposes.

The general MPLS architecture is shown in Figure 10-8. Label switching happens within the MPLS domain only. Outside of the MPLS domain, routers route the IP packets that arrive and depart inside Layer 2 frames just as before. This is shown in the figure. The labels shown in the figure (the "1" and "4" and "5") are added, used, and removed by the MPLS devices in the MPLS domain.

Within the MPLS domain, the LSRs perform label switching on paths according to the value of the MPLS label. These are called *label switched paths,* or *LSPs.* The tables in the LSRs can be set up by hand (the MPLS equivalent of PVCs) or using RSVP for MPLS and/or LDP as signaling protocols (the MPLS equivalent of SVCs). The label value changes as the packet flows from the *ingress* LER to the *egress* LER, just as in a frame relay or ATM network. The ingress LER pastes the label on, and the egress LER pops it off. All of the packets with these labels will follow exactly the same path through the MPLS domain. MPLS has more or less transformed many of the routers into switches.

There is much more to the MPLS than the brief section here has outlined. MPLS has standard mechanisms not only for creation of stable paths and therefore stable delays through a collection of routers, but also for reserving bandwidth along these paths and so

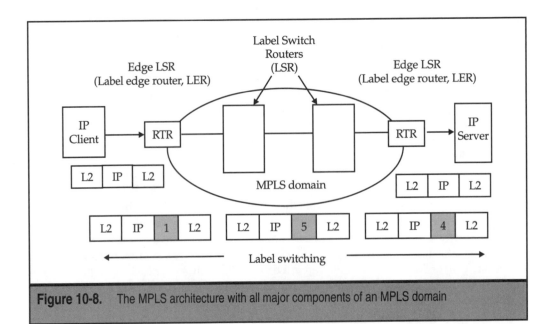

Figure 10-8.   The MPLS architecture with all major components of an MPLS domain

delivering various classes of service to the flows of IP packets. This makes MPLS an ideal addition to IP routers for a mixed voice, video, and data environment.

MPLS AS AN OPTICAL CONTROL CHANNEL

As wonderful as MPLS is, what has MPLS to do with optical networking and the way that IP fits into an optical network? Since optical networking involves the use of switches, not routers, as the network nodes, it is easier to fit IP into an optical networking environment when switching is used than when pure IP routing is used.

IP routing is a rather processor-intensive task. Many decisions have to be made on *each and every* packet before the packet can be forwarded to an output interface on an IP router. And routing tables can be huge, as the number of possible IP network addresses grows larger and larger. In spite of efforts over the years to assign IP addresses more rationally, routers on the Internet can have close to 100,000 routes to examine before deciding where to forward a packet.

Routers have to

1. Remove the frame header and trailer
2. Check the packet header checksum (if invalid, discard)
3. Check other parameters in the packet header (if invalid, discard)
4. Decrement TTL (if 0, discard)
5. Perform any packet security filtering configured (if failure, discard)
6. Look up the next-hop router in the routing table (this could take a while)
7. Discard if fragmentation is needed but the DF bit is set to 1 (then don't fragment)
8. Process any IP option fields present
9. Create new packet header and field values
10. Compute a new header checksum (since fields have changed)
11. Create a new frame header and trailer
12. Deliver to the output interface for transport

On the other hand, an IP router with an MPLS path set up only has to

1. Remove the frame header and trailer
2. Decrement TTL (if 0, discard)
3. Look up the next MPLS hop in the switching table (very fast)
4. Create a new frame header and trailer
5. Deliver to the output interface for transport

This is a much more efficient process. And MPLS offers stable paths and delays, and the promise of always having the bandwidth needed for applications using MPLS.

Recall that MPLS includes its own signaling protocol (in fact, both RSVP and LDP can be used to set up MPLS paths). MPLS sets up paths through the MPLS domain by using the signaling protocol to say to each MPLS LSR "use label 6 from here," "use label 17 from here," and so on.

Now, suppose that each label represented not an arbitrary number, but a *wavelength* on an optical network. Now the MPLS label value would tell each optical switch the wavelength to use on a single DWDM fiber connecting the MPLS devices together. This is sometimes called Multiprotocol *Lambda* Switching, but the point here is that MPLS can easily be implemented on an optical network with DWDM.

The next step is to make MPLS the control channel protocol for optical switches. Recall that SONET/SDH is often used for this purpose. However, IP routers cannot easily generate or interpret SONET/SDH overhead bits and alarms. But IP routers running MPLS can easily employ special extensions to RSVP and/or LDP to set up, maintain, and tear down the MPLS "label tables" forming the paths inside an MPLS domain.

In fact, this makes perfect sense. Why bother having special hardware and/or software just to generate and interpret SONET/SDH frames and overhead and alarms? Just use MPLS for everything.

MPLS has been proposed as the optical control channel for optical networks and DWDM devices. There seems at present to be a lot of support for this proposal from both equipment vendors and service providers.

GETTING IP PACKETS ONTO DWDM

So the approach has been advocated to make MPLS the control channel in an optical network, assigning labels to traffic flows based on wavelength used. This would release MPLS devices (really just IP routers with MPLS capabilities) from the burden of having to use SONET/SDH frames for either transport or control.

But hold on a second. IP packets have to be framed inside of *something*. Frames do provide three things that are difficult to do with pure IP packet transmission. First, frames delimit the packets. That is, a special bit pattern is used as *interframe fill* (an "idle pattern") at Layer 2 so that the Layer 3 process does not have to pay attention to each bit to see if it is the start of a packet. Second, frames allow more than a single type of packet to be sent on the network. This might not seem like a big deal when almost all packets are IP, but there are cases where frames carry packets other than IP, if only when one Layer 2 process talks to another. (If there were no Layer 2 protocol units at all, how would anything be able to communicate with Layer 2 at all?) Finally, frames provide error detection that packets lack (the IP checksum is only done on the packet header itself).

So framing IP packets, even when MPLS is used, seems wise. But what kind of framing is best? Naturally, the fewer steps involved in getting IP packet bits onto a DWDM fiber link, the better. At least six ways have been proposed to get IP packets onto DWDM optical channels. These are shown in Figure 10-9.

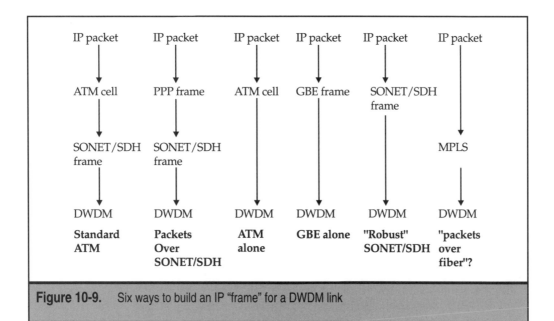

Figure 10-9. Six ways to build an IP "frame" for a DWDM link

At least three of the methods shown in the figure are possible today, and often used. The standard ATM method puts an IP packet inside a stream of cells and then puts the cells inside a SONET/SDH frame before sending the bits as light in a DWDM channel. The Packets Over SONET/SDH (POS) method eliminates the need to form ATM cells first, a welcome relief. But some type of framing is still needed, such as PPP frames (HDLC, or High-Level Data Link Control, frames can be used as well). The ATM alone method keeps the cells but eliminates the need for SONET/SDH. All three of these have been used, and used quite well, on optical networks.

The last three methods are newer and not as common as the first three ways to get IP packets onto a DWDM channel. Gigabit Ethernet, either regular GBE or the new 10 GBE, frames can be used. This method is so important that it is considered more fully in Chapter 12. The "robust" POS method places IP packets directly inside a SONET/SDH frame. Presumably, something has been added to IP to replace or remove the traditional framing roles of delimiting, adding multiprotocol support, and performing error detection (most discussions about this method conveniently leave this "framing" out).

The last method in the figure is of the most interest here. Here, the IP packet gets an MPLS label, and that's that. What about the three framing functions? All that is needed to provide delimiting to add to the MPLS label is a simple delimiter field and an "idle pattern." Only a few bits are needed for this. MPLS provides its own multiprotocol capabilities, and it is simple to reserve a few label values for special purposes. As far as error detection goes, MPLS can follow the lead of ATM and perform no error checking at all inside the network. Bit error rates on DWDM fibers should be low enough to make error recovery strictly an endpoint function and decision.

Once MPLS is used to carry IP packets onto and off of a DWDM link, all that is needed is a truly "optical router" so that labels can be swapped without needing to convert any of the packet or label into electricity. Is there any way that this could be done with currently available technology?

THE OPTICAL ROUTER

The quest for a true optical router has been a goal for a while in the field of optical networking. Research is ongoing, and as a result, there is no real firm architecture for a workable optical router as yet. What follows in this section is necessarily general, but it is based on a number of actual research projects. The architecture described in this section leans heavily on work done at Calient Networks.

Before proceeding, it might be a good idea to summarize some of the advances in packet forwarding technology set forth earlier in this chapter. These are used and expanded upon in the optical router architecture explored here.

MPLS adds switching capabilities to an IP router. The packet now has a *content* consisting of the original IP packet header and information, and a *label* consisting of a short header. No other framing is required. When used in an optical network, the label assigns the packet to a wavelength on a DWDM fiber forming the input and output of the optical router. This means that the optical "router" is more like an MPLS LSR than a pure IP router today, but this should be seen more as an acknowledgment that optical networking requires new methods than a limitation.

This reliance on MPLS techniques also means that initially wavelength/packet routing will be possible only inside an MPLS-like domain where IP switching is supported. Electrical IP routing will still take place at the edges of such a network. However, the development of the optical router will mean that optical network nodes have been optimized for the IP (and therefore Internet and Web) environments.

One of the architectures for an optical router is shown in Figure 10-10. The figure is quite complex, so some words of explanation are in order.

Packets arrive on an input fiber at the left of the figure. The packets already have a label assigned based on wavelength, and there are four wavelengths on one fiber in the figure. This number of wavelengths is low, but the point here is to illustrate the process without making the figure too overwhelming. Since all of the components depicted (splitter, combiner, etc.) have been discussed in detail earlier in this book, only the role of these components in the optical router need be mentioned here.

Wavelengths arriving at the demultiplexer at the left of the figure are split into the four wavelengths and sent on four separate paths through the optical router. Only one path is detailed, but all of the major components on each path are the same. Each wavelength passes through an optical splitter, so there are now *two* copies of each packet flowing through the optical router. One copy goes to the control photonics, and the other passes through the optical router fabric itself.

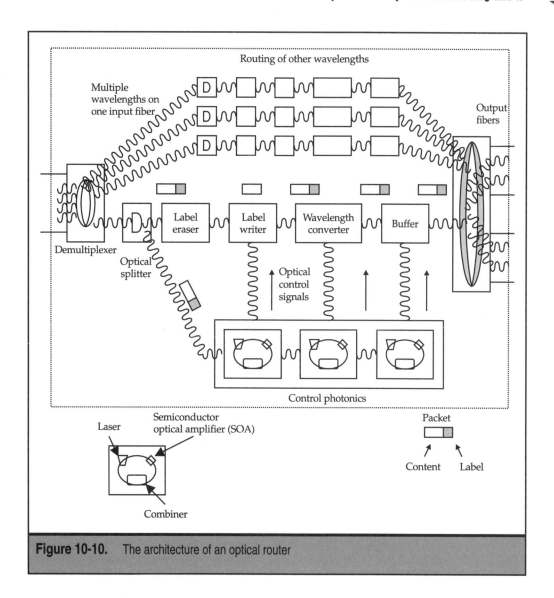

Figure 10-10. The architecture of an optical router

Consider the path of the "fabric copy" first. This packet has its label removed by a "label eraser," as shown in the figure. This is all done optically, leaving the original packet intact. It is somewhat tricky to time the flow of photons to remove only the label, but it can be done. The figure shows that only the packet content leaves the label eraser. The label writer component of the optical router adds a new label based on the information provided to the control photonics by the label of the "control copy" of the original packet. The control photonics requires no conversion to electrical bits either, performing its task through an assembly of driving laser, semiconductor optical amplifier (SOA), and combiner, all linked by waveguides. The output of the first optical control station is used to control the label writer with another waveguide.

Once the new label is created for the packet, another optical control signal drives a wavelength converter. This device shifts the packet onto the proper outbound wavelength corresponding to the incoming label on the packet. This wavelength is related to the value of the new label just written, of course.

So the place of the routing table is taken by the first of the control photonics stations. The control copy of the packet label serves as an "index" to drive the optical control signal to instruct the label writer to construct the new label properly. The place of the switching fabric linking input interfaces to output interfaces is taken by the wavelength converter, again under optical control driven by the control copy label "value."

The last station is a buffer area designed to hold the packet until all other packets are again ready to be output on the same fiber. At high optical speeds, even slight variations in the path taken by the light through the router can distort the timing between and among packets on different wavelengths. This station is also under optical control.

Finally, all of the wavelengths are multiplexed together and sent out on the output fibers. In the figure, two of the wavelengths exit on one fiber and two of the wavelengths exit on another, but this is just an example.

Optical routers, and the optical control devices and optical logic gates that make them possible, are an exciting application of optical networking that remains for now a vision. But work is underway to make this vision a reality.

PART IV

Local Area Networks

Part 3 examined the role of optical networking and DWDM in a WAN environment—the emphasis was on how SONET/SDH, ATM, and IP all fit into the optical networking picture. These popular WAN technologies were chosen to assess the impact of optical networking on each of these three.

Part 4 explores the relationship between optical networking and LANs, although MANs are considered as well. Chapter 11 examines the impact of optical networking on FDDI and Fibre Channel (FC). In spite of the fact that FDDI is most often deployed as a MAN, and FC is most often used in situations even smaller than a traditional LAN to build storage area networks (SANs), both architectures fit a loose definition of a LAN. Both FDDI and FC can take advantage of optical networking to do things undreamed of before.

In order to better appreciate how optical networking and DWDM change the nature of the LAN in many ways, it is necessary to extend the concept of the *power budget*, introduced in Chapter 4, to include more parameters such as bit-error rates and optical return losses. This will involve some math, but nothing more complex than the simple equations and relationships used earlier in this book. It is essential that the performance characteristics of FDDI and FC be appreciated before the impact of optical networking on LANs can be properly evaluated.

Chapter 12 explores the relationship between Gigabit Ethernet (GBE) networks and optical networks. The anticipated support for the GBE standard among equipment vendors and service providers has been astonishing. Such enthusiasm for a standard has not been seen in the networking world since the early days of 10Base-T and 100 Mbps Ethernet. Even optical networking and DWDM, as popular as they are, seldom rally much support for a *standard* version of any aspect of the technology.

Some of the basics of GBE were investigated in Chapter 4 as well. But Chapter 12 extends the brief introduction offered there. Using GBE for optical networking is a given, and GBE was developed specifically for the world of optical networking. And 10 GBE adds speed and reach to GBE. With GBE and 10 GBE, the lines between SAN, LAN, MAN, and WAN blur. Perhaps the use of GBE and 10 GBE as one and the same set of hardware, software, and protocols from desktop to desktop to global reach will fulfill the vision of early ATM proponents. But instead of the WAN (ATM) invading the LAN, with GBE the LAN has invaded the WAN.

CHAPTER 11

Fiber LANs and Optical Networks

It is apparent that the rise of DWDM and optical networking will have a large impact on the way wide area networks are implemented and deployed and used. Whether the WAN is based on SONET/SDH leased lines, or based on ATM switches, or even viewed mainly at the IP router level, great change is inevitable. Higher bandwidths and longer distances always have a major impact in a WAN environment, and optical networking goes beyond even these simple characteristics.

But what about the local area networking environment? What impact will optical networking have on the LAN? And speaking of LANs, what about networks that are traditionally positioned as a kind of "stretched LAN," the metropolitan area network? Large MANs share many WAN characteristics and features. Smaller MANs span a campus or collection of office buildings in an office park and are sometimes called campus area networks (CANs). Today, there are also networks that can be positioned as "shrunken LANs" or "specialized LANs," the storage area network or server area network (SAN). What will be the impact of optical networking on SANs, LANs, and MANs?

This chapter will show that while the impact of optical networking might not be as revolutionary in this arena as in the WAN arena, there are still many issues to be resolved when optical networking comes to the LAN and related networking environments. The reason for this restricted impact is mainly due to the fact that there are many more desktops than buildings and much more inertia closer to the user than on the backbone. It is exceptional that sweeping changes take place close to the user. The success of TCP/IP and Ethernet in this regard is as much an indication of how rare such a sweeping change really is as it is an indication of the need for TCP/IP and Ethernet.

When change does come to the network below the WAN level, it is either because the change is driven by a fundamental change in the way the people live and work (as the Internet and Web drove TCP/IP to the end user) or because the change is driven by an improvement in a widely accepted technology (as 10 Mbps Ethernet drove 100 Mbps Ethernet and beyond).

Ethernet and the traditional LAN environment will be considered in the next chapter. This chapter will explore how optical networking will impact and change the way that FDDI (traditionally positioned as a MAN networking) and Fibre Channel (recently positioned as a SAN technology) will be transformed by the possibilities presented by DWDM and optical networking.

Significantly, both FDDI and Fibre Channel (FC) are already heavily dependent on fiber optics for their basic configuration and, to a large extent, success. Changing an electrical SAN, LAN, or MAN environment to work more closely with optical networking is one thing. The change is abrupt and clean. Adapting existing optical technologies such as FDDI or FC to DWDM and newer optical networking configurations can be much more gradual and results in workable "hybrid" environments that need careful implementation and testing plans. However, the effort might be well worth it.

FDDI

The history and basics of the Fiber Distributed Data Interface (FDDI) were introduced in Chapter 4. There is no need to repeat this information. But there is much more to FDDI than discussed in that chapter. In order to appreciate how optical networking can change the way that FDDI is used, it is necessary to examine more of the operational details of FDDI.

As outlined in Chapter 4, FDDI was finalized in 1990 and in a sense ahead of its time, although work had begun on FDDI as early as 1985. FDDI is often positioned as a "stretched" version of the Token Ring LAN technology, supporting distances of 100 km, a speed of 100 Mbps, and a large frame size. There is a token circulating on the FDDI ring, which was really a *dual counter-rotating ring* with recovery features similar to (and foreshadowing) some types of SONET/SDH rings. The structures of the generic FDDI frame and token are shown in Figure 11-1. Although the address fields are shown as in original FDDI as being either 16 bits (2 bytes) or 48 bits (6 bytes) in length, only 48-bit addresses are supported on LANs today.

FDDI nodes can have two pairs of fiber connecting each node, and such nodes are called *dual attachment stations (DAS)*. There are also *dual attachment concentrators (DACs)* that have multiple ports for linking devices to the dual ring. FDDI also has *single attachment*

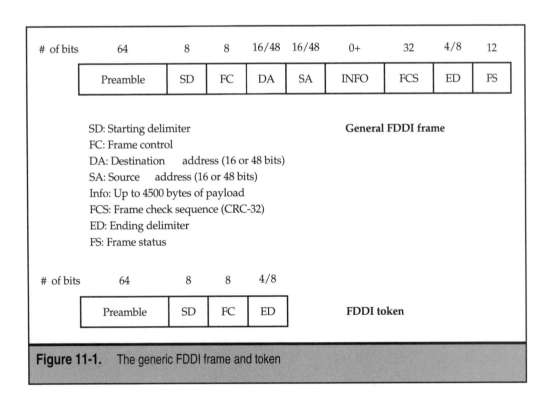

Figure 11-1.   The generic FDDI frame and token

stations (SASs) with two fiber connections and *single attachment concentrators (SACs)* with multiple ports. In this chapter, unless explicitly noted, all FDDI nodes will be dual attachment stations (DASs).

Having found a home as a LAN backbone technology, FDDI is most often used today as a backbone LAN connecting hubs and routers. There are FDDI rings around whole cities and FDDI rings that run from one side of a room to another. But they all are LAN backbones. The "dual attachment stations" have become boards in the hubs and routers. DASs are always used when constructing an FDDI ring.

This technology usually uses multimode fiber and LEDs, but FDDI variations allow for the use of single-mode fiber, and four distinct combinations of lasers and receivers, depending on the distance spanned. With single-mode fiber in use, FDDI nodes can be 40 to 60 km (25 to 37 miles) apart. It is this aspect of FDDI more than anything else that makes FDDI into a MAN technology when used to link LANs.

FDDI rings are dual, or two in number. One of the rings is the *primary*, and the other is the *secondary*. Each ring consists of a single strand of fiber, and the fiber operates in the usual unidirectional manner. The FDDI token ordinarily circulates on the primary ring fiber only, and the token only circulates on the secondary ring fiber only when a "ring wrap" occurs due to link or station failure.

An FDDI "ring wrap" shunts the token arriving on the primary ring fiber and places the token onto the secondary ring fiber *in the opposite direction*. Whether the failure is a link or a station, the intact FDDI nodes closest to the failure will wrap because they lose the signal on one pair of fiber links. Both will pass the token from the primary to the secondary, depending on where the token appears. The general idea behind FDDI "ring wrap" recovery is shown in Figure 11-2.

Normally, only the primary ring fiber has its token and frames passed through the higher-layer process in the DAS. The solid dot indicates where the token is processed by the MAC layer in each DAS. The stations notice a jump in delay but keep on working. In the wrap condition, the failed pair of primary and secondary fibers is bypassed and the token and frames flow onto the secondary fiber. This procedure is shown from the perspective of the DAS in Figure 11-3. Normally, only the primary ring (Pri in and Pri out in the figure) has its frames pass though the MAC layer process. When in the wrapped condition, the DAS puts traffic onto the secondary ring (Sec in and Sec out in the figure).

Although it was not mentioned in Chapter 4, the line coding used in FDDI is a distinctive part of the new technology. The line code chosen is called "4 in 5 encoding" or just *4B/5B encoding* (sometimes *4b/5b* is seen, but the use of the capitals is far more common). This means that every 4 bits sent on an FDDI link are represented as 5 bits "on the wire." Technically, the basic unit sent and received in FDDI is not the bit or the byte, but the 4B/5B *symbol*. The added fifth bit helps with error detection on the fiber link, acting as a kind of "parity check" (the actual mechanism used is much more complex) on the half-byte or "nibble." Naturally, this extra bit can never be used for information transfer. But the presence of this extra bit does mean that while the information rate of FDDI is 100 Mbps, the actual bit rate employed on the fiber is 125 Mbps.

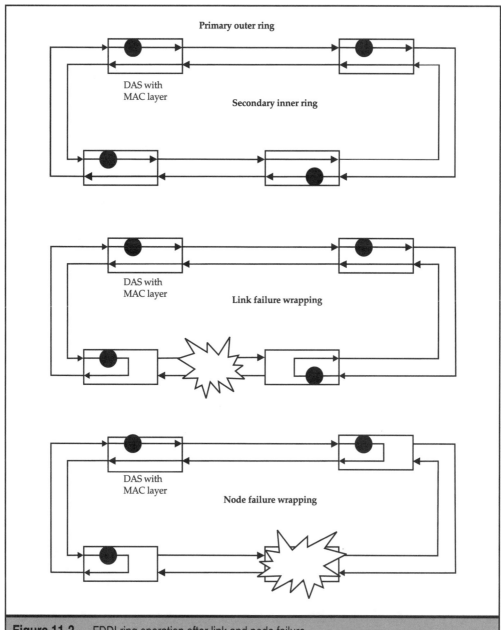

Figure 11-2.   FDDI ring operation after link and node failure

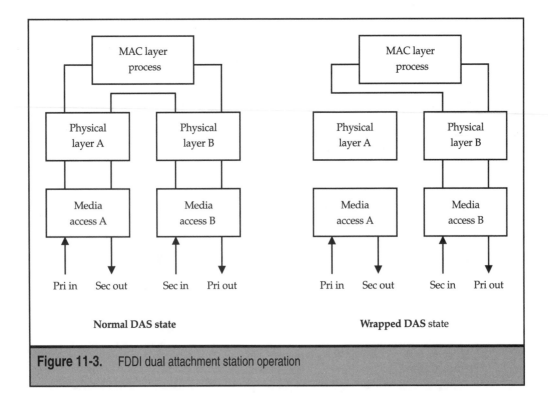

Figure 11-3.   FDDI dual attachment station operation

The significance of this 125 Mbps operation of FDDI on the fiber should never be overlooked. It is due to this simple fact of FDDI life that the impact of optical networking on FDDI might be considerable.

FDDI and DWDM

FDDI is normally deployed today as a LAN backbone connecting sites over a MAN distance, as mentioned several times already. MAN distances are usually defined as 100 km (about 60 miles) or less, more than enough to link sites clustered in almost any major metropolitan area. To appreciate the position of FDDI in the DWDM world, consider the FDDI configuration shown in Figure 11-4.

There is nothing unusual about this FDDI configuration. It is deployed many times with minor variations all over the world, connecting a rather remote site with two closer sites. Although not shown in the figure, each DAS is typically a router or Ethernet switching hub, which would normally connect to each LAN at the site through a series of other routers or LAN switches.

Here is the issue when it comes to the position of FDDI in the world of DWDM and optical networking: Note that the sites connected by the FDDI ring can be up to 40 or even 60 km apart when single-mode fiber is used. Now, the dual rings mean that there can be

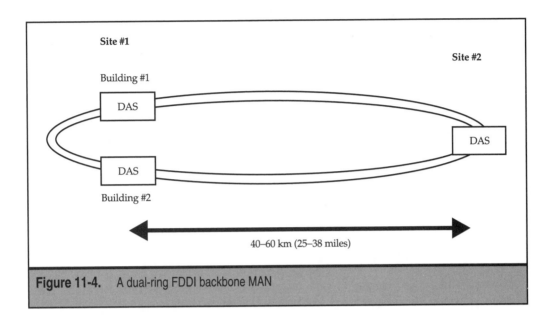

Figure 11-4. A dual-ring FDDI backbone MAN

200 km of fiber, but the total geographical "dual ring span" is not more than 100 km. It is unlikely that the organization using the FDDI ring has right of way to run their own fiber to connect the sites. It is much more likely that the organization will have to buy a MAN fiber service from one service provider or another to span the distance.

What technology is most likely to be used on this fiber service? Probably some form of SONET/SDH. However, even in the United States, where there is some SONET available at the OC-1 rate of 51.84 Mbps, this speed cannot be used to link 100 Mbps FDDI nodes. And although the SONET OC-2 speed slightly exceeds 100 Mbps (103.68 Mbps), OC-2 exists only on paper and cannot be used for the full FDDI line rate of 125 Mbps. So the closest SONET/SDH speed that can be used for FDDI is the OC-3 or STM-1 rate of 155.52 Mbps. In fact, an FDDI payload mapping exists in both SONET and SDH at this speed.

What's wrong with that? Only that a considerable portion of this bandwidth is useless to the organization. Of the entire 155.52 Mbps, only 100 Mbps is used for information transfer between the sites. There is 25 Mbps for FDDI 4B/5B encoding "overhead," but even the bandwidth leftover is useless for anything except an "idle pattern." This is because the SONET/SDH mapping for FDDI is an all-or-nothing proposition: either the entire 155 Mbps is used for FDDI, or none of it is. Nothing else can share the mapping with FDDI bits. (There were some proposals to map additional services onto SONET/SDH when carrying FDDI frames, but no new standards seem to have come of the effort.)

However, when a wavelength is used for the FDDI traffic on an optical network with DWDM, there is no need to "map" FDDI to anything except an optical coding on the assigned wavelength. There is no additional overhead at all. The user pays for—and gets—125 Mbps of bandwidth and uses 100 Mbps for information transfer. Any optical transponder can be used to transfer the normal FDDI wavelength to a wavelength on the DWDM link(s).

This situation of mapping FDDI onto SONET/SDH and onto DWDM is shown in Figure 11-5. The SONET/SDH overhead from transport and path is 1.728 Mbps. The FDDI occupies 125 Mbps for information transfer and 4B/5B encoding, leaving an appreciable 28.792 Mbps of bandwidth paid for but unrecoverable to the organization.

It could be argued that no one would use SONET/SDH to carry FDDI traffic for this very reason. Why pay for bandwidth that cannot be used? But suppose the organization can run its own fiber, or that it purchases (really a long-term, renewable lease) dark fiber to link the sites? Since FDDI transmitters normally use the same wavelengths on primary

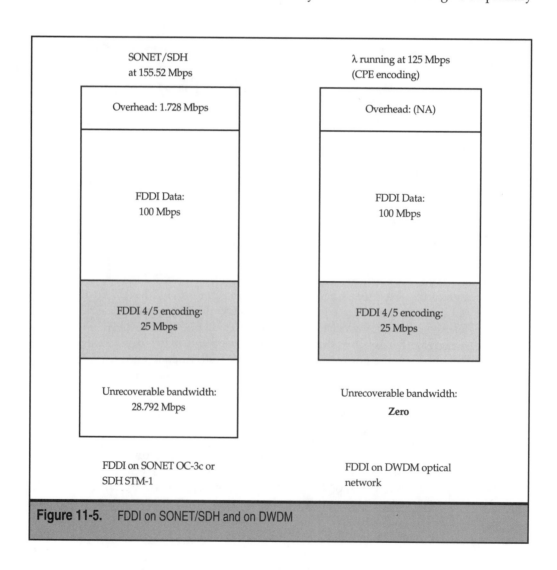

Figure 11-5. FDDI on SONET/SDH and on DWDM

ring fiber and secondary ring fiber, *two* fiber pairs, or four fibers in all, must be used to create the FDDI ring.

When DWDM is used to connect the sites, not only is there no wasted bandwidth on the fibers, only *one* pair of fibers is needed to link the sites. There are still outbound and inbound fibers, but the optical network node can easily cross-connect the FDDI ring signals to preserve the wrap protection characteristics of the FDDI ring as a whole. The DWDM edge for FDDI is shown in Figure 11-6. Note how the FDDI ring fibers are mapped to DWDM wavelengths.

It could be argued that DWDM rings could be relied on to provide the protection that the FDDI rings normally provide, and so the configuration could be simplified even further. However, the economics of service providers deploying MAN-size DWDM rings are still uncertain, and this configuration will not preserve the integrity of the total FDDI ring. Some DWDM ring protection wrapping might actually confuse the FDDI protection process and lead to the sites becoming isolated. And it is unlikely that an organization buying dark fiber would be willing to spend enough to construct their own DWDM ring to provide the same protection that they already have with FDDI.

Using DWDM for FDDI has another advantage to both customer and service provider alike. The FDDI service occupies only two wavelengths in each fiber. There are many more potential wavelengths available for other services between the two sites. Thus there is service provider revenue potential in the DWDM configuration that is totally absent in the SONET/SDH or even dark fiber configurations.

None of this should be interpreted as suggesting that DWDM might lead to a revival of FDDI as a leading LAN interconnection technology. Today, 100 Mbps is a fairly low speed, and the cost of FDDI interfaces remains very high. But it certainly makes more sense to use DWDM to link existing FDDI backbones than it does to use SONET/SDH, or even double spans of dark fiber. FDDI on DWDM uses only a single span of fiber, and the DWDM cross-connect nodes could even be customer owned and operated.

FIBRE CHANNEL

Fibre Channel (FC) was also briefly introduced in Chapter 4. There FC was mentioned as a product of the American National Standards Institute (ANSI). FC is not a traditional networking protocol like Ethernet, but more like the small computer systems interface (SCSI) for connecting scanners or hard drives to a computer. FC runs at 100 Mbps, 200 Mbps, 400 Mbps, 1 Gbps, and even 2 Gbps speeds. FC is defined on coaxial cable, shielded twisted pair (STP), and multimode and single-mode fiber. On single-mode fiber, a maximum link distance of 10 km (about 6 miles) is supported. Longer spans cannot preserve the FC delay of about 10 to 30 microseconds.

Chapter 4 mentioned several topologies for an FC "network." These included a simple point-to-point link, a topology using a central switching hub or FC "fabric," and an odd type of ring called an FC "loop."

FC is most often used today to create a storage area network (SAN) for large Web sites. It is not unusual for a large company to have perhaps 30 or more servers all acting as

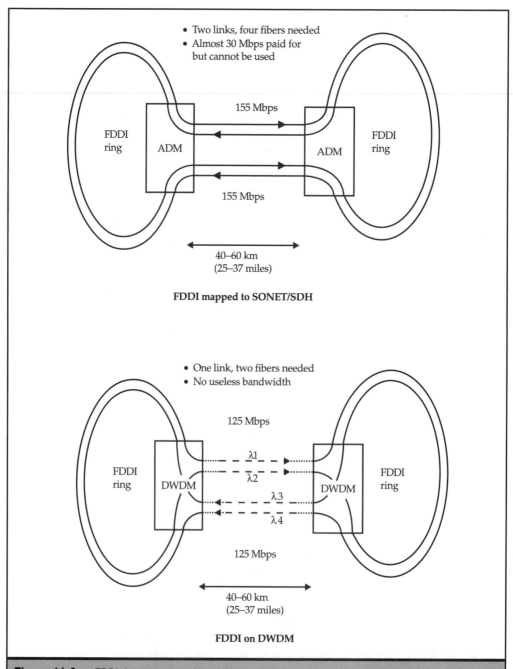

• Two links, four fibers needed
• Almost 30 Mbps paid for but cannot be used

155 Mbps

FDDI ring   ADM   ADM   FDDI ring

155 Mbps

40–60 km
(25–37 miles)

FDDI mapped to SONET/SDH

• One link, two fibers needed
• No useless bandwidth

125 Mbps

FDDI ring   DWDM   λ1   λ2   λ3   λ4   DWDM   FDDI ring

125 Mbps

40–60 km
(25–37 miles)

FDDI on DWDM

Figure 11-6.   FDDI rings connected by two pairs of fibers and one pair of DWDM fibers

a single Web site, all appearing to users as www.example.com. The servers must still be coordinated to the extent that all of the servers look and feel the same in terms of Web content and capabilities. Users that begin transaction such as buying a sweater must be assured that the Web site does not drop the session if a page request goes to a different server in midstream. And if a hard drive fails, the server that has the hard drive is out of commission.

For this reason, the SAN term has been sometimes changed to mean *server* area network. This is because in many SANs, servers can be connected to the router on an FDDI ring (this is only an example). And on the back end of the servers, the redundant arrays of inexpensive disks (RAIDs) make sure that users see a consistent Web site. The server becomes the focus of the SAN.

FC is a layered protocol, but the layers of FC have little to do with the OSI-RM or other layered network architectures such as ATM or the Internet protocol suite (TCP/IP). FC is layered into five layers called FC-0, FC-1, FC-2, FC-3, and FC-4 from bottom to top. The five layers of FC along with their main functions are shown in Figure 11-7.

▼ **FC-0**   This layer defines all of the media that can be used with FC links, the signaling levels to be used on the links, and the overall characteristics of the transmitters and receivers.

■ **FC-1**   This layer defines the encoding used on the fiber. FC uses an "enhanced" version of FDDI's 4-bits-coded-as-5-bits that keeps the bytes together and adds *two* extra bits to each 8 frame bits "on the wire." This is 8B/10B encoding. Low-level link functions are also defined here.

■ **FC-2**   This layer defines the FC frame format and groups of frames called *sequences* and *exchanges.* Here is where FC performs flow control, establishes six distinct classes of service, provides login/logout functions (for FC ports, not users), and defines the basic FC topologies. Fragmentation and reassembly of exchanges into sequences and then into frames and back again is done here also.

FC-4 (ULPs)	SCSI	IP	HPPI	AAL5	IPI-3	SBCCS	VIA	FICON
FC-3	Multiplexing devices with many FC ports work as one port (future)							
FC-2	Frame format, flow control, class of service, login/logout, topology							
FC-1	8B/10B encoding, low-level link functions							
FC-0	Media types, signal levels, transmitters, and receiver specifications							

Figure 11-7.   The five layers of Fibre Channel

■ **FC-3** This layer defines how devices with multiple ports on one FC node (any device with at least one FC port) can act as if they have only one port. This layer is still being defined.

▲ **FC-4** This layer defines various Upper Layer Protocol (ULP) mappings so that a variety of protocols can use FC frames as a transport mechanism. There are currently eight ULP mappings defined or planned.

Together, FC-0 and FC-1 correspond fairly well to Layer 1 (the Physical Layer) of the OSI-RM. FC-2 is similar to a LAN's Media Access Control (MAC) sublayer (the lower portion of the OSI-RM's Layer 2 or Data Link Layer), but FC does not have a true MAC. The FC frame is surprisingly small, only 2,112 bytes of data maximum with a 36-byte header, which is one reason that grouping into sequences and exchanges is done. FC-3 is not really a layer, but a loosely defined concept that FC devices can make a group of ports seem like one.

FC-4 is essentially the "user interface" to FC. The ULPs supported are quite familiar, if an odd mix to those familiar with traditional LANs and WANs. The eight FC-4 ULP mappings are:

▼ **Small Computer System Interface (SCSI)** A popular method for connecting disks internally or externally in servers.

■ **Internet Protocol (IP)**

■ **High Performance Parallel Interface (HPPI)** An older technology for high-performance peripherals.

■ **ATM Adaptation Layer 5 (AAL5)**

■ **Intelligent Peripheral Interface–3 (IPI-3)** A newer interface for disks and tape drives.

■ **Single Byte Command Code Sets (SBCCS)** Based on IBM's ESCON fiber optical channel mainframe storage systems.

■ **Virtual Interface Architecture (VIA)** A new method for bypassing usual network protocol layers and creating *server clusters*. A little like writing an application program that sends bits directly to a device port, as was done before layered protocols became the rule. FC has all the layers needed.

▲ **FICON (Fiber Connectivity)** A version of ESCON used in the IBM S/390 mainframe architecture.

FC runs at a variety of speeds, none of which is particularly well aligned with any former LAN or WAN speed. Oddly, FC speeds are often quoted by name or in megabytes per second (MB/s). Table 11-1 gives the FC speeds in terms of the more familiar Mbps or Gbps, and then it lists them in MB/s and by FC name.

Note that conversion of Mbps to megabytes per second is not straightforward in FC. The conversion not only takes into account the 8B/10B encoding but adds overhead in

Line Speed	FC Speed in MB/s	FC Name
133 Mbps	12.5	Eighth speed
266 Mbps	25	Quarter speed
531 Mbps	50	Half speed
1.063 Gbps	100	Full speed
2.126 Gbps	200	Double speed
4.252 Gbps	400	Quadruple speed

Table 11-1. FC Speeds and Names

the form of frame headers and other "extra" bits. So 25 MB/s is not 250 Mbps, but 266 Mbps.

FC devices all have at least one FC port. FC ports fall into many categories, depending on the topology used and the role of the port in the particular topology. FC ports can also have more than one operational mode.

In it simplest form, two FC devices with N-ports (also seen as N Ports or N_Ports) connect to each other directly with one pair of fibers. In the *arbitrated loop* topology, L-ports are daisy-chained to form a loop in almost the same way that FDDI DASs form dual counter-rotating rings. Until about 1999, the vast majority of FC topologies were loop topologies.

But for the purposes of this discussion, and because of the way that FC is expected to be used in modern SANs, the most important FC topology is using the *fabric*. When a fabric is used in FC, there is a central switching FC device with F-ports used to link a number of N-ports together. F-ports require N-ports to support a login/logout function as FC devices are powered on and off, and the lack of this function limited most FC configurations before 1999 to simple loops of L-ports.

For the sake of completeness, there are also NL-ports that combine N-port and L-port functions into one. There are also FL-ports, E-ports for connecting FC switches, G-ports that combine the functions of E-ports and F-ports, and GL-ports (an E-port, F-port, and L-port all in one).

The basic configuration of a collection of FC devices using a fabric topology is shown in Figure 11-8. All of the N-ports can communicate through the FC switch once they are "logged in." The FC switch can be *blocking* or *nonblocking*. In a blocking FC switch, once a flow of FC frames has begun through the switch from one N-port to another, no other N-ports can communicate. In a nonblocking FC switch, the fabric is fast enough to allow all of the N-ports to communicate at the same time, as long as the N-ports agree to the exchange.

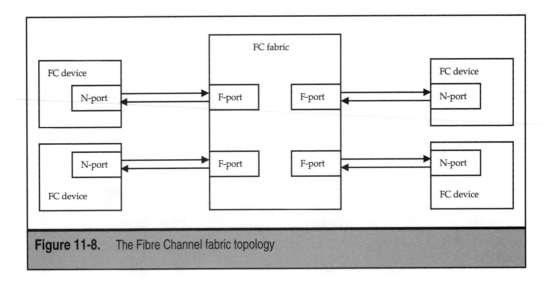

Figure 11-8.   The Fibre Channel fabric topology

How would an organization maintaining a major Web site use FC with an FC switch fabric to build a SAN today? Potentially in two areas: on the storage side of the server and on the *router* side of the server (showing why the term *server* area network is often preferred). This basic configuration is shown in Figure 11-9.

Most large organizations dependent on the Internet today employ four (or more) routers to connect to the Internet. One reason is that such organizations like to employ more than one ISP to reduce traffic loads and increase reliability. Another is because routers frequently need to be restarted periodically, and this process can take many minutes. A totally failed router can be a disaster, costing the organization a lot in terms of lost productivity.

IP is a defined ULP mapping in FC, so a "server farm" of 32 or more servers can connect to the routers through an FC switch, as shown in the figure. Now, FC ports are not common on routers today, but this situation could change rapidly, since connecting many routers to even more servers is an ongoing problem in many organizations.

The figure also shows an FC switch linking the servers to their redundant arrays of inexpensive disks (RAIDs) where the Web information is stored, indexed, and processed in a variety of ways (such as "data mining"). RAIDs often exceed 100 gigabytes of total storage, and some extend into the terabyte (1,000 gigabyte) ranges. So speed is also needed on the "back end" of the server.

Fibre Channel and DWDM

As interesting as the use of FC to create SANs is, what has all of this to do with DWDM and optical networking? Well, when FC is used with single-mode fiber and 1,310 nm lasers, the distance between FC devices can be up to 10 km, or about 6 miles. In many cases, this can be the distance from the organization's site to the ISP point of presence (POP).

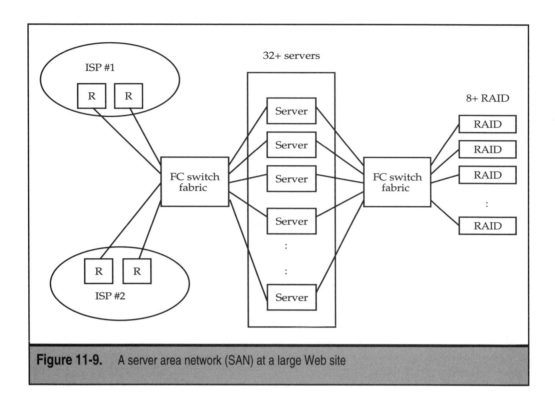

Figure 11-9.   A server area network (SAN) at a large Web site

Also, the storage side of the SAN can extend to 10 km as well. So the RAIDs can be at a site other than the place where the servers are located. In fact, the FC switch can connect to *two* distinct RAID sites, one active and the other acting as a hot standby with the same information. FC can be fast enough to make the necessary multiple updating possible.

So far, all of this could be done without DWDM. But replace the FC switch with a DWDM cross-connect or optical switch. Such a "fabric" would be inherently nonblocking due to the multiple wavelengths that the FC port traffic would be mapped onto (probably with transponders, but pure optical transponders are available today). Nonblocking FC switches, which might need to handle 16 or 24 or even 32 quadruple-speed FC links each running at about 4.2 Gbps, have been prohibitively expensive in the past. But DWDM makes a perfect platform for a nonblocking FC switch fabric.

In fact, DWDM, with its complete independence of bit rate and coding method used by traffic, is about the only technology capable of extending server area networks outside of simple loops based on FC. FC bit rates just do not align well at all with traditional LAN speeds based on Ethernet (10 Mbps, 100 Mbps, 1 Gbps, and 10 Gbps) nor traditional WAN speeds based on SONET/SDH (155 Mbps, 622 Mbps, 2.4 Gbps, and 10 Gbps).

How FC could fit into an optical network with DWDM is shown in Figure 11-10.

The DWDM switch shown could be on the customer premises or in the service provider POP. Either link could be 10 km. The same is true of the storage side of the SAN. Not

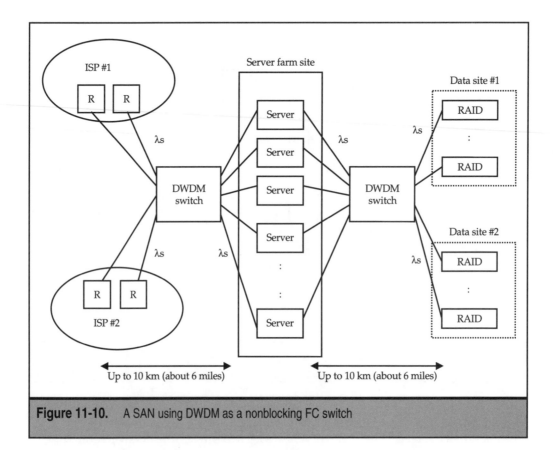

Figure 11-10. A SAN using DWDM as a nonblocking FC switch

only that, but the DWDM switches in the figure could be one and the same device. As long as there are enough wavelengths to go around, what difference would it make? The DWDM/FC switches could even be connected to increase the span of the network, but this would also raise the disk access time.

The DWDM switch can look as much like an FC fabric as desired by the vendor of the DWDM switch. The DWDM switch could have a series of F-ports, or the user devices could appear to be point-to-point connected N-ports with each connection having a distinct wavelength. But it is more in keeping with FC fabric topology if the DWDM cross-connect appears to have F-ports.

So it is possible to let the service provider offer outsourced router *and* storage services in this architecture. DWDM and FC are equally at home with IP and SCSI and more. Let the customers concentrate on their main line of business. Optical networking can be of real help here.

SAN/LAN/MAN?

So far this chapter has shown how optical networking in general, and DWDM in particular, can extend and enhance the capabilities of both FDDI and FC. Not only are these extensions and enhancements attractive to customers, but they offer service providers opportunities to offer new services and earn new revenue drawing on optical networking capabilities.

The DWDM cross-connect or switch has blurred the line between SAN and LAN and WAN in terms of cost, distance, and services offered. A large enough optical switch can support FDDI and FC services, plus other services like ATM connectivity and IP telephony services *all on the same platform*. Optical networking can be used not only to break bandwidth boundaries on networks, but distance boundaries as well.

The only major LAN and WAN technology left out of the equation so far is Gigabit Ethernet (GBE) and 10 GBE. The almost unique position of GBE and 10 GBE with respect to optical networking and DWDM is the topic of the next chapter. But first, this chapter will close with a look at implementing services like FDDI and FC on DWDM links.

TESTING FDDI AND FC WITH DWDM

It can attractive to extend the reach of both FDDI and FC with a DWDM system. But unlike the homogeneous optical network systems put in place by service providers and governed by stable standards, these "hybrid" FDDI/DWDM and FC/DWDM systems require more forethought. This is because such hybrid situations often do not have all of the performance parameters of the end-to-end system spelled out ahead of time. Organizations that want to use a DWDM link to extend FDDI and/or FC have to understand all of the relevant parameters and confirm proper operation of all network components. Extensive testing during installation is needed, and ongoing performance checking is required through maintenance procedures. DWDM adds many more variables to the equation.

This section takes a look at the important parameters when mapping FDDI or FC to a DWDM link on an optical network. This section should be seen as an extension to the basic concept of a *power budget* applied to a fiber link introduced in Chapter 4. There are many more critical parameters to consider with DWDM.

When it comes to DWDM, critical system parameters differ for individual system components. And the effects introduced by each component combine in ways that are difficult to predict and so must be monitored each step along the way. In some cases the effects may add up quite simply; in other cases, the effects are increased in more complex ways; and in still other cases, the effects might cancel each other out. At each step along the way, the environmental conditions, the physical distance, and the choice of connectors and even patch cords can create problems for the simplest network component.

Overall performance of a DWDM network link in determined by the following major characteristics of the link:

▼ **Laser transmitter power level** This is even more critical in DWDM systems, where nonlinear effects can become a factor at much lower power levels than on single-wavelength systems, especially where bidirectional transmission is in use. But this factor is most important for distance, and so power should be as high and stable as possible.

■ **Number of channels** The number of DWDM channels multiplied by the modulation rate of each channel gives the total bandwidth of the link. For example, 32 wavelengths at 2.5 Gbps each (a common enough SONET/SDH speed) gives a total link bandwidth of 80 Gbps.

■ **Channel spacing in gigahertz** The standard ITU-T grid spacing is 0.8 nm, or 100 GHz. There is also a proposed 50 GHz spacing grid for 0.4 nm separation. Some vendors use even smaller spacings, down to 0.1 nm in some cases.

■ **Laser modulation rate in Gbps** Most long-haul WAN links operate at 2.5 Gbps or 10 Gbps. Shorter MAN links usually run at much lower rates.

■ **EDFA gain** Fiber amplifiers should be considered from both an amplitude and spectral width perspective. Signal gains of 30 to 40 dB are common, and spectral widths of 40 nm are typical, except for "extended range" EDFAs.

■ **Receiver sensitivity** The receiver gain should be as high and stable as possible. Period.

▲ **Fiber specifications** Fibers used for DWDM links vary considerably when it comes to data rate/distance, dispersion, and maximum channel capacity.

In addition these factors, overall system performance is affected by other considerations. However, the impact of most of these other considerations is determined by the values of the parameters listed. In many cases, an organization has control over some of the listed parameters (channel spacing, etc.) but little to no control over others (EDFA gain, etc.). So more is needed to make sure the link will operate as intended.

Here are some of the tools and methods needed to make sure that an FDDI or FC link mapped across a DWDM link behaves once it is ready for operation.

Bit Error Rates

It is sometimes difficult for an organization to have access to all of the information about link performance that it would like to have. The link might be "invisible" and removed from direct control or observation. The situation might be as shown in Figure 11-11. But the end-to-end bit error rate (BER) is always available to the end users.

Not so long ago, and in some cellular telephony environments today, an end-to-end BER of 10^{-4} was not unusual. Today, end-to-end BERs of 10^{-12} are common, and the trend is toward ever-lower BERs on optical links.

To see if the proposed link will meet the BER requirement, it is necessary to examine the power budget. This ensures that enough optical power is available at the receiver to deliver the required BER at the designed level of receiver sensitivity. It is not necessary

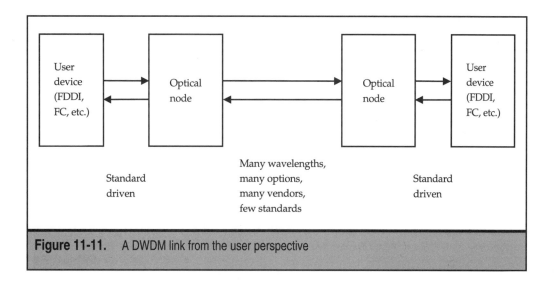

Figure 11-11. A DWDM link from the user perspective

to repeat the power budget computations presented in Chapter 4. The important factor here is the *loss margin,* or the amount that the received optical power exceeds the receiver sensitivity chosen for the BER required. For example, if the receiver chosen delivers a BER of 10^{-12} with a margin of 10 dB, and the loss margin is 15.85, then the design should work as intended. The margin should always be more than 10 dB in any case, since some entries in the power budget might be too optimistic, and others (such as splices) might be forgotten entirely.

Loss Testing

Once the overall end-to-end BER has been established for the link, the next test that should be performed is the loss test. Loss testing, which can be done on individual portions of the entire link, is essential when the end-to-end BER is not what was expected. Only a loss test can find the sections of the link that are not performing as intended.

Losses on an optical link depend on the individual components and the distances involved. Every connector, multiplexer, and stretch or fiber contributes to losses in the power budget. Whenever light is injected into fiber optical equipment, some of the light is reflected back to the source. The optical return loss is a direct indication of the presence of bad splices, defective connectors, or improperly functioning components. There are also unintended air gaps, misaligned splices or connectors, and refractive index mismatches to be dealt with. There are also the multitude of DWDM impairments to consider, such as scattering, that will affect the stability of the light source.

Only a loss test can verify that the losses computed in the power budget are an accurate reflection of reality. In multiple wavelength systems, loss varies with channel, and so a loss test must be done over a range of wavelengths. The lasers used in most DWDM systems are very sensitive to reflected light, which can not only degrade the signal but also damage the source laser. EDFAs are even prone to reflections, and these can raise the signal-to-noise ratio dramatically. All of these factors must be measured for quality assurance purposes both before and after the link is live.

Optical Return Loss and Reflectance

The preceding section used the term *optical return loss* as a measurement of the amount of light reflected back to the source on a DWDM link. This term is sometimes used interchangeably with *reflectance*. However, the two terms really mean two slightly different things. Optical return loss is a "global" or total property of the system as a whole, end to end. Reflectance, on the other hand, is most often used to describe an optical reflection at a single interface or through a single component such as a connector.

Both optical return loss and reflectance represent the ratio between incident power and reflected power. Both are commonly cited in decibels, but whereas optical return loss can be cited as a positive value, reflectance should always be cited as a negative value.

For example, the reflectance of a connector could be –30 dB. This represents perhaps 1 mW of incident optical power and a measured reflected power level of 1 microwatt, a ratio of 1,000 to 1. The greater the negative value, the less reflected power and the better the performance of the component. So a reflectance of –25 dB is *worse* than a reflectance of –30 dB.

Optical Signal-to-Noise Ratio (OSNR)

The BER is the best measurement of optical link performance from a user perspective. But from a pure engineering perspective, the BER is just a result of the underlying *optical signal-to-noise ratio (OSNR)* of the link. When a DWDM link is installed, the OSNR is determined by the choice of components and the work quality of the installers. The OSNR gives the "elbow room" of the link between the peak optical power of the source and the noise floor at the receiver for each channel. This is an indication of how "legible" the signal is at the end of the link. And as links grow longer and longer, the OSNR becomes more and more critical.

In DWDM systems, the OSNR is represented graphically as the ratio between signal power and noise power as a function of wavelength. The OSNR value should be more than 40 dB for each channel. Any optical amplifiers will inject noise, of course, and the OSNR could drop to about 20 dB at the end of the link. The precise amount of this drop will depend on the exact length of the link, the number of cascaded EDFAs, and the bit rate being used for the test on the channel. Each individual EDFA should contribute no more than 3 to 7 dB to the degradation of the OSNR.

Optical noise in EDFAs is mostly a result of the amplified spontaneous emissions (ASEs) that are a natural consequence of EDFA operation. Even though individual EDFAs are tested by the manufacturer, EDFA performance can vary in the field. All intended operational channels should be tested, both in isolation and once cascaded on a link. Variations are a warning sign, since system power flatness is always a goal. And the

OSNR will vary over time due to variations in temperature, local wear and stress, degradation of components, and changes to the network.

A questionable OSNR can only be dealt with by changing the link configuration, or by adding optical amplification where needed. This second approach might be difficult if it is suspected that the cascaded EDFAs are the source of the OSNR issue in the first place. So in these cases, only changing the configuration of the link can help.

OSNR is especially useful at the boundary between service providers or where a customer's equipment interfaces with a service provider's equipment. The OSNR is the best measurement of the overall quality of the signal provided.

Amplifier Gain

An optical system with EDFAs really modifies an optical signal many times as it travels from source to receiver. The EDFA amplifier gain (signal power boost) for an individual channel depends on the number of spacings of channels at that particular EDFA. As optical cross-connects add and drop wavelengths along a fiber link, the wavelength profile presented to an EDFA can vary. So aggregate measurements might not be a true picture of the overall situation.

The OSNR must always remain adequate, no matter how the EDFAs are loaded. So even though the final decision to turn up a link might depend on total power and the power in each channel, it can be necessary to track down weak channels through an EDFA.

There should therefore be a capability to check the gain flatness at each EDFA. This should be coupled with a method to check for nonlinear and polarization effects. All of these measurements depend on having an accurate profile of the signal across many wavelengths.

Central Wavelength and Drift

The goal of DWDM links is to pack many wavelengths as close together as possible. The central wavelength of each channel is the most important characteristic of DWDM systems, since the central wavelength determines the path used by the source to communicate with the receiver.

The exact value of the central wavelength used in each channel must be precisely determined when the optical system is first installed to make sure it meets the design goals. Because central wavelengths tend to drift over time, as channel spacings become smaller and smaller, maintenance procedures must be in place to detect this drift.

Wavelength cross-connecting and other forms of optical switching complicate the issue. The wavelength used on a particular link might change many times from source to destination. Careful planning is needed to avoid unwanted wavelength interactions. Newer systems using 50 GHz channels (0.4 nm spacing) instead of the ITU-T 100 GHz channels (0.8 nm spacing) impose very strict limits on spectral drift. So the spectral drifts that distributed feedback (DFB) lasers are prone to can be devastating, not only for the user whose laser is drifting, but for the adjacent wavelength as well. The issue of drift is shown in Figure 11-12.

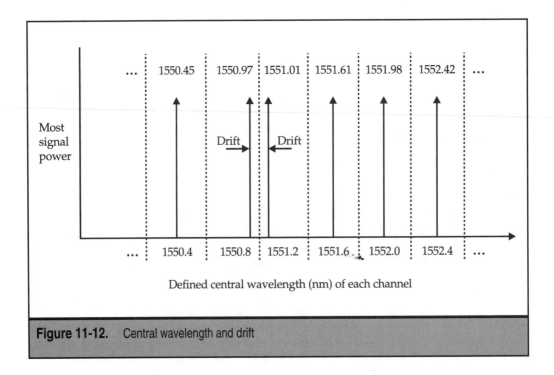

Figure 11-12. Central wavelength and drift

Source stability is very important for this reason. Side lobes at the transmitter are an issue, since they can easily add noise to adjacent channels. And even if the wavelengths were better aligned with DWDM windows, SONET/SDH lasers were never designed for strict stability and so cannot be used directly in DWDM channels. However, SONET/SDH lasers operating at 2.5 Gbps can be used to drive stable DFB lasers directly.

Drift can be as much as 0.05 nm in one direction or another over an hour or two on many operational DWDM systems. This does not sound like much, but if the central wavelength in a DWDM channel has to fit in 0.4 nm, there is only 0.2 nm of latitude on each side of the central wavelength for the channel. So 0.05 nm is a full 25 percent of this center-to-edge value. Spacings of 0.1 nm are even more stringent, of course. Drift is caused by temperature variations, back reflections, and other laser-related variables.

Crosstalk

Crosstalk is just the unwanted appearance of power from one channel on another channel on a DWDM link. Crosstalk is very difficult to predict from design data such as power budgets and the like. Crosstalk only becomes obvious with real or test signals present on many channels.

Crosstalk can be calculated from a detailed examination of the spectral profiles of two channels. A minimum value of 25 dB is normally required between channels. However, undersea links often relax this to about 13 dB, and very long transcontinental links often have adjacent channel crosstalk of about 17 dB.

However, calculations are no substitute for direct measurements. Crosstalk, like many other parameters, should be checked periodically once a DWDM link is in service.

Nonlinear Impairments

Nonlinear impairments have been introduced and discussed in Chapter 6. This section is not intended to repeat this information, but to build on the information presented there. In this section, the importance of nonlinear effects is that they often become apparent only after the DWDM link has been put into operation. For example, it might only be after all components of an optical system are interconnected that nonlinear effects that were unimportant in isolated subsystem testing combine to make the link useless.

Four-wave mixing is a perfect example of nonlinear impairments in action. Careful loading of the possible channels is needed to avoid the power of interacting optical signals from turning up in an channel in use for information transfer. The overall effect is similar to crosstalk but can affect more than one other channel as long as the polarizations and magnitudes of the interacting signals are just right. Only careful channel loading and the use of newer specialized fibers that add chromatic dispersion just to offset four-wave mixing can reduce the effects of four-wave mixing.

Polarization mode dispersion (PMD) is a consequence of imperfect fiber fabrication and the normal stress and bending of fiber during installation that distorts the roundness of the fiber core. Temperature variations play a role as well. These causes are more or less random, making PMD difficult to predict, of course. PMD is seldom a concern in links less than a kilometer or so in length or at speeds below 10 Gbps. Older fibers, meaning fibers made or installed before around 1990, are much more prone to PMD than newer fibers. Newer fibers benefit from better fiber fabrication techniques and more careful installation procedures. PMD usually only comes into play when older fibers are used to upgrade a planned DWDM link for high-speed operation.

A complete discussion of every parameter important for DWDM link operation would include much more than this brief outline of critical test parameters. A full DWDM link measurement suits all of the parameters discussed in this section and would also include:

▼   OSNR degradation

■   In-band and out-of-band crosstalk ratios

■   Ripple and insertion loss

■   Differential group delay (useful in PMD measurements)

▲   Polarization-dependent loss

Interested readers needing more details on DWDM testing are referred to the testing references in the bibliography.

CHAPTER 12

Gigabit Ethernet and Optical Networking

It would be somewhat of an exaggeration to say that Gigabit Ethernet (GBE) and 10 GBE were designed for optical networking, but it would not be much of an exaggeration. The fact remains that GBE and 10 GBE fit into a world where services are provisioned by the wavelength and packets are switched and routed at terabit rates as easily as computers fit into a world of LANs. They just seem to be made for each other, although neither was developed expressly with the other in mind.

This chapter explores the close relationship between optical networking and GBE in all its forms. GBE solves one of the trickiest problems involved with using IP packets in an optical network: the problem of how the IP packets should be framed. Speed and throughput are the most important aspects of the new world of practically unlimited bandwidth that optical networking offers. Thus taking IP packets out of LAN frames such as the lower speeds of Ethernet and placing them in other frames suitable for WAN transport such as SONET/SDH or PPP or ATM cells introduces unacceptable delays into the process of IP packet delivery.

Once framed, an IP packet should be kept inside the frame as far as possible along the path from source to destination. This principle has been recognized for years. ATM cells were once proposed as a way of taking what was essentially a WAN format for framing (although the cell was a small, fixed-sized "frame") and extending it to the LAN. B-ISDN was dependent on ATM cells flowing from desktop to desktop across LAN and MAN and LAN with equal ease. Both B-ISDN and ATM came up short in the marketplace, in no small part because there was no way that ATM was going to replace Ethernet as the interface of choice in the PC.

Now it seems to be Ethernet's turn to try where ATM failed. If the WAN cannot displace the LAN protocol, perhaps the LAN can replace the WAN protocol. In a very real sense, GBE and 10 GBE are intended for the network backbone rather than the LAN. There might come a day when a UNIX-based or Windows-based PC architecture will be able to fill up a sizeable portion of a GBE or 10 GBE link. But this day seems to be far off, and most PC architectures cannot begin to fill a fraction of the bandwidth represented by 100 Mbps Ethernet, even with full-motion video.

It is somewhat surprising that GBE and 10 GBE use the Ethernet name at all. There are many differences between the way that the venerable 10 Mbps Ethernet and 10 GBE function. This is to be expected, since 10 GBE is fully a thousand times faster than 10 Mbps Ethernet. An automobile that ran at 65,000 mph or Kbps instead of 65 mph or Kbps would look and run very differently from the cars driven today. And the roads would have to change as well. This one reason why the relationship between optical networking and GBE is such a close one.

THE MAGIC ETHERNET NAME

In spite of their differences, Ethernet is Ethernet whether it runs at 10 Mbps or 100 Mbps or 1 Gbps or 10 Gbps. It is very important to preserve the Ethernet name in all of these flavors. This is because of the power of the Ethernet name in the consumer's mind. Users do not understand much about LANs and networking in general, and what little they do know is all about Ethernet. It has been said by supporters of everything from token ring to ATM that "the good news is that customers know Ethernet. The bad news is that they don't want to know anything else."

Why is the Ethernet name magic? There are many reasons. Ethernet was the first fully standardized LAN technology, predating even the IBM PC itself. Ethernet was about the simplest way that multiple stations could share bandwidth on a LAN without some form of coordinated control (such as a token). This meant that Ethernet was also the least expensive way to link almost any number of devices. And Ethernet was soon built into most UNIX-based machines by the mid-1980s. Why buy something else when the Ethernet was already there?

Not that Ethernet did not have shortcomings. Ethernet-attached stations just sent anytime they thought no one else was sending (they were usually right). If someone else was sending at almost the same time, the result was a "collision"; neither frame got through and both frames had to be resent. The heavier the traffic load on an Ethernet, the more collisions there were and the more resends (which also generated collisions).

This access method was called Carrier Sense Multiple Access with Collision Detection, or CSMA/CD. The "carrier sense" meant that every Ethernet station listened before sending and if a carrier signal was present due to another station sending, the station that was doing the sensing had to wait. The "multiple access" meant that every Ethernet station could send whenever the medium was idle and needed no one's permission. The "collision detection" meant that Ethernet stations knew when someone else had started sending at almost the same time and the frames overlapped in a collision (due to signal propagation delays).

What Makes Ethernet Ethernet?

CSMA/CD operation became the constant that defined Ethernet. In fact, the IEEE, in charge of international LAN standards, says that Ethernet LANs should be more properly called "CSMA/CD LANs." However, it is difficult to change the way that people talk. Today there are only two types of LANs to most people: Ethernet LANs and non-Ethernet LANs. It seems likely that more than 90 percent of all the LANs that have ever been built are some form of Ethernet.

And Ethernet has evolved over the years. Properly, the Ethernet term describes a particular form of LAN that used a special type of coaxial cable and was pioneered by DEC (now part of Compaq), Intel, and Xerox. This original form of Ethernet is often called "DIX Ethernet" to distinguish it from the other types of Ethernet that have sprung up over the years.

The IEEE changed the original Ethernet frame structure and created what is called 10Base-5. Then came 10Base-2, which daisy-chained PCs together with thinner coaxial cable. Soon 10Base-T appeared, bringing "Ethernet" into the world of unshielded twisted pair (UTP) wiring in the familiar star-wiring to central hub configuration. Fiber versions appeared as 10Base-F. Ethernet then ran at 100 Mbps, and then came GBE. But they all supported CSMA/CD operation. Some of the forms of Ethernet (but hardly all of them) are shown in Figure 12-1.

But whatever the physical appearance of the "Ethernet" LAN, it is CSMA/CD that makes Ethernet Ethernet.

A Case Study: 100VG-AnyLAN

Probably the best way to understand the position of Ethernet in the marketplace and why Ethernet makes so much sense for optical networks is to consider a case study in "not Ethernet." In the early 1990s, another technology was often mentioned in the same breath as 100 Mbps Fast Ethernet or 100Base-T as a way to make LANs run at 100 Mbps. Proposed by HP and AT&T, this originally was to be called 100Base-VG. The "VG" stood for "voice grade" because 100Base-VG was supposed to be able to run at 100 Mbps over voice wiring, as opposed to the special types of UTP required for the 100Base-T proposal. The feeling was that

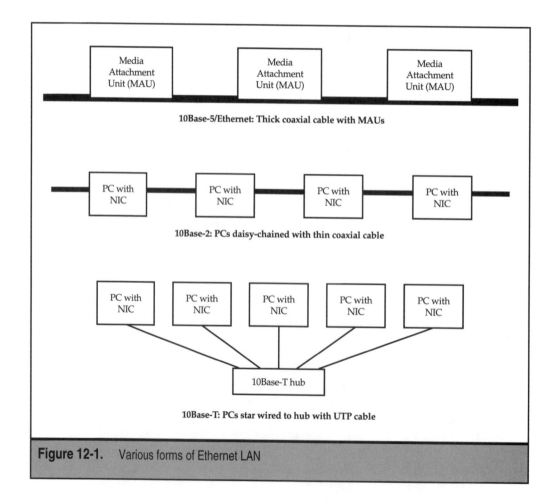

Figure 12-1.   Various forms of Ethernet LAN

many customers would not want to run new cable for 100Base-T (as it turns out, this is exactly what was required when Category 5 [Cat 5] UTP became available). The basic frame structure used in 100Base-VG was the same as that used in Ethernet, so 100Base-VG started out as a proposed variation of Ethernet.

Eventually coming to be known as 100VG-AnyLAN, 100Base-VG got its own IEEE committee (IEEE 802.12) outside of the committee that handled Ethernet (IEEE 802.3). This was because the underlying MAC layer that 100VG-AnyLAN was built on was called Demand Priority by the developers.

The relationship between 100VG-AnyLAN and Demand Priority is the same as the relationship between 10Base-T and CSMA/CD. The 10Base-T standard is the overall specification for manufacturing hardware and software that is compliant with the standard to run 10 Mbps Ethernet over UTP wire. CSMA/CD is the MAC layer protocol used to allow many users to share the single bandwidth available on the LAN. In this same way, 100VG-AnyLAN was the overall standard specification for HP and AT&T's scheme, and Demand Priority is the MAC layer protocol.

The 100VG-AnyLAN specification was developed to address the same limitations that Ethernet-type LAN users were encountering in the early 1990s. HP focused in on the need for imaging and graphics transfers, direct attachment of PostScript (not simple text) printers on the LAN, and video/audio requirements as the main performance problems that 100VG-AnyLAN was designed to overcome.

HP worked primarily on the interface cards and protocols, and AT&T worked mainly on the physical wiring and connections for the network. AT&T's original specification called for the use of Category 3 (Cat 3) UTP wiring, support for all types of cross-connects, and the use of 25-pair cable, 50-pin telco connectors, and RJ-45 jacks. HP found that CSMA/CD could not support 100 Mbps over the distances needed on Cat 3 UTP and turned to a different MAC layer method, which they called Demand Priority. Eventually, this work came to be known as 100Base-VG, the "VG" indicating the emphasis on AT&T "Voice Grade" UTP wire instead of coax or something else.

The distinguishing feature of 100Base-VG was its use of Demand Priority, which eliminated the collisions that CSMA/CD was designed to deal with. The essence of CSMA/CD MAC control was *contention:* try to send whenever there is data. The essence of Demand Priority was *fairness:* only send when it is okay to do so.

Demand Priority required a switch in the LAN hub. The hub switch managed the access to the network. In this system, a station asks (demands of) the hub for permission to send, and it indicates whether the data is either a normal- or higher-priority traffic frame (hence "Demand Priority"). If the hub receives multiple requests more or less at the same time, it services the higher-priority demands first. The hub receives the frame on an input port and immediately switches it to an output port determined by the destination MAC address. (The basic idea of Demand Priority is shown in Figure 12-2.) Traffic delivery is handled by the Demand Priority switching hub according to current load (in terms of buffers) and traffic priority mix. Demand Priority was intended to perform much more efficiently at higher loads than CSMA/CD Ethernet.

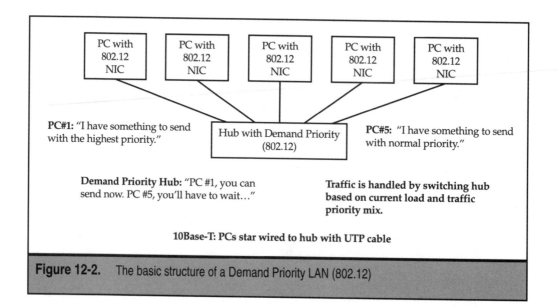

Figure 12-2.   The basic structure of a Demand Priority LAN (802.12)

Note that the Demand Priority protocol had to be present in the NIC cards in the LAN stations as well as in the hub itself. And of course to take advantage of the priorities, end-user application software had to be written to accommodate Demand Priority.

Demand Priority had its shortcomings. All higher-priority demands got serviced first, and normal (never called "low") priority demands always waited. If enough higher-priority frames were always arriving at the hub, the normal frames suffered from "starvation." Starvation would lead to high delays between some workstations on the LAN. But since normal priorities were to be used for traditional LAN traffic such as file transfers and electronic mail, this could have been perfectly okay. Higher priorities were reserved for timing-sensitive applications like video and audio. There was no mechanism for reserving bandwidth either, but at one point adding such a reservation mechanism was considered.

The original 100Base-VG result was submitted to the IEEE 802.3 committee around the same time as the 100Base-T approach and immediately ran into trouble. The debate centered around the absence of CSMA/CD as a MAC layer protocol in favor of Demand Priority. All flavors of Ethernet-type LANs, from 10Base-5 to 10Base-T, preserve CSMA/CD unchanged. But here was a proposal that Ethernet-type LANs include at least one flavor that did not use CSMA/CD. So backers of 100Base-T immediately protested that the HP and AT&T proposal was not a new kind of 802.3, but something else entirely. HP and AT&T maintained that since 100Base-VG still preserved the Ethernet frame structure it should still fall under the 802.3 committee's jurisdiction.

Things soon got so bad that Bob Metcalfe, the Ethernet inventor, was brought in to say in effect "If it does not include CSMA/CD, it cannot be Ethernet." HP and AT&T, for their part, were quite anxious to keep the whole process and result under the 802.3 umbrella.

After all, 802.3 LANs were the most common LANs in the world, and customers and users were comfortable with their use and components. This wide customer acceptance was important; in fact, many observers especially felt that it would be as hard to sell 100 Mbps LAN schemes if they were not "Ethernet" as it had been to sell 802.7 or 802.9 LANs. (Few people have ever heard of these, which is exactly the point.)

In any case, Demand Priority was deemed to be different enough from Ethernet, in spite of the retention of the frame structure, to be given its own LAN committee: 802.12. Once the Ethernet "magic" connection had been severed, most equipment vendors and potential customers quickly lost interest in 100VG-AnyLAN. Attention focused on 100Base-T. Demand Priority had some nice features, especially the priority scheme, which have to be added to an Ethernet LAN and are still uncommon. Demand Priority had only one real flaw, but it was a fatal flaw. Demand Priority was not Ethernet.

OPTICAL ETHERNET: GIGABIT ETHERNET

A brief overview of GBE was presented in Chapter 4. It was pointed out that GBE freely borrowed from Fibre Channel (FC). GBE allows for the use of UTP in what is known as 1000Base-T, but only for very short, 25-meter (about 75 feet), runs for patch cables and the like. There is also 1000Base-CX defined on 25 meters of coaxial cable. Meaningful distances are only possible on fiber optic cable.

GBE defines two physical media interfaces for fiber: 1000Base-LX and 1000Base-SX. Both share a common coding technique and other aspects with Fibre Channel. The specification defines a Gigabit Media Independent Interface (GMII) that plays the same role as the Attachment Unit Interface (AUI) in 10 Mbps Ethernet and the Media Independent Interface (MII) in 100 Mbps Ethernet. But no connector is defined for the GMII, so the optical transceiver (the E/O conversion device) cannot be an external device. This was done because of the parallel nature of GBE data transfers. External cables would be hard to control with such high frequencies and bit rates. GMII can be bypassed in cases where the GBE MAC transfers directly to the 1000Base-X encoder and decoder. The coding is the same as Fibre Channel's "8 data bits as 10 bits on the wire" (8B/10B) coding scheme.

The 1000Base-LX (the L is for "long") GBE interface uses long-wavelength lasers operating in the 1,270 to 1,355 nm range. Multimode (62.5/125 or 50/125) and single-mode fibers can be used. Half duplex mode with CSMA/CD collisions is supported, but most GBE operation is full duplex from switched hubs. Half duplex mode supports a 316-meter (1,036-foot) segment, and full duplex supports 550 meters (1,804 feet) on MMF and 5 km (about 3 miles) on SMF. SC connectors are used. But since CSMA/CD can be present, GBE is definitely Ethernet.

The 1000Base-SX (the S is for "short") GBE interface uses shorter-wavelength lasers operating in the 770 to 860 nm range. These lasers are less expensive than longer-wavelength transmitters, but only MMF is supported. Half (CSMA/CD) and full duplex (no collisions) operating modes are supported. Half duplex segments can be up to 275 meters (902 feet) on 62.5/125 MMF and up to 316 meters (1,036 feet) on 50/125 MMF. Full duplex segments can be up to 275 meters on 62.5/125 MMF and 550 meters (1,804 feet) on 50/125 MMF.

The IEEE 802.3 committee called the work done on GBE on fiber 802.3z. The work done to enable GBE to run on UTP wire was called 802.3ab. This method does not use 8B/10B encoding and requires *four* pairs of Cat 5 UTP to reach 100 meters (328 feet). How all of the pieces of GBE fit together, which pieces were taken from FC, and which pieces were added to create 802.3ab (GBE on UTP) are shown in Figure 12-3.

When incorporated into GBE, the FC components were stripped down to their bare essentials. The main thing that was preserved was the mature and well-understood 8B/10B line coding. To make CSMA/CD work at gigabit speeds, it was necessary to add a feature called *carrier extension* to the CSMA/CD standard implementation. The details are not important, but this involved the relationship between the smallest allowable frame size and the round-trip delay of the LAN segment. An optional feature called *frame bursting* was added to CSMA/CD as well to improve throughput at gigabit speeds. Frame bursting allows a sender to continue sending more than one frame after a short pause (called the *interframe gap*) if no collision occurs.

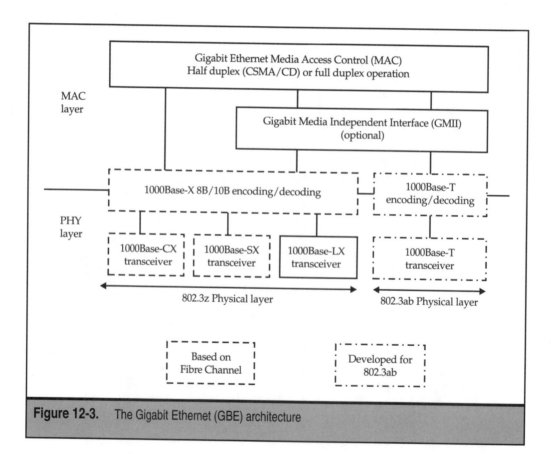

Figure 12-3.   The Gigabit Ethernet (GBE) architecture

There is even a CSMA/CD GBE repeater in the specification. But CSMA/CD in any form is just very inefficient at gigabit speeds and the distances are severely limited. Most implementations of GBE will be full duplex. But the magic Ethernet name is preserved.

Using GBE

GBE will initially be used to connect GBE switches with lower-speed 100Base-T Fast Ethernet ports, routers, and servers with special GBE interfaces. Servers and other forms of hosts have a hard time keeping up with GBE, but it can be done.

There are four things to consider when a GBE interface is envisioned for a server.

▼ **PCI bus bandwidth**   The PCI bus is the most common "Wintel" (Windows and Intel) architecture, and it is also used on some other platforms. All bit transfers from board to board and from board to the CPU and back happen over the PCI bus. In the mid-1990s, the PCI bus was 32 bits wide and ran at 33 MHz, limiting throughput to about 1 Gbps. Naturally, this is a problem when one board has as much bandwidth as the whole PC internally, and bits have to be buffered on the NIC—lots of them. By 1998, the PCI bus was 64 bits wide and still ran at 33 MHz, doubling the speed to 2 Gbps. Today, the PCI bus is usually 64 bits wide and runs at 66 MHz, giving 4 Gbps bandwidth. Make sure the server bus can keep up with the GBE interface. That is, do not use an older server for GBE.

■ **Network Operating System (NOS)**   Most GBE interfaces will still be using TCP/IP inside the GBE frames. Single-processor UNIX-based systems a few years ago could only handle about 500 to 800 Mbps through the whole protocol stack. Windows ran about 600 Mbps maximum. Today, both UNIX and Windows do better, but it still might be necessary to run dual processors to keep up with the GBE packet arrival rate on a server.

■ **CPU utilization**   Even simple TCP/IP checksum calculations can tie up a CPU at GBE speeds. Other header processing demands even more attention. Now, some of this processing can be offloaded to the GBE NIC, but this raises the cost of the board. Another way is to cut down on the number of headers by using *jumbo frames*. Jumbo frames allow "Ethernet" frames to be much larger than the usual 1,500 bytes or so, up to 9,000 bytes in most cases. However, jumbo frames are not standard Ethernet frames and so cannot interoperate with other switches or NICs that do not also support jumbo frames.

▲ **Memory subsystem**   System buses (not the same as the PCI bus) connect CPU to memory. Older buses ran at 66 MHz and were overwhelmed by caching requests and the like at GBE speed. Newer system buses at 100 MHz should always be used with GBE. Configurable memory cache size is a nice feature also.

It is most likely that GBE will be used in a *mixed speed* Ethernet environment. Older machines can still run at 10 Mbps. Most client and server machines will run at 100 Mbps.

Hubs that can auto-sense and operate ports at either 10 Mbps or 100 Mbps will link these clients and servers together. Newer servers can run at GBE speed with a GBE NIC, and a special 10 GBE switch will connect them to the slower ports. The GBE hubs will also connect to each other at 1 Gbps, or to new 10 GBE switches. Figure 12-4 shows this environment. Use of the 10 GBE links *between* sites (shown in the figure) will be considered later in this chapter.

This figure represents just the overall architecture, so many hubs are not shown with any clients or servers. But they are there.

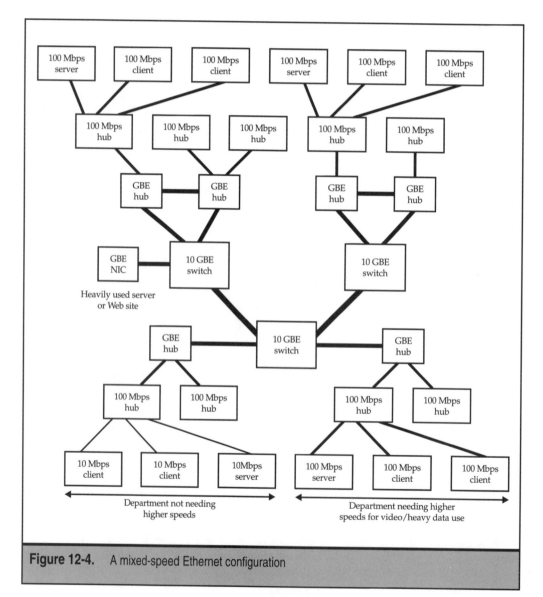

Figure 12-4.   A mixed-speed Ethernet configuration

GBE is best suited for an environment that is experiencing network bottlenecks in three main areas:

▼ **Switch-to-switch connections**   Many Ethernet hubs today are switches. That is, the hubs do not simply send traffic out every port, but rather they know which MAC address is on each port. The hub examines the destination MAC address on each frame and forwards it only to the proper port. This makes full duplex operation possible in the first place.

■ **Connections to high-speed servers**   Heavily used servers can take advantage of GBE speed to increase throughput and decrease user response time, as long as the server can keep up with the GBE, as mentioned previously.

▲ **Backbone connections**   Given the reliance on Internet connectivity today, it is tempting to use GBE to connect a site to an ISP. DWDM could be used, but probably not on this link. This possible use of GBE will be considered at the end of this chapter.

Migrating to GBE does not have to be difficult. The starting site LAN can be 10 Mbps Ethernet. Several Ethernet segments or workgroups can then be "collapsed" onto a 10/100 Mbps switch, which in turn has several 10 Mbps Ethernet server connections. Next, the LAN backbone between switches can be upgraded to 100 Mbps Fast Ethernet, and power users can be given 100 Mbps connections as well.

Some of the switch links can be upgrading to GBE as traffic grows. Key servers are upgraded with GBE NICs at the same time. Desktops still use 10/100 Mbps switched Ethernet. Finally, the Fast Ethernet backbone switch that connects the 10/100 switches can be upgraded to a GBE switch, supporting multiple 100/1000 switches. Once the backbone is upgraded to a GBE switch, high-performance server farms can be connected directly to the backbone with GBE NICs. This increases throughput for the servers with users running high-bandwidth application files or bandwidth-intensive data warehousing and backup operations.

GBE Frames

In might seem odd that to this point no mention has been made of the GBE frame structure. This is only because the GBE frame structure is exactly the same as it is for a basic Ethernet frame running at 10 Mbps. The basic Ethernet frame structure is shown in Figure 12-5.

The preamble is a seven-byte sequence that tells the receiver to "get ready for a frame." It is just an alternating string of 1's and 0's ending in a 0 bit. The start of frame delimiter (SFD) is one byte that is used to continue the alternating bit pattern of the preamble and ends in two 1 bits: 10101011. Then come two MAC layer address fields, destination (DA) and then source (SA), that are six bytes (48 bits) long. Before 1997, two-byte address fields were supported. This is the address of the NIC in the device, and *not* the IP address (which is normally 32 bits long). The length or type field depends on whether the frame follows the original Ethernet frame structure (most LANs do) or follows the structure detailed in the IEEE 802.3 specification. In Ethernet, this two-byte field

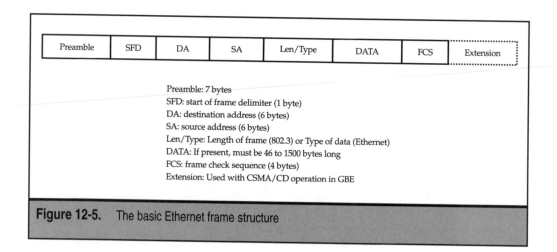

Preamble	SFD	DA	SA	Len/Type	DATA	FCS	Extension

Preamble: 7 bytes
SFD: start of frame delimiter (1 byte)
DA: destination address (6 bytes)
SA: source address (6 bytes)
Len/Type: Length of frame (802.3) or Type of data (Ethernet)
DATA: If present, must be 46 to 1500 bytes long
FCS: frame check sequence (4 bytes)
Extension: Used with CSMA/CD operation in GBE

Figure 12-5. The basic Ethernet frame structure

is a frame *type* code. In IEEE 802.3, this is a two-byte frame *length* field. The two forms are not interoperable, but few LANs run with 802.3 frames anyway.

The data field can contain up to 1,500 bytes of data (if 802.3 frame formats are used, not all of these bytes can be *user* data). If the data field contains less than 46 bytes of data, it is necessary to *pad* this field to create 46 bytes. The frame check sequence (FCS) is a four-byte (32-bit) field that provides error detection for the frame. The extension field is present when GBE is used with CSMA/CD. This extends the "length" of the frame to occupy a normal CSMA/CD "time slot" and allows CSMA/CD to function even with short frame sizes.

Lately, a lot of GBE vendors have begun to support *jumbo frames* for GBE. These extend the allowable data field size far beyond 1,500 bytes. Most hardware and software vendors support 9,000-byte jumbo frames, which increases throughput slightly and cuts down on the number of headers (both frame and packet headers) that a GBE switch or server has to process. One major router vendor supports a configurable jumbo frame size of up to 10,240 bytes.

There is no real standard for jumbo frames in GBE, so interoperability issues are always present. And jumbo frames do little to increase bandwidth efficiency in GBE. After all, 1,500 data bytes out of a 1,518-byte total is already 97.5 percent efficient. With jumbo frames, 9,000 data bytes out of 9,018 is 99.6 percent efficient, but this lowers the transfer time of a one-megabyte file by only 0.1 milliseconds.

But the real attraction of jumbo frames is for devices with GBE interfaces. Testing has shown that jumbo frames can increase throughput by 50 percent and at the same time decrease CPU utilization by 50 percent. And many common LAN applications use large packets. Sun's Network File System, for example, uses an 8,192-byte block of data. Fragmenting this across six frames adds more processing overhead. If interoperability is an issue, ways exist to partition a LAN so that jumbo frames are kept isolated on one part of the LAN.

To many people, the move to jumbo frame support is long overdue.

10 GIGABIT ETHERNET

No sooner did GBE come along than there was talk of extending GBE to an even higher speed. So 10 GBE was born. It made perfect sense. The highest speed of SONET/SDH was about 10 Gbps, so plenty of lasers and receivers ran at that speed. And since GBE ran over long enough distances on single-mode fiber (SMF) to leave the building (up to 3 km), why not try to aggregate GBE signals onto one higher-speed link for MAN use? (By the end of this chapter, GBE and 10 GBE as a WAN technology will be considered.)

Some information on 10 GBE was also introduced in Chapter 4. There, it was pointed out that 10 GBE is very much still a work in progress; nonetheless, 10 GBE extends Ethernet closer to the limits of state-of-the-art fiber optic speeds for serial transmission, which currently tops out at 40 Gbps. This standard comes from the realization that the vast majority of LANs linked by WANs today run Ethernet in some shape, manner, or form. It would be more efficient to minimize the amount of reframing, conversion, encapsulation, and other forms of overhead processing that must take place as a "low speed" Ethernet frame makes its way from LAN to MAN to WAN and back again.

What can be said with confidence about 10 GBE at this stage of its development? Well, 10 GBE will be used to tie GBE switches together, perhaps even over MAN and WAN distances. When that day comes, most end systems on the network will run at 100 Mbps at least. This vision of 10 GBE is shown in Figure 12-6.

The 10 GBE specification will feature full duplex operation on fiber optic cable. Support for CSMA/CD and other media is unlikely. But there will be as much consistency with 10/100/1000 Mbps operation in the sense that the 10 GBE frame will be the same (perhaps with standard jumbo frame support), and other aspects of Ethernet such as management will be the same.

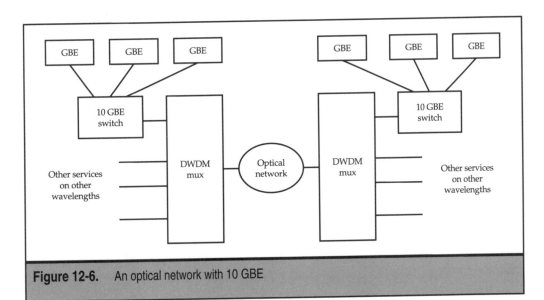

Figure 12-6.   An optical network with 10 GBE

Block coding such as 8B/10B will be used in at least some form of 10 GBE, and there will be a media-independent interface called XMII between the MAC layer and the line coding used in 10 GBE. However, there are proposals to use another form of block encoding, called 64B/66B, for running 10 GBE over WAN distances. This will provide great flexibility in implementation, and 10 GBE can run over a physical interface that operates in a serial or parallel fashion, or over "coarse" WDM.

Work is being done by the IEEE 802.3ae task force, and it seems that at long last the worlds of data communications (LAN connectivity above all) and telecommunications (voice above all) will finally come together. The highest speed of SONET/SDH is OC-192/STM-64 running at about 10 Gbps. So the potential exists to use at least some of the same components used for OC-192/STM-64 and migrate these to 10 GBE. Alternatively, 10 GBE links could be added with DWDM to existing OC-192/STM-64 links.

There are three main areas of concern when trying to scale GBE to 10 GBE. These are speed scaling, power consumption, and density issues.

▼ **Speed scaling**   Increasing speeds for optical components is not just a question of generating bits 10 times faster or making a clock tick 10 times faster when going from GBE to 10 GBE. Circuits operating at 10 Gbps are now running at microwave communications frequencies. Circuits must be laid out differently. Even when shrinking things down to a package 1.5 mm on a side, it is necessary to think of many of the internal circuit paths as transmission lines all by themselves. Techniques used at 2.5 Gbps and below cannot be used at 10 Gbps. However, one benefit that 10 GBE has is that the strict limits on timing jitter needed for SONET/SDH chipsets intended for WAN use in telephony applications are not needed in 10 GBE. Voice support in 10 GBE will be through voice over IP (VoIP), which handles jitter in other ways.

■ **Power consumption**   All high-speed optical interfaces need to consider power requirements. But as the size of components shrink, and higher port densities are needed, every watt of power adds to the complexity of issues regarding power supplies, cooling fans, heat sinks, and the total size of the device. Fortunately, the use of vertical cavity surface emitting laser (VCSEL) diodes has allowed great improvements in these areas. Less drive current is needed for these devices.

▲ **Density issues**   Minimizing size always has a way of minimizing cost in communications systems. But in 10 GBE, size not only refers to the physical dimensions of the component, but also the space needed to interconnect components. The "size penalty" doubles as increasing the chipset rate tenfold requires more isolation of signal paths. Integration of function will help, and this will in turn help with power issues.

The future of 10 GBE is bright. All in all, 95 percent of all Internet traffic comes from or goes to an Ethernet-based LAN. Uniting separate LAN/MAN/WAN structures with various speeds of Ethernet is logical and probably inevitable. New networks based on

10/10/1,000/10,000 Mbps Ethernet will minimize the number of reframing steps that need to be taken by routers and switches and will be more efficient than any other infrastructure in use today.

The LAN Invades the WAN?

How can it be claimed that GBE and 10 GBE somehow allow LAN technologies to invade the MAN and WAN arenas? GBE on single-mode fiber (SMF) only supports up to 5 km (a little more than 3 miles) spans between GBE hubs. That might be far enough to link a site to an ISP, but hardly far enough to call this a MAN or LAN. And 10 GBE would be expected to follow the general rule that going faster means going shorter on links between active devices.

Clearly, something is missing if GBE and 10 GBE are being proposed as all the networking needed for LAN and MAN and WAN. Here is the missing piece.

The simple fact is that full duplex Ethernet at any speed has no fundamental distance limit. Half duplex Ethernet using CSMA/CD must respect propagation delay limits, but the simple serial links used in full duplex Ethernets have no such limitation. Link lengths are only limited by the capabilities of the particular physical layer topology and the performance characteristics of the medium used.

For example, IEEE 802.3z guarantees that all 1000Base-LX-compliant links using SMF will *interoperate* over links up to 5 km in length. But this is just a baseline specification for worst-case scenario interoperability when there is no control over the design of the two end-points of the link. This ensures than low-cost, standard components are always available. If interoperation is not of paramount concern, nothing stops a vendor or customer from extending the distance spanned, as long as the overall power budget is respected, of course.

With GBE, link distances can be greatly extended and still be in total compliance with the GBE standard. In particular:

▼   If the performance of end equipment using 1000Base-LX is carefully matched, link spans of up to 20 km (about 12 miles) are possible.

■   With currently available link extenders, link lengths in GBE of up to 120 km (75 miles) are possible, even without optical amplifiers.

▲   When optical amplifiers are coupled with the link extenders, link lengths of up to 1,000 km (about 620 miles) have been demonstrated for GBE.

The extended GBE link possibilities are shown in Figure 12-7.

Many service providers supporting GBE envision links with a standard span of up to 40 km (25 miles), the same as in SONET/SDH. Current plans call for 10 GBE at 1,310 nm to support up to 10 km and at 1,550 nm (the normal DWDM window) up to 40 km without optical amplifiers. If SONET/SDH running at 10 Gbps can do it, then so can 10 GBE.

GBE and 10 GBE and Optical Networking

A new breed of service providers such as Telseon and Yipes Communications are already using GBE switches from vendors like Extreme Networks, Foundry Networks, and

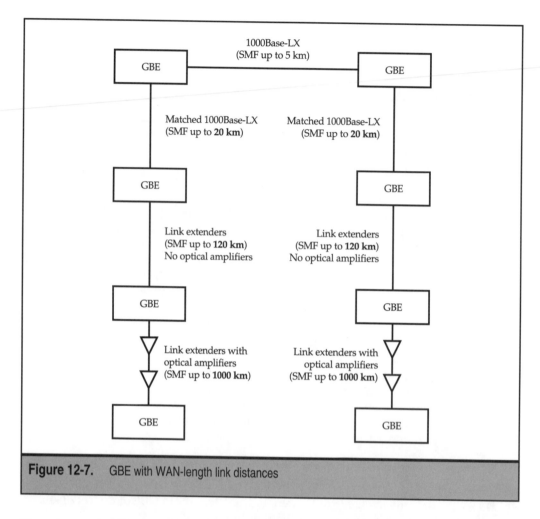

Figure 12-7.   GBE with WAN-length link distances

Riverstone to build networks that are based entirely on Ethernet. In some cases, users can use a simple browser interface to provision their own bandwidth in 1 Mbps increments up to a full 1 Gbps. Once the fiber is run to the customer premises, Ethernet allows this to happen almost naturally. Technologies that are as "old" as OC-3 or STM-1 at 155 Mbps just do not have this flexibility.

These new GBE service providers are more than ready to replace traditional leased-line services and SONET/SDH with Ethernet. If IP over Ethernet has been so successful on the LAN, that must mean something. Why not extend IP over Ethernet into the DWDM core of the optical network? If nothing else, MAN-size GBE networks can form the basis for ISP peering points.

To this point, conserving fiber pairs in a GBE network has not been a problem, mainly due to the enormous bandwidth involved. But there are already products on the market to combine eight full duplex GBE channels onto one pair of fibers using "coarse" or wide WDM (WWDM). However, use of GBE with DWDM has not been very popular so far, mainly because it makes much more sense to put multiple SONET/SDH links running at 2.4 Gbps (actually 2.48832 Gbps) onto a DWDM link than many GBE links with less than half the bandwidth. And once the 8B/10B encoding is factored in, GBE really takes up 1.25 Gbps. So one 2.4 Gbps channel cannot carry two GBE links. One nice solution is to map two GBE channels onto a single 2.4 Gbps SONET/SDH channel by dropping the GBE 8B/10B overhead and using some other encoding scheme (several are possible) and then mapping this SONET/SDH channel onto DWDM. This is shown in Figure 12-8.

Products to this point have been proprietary, of course. But the rest of the DWDM channels can contain almost any other type of traffic necessary to preserve compliance with standards.

When it comes to 10 GBE, the IEEE group plans to make 10 GBE capable of being used for a LAN with direct links to devices running at a full 10 Gbps. But 10 GBE will also be compatible to the greatest extent possible with the SONET/SDH rate of 9.58464 Gbps, usually rounded to 10 Gbps. So existing SONET/SDH ADMs and DWDM equipment should be able to gracefully migrate to 10 GBE. However, 10 GBE will not be SONET/SDH compliant, just SONET/SDH "friendly." Thus 10 GBE will not adhere to the ITU-T DWDM grid, nor will it support SONET/SDH jitter and timing mechanisms.

Nevertheless, the proper mix of transponders should be able to allow 40- or 160-wavelength DWDM systems (for example) to mix SONET/SDH 10 Gbps channels with others running 10 GBE channels over the same distances. This would maximize flexibility and service offerings. Also, 10 GBE ports will appear on optical switches and routers. Some servers will have direct 10 GBE interfaces to hook directly into optical networks with DWDM rings, eliminating extra router hops.

However, it takes more that just a data transport mechanism to make a technology into a network. GBE and 10 GBE must also become a "carrier-class" offering if the two GBE offerings are to replace the existing fiber infrastructure based on SONET/SDH. There are issues of reliability, scalability, OAM&P (operations, administration, maintenance, and provisioning), and quality of service (QoS).

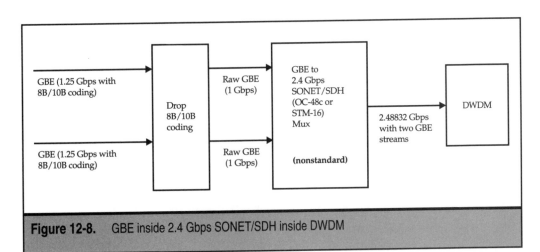

Figure 12-8.   GBE inside 2.4 Gbps SONET/SDH inside DWDM

Some early tests with GBE claim a 99.99 percent uptime, which is not bad for new products, and GBE can always rely on DWDM rings to provide the same protection SONET/SDH enjoys, and in the same timeframes of about 60 milliseconds. OAM&P could be handled by MPLS, and some early QoS issues could be resolved the same way. Of course, at 10 Gbps, it would hard to argue that delay-sensitive services like voice would in any way have service quality degraded under the direst of circumstances. Minimal and stable delays should be easy to achieve with GBE and 10 GBE, especially with multiple DWDM channels available.

But at some point, aggregating traffic will tend to fill up even the fastest pipe. Not all traffic should be handled exactly the same. No one would suggest that a bulk file transfer should have priority over a Web page or that a Web page should have priority over a voice call or video stream. ATM struggled to provide proper QoS for these services because all forms of traffic started appearing inside IP packets, and all IP packets were essentially treated the same by ATM. Now the situation is potentially even worse: not only are all the packets IP, but all of the frames are Ethernet! In frame relay and ATM, QoS parameters are supported by mapping them to virtual circuits, which form stable paths across the network for traffic to follow. The LAN legacy of Ethernet provides no support for virtual circuits and so nothing to "stick" QoS to as traffic flows across the network. How can new Ethernet WANs provide adequate QoS and virtual circuits for all forms of traffic?

One way is to use the 1998 IEEE standard called *VLAN tagging* with GBE (and 10 GBE). Many vendors of GBE and 10 GBE products rely on VLAN tagging to add WAN features to Ethernet.

The concept of a virtual LAN (VLAN) was introduced in Chapter 9 in the ATM LANE section. VLANs allow clients and servers attached to different hubs to act as if they were all part of the same physical LAN segment. What this boils down to is that the network administrator can define a separate *broadcast domain* for each VLAN. The broadcast domain establishes the group of devices that receive an Ethernet broadcast frame sent to all devices on the LAN. Many protocols rely on Ethernet broadcasts to function properly, most prominently the IP Address Resolution Protocol (ARP). Ordinarily, all devices linked to the same physical hub make up a broadcast domain. VLANs make the idea of a LAN into a logical object, not a physical object.

Now, Ethernet frames contain MAC layer source and destination addresses. The broadcast address is a form of Ethernet destination address. How can the LAN hub possibly know which VLAN an arriving frame is to be distributed to? The hub could maintain a list of all the ports that make up a VLAN, and all hubs do, but there is more that needs to be done. It turns out that devices can belong to more than one VLAN. Not only that, but a single hub could have links to two or more other hubs also having devices that belong to the VLAN. How would the first hub know which inter-switch ports to send the broadcast frame out on? The hub certainly does not want to blindly send these frames everywhere.

Both the VLAN identification problem and the QoS prioritization issue are neatly solved with VLAN tagging. VLANs are defined in IEEE 802.1Q. VLAN tagging uses 802.1Q formatting to add four bytes (called the Q Tag or 802.1Q header) between the source address and Length/type fields in an Ethernet frame. The Q Tag has two components, both two bytes long. The first two bytes are the Tag Type field, and the second two bytes are the Tag Control Information field. The position of the VLAN Tag in the Ethernet frame, along with one possible use for these Q tags, is shown in Figure 12-9.

The Tag Type (or Tag Protocol Identifier in some documentation) field is just a value of 81-00 in hexadecimal that lets devices know that the next two bytes are the Tag Control Information field and that the Length/type field and the rest of the Ethernet frame will follow. The Tag Type field is really just a special value for the Ethernet Type field, which would normally be in this position in the Ethernet frame, saying that a VLAN is in use.

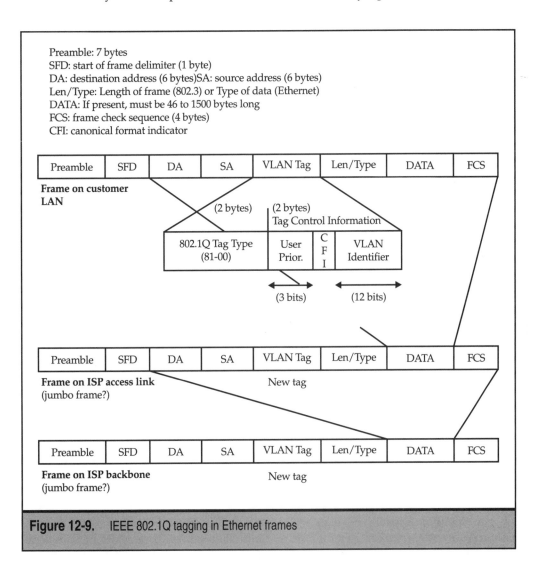

Figure 12-9.   IEEE 802.1Q tagging in Ethernet frames

The real power of the 802.1Q header is revealed in the second two bytes, the Tag Control Information. Two formats are allowed for this field, called "802.1Q (part a)" and "802.3 (part b)." The 802.1Q format can be used with any LAN frame, but the 802.3 format is to be used with Ethernet. The figure shows the 802.1Q format (the only reason for the 802.3 format is because of the odd way the Ethernet puts bits on the wire compared to other LANs), but the same fields are present in both formats.

There are three subfields in the Tag Control Information field. One of them, the Canonical Format Indicator (CFI), is a single bit used primarily for token ring and so not a concern here. The first field of importance is the VLAN Identifier (VLAN ID) field. This 12-bit field is used to associate the Ethernet frame with a particular VLAN whether the frame is a broadcast frame or not. Now, VLANs are mostly campus-wide or even confined to a single site. When used with GBE or 10 GBE, however, the VLAN ID can serve as a virtual circuit identifier just as the DLCI in frame relay or the VPI/VCI in ATM can.

The VLAN ID is useful in a WAN-scale Ethernet because now Ethernet "circuits" can be provisioned and switched as easily as frame relay or ATM virtual circuits. And of course Ethernet includes its own speed specifications, which frame relay and ATM lack by themselves.

The value in the VLAN ID can be assigned by the service provider completely independent of any other considerations. But the VLAN ID can also be chosen to reflect the value of the MPLS flow identifier if MPLS is carried inside the Ethernet frame, or the value of an IPv6 header flow field, or something else entirely. Now, all these fields have different sizes, but the point is that they can be mapped to one another. So for part of its path across a WAN, an IP packet uses MPLS to traverse a former purely ATM portion of the network, and then some interworking device could map this MPLS flow value onto the GBE portion of the network if VLAN tagging is used.

The other field of interest is the User Priority (usually just seen as P) field, based on the IEEE 802.1p specification. This is a way to add user priority values to the Ethernet frame. The P bits are a three-bit field, so there can be eight different priorities for the Q-tagged Ethernet frame. This should be plenty for most WAN applications, and B-ISDN/ATM has only four distinct service classes. A value of 0 has the lowest priority, and a value of 7 the highest. In spite of the presence of the "user," 802.1Q does not really expect end users to decide what priority their traffic should have. When it comes to 802.1Q, the end system application program is the "user" of the network. These standardized applications will have to decide if Web pages rate a 4 or a 5 when it comes to priorities.

Like the VLAN ID field, the User Priority field can be set independently of other considerations. That is, voice could always be a 7, FTP file transfers always a 0, error messages a 1, and so on. Or the bit pattern in this field could be drawn from the value of priority bits drawn from somewhere else. This source could be the type of service (TOS) field in the IP header, the DiffServ interpretation of these bits, the MPLS experimental field, and so on. In contrast to the VLAN ID, all of these priority fields are also three bits, so mapping is not really an issue at all.

Another nice thing about VLAN tagging is that it is possible to implement hierarchical layers of tagged Ethernet frames. So an internal GBE LAN connecting buildings can use one set of headers and VLAN tags. The locally tagged frames can be handed off to an

ISP over a GBE link, who encapsulates this frame and retags it for transport on its own GBE backbone. Finally, the ISP frames can be handed off to a national backbone provider who tags them once again. The process is especially attractive when combined with the use of jumbo frames. This layering of tags in combination with jumbo frames is also shown in the figure.

How would this series of Ethernet frames with VLAN tags be used in an optical network? Figure 12-10 shows one way, with edge and core routers now handling the mapping of VLAN identification labels and the packing and unpacking of jumbo frames (if used). This architecture varies slightly from the basic architecture introduced in the previous figure. Here, VLAN tags are only used on the access link, since it is unlikely that most LANs will need such fine QoS distinctions. Another VLAN tag is added at the service provider POP for internal handling, and then a third VLAN tag is added when this POP's traffic is sent onto the DWDM backbone ring.

There will still be plenty of routers used with GBE, but this chapter has emphasized the role of GBE in optical networking, so there have been few routers to be seen up to this point. However, there is no reason that a GBE switch cannot also have IP routing capabilities, or that an IP router could not have a GBE switching fabric as well.

It should come as no surprise that the end result starts to look a little like MPLS. GBE equipment is as much of a switch as MPLS devices are. And GBE nodes can route the packets inside the Ethernet frames as easily as a router. But with VLAN tagging, a lot of Ethernet frame delivery can be done without examining the IP packet inside at all.

Of course, SONET/SDH proponents always point to the protection switching provided by SONET/SDH rings as a feature that cannot be duplicated in an "Ethernet WAN" made up of point-to-point links. On the LAN level, Ethernet LANs are linked by bridges running something called the Spanning Tree Protocol (STP). This allows bridges to form rings for redundancy purposes, but it allows only some of the bridges to operate to eliminate "loops" that cause traffic to circle around endlessly. Now, GBE could always rely on DWDM rings for protection, but it would still be the case that Ethernet had no protection switching of its own.

But this may not be the case for long. Ethernet nodes connecting VLANs are not routers, but more like traditional LAN bridges using switching and running STP. STP will reconfigure a collection of Ethernet bridges to find an alternate path without loops if a bridge fails, but only after a minute or so. A proposal to speed this process up called Rapid Reconfiguration of Spanning Tree, a product of the IEEE 802.1w group, is almost ready to go. This will make STP as quick as SONET/SDH when it comes to finding an alternate path through a collection of GBE bridges when a link or node fails.

The nice thing about the new STP is that it works with almost any collection of bridges forming loops, no matter how large or interlocked. So there are fewer design constraints than there are in SONET/SDH rings.

What about even higher speeds for GBE? The current state of the art for serial transmissions of bits on a fiber optic link is 40 Gbps. SONET/SDH will soon make this leap as OC-768, but this is essentially just four 10 Gbps SONET/SDH frame structures extended to run faster. No breakthroughs with SONET/SDH are required. It might soon make sense to extend 10 Gbps to 40 Gbps as well. Once it is running in full duplex mode, there is really no limit to the speed or span of Ethernet.

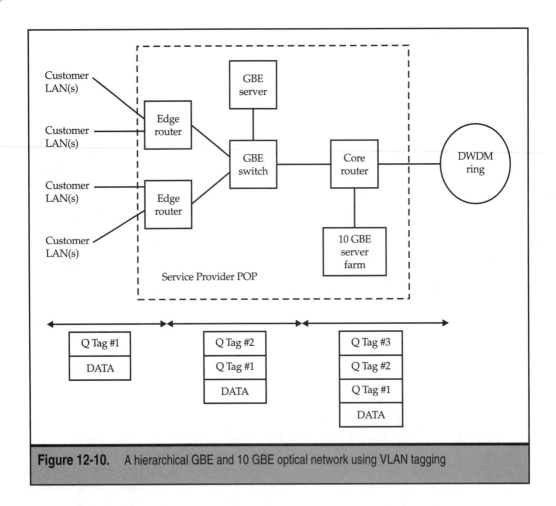

Figure 12-10. A hierarchical GBE and 10 GBE optical network using VLAN tagging

GBE AND 10 GBE: IS ANYTHING ELSE NEEDED?

Perhaps someday ISPs will be known as ESPs: Ethernet service providers. If GBE and 10 GBE make it possible and attractive for ISPs to take advantage of extending Ethernet to WAN distances, then why should any other form of framing be used for IP packets besides Ethernet? The VLAN tagging could be used with, or even replace, the MPLS method of switching IP packets along stable paths with the proper quality of service for all applications.

The issue of the "best" framing for IP packets has already come up in Chapter 10 when IP was considered with DWDM. There, all roads basically led to DWDM, and one of the framing methods for IP packets discussed was GBE. Recall that SONET/SDH framing was required for legacy voice applications, where digital voice samples at 64 Kbps lacked any form of framing that could be used to separate voice channels on a voice backbone. Framing is required for IP packets as well, since without a suitable frame structure only IP packets could be used, there would be much less robust error detection, and a more complete form of packet delineation would be needed.

But what about networks, or portions of networks, that carry IP packets inside ATM cells (or frame relay frames) on virtual circuits? What about IP packets flowing on PPP (Point-to-Point Protocol) leased-line links between routers? And what about IP packets already given MPLS labels?

As it turns out, there are already ways to carry each one of these other IP packet network framing techniques inside Ethernet frames. Specifically, a way to carry a string of ATM cells inside an Ethernet frame is called "cells in frames," or sometimes CIF. CIF was pioneered at Cornell University in New York when it was realized that the school had to teach ATM, but there was no realistic way to convert all of their Ethernet LAN-attached PCs to ATM. The solution was to carry a series of ATM cells inside the Ethernet frame over a 10Base-T LAN. The cells all shared a common header, and the cell payloads that followed filled an Ethernet frame. ATM cell payloads are 48 bytes long, so 31 of them could fit inside an Ethernet frame ($48 \times 31 = 1,488$) with enough bytes left over for the common 5-byte header that all of the cell payloads had to share. The concept of CIF could easily be used on GBE links to carry ATM cells between ATM switches on a GBE backbone.

PPP frames can be carried inside Ethernet frames using PPPoE (Point-to-Point Protocol over Ethernet). PPPoE is frequently used with Digital Subscriber Line (DSL) modems to carry IP packets to and from Ethernet NIC-equipped PCs. But if there are areas of a network still using PPP to carry IP packets between routers, these PPP frames can be carried over a GBE backbone inside Ethernet frames using PPPoE.

And MPLS label information can easily be mapped to an Ethernet 802.1Q VLAN tagging header. This would allow GBE to be used between MPLS switches as well. Frame relay virtual circuit identifiers (DLCIs) could be mapped the same way, and frame relay could ride the GBE backbone along with all of these other forms of traffic.

This philosophy of "all roads lead to Ethernet" is shown in Figure 12-11.

ATM uses CIF to ride the GBE or 10 GBE backbone, and PPP uses PPPoE. MPLS is mapped using VLAN tagging (MPLS labels can also be attached directly to Ethernet frames, of course). Frame relay can be mapped this way also, but in the figure frame relay is shown with a dotted line, since it is more likely that frame relay frames would enter a GBE backbone through an ATM switch or inside MPLS or just be removing the frame relay frame. The "odd technology out" is SONET/SDH. There is just no easy way to map SONET/SDH frames to GBE frames. In fact, as discussed previously, it makes much more sense to map two GBE channels to one SONET/SDH 2.4 Gbps link. Most likely, SONET/SDH frames will continue to occupy channels of their own on DWDM backbones independent of GBE.

Naturally, none of this rules out mapping the GBE links to DWDM channels, with or without the intermediate 2.4 Gbps SONET/SDH mapping discussed previously. This is also shown in the figure.

By now, it should be obvious that FDDI, Fibre Channel (FC), and both forms of GBE work quite well with optical networking. Any organization with FDDI, FC, or GBE in

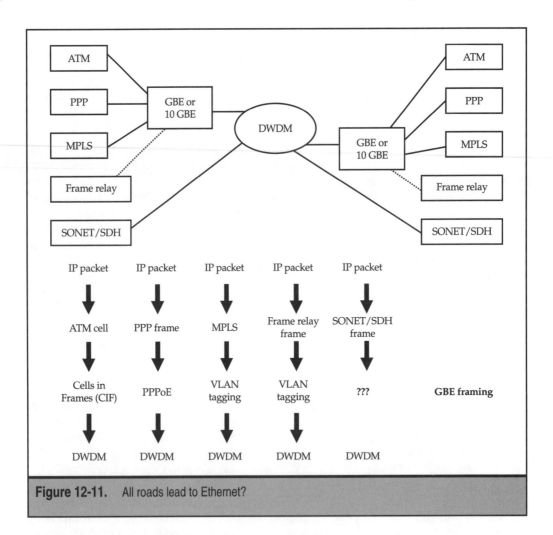

Figure 12-11.   All roads lead to Ethernet?

their past or future will find themselves in a good position to take advantage of optical networking. This is as true of end users and customers as it is for the service providers.

However, no one would propose that FDDI, FC, or GBE should be used as the basis for a residential service. So what about this vast pool of residential users that not only have not seen FDDI or FC or GBE speeds yet, but often do not even have access to DSL or cable modem services? Many still rely on dial-up modem speeds to access the Internet. Is there any hope that optical networking will someday make residential services as fast as organizations enjoy at major sites? This is the topic of the next major section of this book.

PART V

Residential Optical Networks

Parts 3 and 4 considered the impact of optical networking and DWDM on WAN architectures such as SONET/SDH and ATM and LAN architectures such as Fibre Channel and Gigabit Ethernet. Chapters 13 and 14 will apply the concepts of optical networking to the two major *residential broadband* (i.e., high bandwidth) architectures for bringing high-speed services to the home. The two architectures are the cable TV industry's hybrid fiber coax (HFC) architecture and the telephone companies' fiber to the home (FTTH). There is considerable variation in each of these architectures, and nothing prevents a cable TV company from embracing FTTH or a telephone company from building (or buying, in many cases) an HFC network. But the support for each by their respective industries has been remarkably consistent (or maybe not so remarkable, considering the competition).

The two architectures share many features in common. For example, both envision high-speed Internet access a key service offering. Video services are forecast to be the primary source of revenue, especially *digital* video services. Both will support voice services as well. Both architectures can employ a component called a passive optical network (PON), which requires no power between service provider and customer premises.

There are significant differences between the two architectures. For instance, cable TV networks have always had plenty of bandwidth, but usually only one way to the customer, or *downstream*. There have been no real *upstream* channels from customer to service provider *headend* office where the services originate until relatively recently. On the other hand, the telephone network has always been two-way, but only for enough bandwidth to support an analog telephone conversation. The cable TV companies have been busily adding upstream channels while the telephone companies have concentrated on adding bandwidth. This is one of the main reasons that the two approaches differ: they are both evolving from very different starting points.

Chapter 13 looks at the cable TV network infrastructure. The current HFC architecture is detailed, as well as the fact that the cable TV networks of today still mainly distribute analog TV signals. This is changing, however, so the chapter quickly moves on to consider the impact of digital TV (DTV) on this HFC architecture with PONs. Finally, the chapter closes with a look at how optical networking with DWDM will change DTV and HFC.

Chapter 14 looks at the public switched telephone network (PSTN) infrastructure. The emphasis here is on the use of fiber in residential areas. After a quick look at copper-based technologies such as digital subscriber line (DSL) technologies, the discussion turns to FTTH and its variations. Fiber to the neighborhood (FTTN), fiber to the curb (FTTC), and the ultimate fiber destination—the end-user device—are all considered in turn. This last variation could be called fiber *in* the home (FITH?). Chapter 14 closes with a look at how optical networking with DWDM will change the way that DSL and the PSTN operate.

CHAPTER 13

Digital Cable TV Networks

Cable TV networks were once used to bring TV signals to people in remote areas. Now, cable TV networks can distribute more than 100 channels of analog TV, and many include a number of digital TV channels. By 2006, says the United States government, all TV signals will be digital and consist of a string of 0's and 1's rather than the various intensities of red-green-blue (RGB) that characterize analog TV signals. The conversion effort will be substantial, because there are currently more than 10,000 cable TV *headends* (the offices where the signals originate on a cable TV system) in the United States alone. This is about 200 headends *per state.*

Cable TV systems can now offer both telephony and high-speed Internet access services, and many do. This chapter will highlight the changes that have been and will be necessary to the cable TV infrastructure as new services and technologies become more of a necessity than a luxury. The chapter ends with an assessment of the impact of optical networking and DWDM on cable TV in general and digital cable TV in particular.

As the cable TV industry is poised for a great leap into the future, this is a good place to review the history of the cable TV industry. Since the change to digital TV in the United States has been mandated by government decisions, the role of the U.S. government in the cable TV industry is stressed.

HISTORY OF CABLE TELEVISION

In the late 1940s in the United States, television was starting to become an important new medium for news and entertainment. Yet there were many communities in rural areas where regular broadcast signal reception was of poor quality. In many cases, the audio and video was next to impossible to hear and see. Cable TV was developed at this time for communities unable to receive TV signals because of terrain or distance from broadcast TV stations. Cable TV was called CATV (community antenna television) at the time, and the acronym is still seen sometimes, especially in the United States. Lately, the CATV acronym has even been interpreted as cable-assisted TV, since use in rural areas only has been expanded to almost everywhere. But in those days, cable TV system operators located antennas in areas with good reception, picked up broadcast station signals, and then distributed them by coaxial cable to customers for a fee.

There is a lively debate between Pennsylvania and Oregon over the origin of cable TV. Oregon claims that cable TV began on Thanksgiving Day in 1948, and that the first commercial headend went into service in February 1948 in Astoria, Oregon. Pennsylvania claims that cable TV began in June of 1948 in Mahoney City, Pennsylvania, in an effort to bring TV to places blocked by the Allegheny Mountains. In any case, it is clear that cable TV was a 1948 invention.

By 1950, cable TV systems operated in 70 communities in the United States. These early systems served about 14,000 homes combined. In contrast, in October 1998 there were more than 10,700 systems serving more than 65 million subscribers in more than 32,000 communities, and by now, large systems serve more than 20,000 homes from a single headend. Cable TV systems are operating in every state in the United States and in many other countries, including Austria, Canada, Belgium, Germany, Great Britain, Italy, Japan, Mexico, Spain, Sweden, and Switzerland.

Even the most basic cable TV systems are technically capable of offering between 36 and 60 channels on older coaxial cable. Channel capacity and offerings in the industry have increased dramatically in recent years. Many systems now offer in excess of 100 channels, and some of the channels are digital (soon they will all be digital). Most cable customers receive service from a system offering more than 54 channels.

The current structure of the bandwidth used on the coaxial cable for a cable TV network is shown in Figure 13-1. There are four main sections to the spectrum, which today can extend up to about 1 Gigahertz (1,000 Megahertz, or MHz). In the United States, the frequency range below 52 MHz is not used for downstream traffic signals to the customer. The frequency range between 5 and 40 MHz is used for upstream traffic from the customer to the headend, such as for cable modem operation. These analog channels take up about 6 MHz, as do the downstream analog TV channels.

All of these analog channels share the cable with frequency division multiplexing (FDM), of course. Until digital TV becomes the rule, the channels are all 6 MHz analog channels, and these are much more efficiently multiplexed with FDM than anything else.

Outside of the United States, this system of upstream channels to 40 MHz, downstream channels starting at 52 MHz, is not always used. The 40/52 split is used in North and South America, and many places in the Far East outside of Japan such as China, Korea, the Philippines, Thailand, and Singapore. Some systems do have a 42/54 split, however, and before 1994 the North American upstream or return band ended at 30 MHz. Australia uses a 65/85 split system; Japan and New Zealand use 55/70. India, Malta and Eastern Europe use 30/48, and Western Europe, Ireland, and the United Kingdom use 65/85. On the low end of the spectrum, most systems avoid 5 to 10 MHz because there is a lot of "noise" (unwanted external signals) in that frequency range.

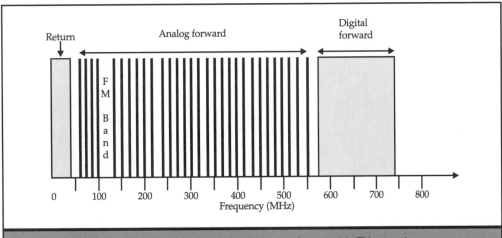

Figure 13-1.   The typical spectrum allocation of a United States cable TV network

The second area of interest in the spectrum is the broad range from 52 MHz (in the regions previously noted) to about 550 MHz. These are known as the *analog forward channels* and make up whatever number of 6 MHz analog TV channels the customer gets to see. Note that there is a space for FM distribution around 100 MHz, and this is the third area of interest. Many cable TV operators distribute FM signals, mostly music. The fourth and final area is for the *digital forward channels* above 550 MHz. These are the newer digital TV signals to the customer. As time goes on, this portion of the spectrum will displace the analog forward channels.

The channel capacity of a cable system makes it possible for a cable TV system operator to provide many services in addition to over-the-air television broadcast signals. Most systems also offer diverse program services, including (for example) news, weather, business information, movies, sports, and special entertainment features, as well as programming designed for specific audiences such as children, women, and ethnic and racial groups. Within the past few years, some cable systems have begun offering a full range of telecommunications services, including high-speed Internet access and local telephone service. High-speed Internet access allows customers with cable modems (sometimes seen as *cmodems*) to connect to the Internet more than 100 times faster than with the fastest standard analog modem.

A real milestone in the cable TV industry was reached in November 1972. This was the birth of Home Box Office (HBO) in Wilkes-Barre, Pennsylvania. HBO was the first real service that was made just for cable TV. It took a while for HBO to become an essential part of a cable TV offering, but once it did, HBO and cable TV each fueled the growth and acceptance of the other. When MTV (Music Television) and ESPN (Entertainment and Sports Programming Network) came along, also lacking broadcast outlets because they were cable-only services, these popular offerings made cable TV that much more nearly indispensable for many people in the United States and around the world.

Some cable TV operators also create their own local programming and provide access channels for public and institutional uses. They also provide leased access channels for those wishing to show specific programs. And the service is no longer just a one-way broadcast to the customers. Electronic banking, shopping, utility meter reading, and home security are some of the home services that are possible using the newer two-way transmission capabilities of cable TV systems.

Cable TV in the United States slowly but surely gained momentum over the years. When Ted Turner launched his "superstation" TBS in December 1976, the programming was seen on only four cable TV systems. Soon, TBS became the first local station broadcast everywhere, and people in places like Vermont could become baseball fans after watching the Atlanta Braves play night after night.

ESPN began on September 7, 1979, when most sports news on TV was limited to three-minute segments on the local news. ESPN was first limited to filling 24 hours of sports with such odd content (to people in the United States, at least) as Australian football.

Ted Turner and Reese Schonfeld, a noted TV and print journalist, started the Cable News Network (CNN) on June 1, 1980. CNN's "fluid news" style was pioneered by Schonfeld, CNN's president for three years before a clash with Turner drove him out.

MTV was launched on August 1, 1981, with a video called, appropriately, "Video Killed the Radio Star" by the Buggles. An unknown "video DJ" named Martha Quinn first brought music to television. That same year, TBS qualified for tracking by A.C. Neilson's rating service, a cable TV first.

Cable TV really took off in the United States in the 1980s, as momentum made cable TV a mainstream service. MTV gave birth to a "classic" version of itself called Video Hits 1 (VH1), and by 1987, National Football League (NFL) games were on ESPN (fans of Australian football in the United States were not happy). MTV even moved itself onto the Web in 1995, a case of an enterprise becoming an innovator in two arenas.

A short (and subjective) history of cable TV landmark events is shown in Figure 13-2. All of this growth and success in the cable TV industry was closely watched in the United States by a very interested body charged with establishing rules for the communications industry: the FCC.

The Federal Communications Commission (FCC) first established rules for cable TV systems that received signals by microwave antennas in 1965. In March 1966, the FCC established rules for all cable TV systems (whether or not served by microwave). In those days, there were frequent fights over federal versus local regulation of cable TV systems. Many broadcasters (which offered "free TV") were not happy that the cable TV companies took something they got for free (the broadcast TV stations) and charged for it ("fee TV"). The fights went all the way to the Supreme Court, which found that the FCC "needed" authority over cable TV systems to handle these and other issues.

New FCC rules regarding cable television became effective in March 1972. These rules required cable TV operators to obtain a certificate of compliance from the FCC prior to operating a cable TV system or adding a broadcast television channel. The 1972 rules covered many broad subject areas, from franchise standards to network program nonduplication (broadcast stations should not have to compete with themselves) and syndicated program exclusivity, from cross-ownership rules to technical standards. Cable TV operators with original programming were subject to equal time, the Fairness

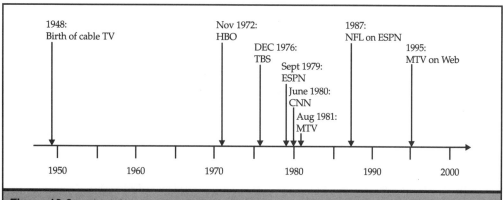

Figure 13-2. A brief history of cable TV

Doctrine, sponsorship identification, and other provisions similar to rules applicable to "free TV" broadcasters.

Over the years, the FCC modified or eliminated many of the rules. Among the more significant actions, the FCC got rid of most of the franchise standards in 1977, substituted a registration process for the certificate of compliance application process in 1978, and eliminated the distant signal carriage restrictions and syndicated program exclusivity rules in 1980.

In October 1984, the U.S. Congress amended the Communications Act of 1934 by adopting the Cable Communications Policy Act of 1984. The 1984 Cable Act established policies in the areas of ownership, channel usage, franchise provisions and renewals, subscriber rates and privacy, obscenity and lockboxes (to prevent access to some programs), unauthorized reception of services, equal employment opportunity, and pole attachments. The new law also defined jurisdictional boundaries among federal, state, and local authorities for regulating cable television systems.

The 1984 Cable Act was followed by an explosion in cable TV systems in urban areas. The number of households subscribing to cable TV systems increased, as did the channel capacity of many cable TV systems. One thing that did not increase was competition among distributors of cable TV services. In many communities the rate increases for cable TV services far outpaced inflation. Responding to these problems, Congress enacted the Cable Television Consumer Protection and Competition Act of 1992. The 1992 Cable Act mandated a number of changes in the manner in which cable television is regulated.

The 1992 Cable Act had many perceived flaws, mostly relating to how the Act seemed not to stop the increase in cable TV rates in any appreciable way. By 1996, the United States Congress was eager to make a sweeping change to the telecommunications industry as a whole in the United States, not just the cable TV industry.

The Telecommunications Act of 1996 (TA96) was intended to provide an environment to promote competition, deregulate many industry segments, and accelerate private sector deployment of advanced telecommunications and information technologies and services. The key to making all this happen was competition. The FCC has adopted a host of regulations to implement the requirements of the TA96, not all of which have been universally applauded in the United States.

WHEN CABLE TV MEANT COAXIAL CABLE

Cable TV was first introduced in Chapter 4 along with the other major LAN and WAN technologies mentioned in previous chapters. Only the barest outline of the cable TV network architectures was given, however. There was brief mention of cable TV networks based wholly on coaxial cable and then cable TV networks based on coaxial cable combined with fiber optic cable to create a hybrid fiber-coax (HFC) cable TV network. This is the place to fill in the details and position cable TV networks for the new world of cable modems for high-speed Internet access and digital TV (DTV) offerings.

The basic architecture of a coaxial cable TV system—one that contains no fiber optic cable at all—was shown in Figure 4-9 of Chapter 4. The architecture consisted of a single headend with coaxial cable deployed in a *random* or *arbitrary branching bus* topology. It is a branching bus in the sense that all of the signals go everywhere they can (making it a bus) and branches split off the signal at specific points to distribute the signal to as many *homes passed (HP)* as possible. The branching bus is *arbitrary* or *random* in the sense that the system designer can split off branches almost anywhere the geography of the distribution area dictates. The only requirement is to have as many analog *amplifiers* as necessary to keep the received signal quality as high as it should be. So amplifiers are *cascaded* one after another between headend and home passed. The HP metric is better than simply counting customers, since subscribers come and go, and not everyone whose home is passed by the cable TV network will subscribe to the service.

Figure 13-3 shows some of the details of the coaxial cable infrastructure. The emphasis here is more on the details of the amplifiers. There are large distribution amplifiers and smaller line extenders on the branches. The distribution amplifiers also allow the signal to branch, so there are many outputs (typically up to four) and one input on a distribution amplifier. Line extenders have only one input port and one output port. On larger systems, there can be as many as 30 or more amplifiers between headend and home. *Taps*, which do not amplify the signal and are distinct from line extenders, carry the signal from branch to home over a *drop cable*, also made of coaxial cable, of course. Taps have four, eight, or even more ports depending on whether the branch services a residential block of single homes, a group of townhouses, or a large apartment complex. Apartments and townhouses are called *multiple dwelling units (MDUs)* in the residential services industry.

Some cable TV services have very high *penetration rates* or *buy rates*. These rates mean that if a customer has the opportunity to buy the service for an extra cost, what percentage of the customers will pay for the service? Buy rates for some services, such as premium movie channels, are quite high, usually over 80 percent. Some buy rates remain distressingly low for the cable TV operators, such as for ordinary dial-tone telephony services, which seldom exceed 5 percent or so.

Service demand above and beyond basic TV channel offerings in the cable TV industry tends to be highly elastic, meaning that the less expensive the premium service is, the greater the buy rate. Of course, the lower the cost to the consumer, the lower the revenue

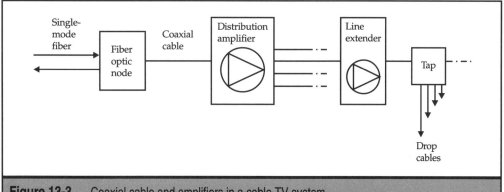

Figure 13-3.   Coaxial cable and amplifiers in a cable TV system

from each customer to the service provider. Only a higher buy rate can generate increased revenues for the cable TV operator as prices drop.

Competition from the telephone companies for voice and Internet access, and satellite TV companies for video services, keep a tight cap on the amount that a cable TV operator can charge for any individual services. So *bundling* of services is a popular way to achieve acceptable buy rates and revenue levels. For instance, a cable TV operator might bundle premium movie channels with high-speed Internet access through a cable modem. The package price would be less than for each service sold separately, but the package price effectively lowers the price to the customer while at the same time, it is hoped, establishing a higher buy rate for one (or both) of the services.

These service considerations are important when it comes to assessing the impact that optical networking and DWDM will have on cable TV networks. But most of this impact concerns HFC cable TV networks, cable modems, and digital TV services. So the next step will be to consider HFC cable TV networks in more detail.

HYBRID FIBER-COAX (HFC) CABLE TV

The basic architecture of a newer hybrid fiber-coax (HFC) cable TV network was also outlined in Chapter 4. Figure 4-10 showed the general structure of a cable TV network using HFC, and there is no need to repeat that information here. It is more important to consider more fully the reasons that fiber optic cable was added to the coaxial cable in a cable TV network in the first place.

The weak link of the all-coax cable TV architecture was the analog amplifier. These analog devices, which were very numerous, each added noise to the signal as they were cascaded to reach more and more customers. Distribution amplifiers were typically needed every mile or so (about every two km), and line extenders every half mile (about one km) or so on the branches. Customers that ended up 30 or more amplifiers away often had worse reception than they had from regular broadcast TV channels delivered with an antenna.

The analog amplifiers required huge amounts of power as well. Usually the local cable TV operator was one of the power company's largest customers, if not *the* largest customer the power utility had. And the branching bus nature of the architecture meant that if a single large distribution amplifier failed, a large proportion of the customers "downstream" from this single device lost service.

Finally, once two-way services became common, having dozens of amplifiers all "pointed outward" meant that weaker upstream signals from the home could not fight their way back to the cable TV headend. Not only did all-coaxial cable TV amplifiers not amplify upstream signals, many of them actually damped out (weakened) upstream signals. After all, no one had a TV set-top box that talked back to the cable TV company's headend directly in those days. Any signals appearing at an amplifier heading upstream had to be "noise." Not only that, but as many small sources of "noise" traveled the branching bus in reverse, from many sources to one headend destination, the upstream sources got more and more concentrated as they approached the headend. At the headend, the additive upstream "noise" could affect the quality of the downstream signals or disrupt the sensitive equipment in the cable TV headend. So many cable TV operators placed a *noise filter* at the end of the coaxial cable at the headend to prevent any upstream signals at all from entering the building.

Two-way services often posed a capacity problem as well. There were sometimes more than 100 channels capable of delivering high-speed Internet data (for instance) downstream to a customer. But the upstream channels were quite limited given the 5 or 10 MHz to 40 MHz frequency range they had to live with in many places around the world. Only four or five 6 MHz upstream channels could be jammed into the 30–35 MHz upstream bandwidth. If cable modems enjoyed a high enough buy rate, these upstream channels might become overwhelmed and be a bottleneck for throughput over the network.

Analog amplifiers in an all-coaxial cable TV network had a number of drawbacks:

▼ Additive noise accumulated through many amplifiers.

■ Power consumption was high individually and collectively.

■ A single amplifier failure caused massive service outages.

■ Upstream "noise" was damped and filtered.

▲ Two-way services had limited upstream capacity.

Once fiber optic cable was added to the cable TV architecture, every one of these issues related to amplifiers was lessened or eliminated. HFC networks remedied many aspects of these issues because:

▼ There were fewer amplifiers in the HFC networks, so there was much less additive noise to affect service quality.

■ The few fiber optic amplifiers consumed much less power individually and collectively.

■ There were more "trunk" fibers, so a single amplifier failure was much less disruptive.

■ The few fiber amplifiers were easier to push upstream signals through, and upstream "noise" at the headend was less.

▲ Two-way services enjoyed much higher upstream capacity.

Most of these improvements arose from the simple fact that fewer amplifiers were required in an HFC cable TV network. In most cases, the entire fiber "trunk" to a neighborhood branch could be handled without any amplifiers at all. On the other hand, splitting the fiber optic signals as the coaxial signals are split makes no sense, because the losses through a 4-way splitter, very common on coaxial distribution "trunks," are much higher because of the way optical power is split. This is why HFC cable TV networks retain coaxial cable for at least the neighborhood portion of the HFC network. What amplifiers remain in HFC as usually only the coax amplifiers.

In Chapter 4, it was pointed out that HFC cable TV networks typically have many point-to-point fiber "trunks" in place of the single coaxial cable. Understanding the power loss implications when attempting to split optical signals now explains this need for many fiber optic cables. Of course, as Chapter 4 also pointed out, it is up to the individual cable TV operator just how far the fiber is to penetrate into the network.

Most cable TV networks today employ a fiber to the feeder (FTF) approach that leaves perhaps 500 to 1,000 homes serviced on a fiber optic link. This means that a large cable TV system with 20,000 homes passed will have to have anywhere from 20 to 40 fiber optic links to connect to each and every neighborhood, townhouse group, and apartment complex on the cable TV system. And as has been pointed out, the older coaxial cable is usually retained for the purpose of distributing electrical power to the optical components. So things are a little more complex with HFC than they seem to be.

Figure 13-4 shows the general structure of an HFC network. The emphasis here is on the headend, where the services are combined onto the fiber optic links. Today, a cable TV headend will supply analog TV signals, digital TV signals, local ads that are inserted to the main feeds, two-way Internet access, and perhaps even landline and wireless public telephony services. Two-way services require an outbound (downstream) and inbound (upstream) 6 MHz channel of their own. An upconverter loads outbound channels. The fiber optic node at the right of the figure is the same as the fiber optic node at the left of Figure 13-3.

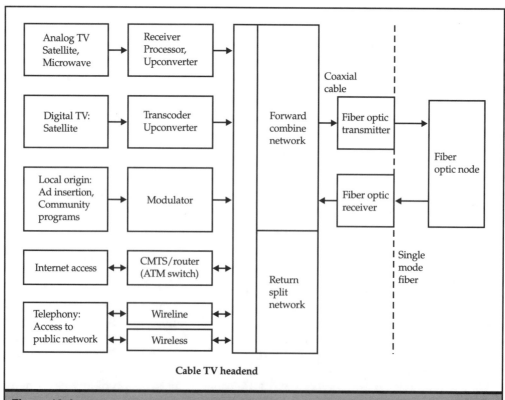

Figure 13-4.   The overall structure of an HFC cable TV network headend

Note that all of these services have to be combined and delivered over the HFC infrastructure. Any upstream signals, especially for the Internet and telephony, have to be demultiplexed and distributed as well. But there is only one fiber optic cable outbound—and one fiber inbound—to handle everything. Of course, the all-coaxial cable network had only one coax cable to handle everything both upstream and downstream. But fiber is different and needs one fiber in each direction.

This last point about analog fiber cannot be overemphasized. All that has been done in an HFC is to replace the coaxial cable with two fiber optic cables. The fiber is not carrying a stream of 0's and 1's (except when used for digital TV, as is discussed in the "Digital TV" section), but just a frequency division multiplexed (FDM) collection of 6 MHz analog TV channels. The laser transmitter usually operates at 1,310 nm, just as in SONET/SDH, but the modulation technique is the purely analog amplitude modulation (AM) method.

In the cable TV headend, all of the downstream services are stacked using a carrier frequency onto a single coaxial cable before conversion to an optical signal. This coaxial cable becomes input to the fiber optic transmitter, which in turn sends the entire package on an optical carrier at 1,310 nm. The optical receiver in the neighborhood takes this analog optical signal and pipes the channels onto the branch coaxial cable. Upstream channels from 100 to 200 or so homes get the same treatment at 1,310 nm, except that there are fewer of these upstream channels.

The way that the fiber optic cables interface with a four-way distribution amplifier and splitter is shown in Figure 13-5. The figure shows that usually fiber and coaxial units are in the lid and bottom of a case that folds shut. The arrow symbols are outbound signal amplifiers.

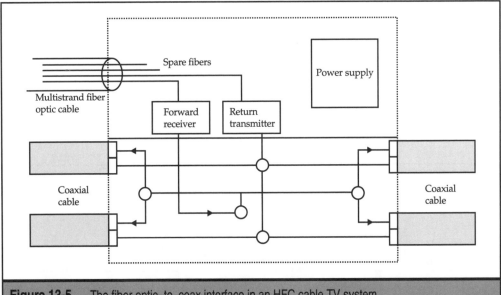

Figure 13-5.   The fiber optic–to–coax interface in an HFC cable TV system

Note that each one of the two fibers will handle only upstream or downstream traffic, while the four coaxial cables handle traffic in *both* directions. This use of two fibers makes a huge difference when it comes to two-way services such as high-speed Internet access using cable modems.

THE CABLE MODEM

A cable modem (sometimes *cmodem*) is a true analog device. The cable modem usually takes IP packets and puts them inside Ethernet LAN frames (although ATM cells can also be used instead of Ethernet) and sends the packets over the cable TV network. The bits can be sent by a number of analog modulation techniques, but the bit stream must fit into an ordinary 6 MHz channel. The channels can be shared among a pool of users (often hundreds of them), and there can be many of them (usually the higher-numbered channels for downstream). Common analog modulation techniques for the IP packet bits and the frames that carry them are quadrature amplitude modulation (QAM, a whole family of methods such as 16 QAM, 64 QAM, and so on) and quadrature phase-shift keying (QPSK).

A simple cable modem protocol stack is shown in Figure 13-6. Note that all of the cable modem users are *bridged* to the cable TV headend behind a single IP router. This means that all of the users share an IP network identifier and all broadcast packets and frames go everywhere. This was once a real security and privacy issue, but the rise of inexpensive "firewall" products for home use has made this much less of a problem. The cable media access control (MAC) layer is just how the cable modems sharing the channel decide when they can send, and the transceivers (transmitters and receivers) operate at radio frequencies (RF) even though the signal remains in the cable.

Early cable modem systems used a simple Ethernet frame without the preamble. Contention was a problem, especially with the shared upstream channels, which were few in number compared to the potential number of downstream channels that could be used for data instead of video signals. These early upstream channels were easily overloaded, and the whole architecture was unsuited for deployment with an anticipated high buy rate for Internet access.

Modern cable modem systems are distinguished by the addition of an *intelligent controller* to prevent contention and sharing from "freezing out" some users. The intelligent controller adds a *scheduler* to the normal media access control (MAC) processor and

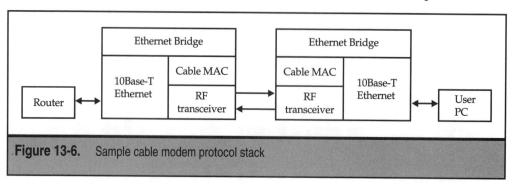

Figure 13-6. Sample cable modem protocol stack

WAN interface to the Internet. With a few enhancements, the scheduler can even guarantee quality of service (QoS) parameters such as bandwidth and delay, making the system much more suitable for voice over IP (VoIP) and eventually video over IP.

The overall architecture of a modern cable modem cable TV system is shown in Figure 13-7. The upstream and downstream channels are logically separated by frequency and might also be separated onto upstream and downstream fibers in an HFC. Two-way operation is also possible in all-coaxial cable systems, but most cable modem services are offered over HFC networks.

By the late 1990s, cable TV operators had determined that certain characteristics were necessary to deploy a successful cable modem termination system (CMTS) at the headend to regulate traffic to the cable modems at the customer premises. These were the features:

▼ Cable modems had to share a single 6 MHz downstream channel. The speed needed to make sure delays were not too high was in the 30–40 Mbps range, depending on which QAM method was used. QAM versions possible were 64 QAM and 256 QAM. Any channel between 54 and 860 MHz could be used, and there could be no interference with adjacent video signals.

■ There had to be one or more shared upstream channels using QPSK or 16 QAM. The range was to be 5–42 MHz. Upstream channels could be from

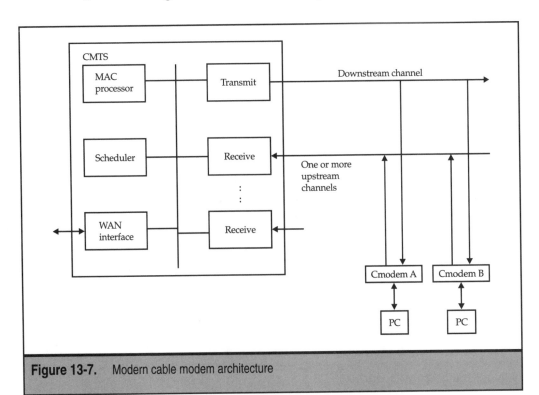

Figure 13-7. Modern cable modem architecture

180 KHz to the full 6 MHz (actually 6.4 MHz) for flexibility, since most cable modem traffic flows in the downstream direction (assuming there are no servers on the customer premises).

▲ The user interface had to be Ethernet on the premises, and at the CMTS to connect to the Internet.

There were additional features concerned with privacy, QoS, voice support, management, and so on. The cable operators worked with many standards groups to come up with CMTS and cable modem products to fulfill these characteristics. The IEEE 802.14 Cable TV Working Group (known as "IP over cable TV") got involved, as did the ATM Forum, and their work involved using ATM on a cable TV network (an idea that seems almost quaint today). A group called the European Cable Modem Consortium, and another European group known as the DAVIC (Digital Audio Video Council) industrial consortium contributed as well. In the United States, most of the work was done by a group of vendors forming the multimedia cable-network system (MCNS) group. Their work was known as DOCSIS (Data Over Cable Systems Interoperability Specification), and now the group is simply known by their work: DOCSIS. DOCSIS now adds a Baseline Privacy Interface (BPI), which supplies a lot of security to cable modem implementations.

Although DOCSIS was originally intended for use in the United States, today there is EuroDOCSIS. DOCSIS now guarantees data rates, even with shared channels. Typical guaranteed rates are 512 Kbps downstream and 128 Kbps upstream. What DOCSIS does is to try to guarantee interoperability by establishing standards for carrying IP packets over an HFC cable TV network, while at the same time allowing and acknowledging the differences in detail in many of these HFC systems. The general idea of DOCSIS is shown in Figure 13-8.

DOCSIS is not the only way to do this, of course. DAVIC is still around and has combined with the Digital Video Broadcasting (DVB) group to explore standards for cable modems, known as DVB-RCCL (DVB Return Channels for Cable and LMDS). LMDS (Local Multipoint Distribution System) is a type of "cellular cable TV" that delivers cable TV services wirelessly. Today, there are significant differences between the way that DVB/DAVIC and MCNS/DOCSIS cable modems operate, mostly at the lowest, physical layer. Standards will help, and DOCSIS has been accepted by the ITU as J.112. DVB still has wide support in Europe, where it is seen as better for European cable TV systems and more of a "home grown" solution. The DVB 2.0 specification is now a European Telecommunications Standards Institute (ETSI) specification, ETS 300800. This now has also become DAVIC 1.5, so all three (DVB 2.0, ETS 300800, and DAVIC 1.5) are essentially the same.

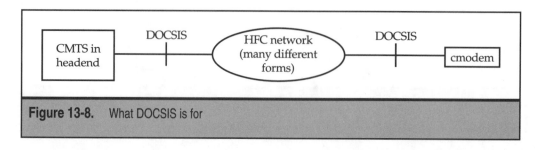

Figure 13-8. What DOCSIS is for

The main thing that DOCSIS adds to a cable TV system is the intelligent controller, as mentioned previously. This addition creates a standardized cable modem termination system (CMTS). The scheduler divides the total available bandwidth on each link into numbered *minislots* that can be used through contention or by a type of reservation called a *direct grant*. Minislots are 2^k times 6.25 microseconds in length, and the allowed values of k are 0 to 7. The minislots, therefore, vary in duration from 6.25 microseconds to 800 microseconds (128×6.25), depending on the needs of the cable TV operator. Bigger slots allow for more bits per grant to be sent, but they increase delays for other users waiting for a grant. Grants assign numbered slots to the cable modem and also set the number of slots the cable modem can use before stopping the send process.

The cable modem has a smaller version of the CMTS scheduler. Figure 13-9 shows how the schedulers handle an upstream packet flowing from cable modem to CMTS. This is the trickier part of the operation, since upstream bandwidth is typically less than downstream.

This is what happens when a cable modem sends a packet upstream:

1. A user's IP packet inside an Ethernet frame arrives in the upstream queue in the cable modem.

2. The cable modem scheduler figures out the number of minislots the packet will take and prepares a bandwidth request message (part of the cable MAC).

3. The cable modem waits for a silent period, then sends the request to the CMTS using ordinary contention mechanisms.

4. The cable modem enters *contention mode* and figures out when to send another bandwidth request if there is still data to send.

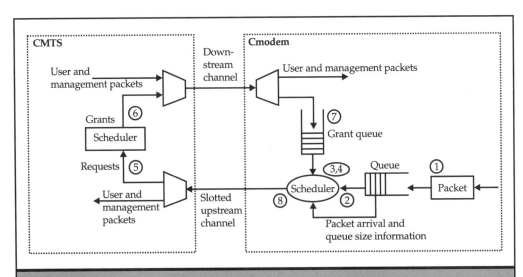

Figure 13-9.   The cable modem and CMTS scheduler

5. The CMTS gets the cable modem's bandwidth request and figures out when it will be able to schedule (or reserve) upstream bandwidth for that cable modem.

6. If the load is light enough, the CMTS can immediately issue a *direct grant* to the cable modem and send it downstream. If the load is heavy, the CMTS schedules a grant and sends an acknowledgment to the cable modem.

7. The cable modem receives either the immediate direct grant or the acknowledgment. Once either is received, the cable modem exits contention mode and either sends (immediate direct grant) or waits (acknowledgment) for the direct grant.

8. Once the direct grant is received, the cable modem can send when the current slot number matches the slot in the direct grant.

The process is complicated by the addition of *embedded bandwidth requests* that do not require separate messages. The whole idea is that the CMTS scheduler can parcel out bandwidth fairly to a large number of cable modems and keep traffic flowing as the number of users grows. The process is a great improvement over pure contention methods that quickly bog down under heavy loads

By adding queues and distributing traffic according to priorities, QoS guarantees can be added to this basic mode of operation. Cable modems have come a long way in just a few years. The security and privacy issues are no worse than in many other technologies, and the capacity bottlenecks have been somewhat eased by running many, slower upstream channels rather than fewer, faster channels. As long as most users stick to short client-to-server traffic patterns, and do not deploy servers in their homes as a matter of course, there should be plenty of capacity on a cable modem system, at least for now. Even with 500 homes per fiber, and 1 Mbps upstream and downstream, there should be no problems even with 25 percent service penetration. However, newer optical network methods might be needed when such Internet access through cable TV networks becomes nearly universal.

DOCSIS also adds security to cable modems through the Baseline Privacy Interface (BPI), as mentioned previously. The BPI adds encryption for the users' traffic, some features to prevent neighbors from browsing each other's computers, and so on. Newer cable modems have as many security features as a personal firewall or a decent router.

PASSIVE OPTICAL NETWORKS (PONS)

HFC networks preserve coaxial cable beyond the feeder level of the cable TV network, usually to about 500 homes. One of the main reasons for doing this is that homes have only a coaxial cable interface and coaxial cable run inside the home. Extending fiber to the home would not be cost-effective, since at some point conversion to coaxial cable electrical signals must be done. It is far more cost-effective to convert optical signals for 500 homes at a centralized location at present.

Also, the *power penalty* for splitting optical signals is substantial. It is much more cost-effective to preserve the line extenders and tap boxes on the coaxial portion of the network than to try to somehow split the optical feed to 500 destinations. Not only is optical power a concern, but *electrical* power distribution to all of these optical components now becomes a concern.

However, what if it were possible to run fiber directly to the end user cost-effectively? What if the optical signal had to be split not 500 times, but only 5 or 10 times? What if the optical signals required no electrical power beyond a certain point of the network? This would describe what is known as a *passive optical network (PON)*.

PONs will be described more fully in the next chapter, since PONs are much the same whether they replace coaxial cable or twisted-pair telephone wire in the local loop. For now, it is enough to point out that a PON can extend fiber in an HFC right up to the door of the customer.

DIGITAL TV (DTV)

Someday soon, analog TV will be a thing of the past, like steam locomotives and horse-drawn trolley cars. In fact, analog TV might suffer a fate worse that either of the other two "technologies," which have a certain nostalgic value. Historical parks might preserve steam trains and horses-drawn trolleys, drawing paying crowds when they do so, but it is hard to imagine anyone paying to see the analog TV of the past.

Digital TV (DTV) is here and will be everywhere soon, if current plans unfold as expected. It is sometimes easy to forget that changes to cable TV architectures are not just about cable modems for fast Internet access. The primary cable TV service is TV. And DTV will enhance TV service offerings in a number of ways.

DTV is an idea, and there are many ways that the specifics of DTV vary depending on the area of the world. There were and are differences in analog TV systems around the world, so this is nothing new. But DTV will not be a monolithic standard with only one way of doing things.

In North America and Argentina, DTV will arrive as ATSC from the Advanced Television Systems Committee. In Australia, India, Korea, Malaysia, New Zealand, Taiwan, and Western Europe, DTV will follow DVB standards. Japan will see DTV according to the rules of ARIB (Association of Radio Industries and Businesses), also known as ISDB (Integrated Services of Digital Broadcasting). All of these systems have many similarities, but also significant differences as well.

What is so special about DTV? Television is just television. But maybe not when it comes to DTV. DTV offers:

▼ **Unprecedented picture quality**   When audio CDs brought digital music to the world, people went out and bought music they already had just to hear it all over again with "digital quality." DTV does the same to television. There is better screen resolution (more scan lines), and an option for a better screen image (progressive scan).

■ **Stereo quality sound**   The sound in DTV is also digital audio, of course. There is no static or distortions. DTV also offers multichannel sound, either two-channel stereo or six-channel "surround sound." The subwoofer channel can literally shake the whole house.

■ **Movie screen format**   Most TVs employ a 4:3 width-to-height ratio, called the *aspect ratio.* So a TV screen 20 inches wide is 15 inches high. DTV supports this format, but it also has an aspect ratio of 16:9, the same as a movie screen. No more "this movie has been formatted to fit your screen" pan-and-scan editing is needed for movies shown in 16:9 format.

■ **Data Services**   DTV can embed a stream of data inside the DTV picture. This information is available on demand and can include product information, sports statistics, news transcripts, and more. Many of the tie-ins currently done through Web sites can now be integrated right into the DTV programming.

■ **Interactivity (Web access?)**   DTV will have upstream channels built in, not added on as with cable modems. The DTV operator could supply services similar to the Internet and Web for "browsing" e-mail, ordering goods and services, responding to opinion polls, and so on. Or the DTV operator could just link this service into the Internet and be done with it. Either approach is possible.

■ **More programs and program choices**   Digital information can be compressed if the service provider chooses in order to make more effective use of channel bandwidth. Programming can be done in standard-definition TV (SDTV) or high-definition TV (HDTV). Instead of one HDTV broadcast, service providers can multicast several SDTV channels. And subscribers can even decide what channel mix they want by time of day or day of the week. It all adds up to more flexibility with DTV.

■ **Electronic programming guide**   Similar to the "program channel" that satellite TV and some cable TV operators offer now, the DTB electronic program guide (EPG) will be more sophisticated. For instance, since the channel mix and content on a DTV system can change rapidly, the EPG will be updated every three hours.

▲ **A better viewing experience**   All in all, watching DTV will be much different than watching analog TV. There will be no more analog interference causing "ghosts" and "snow."

The digital coding technique currently used for DTV is MPEG-2 from the Motion Picture Experts Group. Digital video has many attractions for producers as well as viewers. Digital video can be instantly edited right at the frame level, with no rewind. Digital video can be stored on a server and compressed into less bandwidth than one analog TV channel. Local content can be inserted much more nearly seamlessly than in an analog cable TV system.

DTV will come in broadcast and cable TV system versions. The rest of this section considers DTV as a cable TV offering delivered on an HFC network.

In DTV, the MPEG-2 picture is delivered as a *program stream (PS)* flowing inside a *transport stream (TS).* The program stream is made up of an MPEG-2 video and audio *elementary*

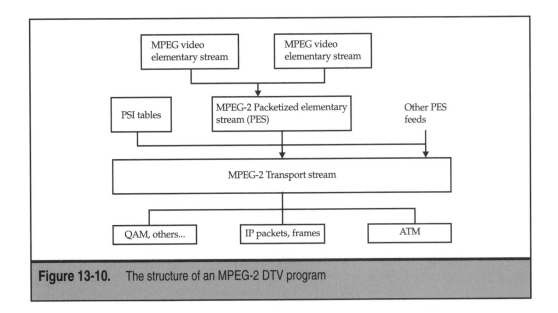

Figure 13-10. The structure of an MPEG-2 DTV program

stream (ES) combined into an MPEG-2 *packetized elementary stream (PES)*. This is done because the ES is as long as the program itself. The packets are pieces of the ES. The PESs are multiplexed onto a single MPEG-2 transport stream at the DTV headend. The transport stream also contains packets with program-specific information (PSI) tables that tell the receiver how to take the transport stream apart into its constituent ESs.

The PSI tables are very important in DTV. Some of the tables are established by MPEG-2 and are common to all forms of DTV. The differences in field values and the other tables used with the main PSI tables define the real differences between the ATSC, DVB, and ARIB versions of DTV. The basic architecture of a DTV cable system using MPEG-2 is shown in Figure 13-10.

As shown in the figure, there are several ways to distribute the transport stream. The viewers can see it over the HFC system using an analog encoding method such as QAM (just because the picture is digital, that does not mean the transport network has to use digital encoding!). Or the TS can be sent using ATM cells, or inside IP packets using frames.

CABLE TV AND OPTICAL NETWORKING

This chapter has spent a lot of time describing HFC cable TV networks, cable modem operation, and digital TV. The point of all this groundwork is to better and more fully appreciate the impact of optical networking on a cable TV network delivering cable modem and DTV services. Optical networking with DWDM has a lot to offer a cable TV network with cable modems and DTV, but less to offer an HFC cable TV network with only analog television delivery.

However, it is up to the individual cable TV operator to decide if the benefits of optical networking and DWDM outweigh the expenses and conversion efforts even when only basic analog services and/or cable modem service is offered. It might be better to add optical networking now rather than wait and try to add DTV *and* optical networking to the cable TV network at the same time.

Optical networking can be added in two places to a cable TV network. It is easiest to add optical networking to the cable TV "backbone." For a number of years, cable TV operators (especially those with very large systems) have been linking local headends to larger "super headends" or regional headends. The resulting headend network is easier to manage and to use for distributing analog TV channels, since each headend no longer has to have a complete set of channel receiving equipment.

In the United States, SONET rings have been used for this purpose, since SONET equipment is readily available and the rings offer protection switching in case of link failure. Ordinary analog TV channels can be digitally encoded and compressed and transported inside SONET frames. The usual digital encoding requires 25.9 Mbps for the 6 MHz analog TV channel. So a ring running at 155.52 Mbps (the OC-3 or STM-1 rate) can transport 6 TV channels. At 622.08 Mbps (OC-12/STM-3), 24 channels can be carried. And at 2.5 Gbps (actually 2.48832 Gbps, the OC-48/STM-12 rate), the ring can carry 96 analog TV channels. This method of distributing 96 analog TV channels between cable TV headends is shown in Figure 13-11.

There is nothing wrong with using a SONET or SDH ring this way, and the protection switching provided by the ring is a big advantage. However, SONET/SDH rings running above 2.5 Gbps are not plentiful, nor are they inexpensive even if available. And there are many issues that are not addressed in the SONET/SDH solution.

One of the main problems is that many cable TV systems have already exceeded the 96 channel "limit" on the ring. And not all of these channels today are analog TV channels. There are now digital channels, such as HDTV, that do not fit as nicely into SONET/SDH. Channels used for cable modems already contain digital information and do not compress with the same methods and with the same efficiency as analog TV channels.

Even when packed with analog channels, the ring does not leave room for much of anything else, even simple voice or data channels between the headend offices. There are only 1.92 Mbps left over when 96 analog TV channels are stuffed onto a ring running at 2.48832 Gbps (96×25.9 Mbps = 2.4864 Gbps). There is always the SONET/SDH overhead, of course, but these bits have predefined uses in many cases and are forbidden for other uses in many other cases.

Ironically, there is plenty of bandwidth that is not used on the ring. Most of the traffic is "outbound" distribution traffic from the regional headend to the local headends. There is very little bandwidth used on the "inbound" side of the ring, flowing from local headend to regional headend. Channels could be picked up by the local headends and distributed, but this is mostly the role of the regional headend. The return traffic is most likely to be the upstream cable modem traffic heading for the routers at the regional headend, if this Internet access is not handled locally (which it often is). If telephony services are offered by

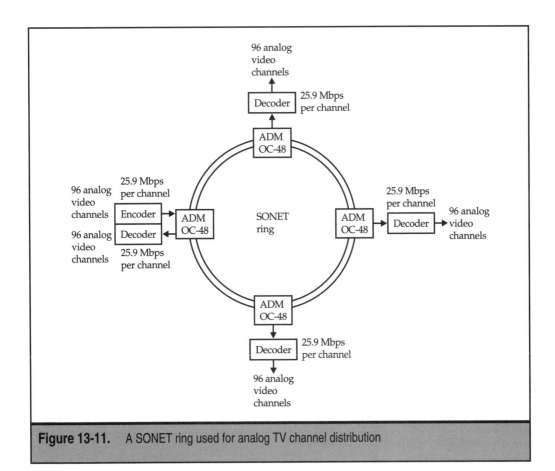

Figure 13-11. A SONET ring used for analog TV channel distribution

the cable TV operator, then this two-way service is also mapped both ways on the ring. In fact, one of the main reasons that SONET/SDH rings came to the cable TV headend was that cable TV operators could more easily link their system up to the local telephone switching office.

Now consider the same backbone situation with optical networking. In this case, a DWDM ring would be used in place of the SONET/SDH ring to connect local headends to the regional headend. The analog TV channels could still be encoded and compressed onto a 2.5 Gbps SONET/SDH channel, of course, and this is a good idea for preserving backward compatibility. But now the SONET/SDH signal is just one of many wavelengths on the ring. Adding another wavelength to add more channel capacity is a relatively trivial provisioning step.

With optical networking there is no requirement for the bandwidths running in opposite directions to match. That is, λ_1 from regional headend to local headend does not have to have the same bandwidth as λ_2 running from local headend to regional headend. This feature alone makes optical networking attractive for this function. The bandwidth can also be changed as channel requirements grow. Also, other wavelengths can be added for digital TV channels, cable modem traffic, telephony services, and internal communications needs as the need arises.

This role for an optical network with a DWDM ring is shown in Figure 13-12.

The second place that optical networking and DWDM can be used in a cable TV network is on the main "trunk" on the HFC distribution network. One of the main reasons to do this is to increase the capacity of the HFC network through channel reuse. Now not only is each *fiber* capable of carrying the full array of offered services, but each *wavelength*

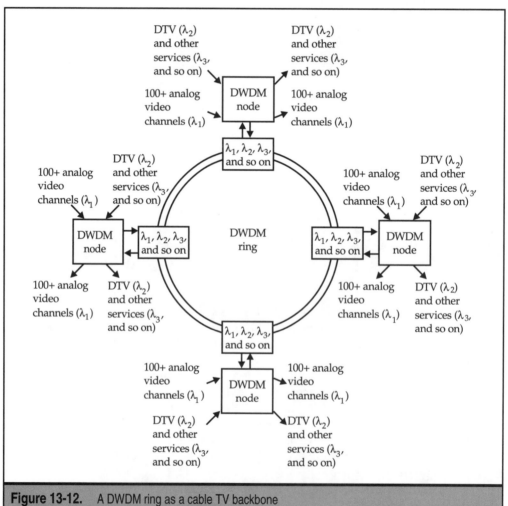

Figure 13-12. A DWDM ring as a cable TV backbone

is capable of carrying this full service array. This is sometimes called *spatial division multiplexing,* because the signals are multiplexed even though they are separated.

When there are 500 homes on a fiber, the downstream bandwidth use can exceed the upstream bandwidth use by as much as 20:1. As mentioned previously, DWDM channels can be flexibly configured to deliver precisely the bandwidth needed in each direction.

Another problem solved by DWDM use is the issue of signal piracy. Due to the shared nature of HFC systems, the same channel mix appears on all of the fiber trunks. Premium channels are scrambled, but they are still there. This presents an opportunity for customers to attempt to descramble signals with a variety of methods, mostly by adding "pirate" boxes to the network. With DWDM, each wavelength can have a different mix of channels per group of customers serviced by the wavelength. If optical networking is used to extend the reach of fiber toward the customer, as seems likely, fiber hubs will not service 500 homes as before, but perhaps groups as low as 12 or so. Extending the reach of the fiber will help to cost-justify the migration from HFC to optical network, since it makes little sense just to replace the functional HFC fiber with DWDM fiber doing exactly the same thing.

An optical cable TV network with DWDM is shown in Figure 13-13. Note that other types of fiber can still be used with DWDM before reaching the coaxial portion of the network. The remote DWDM node can even convert to coaxial cable electrical signals and

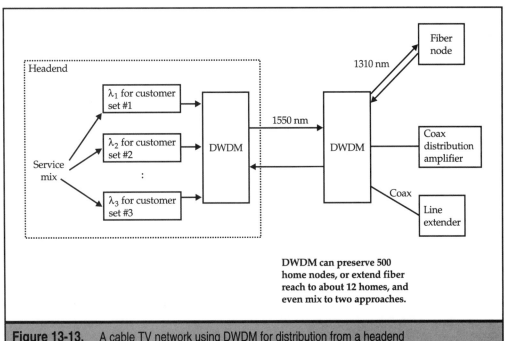

Figure 13-13. A cable TV network using DWDM for distribution from a headend

so extend the reach of the fiber to the line extender level. And DWDM even allows the mixture of these two approaches due to the flexible nature of the wavelength speeds and assignments.

There are even other advantages to using DWDM to reach HFC customers besides more reusable bandwidth per user and two-way capacity. Prices for 1,550 nm equipment are dropping much faster than for 1,310 nm equipment. It is even possible to retain the existing HFC network for two-way and analog services and use the new high-capacity DWDM network for digital TV channels.

When all is said and done, optical networking adds a lot to a cable TV network.

CHAPTER 14

Telephony and Fiber to the Home

Most people have two types of networks running to their homes already. The first is cable TV, and the second is the global public switched telephone network (PSTN). Either or both can be used to access the Internet and Web, and there has been some fairly fierce competition between the two industries, since both want to be the "network of choice" for advanced services of all types.

The last chapter explored how the cable TV networks are preparing to enter the world of optical networking with digital TV. Some of the pieces are already in place, such as some fiber in the infrastructure thanks to HFC and Internet access and data services with cable modems. Telephony service has been added on in many cases and can be done so rather painlessly in other cases. Once cable TV can deliver voice, video, and data services, who needs another service provider?

This chapter will investigate the other major residential service architecture: the telephone network. Just like the cable TV companies, the telephone companies are preparing to enter the world of optical networking. However, the telephone network architecture is almost 100 years older than the oldest cable TV network, and the main telephone service—voice communication—is not the most exciting thing to base a new optical network on. There are some pieces in place, as with cable TV. There is more fiber in the telephone network than in the cable TV network, and the whole idea of using modems to access the Internet and Web began with the telephone network in mind. But it will be much more of a task to add video services to the telephone network, because video requires much more bandwidth on the access network than telephony does.

Not that the telephone companies have not tried to add video bandwidth to the modest bandwidth required by voice. One of the key aspects of this chapter will be to detail the efforts made in this arena during the 1990s. But of course the story will always come back to the role of optical networking and DWDM.

Unlike many of the data and other networks outlined in Chapter 4, not much about the existing structure of the PSTN was presented there. So a good place to start is with an overview of the current PSTN architecture, with the emphasis on where the fiber currently is deployed. This will lead to a more detailed discussion of fiber on the access side of the networks. More specifics about various plans to bring fiber into the access line or local loop as fiber in the loop (FITL) will be investigated. Then a quick look a digital subscriber line (DSL) technology will be presented, but only to illuminate the issues of adding bandwidth to local loop. The idea of a passive optical network (PON) to deliver fiber directly to the home (FTTH) is explored, and finally the impact of optical networking and DWDM on residential telephony is assessed.

THE PSTN IN THE UNITED STATES TODAY

Although there is only one global, public switched telephone network (PSTN) today, it is hard to characterize all of the variations with one short section on architecture. Enough national and local variation exists in all parts of the world to make generalization difficult. Not only that, various degrees of government regulation and control make the simplest architecture more complicated. So this section will describe the PSTN in the United States today, not only because it is essentially the oldest, but also because of its vast size. For example, there are more local access lines on New York's Manhattan Island than on the entire continent of Africa.

This is not the place to discuss the merits of the deregulation process in the United States that led to the creation of local exchange carriers (LECs, now called *incumbent* LECs or ILECs) and interexchange carriers (IXCs) in 1984. Since 1996, any company should be able to be both. Added to the mix now are competitive local exchange carriers (CLECs) and Internet service providers (ISPs). These entities can also offer almost any service to residential customers in most places in the United States, and many do. However, the main point here is that, with few exceptions, the *local loop* or access line between the service provider *central office* (*CO*, the official international term for this switch location is *local exchange*, or *LE*) and customer premises is built, owned, operated, and maintained by the ILEC. Other business units such as ISPs or CLECs can and do lease these lines from the ILECs on a long-term basis (usually for 3–5 years), but only the ILEC can decide when and where to add fiber to a local access infrastructure.

The PSTN

Figure 14-1 shows the basic structure of the PSTN. There are several details that should be discussed before introducing the players that make this simple figure much more interesting today.

Figure 14-1 shows a purely domestic telephone network. International calls would go through an *international gateway* switching office. All telephone network customers (often called *subscribers,* a holdover from the days when customers could not own their own telephones) interact with the voice network using an *instrument.* The most common

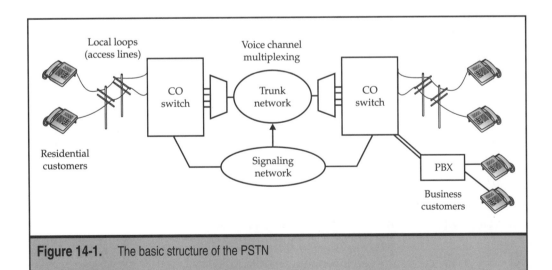

Figure 14-1. The basic structure of the PSTN

instrument is the telephone, as shown in Figure 14-1, but other common instruments are modems and fax machines. The telephone instrument is commonly connected to a central office switch, the network node of the PSTN, by means of a local loop or (more properly) an *access line*. The access line is typically a single pair of twisted pair copper wire. However, many twisted pair voice channels in an housing complex are usually multiplexed together onto fewer pairs of copper wire in an arrangement known variously as a *pairgain system* (since it can "gain pairs" for other uses), a *digital loop carrier (DLC)*, or a *carrier serving area (CSA)*. These are not shown in Figure 14-1, since the main goal of these *distribution network* strategies is still just to bring a number of access lines to the central office (CO). However, the DLC system will be much more important later in the chapter, when fiber is added to this basic architecture.

The main function of the CO is to connect access lines to each other, according to the network address (telephone number) of the destination telephone instrument. The PSTN is a connection-oriented, circuit-switched network in which no information can flow from source to destination until a connection is established between a source and a destination. If the destination instrument is attached to the same central office switch, this job is easy. It is a more difficult task if the telephone number dialed is at the end of an access line on another switch, and in actual practice the destination can be almost anywhere in the world. If the access lines that need to be connected are on different switches, the voice connection must be routed onto a *trunk* voice channel through a trunking network. The trunking network links telephone switches, not telephone instruments. Figure 14-1 shows that many voice channels are routinely multiplexed when sent over the trunking network, although this is not strictly necessary. Trunks can be individual twisted pairs indistinguishable from access lines, although trunks are more likely to be carried on coaxial cable (fiber optic cable most important here), or microwave towers.

Figure 14-1 also shows a *signaling network*. The signaling network is how the source and destination network addresses (telephone numbers) are sent to the destination switch, and also how the voice channel to be used for the call on the trunking network is set up (shown by the arrow). This signaling network is important in understanding the differences and similarities between the PSTN and Internet. This is because the signaling network is basically a connectionless, packet-switched network with routers as the network nodes, just like the Internet. The signaling network *must* be connectionless due to the simple fact that the messages that set up connections, known as *call setup* messages, cannot follow connections if their function is to set up connections (calls) in the first place. Optical networking offers the possibility that the PSTN and the Internet might some day become one, and central office voice switches might be indistinguishable from routers. So it is always good to keep in mind that routing is not an unknown concept when it comes to the PSTN.

Figure 14-1 also shows a business *private branch exchange (PBX)*. The PSTN always distinguished between residential and business customers, and businesses typically have a little voice switch all their own on the customer premises, as a "private branch" of the local exchange. The PBX is there only to show that the same PSTN structure can service both residential and business customers. The emphasis here is on the residential services, but the business aspect of the PSTN should not be forgotten.

In the same way that backbone routers form the global Internet, the PSTN trunking network forms the backbone of the PSTN. The trunking network consists of more voice switches and multiplexed links, but no access lines are attached to the switches on the trunking network. This connectivity is strictly trunk-to-trunk. It would be prohibitively expensive to connect every CO switch with a direct point-to-point link, multiplexed or not, so the trunking network provides this connectivity. Some CO switches do have direct trunks to their closest neighbors, depending on actual calling patterns. But calls are more likely to be routed by the signaling network onto trunks leading to a *local tandem office*. The local tandem is most likely a centrally located CO switch with a second, trunk-only switch housed in "tandem" with the CO switch, which is where the name comes from.

A whole hierarchy of trunk switches were formerly used in North America (the United States and Canada). Today, the trunking network is much "flatter," and most of the calls between trunking switches are handled by IXCs.

Meet the Players

In 1984, after years of court actions, AT&T, the largest telephone company in the world, was broken up in a process called *divestiture*. AT&T's local monopolistic Bell Operating Companies (BOCs) were organized into seven larger Regional Bell Operating Companies (RBOCs). AT&T Long Lines basically became AT&T, the long distance company free to compete (with some restrictions) in a deregulated long distance industry with MCI, Sprint, and a host of others. AT&T retained Western Electric (manufacturing) and Bell Labs (research and development, now Lucent Technologies).

The RBOCs were firmly under state regulation and retained existing monopolies on local telephone services. The RBOCs could carry calls on their own facilities (lines, trunks, and switches) if both ends of the call were within the same *local access and transport area (LATA)*. These were *intraLATA* calls. All other calls were *interLATA* in nature, and callers could choose to use any one of a number of long distance companies to carry part of the call, as long as the long distance company had a switch (called a *point of presence*, or *POP*) within the originating LATA. All LATAs were artificial constructs applied to the RBOCs to make sure that there were enough long distance calls to support the new interexchange carriers (IXCs) now challenging AT&T. Independent telephone companies that were never part of the Bell System were unaffected by the LATA rules. These independents could join an RBOC LATA (many small independents did), establish their own LATAs, or ignore the LATA structure altogether.

So the local exchange carriers (LECs) had a monopoly on local service, but interLATA calls had to be handled from POP to POP by the competitive IXCs. These interLATA calls were listed on the bill usually sent to customers by the LEC (the IXCs paid the LECs to do so), which collected the entire amount on the bill. LECs recovered the costs of the IXC using the LECs facilities through a system of *access charges*, which are still an item of discussion with regard to Internet access and the ISPs.

LATAs and Access Charges

A local call is carried over the facilities owned and operated by the local carrier for just this purpose. A local call may be a toll call, but that just means it costs more. Local calls are precisely defined in the United States: both endpoints must be within the same LATA. Whenever a LATA boundary is crossed, this is a *long distance call*, even if it is only a few miles from end to end across a state line (very few LATAs cross state boundaries). These are handled from IXC POP to IXC POP on interexchange carrier facilities.

So all long distance calls included two local calls. There was a local call on each end, and an IXC call in the middle. The local calls went to the POP on the originator side, and from the POP to the destination in the other LATA. But if the IXC in the middle gained all the revenue from the long distance call (even though the call was billed by the originating LEC in most cases), how could the LECs at each end be compensated for the use of their lines, trunks, and switches? After all, these were in use for the duration of the call and could not be used for local calls or to make money for the LEC in any way.

The answer was to install a system of *access charges*. The IXCs pay the LECs for the use of their facilities to complete local calls. Each call that is carried to and from the POP consumes LEC resources and is subject to a per-call access charge system.

All entities that can cross LATA boundaries for services must pay access charges except for one major category: the ISPs. Data services were explicitly exempted from paying access charges according to the 1984 divestiture agreement. This was done to make sure that the data services segment, newly emerging in 1984, would not be crushed by costs that they could not afford. This also reinforced the basic split between *basic transport* services provided by the protected LECs and *enhanced services* provided by the ISPs. This principle meant that the monopolistic LECs could only transport bits through their network, not store, process, or convert them in any fashion. Changes to bits were enhanced services, reserved for competitive entities like the ISPs.

If anything seems clear, it is that sooner or later the ISPs will have to pay some form of what most of the ISPs see as an "Internet tax" to the LECs in the form of access charges. The only question is how much.

The Modern PSTN

The architecture of the PSTN since 1984 has taken on a structure of LEC, LATA, POP, and IXC. Instead of a hierarchy of switches, the system of IXC backbone switches is linked in a flatter structure.

The architecture of the modern PSTN is shown in Figure 14-2. All of the blocks are voice switches, distinguished primarily by their size and function (tandems perform trunk switching, and so on). There can be direct trunks between COs (dotted lines), and trunks can run from an IXC POP to toll office switches, access tandems, or even central office switches themselves. This is just a representative figure, so not too much should be read into the lack or presence of a trunk here or there (toll offices do not really link to only one CO).

In Figure 14-2, IXC A can carry calls originating from LATA #1 and LATA #2, but not from LATA #3, since there is no IXC A POP in LATA #3. IXC A can complete calls to

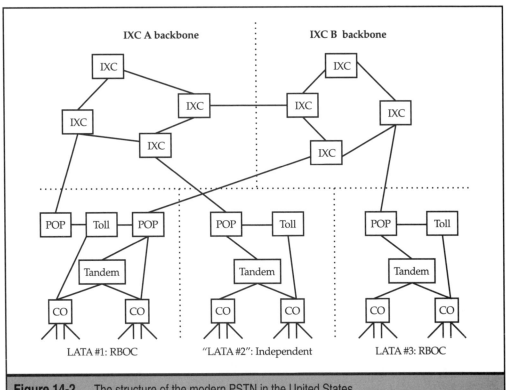

Figure 14-2.   The structure of the modern PSTN in the United States

LATA #3, however, but only by passing the call off to IXC B, which has a POP in LATA #3. Naturally, IXC A pays IXC B for this capability, and it all depends on whether IXC A and IXC B have trunks between them (as they do in this example). Note that only callers in LATA #1 have a choice of long distance service providers, since the other two LATAs only have one POP from one IXC.

There is now no longer a clear distinction between local and long distance calls (the distinction was never as clear-cut as people imagined anyway). There are instead intraLATA local calls (the lowest rates apply), intraLATA *toll calls* (still carried on LEC facilities, but they cost more), and interLATA tolls calls (which might be intrastate or interstate, but are carried by an IXC). Sometimes, intraLATA toll calls are called "local long distance," but this term is not common. The artificial nature of the LATA construct is reinforced by the possibility of interstate, but intraLATA, toll calls, and special "corridors" between some LATAs where LECs can use their own facilities. However, these are just special cases that do not invalidate the general scheme of things.

TA96 and the Local Loop

By the early 1990s, the rise of competitive long distance services was seen as an unqualified success by most people. Rates were down, people could choose their own long distance service providers on a per-call basis (as long as there was a POP reachable in the LATA), and a free-market atmosphere was well established. Attention then turned to the local telephone service market.

LECs claimed that they should be able to offer long distance on their own facilities, without sharing the wealth with the IXCs. LATAs were established to make sure there were plenty of long distance calls, and there certainly were. LECs also felt that they were in a competitive environment anyway, since many states had certified local competitors in the form of competitive LECs (CLECs) who often purchased "unbundled" local loops from the incumbent LEC (ILEC) and ran them to their own voice switches.

So with much fanfare, the Telecommunications Act of 1996 (TA96) was passed by the United States Congress. TA96 proved to have such a complex structure of clauses and rules that no one is quite sure what difference, if any, TA96 has had on local service competition. TA96 allowed ILECs to offer long distance if and only if their local territories were open to competition. Much debate focused on the resale price of unbundled loops offered to the CLECs. What amount should be charged monthly to a CLEC by an ILEC that had installed the wire in 1940? Wasn't it effectively paid for many times over from ILEC services? But what about the incremental costs of maintaining, upgrading, and otherwise making sure the loop was available?

There have been no easy answers to the questions TA96 posed. Depending on the source, CLEC services are either more plentiful than they ever have been before, or the ILECs are still in the habit of placing roadblocks in the path of any serious competitor. In fact, both statements may be true.

ILECs, CLECs, and ISPs

Today, the structure of the PSTN at the local level finds three types of service providers offering connectivity services for voice and data. First there are the incumbent local exchange carriers (ILECs), which formerly had a monopoly of local services, whether RBOCs or not. Then there are the state-certified competitive local exchange carriers (CLECs), which compete either by building their own facilities or leasing "unbundled" facilities for resale from the ILECs.

Finally, there are the ISPs that provide Internet access as well. Sometimes the ISP is a subsidiary of the ILEC, but state regulation means that these ISPs must be customers of the ILEC, just like everyone else. This arrangement retains the split between basic transport and enhanced services.

In the near future, ISPs of all kinds are expected to have to pay access charges to the ILECs just as the IXCs do. Most ISPs just interpret this as a type of "Internet tax" and chalk it up to the cost of doing business. The ISPs handle the needs for local access to the Internet for data. But perhaps the ISPs might handle the local access needs for *voice* as well. This is what voice over IP (VoIP) is all about.

FIBER IN THE LOOP

The previous sections explored the structure of the modern PSTN voice network. The emphasis was on overall structure, not details. So just where is the fiber in the PSTN today?

Much of the fiber is on the trunking network backbone. The large switches operated by the IXCs and LECs are linked in most cases by some form of SONET ring (outside of the United States, it is some form of SDH ring, of course). In many cases, LECs will use SONET rings to link their switching offices in a major metropolitan area. However, not many SONET rings link LECs to IXC POPs (the main issue is sharing control and expenses), but many point-to-point fiber links are used. In some cases, LECs will use fiber to connect digital loop carrier (DLC) equipment to the CO switch. Hardly anyone has installed fiber directly to a home. According to the U.S. Department of Commerce, only some 3 percent of the more than 750,000 commercial office buildings in the United States have direct fiber access, ring or not, after more than 10 years of furious deployment. And there are more than 200 million total access lines in the United States, 70 percent of them residential, and not many of them easily aggregated onto multiplexed links as on the trunking network. The situation is summarized in Figure 14-3. It is hard to be more precise because the situation is rapidly changing as more and more fiber enters the PSTN.

The rest of this chapter considers more fully fiber in the loop (FITL) itself. For many years, it made little economic sense to deploy bandwidth-rich fiber optic cable in the local loop. The main service—analog voice telephony—required only about 4 KHz of bandwidth, so the entire network, even the high-capacity trunks on SONET rings, were engineered to

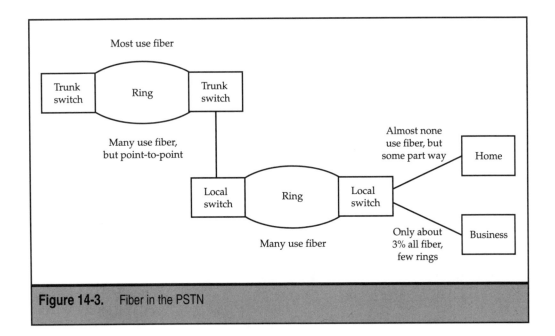

Figure 14-3.   Fiber in the PSTN

just carry a large number of voice channels limited to 4 KHz. When the use of PCs and analog modems became popular in the early 1990s to reach such services as AOL and CompuServe, everything had to funnel through this 4 KHz bottleneck. This limited digital information speeds to what could fit inside 4 KHz, which eventually evolved to about 33.6 Kbps (56 Kbps modems actually require some digital components along the way).

Local loops were made of twisted pair copper wire, twisted three or four times per foot mainly to keep out crosstalk noise from adjacent pairs. There were various thicknesses, or *gauges*, of copper wire used, depending on how far the customer was from the local switching central office (CO). In the United States, the most popular gauges for local loops are 24 AWG (American Wire Gauge) and 26 AWG copper wire, although gauges are often mixed and other gauges are used.

There are more than 20,000 COs in the United States, and more than 200 million local loops. On average, a CO serves 7,500 lines, but some urban COs can have up to 100,000 local loops. Around 20 percent of homes in the United States have second access lines for PCs, fax machines, or home offices in general. Because customers tend to gather into neighborhoods of private homes, apartment buildings, or townhouse complexes, local loops are gathered into *binders* of 50 pairs of twisted pair wires. These *binder groups* are put inside a common sheath and then split out to their individual destinations from a *distribution terminal* closer to the customer. Twisted pair cables normally come in 100- and 200-pair cables, but pair counts as high as 2,700 or 3,600 have been deployed.

When it comes to distance, various architectures and rules evolved to service customers. The typical range of 24 gauge wire is 18,000 feet, and for thinner, 26 gauge wire, 15,000 feet (thinner wire requires less copper but does not carry the signal as far). These figures are for *home run* local loops, which connect a customer directly to the CO. In practice, the figures for the reach of local loops are highly variable, depending on local conditions.

The standard architecture for home run local loops is shown in Figure 14-4. The feeder pairs are taken to distribution points where they are broken out to individual customers. Note that once a "tap" is used for one customer, it cannot be used for another. There is no multiplexing. Furthermore, the pairs "downstream" of the distribution terminal are not cut, as implied by the figure, just tapped into by an arrangement called a *bridged tap*. All of the lines are analog, and the total length usually cannot exceed 18,000 feet even in the best of circumstances.

It is often necessary to reach a remote customer or cluster of customers that is more than 18,000 feet (a little more than three miles) away from the CO. In these cases, the standard architecture in use is the carrier serving area (CSA) with a remote terminal (RT). Today, this is called a digital loop carrier (DLC) because the link from the RT to the CO terminal (COT) is digital. The link might still be copper, but the signals sent are multiplexed digital signals at 64 Kbps, the usual digital rate used to represent an analog voice call. These were also called *pair gain* systems, since the link from RT to CO required fewer pairs of wire, so pairs were gained for use in other places along the way.

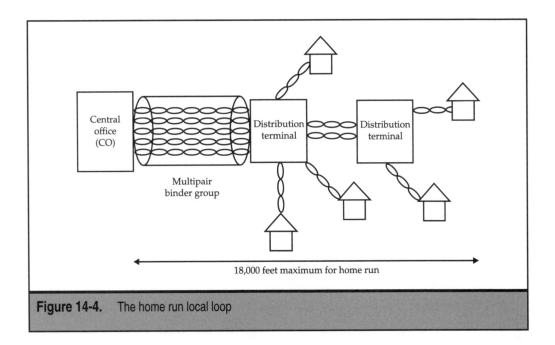

Figure 14-4.   The home run local loop

The overall architecture of CSA with DLC is shown in Figure 14-5. Note that even the "close in" customers are in a CSA, even though there is no DLC with an RT. In practice, clusters of customers more than 12,000 feet from the CO are serviced from a DLC, even though in theory the distance could be much higher on the digital portion of the link. Analog signals on twisted pair are on one side of the DLC, and multiplexed digital signals are used on the feeder side of the DLC.

There is no reason that the link from CO to RT has to be copper. In fact, once DLCs are in use, there is now a very good economic argument for using the high bandwidths fiber offers to replace the copper to the RT. Aggregating even a few hundred 64 Kbps digital voice channels can build up. Two hundred voice calls make up 12.8 Mbps, nowhere near the fiber limit, but more than can be easily handled on copper. And where fiber has been used in the local loop, this is exactly where fiber has been deployed. SONET rings can even be used to link multiple DLCs. This makes more efficient use of the fiber bandwidth, since 155.52 Mbps can carry 672 voice channels. So six or seven DLCs can be placed on a ring.

DSL AND DLC

It is important to note that the analog bandwidth available on twisted pair copper is not limited to 4 KHz. That is just the *useful* bandwidth if people use the network for analog voice calls. Most of the power of the human voice is in the 300–3,300 Hz range, and most

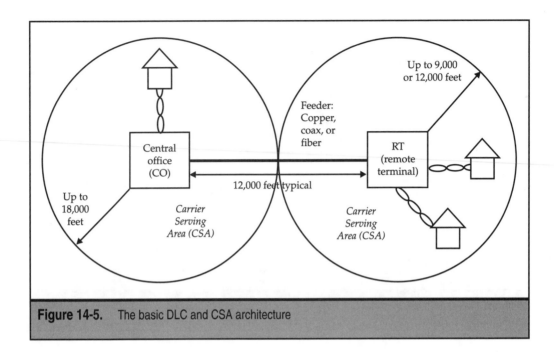

Figure 14-5.   The basic DLC and CSA architecture

telephone networks simply ignore ("bandpass limit") any sounds outside of this frequency range. Since no one could use more than 4 KHz when talking, the whole network from loop to trunk was optimized for 4 KHz analog voice channels and the 64 Kbps digital channels that represented them. The idea behind digital subscriber line (DSL) technologies is to increase the bandwidth used by modems on the local loop, while at the same time preserving the ability to use analog voice at the same time.

Once a large number of residential loops began to be used for modems and access to the Internet and Web, it became more critical to find a way to increase the useful bandwidth on the local loop. Web sessions were longer than telephone calls; in 1996, a US West study claimed 5.64 minutes for voice calls and 32.47 minutes for modem dial-ins to ISPs. Naturally, whenever local facilities are tied up for use, potential LEC revenue is lost. At least the IXCs paid access charges. But ISPs did not. Not only was 32.47 minutes "lost" to the LEC for other revenue purposes (other than the "revenue" generated by a flat-rate local call, in most cases), but no other service provider could make money from someone else calling (the line was busy). Why did Internet traffic have to go through the voice switch? Not a happy situation.

There was much more going on with rate structures and regulation as well. But these discussions are far beyond the scope of this chapter and book. The point here is that the LECs increased the available bandwidth on the local loop (up to 1 MHz and beyond) and took the data traffic off the voice switching network with the technology called *digital subscriber line (DSL)*.

There are many different types of DSL. Some of them are *asymmetrical*, meaning they offer more bandwidth *downstream* to the home than *upstream* from the home. This presumes that the home user will be the client and that the server will be elsewhere in the Internet. Some of the DSLs use analog line codes, meaning they can easily be added to an existing analog voice loop and still allow the loop to carry 4 KHz analog voice. Other DSLs use digital line codes, so no analog voice can share the line (a local loop cannot be analog and digital at the same time). Some DSLs are strictly proprietary, and others are at least based on international standards. All DSLs increase the amount of bandwidth available to the home. Figure 14-6 shows the current types of DSL most often seen today. Some of the most obscure DSLs do not appear in the figure.

A few words about the standard DSLs is in order. HDSL and HDSL2 (now G.SHDSL) are ways of deploying standard E1 (2.048 Mbps) and T1 (1.544 Mbps) links more cost effectively. IDSL delivers 128 Kbps, and SDSL is a form of "fractional" E1 or T1 service, although variations allow for speeds as high as 2.3 Mbps and a reach as far as 22,000 feet if there is no DLC present. ADSL is the most popular form of asymmetric DSL, and G.lite is a variation requiring less equipment. VDSL is basically ADSL with support for digital television (DTV) and made for a fiber feeder environment.

Some of the DSLs, especially those that use digital line codes, are much more clearly intended for business use. Few home users will want to buy a second line just for Internet access. But nothing prevents any service provider—CLEC, ISP, or any other—from doing whatever they want on a line they have leased (although no one but the controlling LEC can physically alter the line, of course).

For the rest of this chapter, the DSLs of most interest are ADSL and VDSL. This is because ADSL is a very popular offering, and VDSL requires the use of a fiber optic feeder.

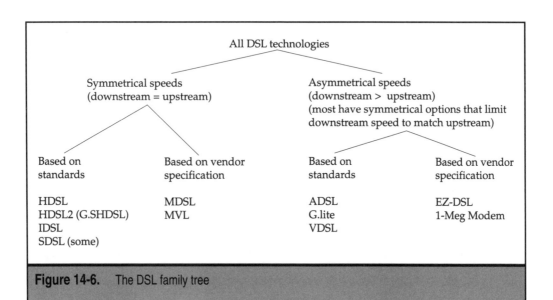

Figure 14-6.   The DSL family tree

Most versions of ADSL require a home run local loop and struggle with the presence of a DLC between CO switch and home. VDSL, on the other hand, *requires* the presence not only of a DLC between switch and home but also of a special type of DLC using fiber optic links or rings as a feeder network. This fiber DLC becomes either a next-generation digital loop carrier (NGDLC) or an optical network unit (ONU), depending on the exact architecture followed. VDSL documents call this device an ONU.

As popular as ADSL is, there remain annoying complications when a telephone company (ILEC) or CLEC or ISP is asked to provide ADSL. The process can take months and is often flatly ruled out for customers more than 18,000 feet from the CO (and the software used to detect this distance is notoriously inaccurate). Some ILECs are reluctant to even try to provide service for customers more than 15,000 feet away, and performance is hardly ideal beyond 12,000 feet (only a couple of miles). The ideal ADSL customer seems to be one located about 5,000 feet from a CO on a home run consisting of a single wire gauge and no bridged taps on the loop. Unfortunately, there are few customers whose local loop fits this profile.

ADSL can usually offer line speeds of 1.5 Mbps downstream and 256 Kbps upstream, but most service offerings promise only about 512 Kbps downstream and 128 Kbps upstream. This is because any ADSL service provider has to aggregate the DSL traffic from many DSL links at a DSL access multiplexer (DSLAM) in the central office. This traffic all has to be passed to and from the ISP, of course, through a series of routers. The faster the speeds promised the users, the more capacity on the backbone between DSLAM and ISP is needed. All this makes it very hard to characterize the throughput speeds actually experienced end-to-end by DSL users. The bottleneck is not the access line, as with the shared channels of the cable TV cable modem architecture, but behind the DSLAM. Surveys show that ADSL throughput varies widely at different times of day and different days of the week, with about 240 Kbps (30 kilobytes per second) being a good expected average. However, ADSL remains as popular as cable modems, and some industry observers even give DSL forms an advantage over cable modems in the long term . All in all, cable modems and DSL have comparable throughputs, with considerable local and time-of-day variations.

ATM is popular as a backbone technology connecting DSLAMs to ISP routers with ATM interfaces. This use of ATM means that *permanent virtual circuits (PVCs)* can be used to connect the ISPs to their customers on the DSLAMs over one physical link instead of requiring an individual point-to-point link to each DSLAM. At some point, the traffic from many PVCs from several DSLAM are in turn consolidated onto another network, usually another ATM network. But the DSLAM network is part of the DSL service, and the second ATM network leading to the ISP is part of an overall data service.

Many protocols are used on DSL service offerings. But one of the most popular architectures is to use Ethernet (technically, 10Base-T) to connect the PC to the DSL modem (the DSL modem can also be a board in the PC itself). This does not have to be a full-blown LAN unless more than one PC is using the DSL link. For one PC, a simple crossover

Ethernet cable will do. Before IP packets are placed inside the Ethernet frames, however, the IP packets are put inside a PPP frame. This protocol structure is called *Point-to-Point over Ethernet*, or PPPoE (sometimes seen as just PPoE). PPP is a WAN protocol that has many attractions for the ISP. Ethernet is a LAN protocol most suited to connect multiple PCs over a limited distance. PPPoE allows Ethernet to be used to reach the DSL modem, and at the same time allows the end-to-end connection to the ISP to appear to use PPP. PPPoE also allows ISPs to make better use of their IP address space, but this is a separate issue. PPPoE has its own Internet specification, RFC 2615.

The Ethernet frame is put inside an ATM cell stream by the DSL modem. The Ethernet frame is examined by the gateway router between the DSLAM ATM network and other ATM network. The gateway router makes router decisions based on where the Ethernet frame came from and where it is going to (so technically this gateway "router" is part ATM switch and part Ethernet bridge). The PPP frame and IP packet inside are not processed until they arrive at the ISP's router. This basic ADSL architecture is shown in Figure 14-7.

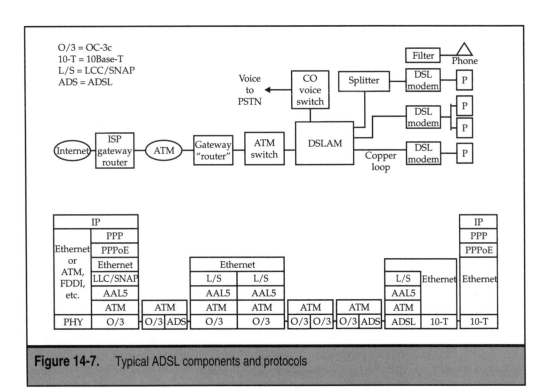

Figure 14-7.   Typical ADSL components and protocols

Figure 14-7 is very dense but is very important when it comes to understanding where optical networking might fit into this structure. At the right of the figure, three common home ADSL arrangements are shown. First, a *splitter* device at the home is used to split the signals to the existing home voice wiring and to a wire running to the DSL modem (this new wire must be installed by the service provider or customer). The filter or *microfilter* allows the phone to be used without too much noise from the DSL line data (there is some hum or buzz on the line). Second, a home might have a full LAN connecting many PCs. Some DSL modems have multiple ports like an Ethernet hub, while others require a hub or router (for security) on the premises. Finally, a single PC can use the line with a simple crossover 10Base-T cable. The protocol stacks below each device shows the flow of information through it. The PCs all use PPPoE in this example.

The DSL modem takes the Ethernet frame and adds some headers called the Logical Link Control/Subnetwork Access Protocol (LLC/SNAP) header and the ATM Adaptation Layer 5 (AAL5) header. It then sends the information out as a stream of ATM cells over an ADSL physical layer. The DSLAM aggregates traffic from many loops, changes the ATM connection identifiers, and sends the cells through the ATM switch network. The DSLAM also can split off the voice traffic and send it to the PSTN voice switch, but this service separation is often done before the traffic hits the DSLAM.

The ATM switching network in Figure 14-7 uses SONET OC-3c links running at 155.52 Mbps to send traffic to a *gateway router*, although it is really a bridge, since it only looks at the Ethernet frames. The function of this device is to form a gateway between the DSL service provider's ATM network backbone to the DSLAMs and the ATM network connecting to the router of the DSL customer's chosen ISP. Often these ATM networks are one and the same, but regulation in the United States still differentiates between DSL service providers and ISPs.

At the ISP's gateway router, which can now gather customer traffic from any DSLAM on the ATM network, the PPPoE is processed and the IP packet is sent onto the Internet inside an Ethernet frame, an ATM cell stream, and something else. This is the logical endpoint of the DSL link.

There are many variations on this scheme in terms of protocols used and even physical arrangements. But this architecture is very popular for ADSL service offerings.

The reason that ADSL (and most other DSLs) cannot run through a DLC is that existing DLCs are optimized for analog voice and will ignore anything that is not 4 KHz analog or 64 Kbps digital voice. However, some 30 percent of the 200 million access lines in the United States are on DLCs already, and the number is rising sharply. Another 20 percent of these lines are more than 12,000 feet long, and this number is increasing as well. This is because in the United States, people are moving further away from congested urban and even suburban areas. As remote clusters of townhouses and apartments spring up almost overnight, the CO remains where it was, just further away.

VDSL adds even more bandwidth to ADSL, enough to support digital TV (DTV). VDSL is in a real sense the telephone companies' answer to the cable TV. (Early trials to deploy *switched digital video [SDV]* over ADSL one channel at a time failed in large part because it took too long to change channels!) VDSL is essentially a fiber optic network with a short copper tail end. VDSL increases the copper bandwidth to 10 MHz, and some early VDSL networks have delivered three or more DTV channels at the same time, along with Internet access and analog voice.

VDSL requires an ONU as a type of DSLAM. For ADSL services, it seems like a good idea to move the DSLAM into the DLC to prepare the way for VDSL. VDSL is sometimes seen as the evolutionary path for ADSL, so moving the DSLAM into the DLC seems to make perfect sense.

Moving the DSLAM into the DLC is not always an ideal solution. DSLAMs are mostly too big for small DLC cabinets and enclosures. Related issues are powering the remote DSLAM, the heat given off, the need for battery backup in case of commercial power failure, the port density, and so on. Integrated DSLAM functions will come to the DLC world, it seems, if only because they must as DSLs and DLCs grow up together.

THE PASSIVE OPTICAL NETWORK (PON) AND FIBER TO THE HOME (FTTH)

One of the main issues regarding fiber in the loop (FITL) is the question of distributing power to the remote optical network units (ONUs) needed for true fiber to the home (FTTH). At the ONU, optical signals are converted to and from electrical signals for compatibility with user equipment. In fact, sometimes the term "PON" is used interchangeably with "FTTH," and this makes a lot of sense. This section explores some of the implications of using a PON to cut down on the number of users serviced by an ONU.

First of all, the telephone service providers' architecture of fiber to the neighborhood (FTTN) looks a lot like HFC. The tail end is twisted pair copper instead of coaxial cable, but the node sizes are quite similar (500 to 1,000 homes). The two more ambitious residential fiber architectures are fiber to the curb (FTTC) and fiber to the home (FTTH). FTTC serves small clusters of anywhere from 10 to 100 subscribers, depending on vendor. FTTH, by definition, serves one home per fiber. However, this does not imply that there is one fiber per home running all the way back to the central office. That would make no sense at all given the enormous bandwidth available on fiber.

Both PONs and FTTH use a feeder fiber to carry optical signals to a distribution device. But in contrast to the HFC or FTTN or FTTC, a PON or FTTH uses fiber optical tail ends to reach all of the homes. A PON, by definition, sends signals on fiber with no need for intermediate power between service provider office and home. The home's electrical power supply is used to power the remote ONU. So the main difference between a PON and FTTH is that a PON does not allow the splitters on the optical distribution network (ODN) to be powered devices.

Most FTTH and PON schemes share the fiber feeder bandwidth, much as HFC cable modems share a few channels on the cable TV network. The role of the cable modem terminal server (CMTS) is played by an *optical line terminator (OLT)* in the central office. The role of the cable modem is played by the ONU at the home, of course. The splitter at the end of the feeder fiber can be powered (FTTH) or not (PON). The basic architecture of a PON is shown in Figure 14-8.

One problem with the PON architecture is that there is nothing in the home right now that can support a fiber interface. So the remote conversion unit must be either included with the service or paid for by the customer. This is a real problem, since conversion for 500 to 1,000 homes can be very efficient. FTTC can be done as well, but at greater cost per home. A PON requires very cost-effective ONUs to be economically feasible.

Home ONUs, which must usually be mounted outside in the heat and snow and rain, are not yet environmentally hardened enough to be both reliable and at the same time affordable. Putting them inside the home raises regulatory issues that have yet to be addressed. Home ONUs are often called *residential gateways* in many PON discussions.

THE TELEPHONE NETWORK AND OPTICAL NETWORKING

Whether they are called FTTH or PON, with powered or unpowered ONUs, a distinguishing characteristic of both architectures is the requirement for wavelength division multiplexing (WDM) or dense WDM (DWDM). This is because, although each home

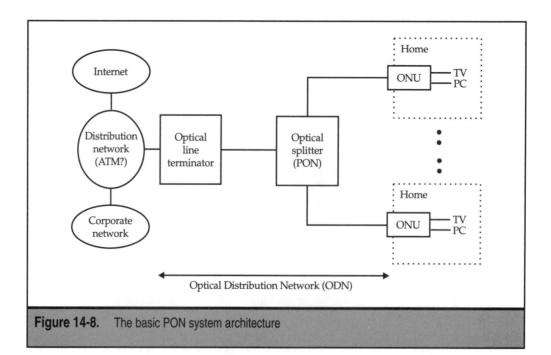

Figure 14-8.   The basic PON system architecture

might have a fiber drop cable running to it, the intention is not to run these fibers individually all the way back to the CO, replacing separate twisted pairs with separate fibers. This architecture might come some day in the future, but no home could ever use more than a small fraction of the bandwidth available over a modern fiber.

So individual *drop fibers* will be gathered onto *feeder fibers* and the multiplexed signal sent back to the CO. Now, there is no requirement to use WDM or DWDM to perform this multiplexing task. The customer optical signals could be converted to electrical signals and use time division multiplexing (TDM); the electrical signals would then be reconverted to optical signals for the feeder. But this would require a powered ONU, just as does having a twisted pair as the drop wire. Analog frequency division multiplexing (FDM) could also be used, as in the cable TV HFC architecture, but this would require electrical conversion (and power) as an intermediate step as well. As has been pointed out several times already in this book, WDM and DWDM are multiplexing methods that essentially apply FDM to fiber wavelengths without the need for any electrical conversion.

Many of the characteristics of the optical splitter used in FTTH or a PON are being standardized. The ODN can consist of one or two single-mode fibers. WDM is used in both, with the one-fiber system carrying optical signals in both directions. The frequency range proposed for one fiber is 1,530–1,570 nm downstream to the home and 1,280–1,340 nm upstream from the home. DWDM might collapse both of these into the 1,530–1,570 nm range. When two fibers are used, one for downstream and one for upstream, both can operate in the 1,530–1,570 nm range.

The distance from the OLT in the service provider's office and the ONU at the home can be up to 20 km (about 12.5 miles, or 66,000 feet). This is more than three times farther than copper loops, as expected. Supported upstream speeds are supposed to be at least 25.92 Mbps with ATM, up to 155 Mbps. Downstream speeds must be at least 55 Mbps but can be as high as 622 Mbps (perhaps not coincidentally, the B-ISDN speed proposal).

The ONU is supposed to be an ATM switch and gateway for home networks not using ATM (few will use it). The ONU can operate with a home LAN built on phone wire, coaxial cable, or even wireless. With current OLT/PON/ONU proposals, the downstream traffic is a shared broadcast and the upstream traffic channel is shared using an arbitrated contention scheme similar to that used in cable modems. This is done so that a single receiver wavelength in the OLT can receive upstream traffic from all of its ONUs easily.

VDSL should be able to ride very easily on this architecture. Analog voice telephony will have to go, however, and the ONU will also have to convert analog telephone conversations to optical signals and back. There are several ways being considered as the best way to do this, including voice over IP (VoIP) and voice over DSL (VoDSL). The only real difference between them lies in the details of how the digital voice packets are encapsulated for their journey across the optical network.

THE TELEPHONE NETWORK AND DWDM

The PON and FTTH bring fiber right to everyone's doorstep. However, there is more that optical networking can do besides just banishing copper from the access lines. Once DWDM is added to the mix, some very interesting things can be done to the basic DSL and PON architectures.

DWDM and optical rings can be used in the same places that DWDM and optical rings can be used in a cable TV network using HFC. That is, DWDM can improve the efficiency and performance of the backbone connecting the DSLAMs to their service providers and distributing traffic to home ONUs.

Consider the backbone situation first. Right now, ATM virtual circuits are used to connect DSLAMs to ISPs. Now, due to local regulation in the United States, the ILEC that controls the local loop (a basic transport service) cannot also be an ISP (an enhanced service). Even when the DSL service provider is part of the same corporate entity as the ISP, they are technically different companies. A customer hooked up to the same DSLAM as its neighbor should be able to choose a different ISP, as long as both ISPs are hooked up to the same ATM backbone network as the DSLAM. ISPs should also have to link only to those DSLAMs where they have customers, of course.

One of the reasons that ATM is used at the DSLAM backbone is that ATM is considered to be a basic transport service. However, as has been pointed out several times already in this book, ATM use involves at lot of overhead, adding a 5-byte header to every 48 bytes of information.

The major advantage of using ATM, besides the regulatory issue, is that ATM allows each DSLAM to be reachable by any ISP with an ATM interface on their gateway router without the need for dedicated, point-to-point private lines running to each DSLAM. The ATM virtual circuits take the place of the point-to-point lines. But having to buy the ATM hardware and overhead just to get the virtual circuits seems like a high price to pay.

If some way could be found to create logical connections from any ISP to any DSLAM customer without requiring ATM, that would be a big advantage over the current way of doing things. ATM just complicates the number of headers required and adds overhead. But DWDM wavelengths can take the place of ATM virtual circuits and provide the needed DSLAM-to-ISP connectivity.

This role for DWDM on an optical ring is shown in Figure 14-9. The interface to the DSLAM is cleaner, the protocol stack much less complex, and the backbone network much more efficient.

There are two ISPs on the DWDM ring in Figure 14-9. Naturally, both the DSLAMs and the ISP routers must be capable of optical node cross-connecting or switching. The optical nodes might be separate boxes so that the routers and DSLAMs can continue to operate as before, but these devices are consolidated in the figure.

Note that each ISP reaches its customers on the DSLAM by wavelength. This does not imply that every customer has its own wavelength, of course, although that is possible. For example, in the figure a small ISP reaches its customers on all DSLAMs over a separate wavelength. A larger ISP can have more than one wavelength carrying traffic to and from another DSLAM. Presumably, this is because the larger ISP has many more customers

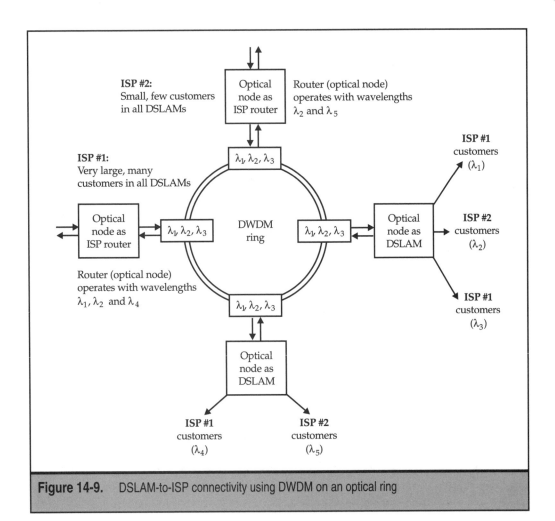

Figure 14-9.   DSLAM-to-ISP connectivity using DWDM on an optical ring

and therefore much more traffic than the smaller ISP. The most important point is that the traffic to and from the DSLAMs is routed by wavelength and not by virtual circuit.

But what about voice? VoIP can easily ride to and from the ISP along with all of the other IP packets. Or the LEC voice switches can connect to each other and to the IXC POPs over the same DWDM ring. There are even other ways to provide for voice on an optical network, as will be seen in the next chapter.

As expected, provisioning is now done according to wavelength. For example, all new customers on a DSLAM can connect to their chosen ISP on the ring over the same wavelength. Upstream and downstream bandwidths can now be assigned asymmetrically. This is a major advantage to linking DSLAMs by wavelength and not by ATM with the same SONET/SDH speeds in both directions.

The bandwidth between ISP and DSLAM is still shared, but more bandwidth can be added easily to the individual wavelengths as the number of customers grows on the DSLAM. And this can be done much more easily than trying to add bandwidth to an ATM backbone using fixed, incremental SONET/SDH speeds. A speed of 622 Mbps might be too slow, but a speed of 2.5 Gbps might be much too high and therefore just cost-justifiable.

The other place that DWDM can be used in the new telephone optical network is on the distribution side of the network from a DSLAM (now an OLT) to the home ONU.

The usual PON architecture allows for an ATM network to distribute up to 622 Mbps of data downstream to up to 32 homes. These 32 homes will also share up to 155 Mbps of upstream bandwidth. This might sound like a lot, but remember that this has to cover all digital video, associated Web traffic, telephony, and whatever other services come to be popular in the next few years. There are already plans for home security, intelligent home heating systems, refrigerators that warn their manufacturers when they are about to break, and so on.

The net result of this sharing is that if all homes are active at the same time, which seems likely given current Internet usage patterns, there is a lot less bandwidth to go around than with HFC systems. The downstream bandwidth works out to about 19.44 Mbps, and the upstream bandwidth translates to about 4.8 Mbps. Of course, if the number of homes serviced from a PON splitter drops from 32 to 8, this is a fourfold increase in the amount of bandwidth available at any one time. There is also the possibility of increasing the upstream and downstream bandwidths beyond the initial target values, of course.

Neither the fewer-homes approach nor the more-bandwidth solution would be easy or painless to implement. But DWDM especially offers an easier way both to provide dedicated bandwidth to a home and to provide for a virtual circuit right through the OLT (now the DSLAM) to the service provider (now the ISP).

In this scenario, information content to one home is sent on one wavelength, while the information content to another home is sent on a separate wavelength. Both frequencies are delivered on the same feeder fiber with DWDM. Separate wavelengths are used for the upstream traffic as well. If it is deemed desirable to increase bandwidth in either direction, these upgrades can be made at the endpoints of the network, at the ONU and OLT, without changing the optical distribution network (ODN) in between. This use of DWDM is shown in Figure 14-10.

Figure 14-10 shows eight wavelengths in each direction, but many more are possible, or course. Note that the OLT now supports DWDM, as do the PON splitter and the ONU. And although the ODN now operates on a wavelength basis, the fiber did not change when DWDM components were added. The mapping of the wavelengths to inbound and outbound traffic to and from the homes shown is purely arbitrary. The only requirement is that the wavelengths chosen be dedicated to a particular user.

The use of DWDM in the PON architecture has the following advantages:

▼ **Bandwidth** Residential users will have more bandwidth with DWDM. A PON with DWDM offers more upstream bandwidth than HFC, VDSL, or any

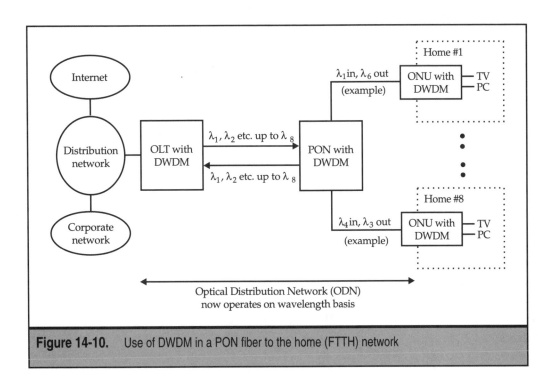

Figure 14-10.   Use of DWDM in a PON fiber to the home (FTTH) network

other architecture. So DWDM is the best choice for the days when most homes will have their own Web sites, as most observers think is more or less inevitable. More and more workers can work at home and still enjoy the high-speed networking that characterizes the corporate network.

■ **Remote provisioning**   As mentioned previously, service providers are reached by wavelength. And services can be turned on and off by wavelength as well. Speeds can even be adjusted by the service provider, but this requires that the ONU have an agile optical transceiver capable of operating at several wavelengths. Bandwidth variations can even be allowed by time of day, day of week, or some other criteria such as family vacations. Management standards to help in this process would be nice to have, but these will probably be slow to come.

■ **Cost reductions**   The lowered need for powering, the self-provisioning aspects of a DWDM PON, and the reduced costs of outside cabling all translate to reducing service provider costs. If the consumer must provide the power to the ONU, as seems likely, costs will be lower still. And the transition from ADSL to VDSL to a PON with DWDM will be smoother than any other residential networking evolutionary path.

■ **Reliability**   Optical fiber networks do not have to face the risk of radio frequency interference. Fibers do not rust, which extends the life of optical components. Optical networks also have fewer active components that might not only fail but be rendered useless by having the power supply disrupted. All this contributes to the greater reliability of optical networks.

■ **Security**   Shared DSL bandwidth and even shared PON bandwidth pose the same security risks as shared bandwidth on HFC networks. The use of Ethernet on the links to the home effectively makes a bridge out of the DSLAM, so traffic goes everywhere and listening in on the neighbors is relatively easy. Plans to add standard security are starting to appear, and many security schemes can be used already. However, above and beyond even that, the use of separate wavelengths adds a lot of security to a basic network.

▲ **Full array of network services**   The idea of a *full service network (FSN)* has been a dream for a long time. With an FSN, there would be no need for separate cable TV and telephone networks and access lines. All services from DTV to Internet and Web access to voice would be delivered over the same infrastructure (although perhaps still by different service providers). An optical network can deliver them all and evolve to do anything that is asked of it.

These advantages to optical networking and DWDM must be balanced against the considerable start-up costs for optical networking, and the still sticky issue of ONU power and control. If the ONU is considered to be customer premises equipment, then the user must provide the power. However, the service provider would have little control over the purchase and use of the device. If the ONU is part of the network, this might be seen as going against the almost 20-year-old trend of deregulation and less control on the part of service providers over their customers. Given this, it seems likely that the ONU will be customer premises equipment, perhaps sold in stores as modems are today. Therefore very strict standardization to ensure interoperability will be a necessity.

About the only place that fiber will not be used for some time to come is in the home itself. There is no reason that fiber cannot be deployed instead of coaxial cable or twisted pair telephone wire from the kitchen through the living room and up to the bedroom. It just makes more sense at present to rely on the single ONU as the *residential gateway* for the conversion of optical signals. It might seem like a fine idea to run the fiber directly to the TV or PC, but in some cases the cost of the ONU for conversion, especially when DWDM is used, would far exceed the cost of the device using the optical network.

However, as time goes on more and more end user devices such as PCs, TV sets, and even perhaps toasters and refrigerators, might come equipped with their own fiber interfaces. In this case, the need for an ONU at the home might well be eliminated. Just have the user run their own fiber in the same way that they now run their own coaxial cable to TVs. Once TVs came with coaxial cable interfaces and built-in cable TV tuners, all that was needed to get basic service to every TV in the house was a simple, passive cable splitter. The same splitter concept might work with fiber optical cable. Maybe the ultimate answer to the ONU question is Fiber in the Home (FITH).

PART VI

Putting It All Together

Parts 1–5 explored the world of optical networking from the days when people stared at distance windmill-shaped paddles to the days when every home will have a fiber optical cable. Each part focused on a part of the story, starting with the fiber itself, then to the transmitters and receivers that were added to create true optical networks.

Early fiber systems were investigated in Part 2, starting with the WAN and SONET/SDH, and then creating the concept of WDM, DWDM, and optical network nodes. Part 3 continued this theme to some extent in the WAN arena, showing some SONET/SDH, ATM, and even IP routing, which will be transformed by the rise of optical networking with DWDM. Part 4 extended this investigation to LANs. FDDI and fiber channel can look and be deployed and used very differently in an optical networking environment. The key position of Gigabit Ethernet was detailed at the end of Part 4.

Part 5 turned to the residential sector—starting with cable TV. Then the world of telephony and DSL was explored to understand how optical networking and DWDM will transform these technologies.

What could be left to form the topics for Part 6? So far, the book has been through time and space from the French Revolution to the Millennium, from WAN to LAN to (maybe) the bathroom. Where else is there left to go?

As it turns out, there are some very important things that still need to be explored in Part 6. There are some loose ends to tie up, a look at a possible complimentary (or perhaps competitive) technology, and, finally, an assessment of the future of optical networking.

Chapter 15 takes a look at what life might be like in the optical networking world. What will it mean to link countries with DWDM undersea cables? What will happen to the Internet and Web? Will the PSTN just fade away? It also takes a final technical plunge into the photonic world with a section on DWDM conformance and performance testing.

Chapter 16 examines a topic that hasn't come up yet: wireless communications. This apparent oversight is understandable when the main topic is concerned with fiber optic cable. But there are many industry experts whose enthusiasm for wireless technologies exceeds even the excitement over optical networking. And big changes are in store for wireless networks. This chapter details them and positions wireless and optical networking in a world where one cannot exist without the other.

Chapter 17 looks at general issues of optical networking. There are network, equipment, and service issues that must be resolved as optical networking moves into the mainstream. For example, will wireless or optical networking have a faster evolution, and is a rapid pace of change in technology always a good thing? And what about the tantalizing promise of true optical computing? Someday optics might rule the network, not only between the clients and the servers, but inside them.

CHAPTER 15

Fiber, Fiber Everywhere

There are probably still good reasons to install media types other than fiber optical cable whenever new cabling is required. But as time goes on, it becomes harder to find compelling reasons to use anything other than fiber to connect things together on a network. The trend started on the WAN backbones of both the telephone companies and the ISPs. The trend continued with the creation of metropolitan area networks (MANs). Initial deployments were of straight point-to-point links, and then came the rings. Major urban areas led the way, and now even relatively sparse suburban areas have substantial fiber ring architectures in place.

In the world of large organizations, at the campus and building levels, fiber has been the medium of choice for a while as well. Once seen only on the campus backbone or in the building rise, fiber is now snaking its way from the telecommunications closets on every floor right to the desktop. For now, the trend seems to be to run fiber to heavily used servers, either as FDDI or (increasingly) Fibre Channel SANs, but the day will come when every client in an office has a fiber optic connector, probably some form of Gigabit Ethernet.

Only the residence seems to be immune to the spread of fiber everywhere. This will not change soon, but when it does, it seems likely that large multiple dwelling units (MUDs) such as apartment complexes and townhouse units will lead the way. And even then there will be plenty of coaxial cable or twisted pair at the tail end of these fiber runs.

The reasons for the fiber explosion have been covered throughout this book. It can cost a WAN service provider more than $70,000 per mile to install new facilities along rights of way in the United States. Why bury anything other than fiber? Most fiber links are engineered to last 25 years, but there is no reason that well-maintained fibers cannot function as long as the copper local loops in the PSTN, and some copper loops have been in service since the early 1900s. Anything else has only a fraction of the bandwidth of fiber and has many other environmental risks from physical damage and electrical interference. No one has ever built a copper ring. (At least none that has been publicized in the vast popular literature on networking.)

On the campus and LAN, new conduit systems using air-blown fiber (ABF) make installing new bandwidth as simple as attaching a small nose cone to a fiber reel in the basement and blowing it up to the fifth floor in about a second. The cost of running fibers once the system is installed is minimal. Compare this with the union cost of installing IBM shielded twisted pair (STP) cable in New York in the late 1980s: about $1,000 per cable. Nobody could blow the rigid and hefty STP anywhere, but the bandwidth was an appreciable (for the time) 16 Mbps.

When it comes to the home, only TV signals require a lot of bandwidth, and compression techniques are lowering the minimum required all the time. PCs with Internet and Web access can chew up a lot of bandwidth. But until all TV signals migrate onto an IP backbone (a scenario that is always seriously considered), the PC can live quite comfortably on copper at the end of a cable TV or DSL network offering perhaps 500 Kbps. More than 90 percent of all cable TV companies offer some form of cable modem service.

This brings up the subject of *convergence.* Convergence is used both to describe the merging of different forms of network into one (e.g., the convergence of the cable TV and telephony networks) and the merging of different end-user devices into one (e.g., the PC as television set). Sometimes, the convergence term is used to cover the "merging" of the service providers themselves (although almost all of these are simple acquisitions), as with Qwest and US West. However, distinctions among types of service providers are probably here to stay for some time to come, for marketing and regulatory reasons. People like to know who "the phone company" is, and deregulation does not mean "no regulation at all."

So the best place to start is by taking a look at how the public switched telephone network (PSTN) might evolve into the optical networking future. Along the way, the role of the Internet and cable TV will be outlined as well.

OPTICAL "DIAL TONE"

A joke made the rounds of all the ILECs in the late 1990s. The joke involved the posting of signs in all the central offices (COs). The sign read "Would the last person leaving the PSTN please turn off the lights?" But that wasn't the joke. The joke was that when the person arrived at the CO, the door was locked, there was no one there to let them in, and the building had been darkened already. Everyone had left the CO long ago.

Today, almost all CO switches are just big computers, not so different than the routers that together create the Internet. What differences there are will start to disappear when voice over IP (VoIP) and voice over DSL (VoDSL) become the rule rather than the exception. This does not mean that everyone will have to throw their analog telephones away and buy new ones, or buy a conversion unit for the analog telephone. Users might revolt at that thought. However, no one seems to mind so much that a new cell phone is required every few years, or that after the conversion to DTV, any older analog television sets will be useless or require the purchase of a converter box.

There are already CO "switch" products on the market that essentially make the CO switch into a very large and very fast IP router. After all, there is a lot of IP packet traffic to and from the Internet on those telephone lines, especially the 70 percent of the 200 million access lines in the United States that run to homes. Why not just take the analog or digital voice and packetize it right at the CO? Let the customers keep their analog phones. This plan would converge the voice switch and router into a single device: the voice-enabled IP router.

Not much work is really needed to voice-enable a router. Analog voice is digitized and packetized, usually at about 8 Kbps with compression, and sent as a stream of IP packet between source and destination IP addresses. The audio packets are usually sent at a rate of around 100 packets per second, but this rate is configurable to trade off between delay and processing overhead. So there is not much load on the router due to any individual voice conversation, although aggregation can still make the packets add up quickly. It helps that most VoIP (and VoDSL) methods include *silence suppression*. With silence suppression, no packets need be sent during the silences as one listens, or pauses between sentences and phrases (or even within words). About 60 percent of all calls are silence. So the effective throughput needed in each direction is only about 40 percent of 8 Kbps, or about 3.2 Kbps. A "CO router" with about 32 Mbps capacity could easily handle 10,000 voice calls at once.

Of course, many lines are idle for every line in use. And routers can handle multiple 100 Mbps Ethernet interfaces with ease today.

The real work that needs to be done to voice-enable a router concerns the signaling used to set up voice connections not only across an IP network such as the Internet, but onto and from the PSTN itself. Naturally, for a long time to come, most telephones will still be attached to CO circuit switches, especially outside more technologically advanced regions of the world.

The Signaling Network

Telephony signaling consists of the analog "touch tones" that a user dials to reach another person on the global PSTN, and digital signaling packets that flow on a special signaling network between LEC CO switches. For long-distance calls, these signaling packets are handed off to a national IXC or to an international gateway. The international standard for these signaling packets, in terms of structure, function, and protocol, is called *Signaling System #7*, or *SS7*. The analog tones usually exist only on the analog local loop today, and the PSTN uses SS7 packets almost everywhere else. A conversion software module, part of the service switching point (SSP), is typically part of the modern CO switch and converts between analog touch tones and signaling packets.

So SS7 is a type of routing network to shuttle signaling packets between voice switches. The most important type of packet contains the telephone connection *call setup* message, but there are others almost as important. One of the problems with integrating IP telephone and the existing voice network of circuit switches is that the signaling network uses entire separate clients, servers, links, and network nodes rather than the voice network that carries the circuit-switching voice. The client is basically the SSP, with the role of the router played by the signaling transfer point (STP) and the role of the server played by the service control point (SCP) hardware and software. The relationship between the voice COs and the SS7 signaling network is shown in Figure 15-1.

In the figure, the end users of the network are telephones, and the light lines represent the local loops. The CO nodes are telephone switches, and the heavy lines are multiplexed trunks carrying many voice channels. The dashed arrows represent the path chosen for this particular call. The call path is full duplex once set up, meaning the two people can talk to each other, but the initial call path set up proceeds from the originator to the destination. Naturally, the signaling paths from CO to switching node are full duplex as well.

How was the dashed path for the call in the figure chosen? Through the SS7 signaling nodes at the bottom of the figure. The links between voice switches, and the signaling nodes themselves, are shown as dashed lines. Another dashed line separates the voice network, where the calls are, from the signaling network. The signaling network is physically separate from the voice network. Note that the end nodes have no direct access to the signaling network, only the switches do.

So the voice call is set up with a signaling protocol. This establishes a path for the call to follow through the network. Now, the route selected for the call could not go through CO C for the simple reason that the destination is not reachable through CO C. The signaling

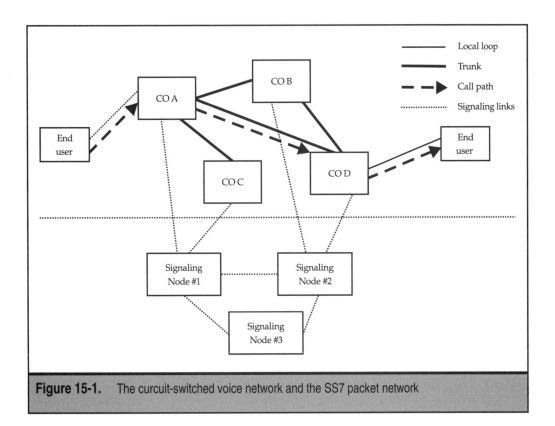

Figure 15-1.   The curcuit-switched voice network and the SS7 packet network

network could route the call through CO B to CO D, but why? There is a direct path from CO A to CO D, which is why the path was used in the figure. Naturally, if all trunk channels are in use between CO A and CO D, it would be possible to route the call through CO B. But this decision can only be made at call setup time by the signaling nodes. Calls in progress are not rerouted, and calls that are *blocked* are simply dropped (the "fast busy" is returned to the user).

The signaling nodes (the STPs) play a crucial part in the call setup process. In the figure, Signaling Node #1 is attached only to CO A and CO C, while Signaling Node #2 is attached only to CO B and CO D. Yet the signaling network must make routing decisions based on *global* knowledge of the network. Information about trunk use and other voice network parameters must be shared by the signaling nodes. On the Internet, routers (which are essentially the functional equivalent of the signaling nodes in the voice network) exchange information through the use of a *routing protocol*. The same thing happens in the PSTN, although the routing protocols used are appropriate for a circuit-switched network. The signaling network in the PSTN is a router-based, client/server, connectionless packet-switched network built on the same principles of operation as the Internet. In the figure, Switching Node #3 can handle communications between signaling

Nodes #1 and #2 if the direct link between them fails. Thus setup requests are routed around failures, and not blocked, as are telephone calls.

Once the call is set up and accepted (the other end answers), the same path is followed for all voice bits. There is always 64 Kbps flowing in each direction, even during periods of absolute silence (e.g., hold). This is not true of the signaling network. Signaling messages are bursty and do not use the same amount of bandwidth to set up a call as to release a call.

SS7 and IP

The SS7 signaling network of the PSTN is a connectionless packet network. SS7 has to be connectionless: one of its main jobs is to set up connections in the form of telephone calls. In the PSTN, the connectionless SS7 network is physically separate from the connection- oriented voice network to minimize hacking and fraud. (In connection-oriented *data* networks like X.25, the call setup signaling messages were sent over the same links as the data.)

SS7 is much more than just a connectionless packet-switched network. But SS7 is a good analogy compared to the Internet, even more so now that VoIP calls that extend to the PSTN must interact with SS7 to set up calls. So a closer look at SS7 is needed to understand how VoIP calls can be made to and from the PSTN, and how optical networking might help the situation.

SS7 is a signaling system that uses out-of-band signaling to send call control information between PSTN switching offices. SS7 includes databases in the network for doing things like 800 number translation and providing other services (such as caller ID) that can be a source of added revenue for telephone companies. Although the SS7 network is designed for call setup–related functions, non-circuit-switching applications, such as connection-oriented or connectionless X.25 packet delivery, can be performed by the SS7 network.

SS7 replaces the older in-band signaling systems that were vulnerable to fraud and slow to set up calls (anywhere from 10 to 15 seconds). SS7 is an example of a common channel signaling (CCS) network and is in fact the international standard for CCS systems. Not only is the signaling moved out-of-band with SS7, but the signaling messages are actually moved off the voice network entirely. Call setup time dropped to about 3 to 5 seconds with SS7, more trunks were freed up to carry calls, and costs were minimized, since less equipment was needed for SS7 than for other signaling methods. Also, CCS networks like SS7 can offer advanced services to users such as caller ID, credit card validation, and so on.

A CCS network like SS7 has a definite architecture, as shown in Figure 15-2. The figure shows the portion of the SS7 that might be used by a LEC.

More details on the signaling network are provided by the figure. The service switching points (SSPs) are the clients of the network, and they are usually implemented as software in central office switches. The signaling messages originate at the SSPs (and usually end up at an SSP as well). The signal transfer points (STPs) are the routers of the network, shuttling the signaling messages around according to routing tables. Finally, the signaling control

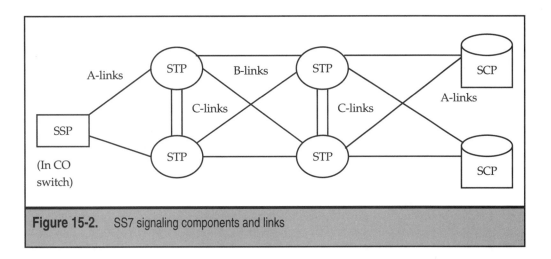

Figure 15-2.   SS7 signaling components and links

points (SCPs) are the databases that store the information needed to provide the customer services, from 800 number translation to credit card validation to call forwarding.

Each of the components is connected by a special link type. The link architecture is designed to provide redundancy to the signaling network. After all, if the signaling network is down, no calls at all can be completed, with a huge impact on revenue. Access links (A-links) connect an SSP or SCP to two STPs. Bridge links (B-links) interconnect the STPs in different regions. Cross links (C-links) connect "mate" STPs and are used when A-links or B-links fail.

There are even three other types of links used for a variety of purposes. These are not shown in the figure but would be used when calls are routed through an IXC, for example. Diagonal links (D-links) connect a LEC SS7 network to an IXC SS7 network. Extended links (E-links) connect an SSP to a second set of four STPs to balance high traffic loads. Finally, fully extended links (F-links) connect two SSP switches or SCP databases directly for load balancing and further reliability.

An SS7 network uses links at 56 or 64 Kbps but can be as fast as 8.448 Mbps if the load calls for it. The signaling messages can be up to 272 bytes in length. The routing of call setups is done based on the address of the destination office, and the routing is purely dynamic, the same strategy used for the Internet.

SS7 is a layered architecture. The bottom three layers of the OSI-RM (a frame and packet structure) are provided by the Message Transfer Part (MTP) protocol. These are called MTP Level 1, MTP Level 2 (the frame), and MTP Level 3 (the packet), although MTP Level 3 does not do all that the OSI-RM Network Layer or IP does. The rest of the Network Layer in SS7, if needed, is provided by a protocol called the Signaling Connection Control Part (SCCP). Some signaling messages do not need SCCP. When the SCP database is queried by an SSP, the database query language is provided by a protocol known as Transaction Capability Application Part (TCAP).

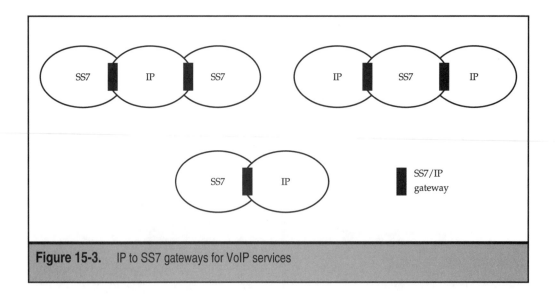

Figure 15-3.   IP to SS7 gateways for VoIP services

So SS7 forms a type of connectionless packet-switched network used to set up telephone calls on the PSTN. If a VoIP call has one end on the PSTN, some interface between the IP network and the SS7 signaling network is needed. There are also scenarios where a call consists of VoIP on each end, with the PSTN in the middle, or the PSTN on each end with VoIP in the middle. In each case, some VoIP to/from SS7 gateway device is needed. These cases are shown in Figure 15-3.

The SS7/IP gateway is easy enough to implement in the IP-to-SS7 direction. SS7 messages have a standard structure that must be observed, and IP packets can trigger the proper signaling messages as an *intelligent peripheral* attached to an SSP. For example, an IP router can be an intelligent peripheral with a direct link to the SSP module of a voice switch, as long as the router can generate the proper signaling messages and packets. This is sometimes called the tail end hop off (TEHO) of a VoIP call.

Things are a little more complex in the SS7-to-IP direction or where two SS7 clouds are linked only by an IP network such as the Internet. There is no standard way (yet) to convert SS7 messages and packets to an IP packet stream. The IP network can be used to *transport* the SS7 messages, of course, but the whole idea is to set up the VoIP call over the IP network. Work continues on standard methods to interface SS7 to IP.

VoIP, SS7, and Optical Networking

So far, it might be hard to see what all of this discussion about the PSTN and SS7 has to do with optical networking. The time has come to bring all of these lines of thought together.

It the near future, the main CO switch might still be a traditional circuit switch. However, this switch might be joined by a smaller (physically, not in terms of processing power) "router" that just handles IP packets. These IP packets could contain voice, or data, or even video, but the point is that data packets will be sent to and from the Internet and Web, and voice packets might be sent to and from the PSTN, depending on the source and/or destination of the voice. There will be a considerable time of transition as circuit-switched voice migrates to VoIP or VoDSL. And the major difference between VoIP and VoDSL is the fact that VoDSL voice packets are sent over a DSL link, while VoIP packets can be sent over almost any kind of link.

VoIP and VoDSL will not use SS7. IP packets are not SS7 signaling packets, and there is no need to reproduce all of SS7 in IP, since SS7 is optimized for circuit-switched 64 Kbps voice. There are several signaling protocols that can be used with VoIP and VoDSL instead of SS7, and it seems likely that more than one might actually be used in various parts of the new voice network. One of the leading IP voice signaling protocols is SIP (Session Initiation Protocol), and another is called MGCP (Multimedia Gateway Control Protocol). SIP is now a proposed Internet standard covered in RFC 2543, released in March 1999. Several other RFCs are important to VoIP and VoDSL besides RFC 2543, but a full discussion of these is beyond the scope of this book.

The hardest thing to do will be to make sure that customers with VoIP or VoDSL can make calls onto the PSTN and take calls off of the PSTN by translating SIP/MGCP signaling messages to and from SS7. This blended PSTN/IP voice world will have to allow for the coexistence of CO circuit switches and voice-enabled routers. When faced with having to handle IP packets with voice inside, there is a standard gateway called a GR-303 gateway that handles the voice conversion between packetized, 8 Kbps, IP voice and digital 64 Kbps circuit-switched voice. The coordination between directory telephone numbers (411) and their IP addresses will be handled by extensions to the IP domain name system (DNS) coupled with another directory service called *Lightweight Directory Access Protocol (LDAP)*.

The relationship among all of these components is shown in Figure 15-4.

In the figure, users can use packet voice over a DSL link, or circuit voice either over a digital loop carrier (remote locations) or direct to the CO switch. Packet voice can be piped into the CO switch through a GR-303 gateway or kept separate through the voice-enabled router. The voice-enabled router links to a gateway "softswitch" that mainly translates between SIP/MGCP packet signaling and SS7. Data is passed on to the Internet, and the voice traffic is passed on to the PSTN once the SS7 network sets up the call. Note that packet voice traffic has two ways to reach the PSTN. Which one is used more depends on how far the service provider is along the circuit voice–to–packet voice migration path.

The role of optical networking with DWDM concerns the connectivity between the CO switches and the gateway softswitch with the Internet, PSTN, and SS7 networks. There are five links shown in the figure, and most arrangements make these separate links. Fiber can be used here, but even if there is a SONET/SDH ring in place, this ring is usually just for CO-to-PSTN connectivity.

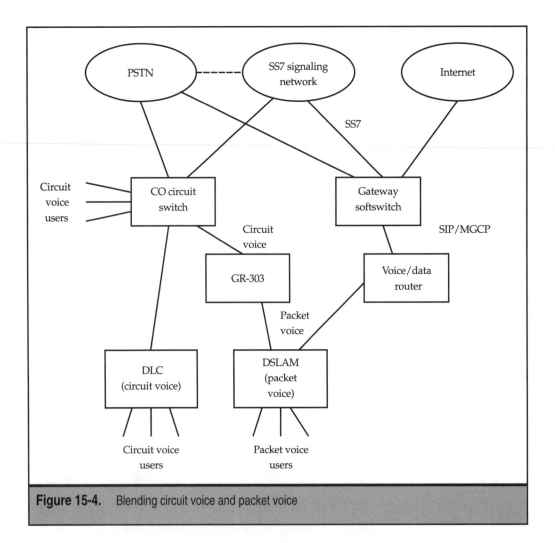

Figure 15-4. Blending circuit voice and packet voice

With a DWDM optical network in place, all of the connectivity needs can be mapped to a single DWDM ring. This role for optical networking is shown in Figure 15-5.

The whole point of this discussion involves the transition from legacy to packet voice. As the circuit CO switch handles less and less of the voice traffic with legacy voice techniques, and the VoIP or VoDSL portion of the network consumes more and more bandwidth, the speed of each wavelength can be adjusted for a smooth transition. There is no longer any need to worry about the relative speed and capacity of the infrastructure between COs for the different services.

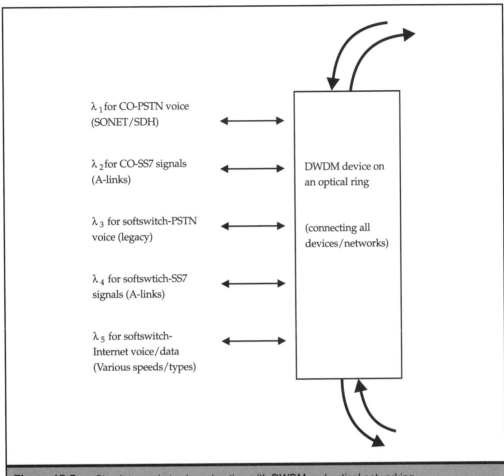

λ_1 for CO-PSTN voice
(SONET/SDH)

λ_2 for CO-SS7 signals
(A-links)

λ_3 for softswitch-PSTN
voice (legacy)

λ_4 for softswtich-SS7
signals (A-links)

λ_5 for softswitch-
Internet voice/data
(Various speeds/types)

DWDM device on
an optical ring

(connecting all
devices/networks)

Figure 15-5.   Circuit-to-packet voice migration with DWDM and optical networking

UNDERSEA DWDM FIBER

So fiber and optical networking can handle the local migration concerns when it comes to moving users from legacy, circuit-switched voice to packet voice. Admittedly, this role will only become significant in the future, but perhaps the not-so-distant future. However, DWDM fiber has a more immediate role as the method of choice today for carrying voice calls and data between the continents. This is where undersea DWDM fiber fits in.

Until recently, most transoceanic messages were carried by satellite, most notably the high-capacity satellites in geosynchronous orbit (meaning they remain stationary, more or less, over a fixed point on the earth's surface). What undersea cables there were had lower capacities in most cases, and they were mainly coaxial cables, not fiber. As recently as 1988, only some 2 percent of the world's transoceanic traffic was carried by undersea cables.

By 2000, however, undersea cables carried about 80 percent of the world's traffic. There are two reasons for this rise of cable versus satellite: channel capacity (a reflection of bandwidth and noise) and delay. The round-trip delays when using geosynchronous satellites, which orbit 1/10th of the way to the moon, are more than half a second. These delays tend to be quite annoying to voice users who are used to lower delays, although there are many voice users who do not complain about the delay at all.

So if the capacity is sufficient, there is a good reason to use undersea cables with a fraction of this delay rather than to continue to depend on satellites. Some service providers actually carry the voice one way over a satellite and carry the return voice path over fiber. This cuts the round trip satellite delay almost in half. Satellites also have a limited capacity, are limited in number due to positioning concerns, and are impossible to repair when a nonredundant component fails.

The newer fiber optic cables have at least 3,000 times the capacity of older coaxial undersea cables. And the cable capacity is always increasing. Current cables carry 42 wavelengths of 10 Gbps each, and 68 wavelength systems are be deployed. Cables offer advantages over satellites in terms of anticipated length of service, signal security, and deployment schedules (satellite booster rockets sometimes fail and have to be destroyed, taking the satellite payload with them). There are now 580,000 km (about 360,000 miles) of undersea cables on the floors of the world's oceans and seas.

Sometimes the cables are actually underneath the sea floor. Satellites suffer from risks due to meteor showers, space debris, and sun spots. The area of greatest risk to undersea fiber cable extends from the shore to the depths, where a lot of undersea activity goes on in the coastal regions. So these cables are typically heavily armored and buried under the sea floor in the relative shallows. The depth of burial varies, but is usually about a foot to eighteen inches (about 1/3 to 1/2 a meter). In the deep ocean the cables simply lie on the ocean floor, where they span the narrower valleys and snake up and down ridges. The environment is calm, and the temperature an almost constant 2 degrees Centigrade (about 34 degrees Fahrenheit).

Fiber optic cables have come so far so fast because of four key developments. The first is improved fiber cable performance, the second is DWDM, the third is the electronics that make higher speeds possible, and the fourth reason is the higher fiber strand count in each cable. Early fiber optic cables for this purpose contained six fibers, two pairs of which were operational, the third pair serving as a backup. Today cables have eight fibers or more, and all pairs are operational.

Most manufacturers of fiber optic cables make special undersea cables. These fibers have a more constant dispersion across a broad range of operational wavelengths. Constant dispersion makes the wavelengths near the "band edges" operate more like the wavelengths in the middle of the range, which allows for higher capacities due to more useful channels. Even when dispersion over long distances can be a problem, fibers with compensating dispersion characteristics can be spliced in periodically to give acceptable dispersion performance end-to-end.

Undersea cables are designed to have an operational life span of 25 years. And when the cables have to be repaired, it is much easier to haul up and fix a fiber cable than to service a satellite. In contrast, geosynchronous satellites usually run out of fuel in 10 to 15 years, and the satellite has to be replaced. The rocket fuel is used to adjust the satellite orbit when slight gravity variations tend to cause the satellite to drift away from its designated position in the sky. And cables generally last beyond their 25-year intended life span. Coaxial cables laid for the United States Navy in 1960 are still functional. Cables can even have a second life after the 25 years have expired. One cable between Hawaii and California, placed in service in 1964, was cut by a trawler in 1989. The cable was not repaired (the 25-year lifetime was up) but was given for research purposes to a group of scientists in Hawaii studying undersea tremors.

Satellite transmissions can be easily intercepted. Encrypted or not, sooner or later the data and conversations will be decoded. The only question is how long it will take. Coaxial cables can be tapped without detection, even undersea cables, but undersea fiber optic cables simply cannot be tapped because it is impossible to penetrate coastal armor (and they are buried as well) or operate tapping devices under the enormous pressures in the ocean depths. Satellites could be shot down or rendered inactive through a number of methods during war. While an undersea fiber could be cut, it has to be found if buried and located on the sea bottom.

No one should ever underestimate the value of meeting an installation date. Undersea cable installation can be delayed for a few days due to weather at sea, but the overall schedule usually allows for that possibility. On the other hand, satellites have been destroyed during launches when the rocket fails and threatens to crash. Some have even failed to reach orbit or do not achieve the proper orbit.

And while it is possible to repair some satellites using the space shuttle, if their orbit is altered to meet the space craft, this can only be done at great expense and is never done routinely for commercial satellites. Undersea cables that have been damaged, have been severed, or are failing can be hauled up by ships, repaired, and placed back into service. This process can take less than a week if the right ship is nearby. The repair crew knows exactly where the cable runs and where the trouble spot is, with the assistance of shore-based technicians and equipment. These technicians have access to information on each component, since a different wavelength is used to monitor each through a loopback coupler.

Oddly, there is much more metal in a modern undersea fiber cable than there is glass. The fibers are organized around a copper-jacketed steel cable called a *kingwire*. The whole assembly is then wrapped in a protective sheath of nylon and hytrel (a kind of "plastic rubber") and hermetically sealed inside a copper sheath that is welded shut. This sheath not only protects the fibers but also can carry electricity to any powered components along the fiber path, such as optical repeaters. The assembly process is carried out in long, narrow buildings hundreds of feet in length. The copper-coated cable is further encased in liquid polyethylene plastic. High-strength steel cables are embedded in the plastic before it can harden. This structure is used for the mid-ocean portions of the cable, to keep them as light as possible as they hang off the back of a ship on their way to the ocean floor two miles (about three km) below.

Closer to shore, the cable is protected by another layer of galvanized steel cables. These portions of the cable can weigh 16 times more than the mid-ocean cables and are much more costly. Yet the cables are surprisingly thin: the typical undersea cable is only about 1 1/2 inches (38 mm) in diameter. Even when another layer of protection is added to the coastal portions, as is sometimes done, the resulting cable is about three inches (65 mm) in diameter. The structure of the modern undersea fiber optic cable is shown in Figure 15-6.

Once the cable has been made, it is wound on huge spools 5 meters (about 15 feet) in diameter. The basic cable segment is 80 km (about 50 miles) long. The spools are unwound as they are loaded onto the ship, and the repeaters or optical amplifiers are inserted between the cable segments. Couplings are attached to the ends of each segment, which weight as much as 300 kg (more than 650 pounds) due mainly to their beryllium-copper alloy housings. The cable is then send down a chute into the ship's hold, where the cable is manually stowed in horizontal arrays (this is exactly the same method used to stow the first transatlantic cable in 1857).

Although repeaters can be placed inside the housings at the ends of the segments, modern fibers use fiber amplifiers. Fiber amplifiers can handle an entire wavelength range, any coding methods, and added channels; they are speed independent as well. The structure of an undersea fiber amplifier unit is shown in Figure 15-7.

In the Figure 15-7, the pair of amplifiers in each unit between segments is shown. Light waves entering on the fiber at the upper left of the figure are amplified by the erbium-doped fiber amplifier (EDFA). The light waves traveling in the opposite direction, shown on the bottom right of the figure, are treated the same way. There are four 980 nm pump lasers, two for each direction. The wavelength division combiner allows the amplifier to continue to function if one of the lasers fails. The filter/isolator removes the 980 nm signal once its job is done.

The function of the loopback coupler is to gather a tiny portion of the amplified light and return it to monitoring equipment on the shore using the light going in the reverse direction. This small signal can be used to determine the overall health of the unit. The monitors can detect a failing amplifier, a totally failed amplifier, or which amplifiers are closest to a cable break. This makes repair chores easier.

The fiber Bragg gratings (FBG) stabilize the lasers and select the light to loop back. The figure shows only one fiber pair, but this is just for clarity. The actual cable will use four or more pairs of fibers in each cable. There will also be some form of DWDM, usually with up to 32 wavelengths, or sometimes even more, on each fiber.

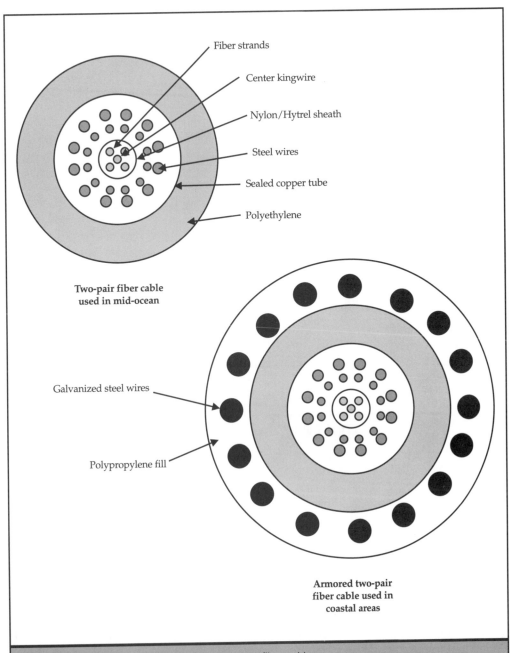

Figure 15-6. The structure of modern undersea fiber cables

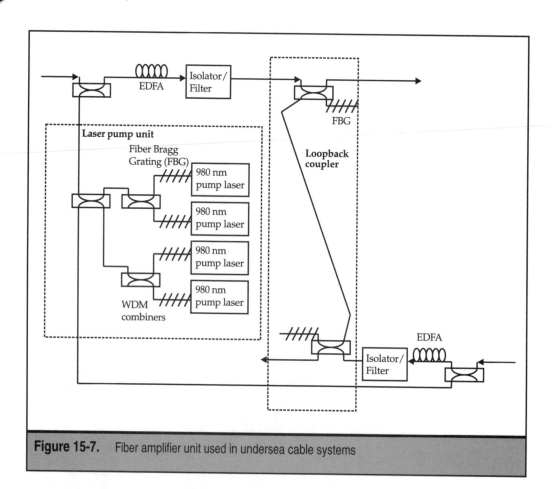

Figure 15-7. Fiber amplifier unit used in undersea cable systems

Laying the Fiber

To prepare the course for the fiber, the route along the ocean floor is surveyed with high-resolution side-scanning sonar. This is done a few months before the actual expedition to lay the cable. The idea is to avoid jagged rocks, shipwrecks, underwater volcanoes, hot water vents, and anything else that might potentially damage the cable.

The actual laying of the cable can begin in one of two ways. One way is to anchor the coastal or in-shore segment to a *beach manhole*. A beach manhole is just a concrete vault buried near the shoreline. Here the buried end of the undersea cable is linked to an underground conduit containing the land-line cable. Once anchored, the rest of the segment is floated out to the ship offshore in a small boat. The second method has the cable-laying ship approach the shore as close as possible to the beach manhole. Careful planning is needed to prevent grounding, and the tides are important. The end of the segment is let out and floated on buoys toward the shore. On shore, a crew of three or so chosen for their strength and clad in wet suits wades out and wrestles the cable end to the landing point.

The cable nearest the shore is often buried by being laid through a hole in a giant steel plow pulled by the ship. The buried and armored cable is protected against fraying from the action of the water on rocks and sand due to surf, tides, storms, and currents. Not only that, but the cable is protected from trawlers' nets and ship anchors. Depending on the part of the world where the cable is deployed, the fishing boat risk can be high or nonexistent. On rocky shorelines that rule out burial, even more heavily armored cable is used.

Some companies deploy undersea fiber cables in pairs in roughly parallel, but widely separated, paths. They use only half of the cables' capacity, so that in the rare event that one fails, the other can pick up the traffic in about 300 milliseconds. This is not as fast as a SONET/SDH ring (about 60 milliseconds), but it is quite fast. Most pairs used in this fashion run east-west, as does most traffic. The north-south cables are usually single cables without any protection.

Undersea fibers are not just for intercontinental traffic. Cities that share a coastline can be linked as well. Run the fiber out to sea, down the coast, and back to land. Brazil has a coastal system, as does Italy. One is planned in West Africa between Nigeria and Senegal. Until this fiber is ready, calls from Lagos, Nigeria, to Dakar, Senegal, will still have to be routed through France.

One reason for coastal systems is to save money on rights-of-way. But in many cases, there are physical reasons not to run the fiber overland. In Brazil, there are few roads than can be used to install the fiber. In Italy, mountains make it difficult to run the fiber everywhere it need be run. However, the cost of repairing a land-line fiber is only a few hundred dollars in many cases. Repairing an undersea fiber can cost millions.

DWDM TESTING

Whether it is used to link PSTN central offices with VoIP routers, or to link continents and island nations with each other, fiber is everywhere. And a lot of the new fibers are designed to handle DWDM wavelengths and channels.

The service providers who install and use this fiber, and the customers that buy or lease dark fiber without any transmitters or receivers, face a much more considerable challenge than the operators and users of light or dark fiber did not long ago. These fibers do not only carry circuit-switched voice but now carry heavy loads of IP packet data, streaming video, and even more exotic payloads. Yet competition and deregulation mean that at the same time operators are expected to overhaul their infrastructures with DWDM fibers, they also have to cut costs. And all of the operators have to provide the proper quality of service (QoS) for the application to satisfy customers and differentiate their services from the rest.

This is what optical networking is all about, of course. The price paid for having a single fiber carry potentially hundreds of wavelengths through totally optical components such as cross-connects is technological complexity. The simplicity of provisioning services by wavelength and adjusting speeds and coding methods without changing the fiber is offset by this complexity of optical networking technology. Precise tuning is needed to make everything work together.

Commercial test and measurement equipment has been struggling to keep up with these needs. Laser wavelengths, once good enough if they were somewhere near 1,310 nm, have to be measured to better than 0.01 nm accuracy. If the concern is the aging of a laser over time, the accuracy needed is 0.001 nm. At the receiver, precise measurements are required to show that the receiver is properly separating the intended signal from the adjacent channels and that any nonlinear channel crosstalk is keeping the power transfer from one channel to another within limits.

In a single-wavelength system, a simple power measurement is all that is ordinarily needed. But the sheer number of parameters that matter in a DWDM system make simple power measurements inadequate. Monitors must operate across a spectrum in order to make sure that the performance of such variables as wavelength, channel power, and signal-to-noise ratio is all correct for each channel in the system. Equipment manufacturers need this information for every component during product development, manufacturing, and system assembly test. Service providers need this information to verify system performance and standards conformance before and after deployment.

A generic optical network is shown in Figure 15-8. Here several access networks feed a collection of DWDM rings making up the network core. Many network elements are not

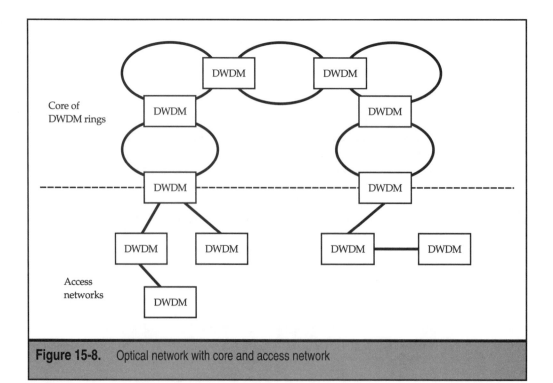

Figure 15-8.   Optical network with core and access network

shown in this figure for the sake of simplicity, and not all service providers or equipment vendors will be equally concerned with core and access network.

The core network is a high-speed IP packet backbone or a backbone handling legacy voice and runs at either 2.5 or 10 Gbps. The access network can also have rings, but running at 155 and 622 Mbps. Signals in the core ring pass from one DWDM ring to another, and on each DWDM access link, SONET/SDH signals can occupy one or more wavelengths.

Each DWDM link that is mapped from access network through the core and then back to an access network has a general structure. Although there are rings everywhere, the basic structure is still the point-to-point link. Source signals at a transmitter operating at the correct wavelength can be sent directly into the DWDM unit. Signals that are at other wavelengths can be fed through transponders to shift their wavelength onto an available DWDM channel. Finally, electrical signals can go through an electrical-optical (E/O) converter unit and emerge at the proper DWDM wavelength. At the receiver, the receiving multiplexer breaks everything down and out again. In the middle, there are rings and cross-connects and optical switches, but the important thing for this discussion is the endpoints. This overall generic structure is shown in Figure 15-9.

However, it would be a mistake to think that every DWDM device marketed and used today could perform all of these conversions and adaptations. The photonics used in these DWDM systems today are of two types. The first type is called an *open* or *stand-alone* system, and the second is called the *integrated* or *terminal* system. In the open system, the device can accept and transport a variety of incoming signals having a wide range of formats and characteristics. For example, an open system should be able to accept not only SONET/SDH feeds, but also Gigabit Ethernet, other IP streams on copper, and so on. The open system will be able to provide all of the signal conversion and wavelength adaptation needed with transponder or remodulation functions. An open system is shown in Figure 15-10. Note the presence and key role of the electrical-optical conversion units.

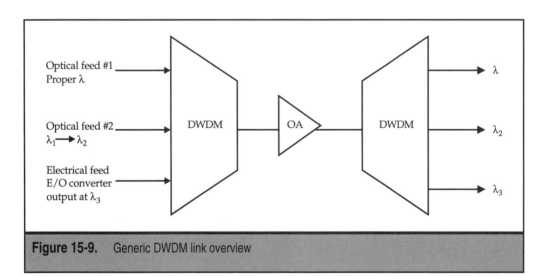

Figure 15-9.   Generic DWDM link overview

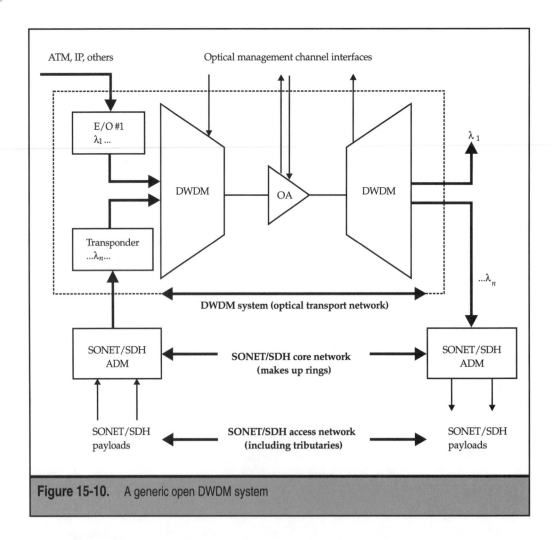

Figure 15-10. A generic open DWDM system

On the other hand, integrated systems are more closely tied to SONET/SDH than open DWDM systems. The SONET/SDH components already have generated a wavelength. A transponder might be needed to shift the wavelength (or not) from 1,310 nm, but no E/O conversion is needed. The integrated system might accept other forms of traffic, but if any conversion from electrical to optical signal is needed, this process is not part of the integrated DWDM system at all. An integrated DWDM system is shown in Figure 15-11. Note the absence of the electrical-optical conversion units.

Any service provider can choose either of these architectures. The decision will be based on economic factors (open systems can be pricey), the installed base of equipment,

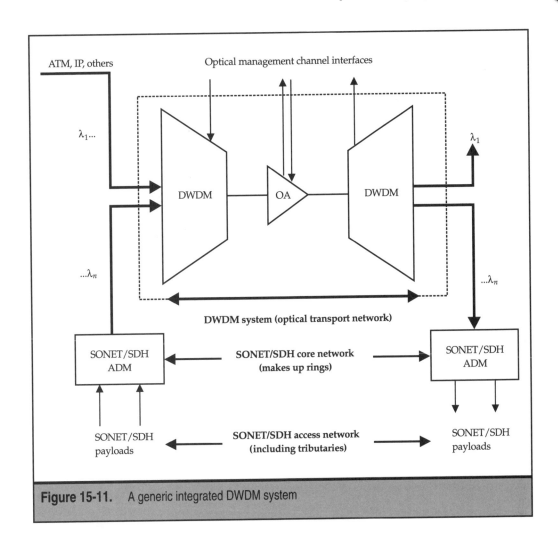

Figure 15-11.   A generic integrated DWDM system

the direction of network evolution (toward more open networking, for example), and services offered. Open systems can use transponders to map SONET/SDH and other signals to the standard ITU-T DWDM grid. Integrated systems can map SONET/SDH and other signals more or less where they like. Open systems are more flexible, since they can transport many different types of legacy and related signals having many different bit rates and a variety of optical power levels. Of course, all of this must be managed and requires additional equipment, which adds expense. Integrated systems can be seen as a long-term solution where customers are supplying the proper photonic signal from an IP router or other type of client equipment.

Testing the Optical Network

The overall reliability of the optical network is a function of the number of components in the system, as has been pointed out earlier in this book. There, the emphasis was on the customer or user perspective. In this section, the emphasis is on the manufacturer of the optical network element, although customers are not ignored.

Optical networks have very high component counts compared to single-wavelength architectures, so the potential for problems is quite high. If any component is not functioning properly, the whole system might be compromised. Performance needs to be measured throughout all stages of development and deployment.

Two general types of testing are of concern when it comes to the complexities of optical networks:

▼ **Performance testing**   This looks at each component and subsystem running at their maximum design limits to fully characterize the performance margins and raw bit error rates.

▲ **Conformance testing**   This is to verify that the behavior of the system conforms to the relevant ANSI, ETSI, ITU-T, or other standards. This is more of a "yes/no" determination than the performance testing.

Both types of test need to be made at all phases of the product and system development and installation cycle:

▼ Design
■ Integration
■ Manufacture
■ Installation
▲ Operation

In the design phase, network modules and subassemblies are subjected to both performance and conformance testing. During the performance test, the design engineers look at the electrical and optical characteristics of the device under many different conditions. The goal is to evaluate the total performance and establish performance margins of the specific component being designed. Conformance testing has less latitude than performance testing, since the module or subassembly must operate within the constraints of the standard specification.

In the integration phase, the same test instruments can be used as in the design phase, but the focus is somewhat different. Here, the manufacturer is trying to integrate multiple modules into a complete network element. The assembled modules must all operate within very narrow margins in terms of power, noise, and so on.

Once the modules and network elements have been designed, the device is ready for manufacturing. The manufacturer must know that each individual module is functioning properly before shipping it, a process known as quality control. In this step, the usual testing is only for standards conformance, not performance. The conformance testing can be done on subassemblies and the entire network elements, even if the device is completely assembled.

Installation should be quick enough so that the customer can generate revenues with the new device as quickly as possible. The customer also is seeking assurance that the system will run reliably and require as little maintenance as possible. Service providers want to make sure that the overall network provides the necessary QoS for competitive markets, but service providers do not want to have to verify QoS in each and every network element. The service providers want to know that the QoS has been designed in and will perform if and when it is needed. This is also where the network can be tested for protection switching mechanisms and to see if troubleshooting procedures are working as planned.

Operational testing can now verify overall system performance. These tests should start with simple bit error rate tests. Although not sufficient in and of themselves, bit error rate tests are always the starting point for overall system performance assessment.

What to Test

Optical networks have a number of basic building blocks. Each basic building block must be tested properly for many different performance parameters. These parameters have been detailed earlier in this book and need not be detailed here.

The five optical networking building blocks that must be tested for conformance and performance are:

- ▼ Transmitters
- ■ Channels
- ■ The optical path
- ■ The optical amplifiers (if any are used)
- ▲ Receivers

Each of these five building blocks is subjected to tests in six different measurement categories:

- ▼ Transmitter output
- ■ Channel interface
- ■ Optical path parameters
- ■ Optical amplifier effectiveness
- ■ Receiver input
- ▲ System test end-to-end

The point where each of these test suites is needed is shown in Figure 15-12.

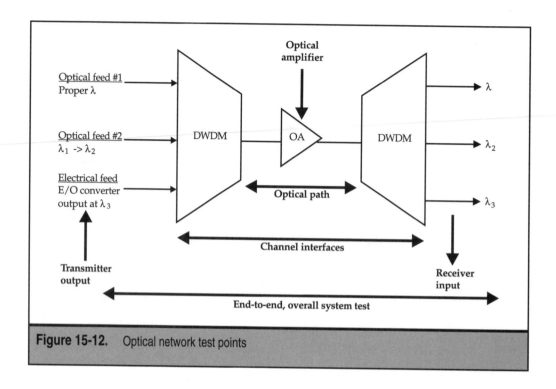

Figure 15-12.   Optical network test points

Each type of interface used in an optical network will require a different test procedure based on the classification of the interface. Timing tests, which are essential for SONET/SDH interfaces carrying virtual tributaries, are not needed at all for Gigabit Ethernet interfaces. Even within a particular interface category, there can be differences in testing for short-haul or long-haul use, for high or low bit rates, and even wavelengths. The requirements for each subcategory determine the expected testing results, naturally. The challenges of system testing for optical networking require careful thought when it comes to test equipment and testing procedures.

CHAPTER 16

Wireless

As this book draws to a close, it might be a good thing to remember that fiber-based optical networking is not the only technology that is growing in leaps and bounds, or almost universally praised, or very cost effective in many situations. There are also many forms of wireless networking. Wireless is about the only alternate technology that might slow the rise of optical networking.

The title of this chapter is simply "wireless." This is because this chapter will explore all forms of wireless communications, from cellular telephones (the major form of wireless "WAN") to wireless LANs to "wireless broadband" systems used in a metropolitan area. Along the way, there will be a section on Bluetooth, the new technology that makes a personal network out of a collection of devices from hand-held games to personal digital assistants (PDAs). There will also be mention of the special protocol for the practice of making special Web pages just for wireless users, the Wireless Application Protocol (WAP).

Obviously, not all of the details about any particular one of these technologies can be presented here, since all of them are subjects of books of their own. But this chapter is intended to be a fairly comprehensive overview of what people are doing and can do with wireless technology today. Optical networking will not be forsaken. The end of this chapter will even explore a type of wireless optical network in the form of the *free space optics,* a kind of "wireless fiber."

The overall conclusion of this chapter will be that it is very wrong to view wireless networking and optical networking simply as alternatives. They can and must work together quite closely to form the networks of the future. Most likely, people will eventually interact with networks through some form of wireless device. The backbones and core will most likely be based on DWDM fiber and optical networking principles. This evolution seems to have already begun today. Many people have used a wireless device. But few people have ever plugged in a fiber connector of any kind. So wireless networking might just be the preferred method for accessing an optical network. When it comes to wireless, it is best to begin at the beginning.

WIRELESS TELEPHONY SYSTEMS

During the late 1970s, people in New York City witnessed something they had never seen before. It was enough to stop traffic and halt people in the streets, even sophisticated New Yorkers. It was then possible to stroll down the sidewalks of New York talking on a telephone. The sight of a wireless telephone was a new one to almost everyone at the time.

There had been radio long before, or course. Police and fire departments had used radios since the 1920s. Many of these systems were a form Specialized Mobile Radio (SMR). But these were "open channel" systems where users had to identify both themselves and the person they wanted to talk to, since the signal was broadcast to all endpoints. Extending the concept of voice telephony to the wireless arena required porting telephone numbers to radio systems, and establishing voice connections for privacy.

There were four huge drawbacks to the early wireless telephony systems. First and foremost, the system was enormously expensive to deploy and use. Second, the equipment was very bulky: the battery alone was so large that it had to be lugged around in a backpack and did not last very long anyway. Third, the capacity of the system was limited, so only a few people could be making calls at any given time. Finally, the voice quality left much to be desired, as signals faded and calls were cut off abruptly.

The first two issues have largely been solved, but the last two problems are still being addressed in many ways. The early wireless telephony systems used a large central transmitter and receiver, so not only did user equipment signals have to be very powerful (adding to the bulk of the equipment), but having only one tower for all the users limited system capacity severely. For instance, the New York system serviced the entire New York metropolitan area, including parts of Long Island, New Jersey, and Connecticut, and could only allow about 50 people at a time to make calls, no matter how many people actually had wireless phones. And wireless voice quality, although getting better with each generation of system and equipment, still all too often left much to be desired compared to land-line voice calls.

The entire history of wireless telephony since 1983 could easily be read as a battle for increased capacity and better voice quality. Fortunately, an advance in one area was also a breakthrough in another. For example, digital personal communications service (PCS) not only made the voice sound better but increased the capacity of the whole system. And advances in battery power and life, antenna sensitivity, component size, and so forth also had an effect on other areas of concern, although not often directly.

It is important to realize that in most cases it is only a *portion* of the entire call path that is carried wirelessly. One end of the call might be a person at home using a wired telephone. Or in between two wireless telephones there might be hundreds of miles of fiber optic cable. Occasionally, the entire call path might be wireless, from phone to microwave to satellite and back, but this is unusual. There is always plenty of wire in any wireless network. This is always good to keep in mind when wireless proponents forecast the eventual use of "wireless for everything." Usually, they just mean "wireless *access* for everything."

Analog Cellular Telephony: The First Generation (1G)

The revolution in wireless telephony that is still continuing to this day began in 1983 with the invention of *cellular* wireless telephony systems. Cellular telephones were introduced in the Chicago area in 1984 by AT&T, then the apex of the monolithic Bell System in the United States. The breakthrough behind cellular involved the use of much smaller coverage areas for the calls, known as *cells*. Idealized for engineering purposes as hexagon-shaped areas, the capacity of each cell was about the same as for an entire central tower system, about 50 calls. But the small cell coverage area meant that users in cars (or any other "mobile phone platform") would often move from one cell to another. So cellular telephony included the use of a call *handoff* system that passed a call in progress from one cell tower to another. This required the coordination of a central Mobile Telephone Switching Office, or MTSO. A single MTSO, usually located in a LEC's central office (CO), could orchestrate handoffs among many cell towers.

The revolution started by the Advanced Mobile Phone System (AMPS) in the United States is shown in Figure 16-1.

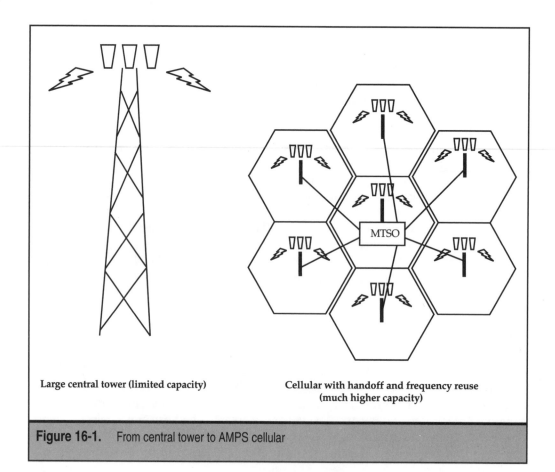

Large central tower (limited capacity)   Cellular with handoff and frequency reuse
                                                        (much higher capacity)

Figure 16-1.   From central tower to AMPS cellular

Calls on the AMPS network were still entirely analog in nature, which left AMPS calls susceptible to the same problems of noise and interference as all other forms of analog communications. Use of analog technology characterized this first generation (1G) of wireless telephony system. The continued use of analog signals limited the application of advanced microelectronics to the systems, and end-user equipment remained bulky and relatively high powered (limiting battery life). But by the mid-1980s, the appearance of people at meetings (especially at AT&T) with briefcase-sized mobile phone battery packs and with handsets the size and shape of landline phones was no longer an object of wonder or even a worthwhile topic of conversation.

In spite of the considerable expense of using the early cellular systems, AT&T foresaw a big future for the new AMPS phones. They confidently predicted hundreds of thousands of mobile phone users by 1990. In fact, the number of users worldwide in 1990 was more like 20 million, growing 30–50 percent per year throughout the 1980s. How could they be so wrong? What AT&T and other early wireless service providers were really saying was "we can't believe that people will pay $7 per minute for the worst voice quality they have ever heard." What AT&T and the other phone companies took a while to find out, and users realized at once, was that cellular service providers were not selling voice, they were selling *convenience.*

AMPS gave ordinary people the ability to make or take calls in their cars (the large size and weight of the units, mostly battery weight, made carrying them around everywhere quite a chore). This ability outweighed all other limitations in terms of price (usage charges remained high well into the 1990s), quality (calls were easily lost when going under bridges or during handoffs), and capacity (there were a lot of "system busy, try your call again later" messages). How else could one call for directions when hopelessly lost, or to say one would be late for an important meeting, or just to find someone during the course of the day? People liked the idea of car phones, wanted car phones, and bought car phones in increasing numbers. Sociologists sometimes pointed out that this meant that people were spending entirely too much time in their cars, but many people liked their cars a lot as well.

The service providers promoted the new technology whenever and however they could. Many people balked at the early cell phone price of $700 or more. So the service providers would rebate a portion of their monthly service charges to the manufacturer of the cell phone until the full cost of the phone was covered. This dropped cell phone prices dramatically, encouraged hesitant users, and helped popularize the service.

The analog 1G systems had many variations. There were AMPS in the United States, Nordic Mobile Telephony (NMT, or two types) in the Scandinavian countries and elsewhere, and Total Access Communications System (TACS) around the world. An AMPS variation known as Narrowband AMPS (NAMPS) also enjoyed popularity. The C450 system was used in South Africa and some other parts of the world.

All of these 1G systems were analog. They differed in terms of the frequency ranges used for sending and receiving, their basic channel bandwidth needed for a voice call (called the *channel spacing*), the number of channels available, and the basic allocation of *spectrum*, or overall bandwidth (usually 50 MHz), given to a service provider. The spectrum was shared through the use of *frequency division multiple access (FDMA),* where each channel to and from the cell tower is given a different channel to operate on. As usual with analog systems, all of them used frequency modulation (FM) to represent voice. These differences are summarized in Table 16-1.

It is tempting to read the table entry for the base capacity, or number of channels, as the capacity of each cell site. But this is rather the capacity of the *entire network's* basic frequency allocation. To increase capacity beyond this number of simultaneous users, cellular systems used a *frequency reuse pattern.* This allowed the total number of users to be in the many thousands, since users on the same frequency channels would be many cells away from each other. Seven- and twelve-cell reuse patterns were common. In a twelve-cell reuse pattern for AMPS, for example, each cell site had a total capacity of about 50 simultaneous calls out of the total base capacity of 832 (some channels were used for control). This is not really all that many, and when a sports event ended late, the departing sports fans easily overwhelmed the local cell site. The only solution was to make the cell sites smaller in areas of very dense use.

Feature	AMPS	NAMPS	TACS	NMT450	NMT900	C450
Transmit MHz	869–894	869–894	935–960	463–468	935–960	461–466
Receive MHz	824–849	824–849	890–915	453–485	890–915	451–456
Channel spacing	30 kHz	10 kHz	25 kHz	25 kHz	12.5 kHz	20/10 kHz
Base capacity	832	2496	1000	200	1999	222/444
Basic allocation	50 MHz	50 MHz	50 MHz	10 MHz	50 MHz	10 MHz
Access method	FDMA	FDMA	FDMA	FDMA	FDMA	FDMA
Modulation	FM	FM	FM	FM	FM	FM

Table 16-1.   Important 1G Analog Cellular Systems

In the United States, deregulation concerns led to the practice of splitting up the frequencies into *A block* and *B block* frequency allocations. This ensured that there would be at least a choice of service providers in each area for competitive purposes.

Today, analog cellular networks occupy the same position in the history of cellular telephony as steam engines do in the history of rail transportation. Steam engines are still admired by some, but passengers hated the soot and noise, and railroads hated having to stop every 50 miles for water and 100 miles for coal. Once diesels came, steam engines were gone in about five years.

Digital Cellular Telephony: The Second Generation (2G)

The overriding concerns of capacity and quality, and to a lesser degree concerns about power and bulk, were addressed with the introduction of *digital* cellular telephony. These systems were sufficiently different from their analog cousins to be grouped together as the second generation (2G) of cellular telephony.

The shift to digital enabled the industry to take full advantage of the breakthroughs in semiconductor technology and lower-powered microelectronics. This last aspect is especially important, as concerns over the long-term exposure to microwave radiation grew with the popularity of cell phones.

The fact is that cell phones operate in the same frequency range as microwave ovens. The simple act of holding a microwave generator against a head for long periods of time came to be seen by some experts as an overall health risk. Perhaps over a long enough period of time, cell phones would "cook" a person's brain. But for every study that suggested brain tumors among heavy cell phone users, there was another study showing that the risks were either exaggerated or acceptable. After all, no one would consider walking everywhere because there were motor vehicle accidents. Life is never absolutely risk free, and even walkers stumble and fall.

However, evidence is slowly building over time that shows that there is an increased chance of health problems with high-powered cell phones. These effects can be reduced by several means, from special filters to choosing low-powered cell phones. Even simply holding the cell phone at a certain angle helps reduced the microwave radiation exposure. The bottom line is that those who are concerned about the health risks can take steps to minimize the risks without giving up cell phones altogether.

There was also a desire among the service providers to offer more than simple voice communications. Service providers wanted to add paging, short text messaging, and so on. There was also a concern over the number of fraudulent phones calls that could be made using the older analog phones, so various supplemental services were added to the new digital phones, such as user authentication and encryption.

The new digital phones and systems were completely incompatible with the older analog phones and systems, of course. Not only that, but there were many variations on the digital theme, even within a country or metropolitan area. To deal with these issues of incompatibility, many of the digital phones could operate in *dual mode*, using digital technology when they were able and reverting to the older analog form when faced with an incompatible digital system or an older analog network. The need for dual-mode operation kept the cost of these digital phones higher than the purely analog phones, which continued to exist.

Digital cellular system worldwide formed a very diverse group. The analog systems generally differed only in channel size and therefore numbers, but the digital system differed in many more ways. Analog systems invariably used FDMA to share spectrum among users, assigning two different mobile platform–to–cell tower and cell tower–to–mobile platform frequency channels for a call. Digital systems also used FMDA in some cases, but they were not limited by FDMA. Some digital systems used Time Division Multiple Access (TDMA), where the same frequencies were assigned to different calls but different time slots were allocated for the bits forming each call. Some digital systems even used TDMA and FDMA in combination. There was also a method called Code Division Multiple Access (CDMA), the details of which are beyond the scope of this chapter. CDMA could be used with FDMA to increase the capacity of a cellular phone network.

While FM modulation was the rule in analog system, digital systems use a variety of modulation techniques to send the 0's and 1's over the airwaves. There are methods called Pi/4 differential phase-shift keying (Pi/4DPSK), quadrature phase-shift keying (QPSK), and Global Mobile Shift Keying (GMSK), as well as a pure digital technique called 16-level quadrature amplitude modulation (16QAM). The coder/decoder (codec) methods used to convert analog voice to 0's and 1's and back included algebraic and vector code–excited linear predictive (ACELP/VCELP) coding, the original CELP, and a technique known as residual exciting linear predictive coding (RELP). Some systems allow multiple methods.

The major cellular digital systems used around the world are listed in Table 16-2.

In the table, IS-136 is listed twice. This might be confusing, but notice that the transmit and receive frequency ranges are different. All wireless regulatory bodies consider services that use different frequency ranges to be fundamentally different services, even

Feature	IS-136	IS-136	IS-95	GSM	iDEN
Transmit MHz	869–894	851–866	869–894	925–960	851–866
Receive MHz	824–849	806–821	869–894	880–915	806–821
Channel spacing	30 kHz	30 kHz	1.25 MHz	200 kHz	25 kHz
Base capacity	832	600	9/10	124	600
Users/channel	3	3	64	8	3/6
Basic allocation	50 MHz	30 MHz	50 MHz	50 MHz	30 MHz
Access method	TDMA/FDMA	TDMA	CDMA/FDMA	TDMA/FDMA	TDMA
Modulation	Pi/4DPSK	Pi/4DPSK	QPSK	GMSK	16QAM
Voice codec	ACELP/VCELP	ACELP	CELP	RELP	--

Table 16-2.   Major Global Digital Cellular Systems

though the services that the customers get might be exactly the same and the service is covered by the same standards document.

In the United States especially, there has been a lot of interest and activity in the personal communications service (PCS). Again, there is a wide variety in terms of access method, modulation, and codec. Many of the same services available on digital cellular systems are also available on PCS systems, and sometimes even more. Major PCS systems are shown in Table 16-3. The only new term is *time division duplexing (TDD)*, which is just a variation on TDMA.

As can be seen, there are a lot of similarities between digital cellular and PCS systems. In fact, the biggest difference is in the transmit and receive frequency ranges for the system. Since most regulatory agencies have always considered services that operate in widely separated frequency ranges to be different, PCS is *not* the same as digital cellular, in spite of their similarities.

None of this should be interpreted as implying that all digital cellular technologies are created equal. There are two that have been overwhelmingly popular in the United States and around the world. IS-136 was developed in the United States, where it is also known as IS-54 or NADC. IS-136 uses TDMA to allow users to share a channel through the use of time slots. IS-136 grew out of IS-54 and has enough additional features to please both users and service providers. Systems based on IS-136 have three times the capacity of analog systems, and much less noise. There is a built-in authentication system and privacy through encryption. A nice feature of IS-136 is that it can be deployed one channel at

Feature	IS-136	IS-95	DCS1800	DCS1900	IS-661
Transmit MHz	1930–1990	1930–1990	1805–1880	1930–1990	1930–1990
Receive MHz	1850–1910	1850–1910	1710–1785	1850–1910	1850–1910
Channel spacing	30 kHz	1.25 MHz	200 kHz	200 kHz	5 MHz
Base capacity	166/332/498	4-12	325	25/50/75	2-6
Users/channel	3	64	8	8	64
Basic allocation	10/20/30 MHz	10/20/30 MHz	150 MHz	10/20/30 MHz	10/20/30 MHz
Access method	TDMA/ FDMA	CDMA/ TDMA	TDMA/ FDMA	TDMA/ FDMA	TDD
Modulation	Pi/4DPSK	QPSK	GMSK	GMSK	QPSK
Voice codec	ACELP/ VCELP	CELP	RELP	RELP	CELP

Table 16-3.   Major PCS Systems

a time, allowing the rest of the cell site to operate analog or with some other technology. And since IS-136 exists as both digital cellular and PCS band services, IS-136 is attractive for service providers that have licenses for both spectrums.

Outside the United States, the Global System for Mobile Communications (GSM) is very popular, especially in Europe. This technology features increased capacity over analog systems, of course, but also international roaming, very high speech quality, better security, and the usual additional features such as text messaging. Some 80 countries had input to the GSM standards process, ensuring integrated cellular system operation across borders and different cultures. GSM is the most widely deployed cellular technology in the world today.

DCS1900 and DCS1800 are also GSM systems. The components are the same, but the frequency ranges are in the PCS bands.

The basic architecture of a GSM system is shown in Figure 16-2. The three major pieces are the Base System Station (BSS), the Switching System (SS), and the Operations and Support System (OSS). A series of Base Transceiver Stations (BTSs) and their Base Station Controllers (BSCs) form the Base Station System (BSS). The ABIS interface connects the BTSs to their BSCs. Several BSCs connect to a central Mobile Switching Center (MSC) through an A link interface. The MSC handles connections between cellular phones and to and from the PSTN.

The BSC(s) can be located in the SS along with the MSC for smaller systems. However, the SS must contain, along with the MSC, the Home Location Register (HLR) for customers

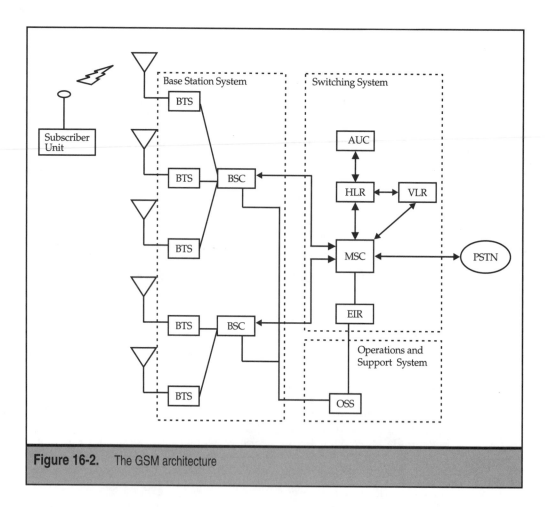

Figure 16-2. The GSM architecture

that are not roaming, and the Visitor Location Register (VLR) for those that are. There also must be the Authentication Center (AUC) to make sure the user is a valid one, and the Equipment Identity Register (EIR) to find the active phones in the first place. Finally, the OSS coordinates everything and makes changes to the overall system configuration.

This section will only briefly mention wireless local loop (WLL) technology. The WLL technology is essentially the same as analog cellular, digital cellular, or PCS, but without the mobility features. There are usually the same types of towers and phones used in cellular systems. In fact, even other technologies can be, and sometimes are, used. WLL is just a way to connect telephone users to the local exchange switch without the need to use copper wires. There are proprietary systems not based on any standard as well as relatively "open" systems using licensed or unlicensed frequencies. Sometimes the home or office is wired but the link is not, and sometimes there are no wires at all between phone and local tower (the link from tower to switch could be wireless or wired).

WLL does not have a specific frequency range to operate in. Local variations and rules will prevail. The details of the WLL system depend on the distance, density, and number of anticipated users. Volume, usage patterns, and bandwidth desired for each user will also play a role in the exact choice of WLL technology.

WLL is especially popular in countries that the United Nations characterizes as having an "emerging" telecommunications infrastructure. It is much more cost effect for these areas to deploy WLL than to run copper wire pairs, or fiber, to every location with a phone or phones. The details of the technology chosen for WLL depend on the services and applications to be supported.

Finally, cordless phones fall into a category all their own. Short range and low power characterize these wireless handsets, which typically just connect to wired base units. Unlicensed frequencies are used. There have even been cordless *public* pay phones, although users tended to wander away from the base unit and not bother to return the handset to its proper place.

Data over Wireless

Both 1G analog and 2G digital systems, cellular and PCS alike, are characterized by an annoying feature: they are very restrictive when it comes to data services. This emphasized the enormous bandwidths available on optical networks, but wireless networks have been sadly lacking in any kind of strides in increasing data bandwidths.

As the quest for voice capacity goes on in the cellular world, the bandwidth needed to send a voice call over the airwaves has actually fallen. Optimized digital voice requires as little as 4 Kbps for decent quality. But data is not voice. Most analog systems could provide about 9,600 bps over a voice channel when used for data purposes, and even GSM does the same. Even with compression and other advanced data techniques, digital systems top out at about 13,000 bps.

CDMA, which employs what is known as *spread spectrum* technology, seems to offer a way out of this bandwidth bind. CDMA channels are very wide, as the term "spread spectrum" implies, but data users would not be able to monopolize all of the bandwidth all of the time at the expense of voice users. One part of the issue is that all of these cellular networks have been chartered and approved by governments as *voice* services. If the data service can be squeezed in, fine. If not, the voice must go through at the expense of the bandwidth available for data.

There have been attempts either to add data services on to a cellular system or even to deploy a data-only wireless network. These attempts have been modestly successful, especially in certain areas of the United States. But a scheme to add "TCP/IP over cellular" with a method called Cellular Digital Packetized Data (CDPD) stalled because it was tied firmly to the older analog cellular systems. Other pure data wireless services suffered from coverage woes, high start-up costs, expensive services, and poor quality (many resends).

GSM digital systems also have an enhancement for data services. This is called the General Packet Radio Service (GPRS). GPRS is a very sophisticated service that handles packet data at a much higher speed than ordinary voice channels attain. There are also various IP-based services available with GPRS. But the big attraction is the considerable (for wireless) 115 Kbps top speed that GPRS can run at. GPRS does require extensive and expensive upgrades to a GSM system. Many service providers might wait until more advanced *wireless broadband* systems (as opposed to cellular *narrowband* systems) are more mature to deploy for advanced IP services such as e-commerce.

Nevertheless, there are colleges and airports where a person with a properly equipped laptop can simply sit down and power up and connect to the Internet at modest speeds. But as with most cellular networks, analog and digital alike, only the access portion of the link is really wireless. However, this is the portion that counts to users.

Wireless Application Protocol (WAP)

The limitation for many forms of wireless Internet access, especially through cellular phones, has led to the development of the Wireless Application Protocol (WAP). WAP is used by many wireless service providers today to strip off much of the motion graphics and other forms of bandwidth-eating elements on a Web page and reduce the page to its information content. Although mainly aimed at cellular/GSM/PCS users, WAP is also used in personal digital assistant (PDA) devices as well.

WAP enables mobile users to access the Internet and Web with limitations. For example, users can send and receive e-mail and view the text content of Web pages. In order for a user to utilize WAP, the service provider must implement WAP in their system, and the user must have a WAP-enabled device. WAP can be used with GSM, IS-95 based on CDMA, IS-136 based on TDMA, and newer third-generation (3G) systems discussed later.

While WAP certainly allows a kind of mobile and wireless access to the Internet, the use of WAP points out a glaring limitation in almost all forms of wireless Internet access. There is simply not enough bandwidth to enjoy the Internet and Web experience to its fullest.

Mobile IP

A related development for using a cellular network for data was the concept of Mobile IP. As expected, this idea focuses on Layer 3 of the OSI-RM, the Network Layer, working with the current version of the Internet Protocol (IP version 4). In the Mobile IP protocol, the IP address of the mobile machine does not change when it moves or roams from a home network to a foreign network. In order to maintain connections between the mobile node and the rest of the network (the Internet), a forwarding routine is implemented in the components of a Mobile IP network.

When a person moves around the physical world, that person often lets his or her home post office know to which remote post office mail should be forwarded. Arriving at the new location, the person usually registers his or her address with the new post office. This same type of thing happens in Mobile IP. When the *mobile agent* moves from its home IP network to a foreign (or *visited*) network, the mobile agent tells a special Mobile IP *home agent* on the home network to which *foreign agent* its IP packets should be forwarded. Moreover, the mobile agent registers itself with the foreign agent on the foreign network. So all packets intended for the mobile agent that arrive at the home network are forwarded by the home agent to the foreign agent, which sends them to the mobile agent on the foreign network. When the mobile agent returns to its home (original) network, the mobile agent tells both agents (home and foreign) that the original configuration has been restored. No one outside the local IP network (i.e., on the Internet at large) needs to know that the mobile agent has moved at all.

The biggest advantage of Mobile IP was and is that every routing table in every router on the Internet does not have to track down the moving user. Users keep their IP addresses everywhere, which is a big plus.

All in all, the best that can be said for Mobile IP is that it works, but this configuration has some drawbacks. Depending on how far the mobile agent moves and how long it takes to get there, there might need to be some store and forwarding of packets while the mobile agent is on neither the home nor the foreign network. Also, if a client PC from the United States is in Japan, and accessing a Japanese Web site, all the packets must journey to the United States, only to be turned around and sent straight back to Japan. Naturally, the return path follows the same tortured path. In addition, Mobile IP works only for IPv4 and does not take advantage of the features of the newer IPv6.

UMTS AND 3G

The twin goals of increased capacity and adequate data bandwidth might actually be achievable in third-generation (3G) cellular networks. The planned successor to GSM is sometimes called the Universal Mobile Telecommunications System (UMTS). UMTS recognizes the need for Internet access and related data services as well as the need for voice services. The plan calls for UMTS to allow for up to 2 Mbps per mobile user, and it includes a global roaming standard.

UMTS is a relatively new term, and the underlying technology, called Wideband CDMA (W-CDMA) is more familiar in the industry. UMTS is meant to unify various 3G proposals, allowing interoperability. UMTS is the cellular portion of a group of 3G wireless technologies known as IMT-2000, or International Mobile Communications 2000. The array of terms is unfortunate, but characteristic of the early phases of any technology and standard.

IMT-2000 systems are intended to be one set of standards for all countries and all services and all applications. IMT-2000 systems will feature:

▼ Worldwide usability (no more carrying several different phones internationally)

■ Support for all applications (voice/data/messaging all in one)

■ Support for circuit-switching (usually for voice) and packet-switching (usually for data)

▲ Data rates up to 2 Mbps (depending on user mobility and velocity)

The "2000" in IMT-2000 officially stands for the general frequency range of operation. IMT-2000 will operate in the 1,885–2,025 MHz and 2,110–2,200 MHz ranges. There will be IMT-2000 standards for both terrestrial wireless and satellite systems.

Data speeds will grow from the 9,600 bps speeds in early GSM systems to 1.920 Mbps with UMTS. GSM systems can build up to this speed, with stops at 57.6 Kbps, 115.2 Kbps, 171.2 Kbps, and 553.6 Kbps. UMTS Phase 1 will add support for wireless laptops onto an existing GSM cellular network. This add-on is called the UMTS Terrestrial Radio Access Network, or UTRAN. The User Equipment (UE) can be a laptop, Palmtop, or other package. Traffic makes its way through a hub to the GSM Switching System. Circuit traffic is handed off to the PSTN as before, while packets make their way through a router onto the Internet (or other packet network[s]). The same locator and authentication steps apply to the data devices as well.

The basic UMTS Phase 1 network is shown in Figure 16-3.

The addition of speeds near 2 Mbps for IMT-2000 will be a welcome relief for users of wireless data services. But these speeds apply to roaming users, often traveling at high speeds in cars, over a WAN or MAN. If users can be convinced to slow down and confine themselves to a smaller area, much higher wireless speeds are possible.

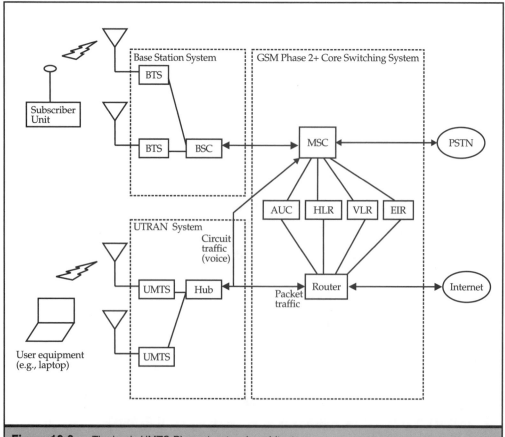

Figure 16-3.   The basic UMTS Phase 1 network architecture

THE WIRELESS LAN

The market for all forms of wireless communications has enjoyed tremendous growth in recent years. In fact, some form of wireless technology, if only a simple pager or beeper, can now reach virtually every location on the face of the earth. Given the tremendous success of wireless telephony and messaging services, it is hardly surprising that wireless communication is now being applied to personal and business computing. Why should a PC have to be linked (chained) to a network by a piece of wire?

This section will explore some of the basics of the terminology and architecture of wireless local area networks and present a brief description of some protocols that have been developed, with emphasis on IEEE 802.11.

The basic technology has been in place for local area networks (LANs) since the success of the Ethernet project at Xerox's Palo Alto Research Center (PARC) in the 1970s and 1980s. Standard LAN protocols such as Ethernet operate at very high speeds and are built from inexpensive connection hardware that can bring networking to almost any computer. Until recently, however, LANs were limited to the physical, hard-wired infrastructure of a building or small group of buildings. Even remote users with phone dial-ups were limited to access through wired, land-line connections. Many network users, especially mobile users in businesses, the medical profession, factories, and universities, to name only a few, could benefit from the added capabilities of wireless LANs.

There is one major motivation and benefit from wireless LANs: increased mobility. Freed from the need for conventional wired network connections, network users can move about almost without restriction and access LANs from nearly anywhere. So medical professionals can not only obtain patient records but have access to a patient's real-time vital signs and other reference data while gathered at the bedside without relying on access to paper charts and requiring other physical paper handling. Factory floor workers can access part and process specifications without wired network connections, which are also impractical or impossible in some factory situations. Wireless connections with real-time sensors allow a remote engineer to monitor and maintain manufacturing equipment, even on the computer-hostile factory floor. Wireless scanners connected to the main warehouse inventory database can be used to carry out and verify inventories quickly and effectively. Eventually, wireless "smart" price tags, complete with liquid crystal displays (LCDs), will allow retail merchants to eliminate discrepancies between stock-point pricing and scanned prices at the checkout lane (a real problem in any store). The list of possibilities is almost endless.

Of course, in addition to mobility, wireless LANs offer increased flexibility. There could be meetings in which employees use small laptop computers and wireless links to share and discuss future products or design plans. This type of "ad hoc" network should be able to be brought up and torn down as needed in a very short time, and either around the conference table or with participants from around the world as circumstances require. Most car rental companies already use a form of wireless networking to help with car returns. Traders on Wall Street are now using wireless terminals to make market trades. Students on university campuses now have access to lecture notes and other course materials while wandering about the campus.

Finally, it is sometimes more economical to use a wireless LAN than anything else. For example, old buildings often require an expensive asbestos cleanup or removal before any new cable can be run. This far outweighs the cost of installing a wireless LAN. In other situations, such as a factory floor, it might not be possible to run a traditional wired LAN, even on fiber. Wireless LANs offer the connectivity of the wired LANs without the need for expensive wiring or rewiring.

The term "wireless LAN" can be defined as a technology that allows two or more computers to communicate using standard network protocols, such as IP, without the need for LAN cabling in between. Of course, wireless WANs could also fit this definition, but this discussion is limited to wireless LANs. Lately, wireless technology has been driven by the emergence of the new IEEE 802.11b standard. This replaces the older IEEE 802.11 (now 802.11a) specification, which was expensive, much slower than an Ethernet LAN, and not very popular. The new 802.11b is truly a "wireless Ethernet." IEEE 802.11b has already produced a number of inexpensive wireless solutions that are growing in popularity with businesses, schools, and even home users.

There are two kinds of wireless LANS: *ad hoc* and *access point*.

An *ad hoc* (or *peer-to-peer*) wireless LAN consists of a number of computers, each equipped with a wireless LAN networking interface card (NIC). Each computer can communicate directly with all of the other wireless-enabled computers. These computers can share files, applications, and printers among themselves this way, but they might not be able to access wired LAN resources unless one of the computers acts as a LAN *bridge* to the wired LAN using special software. This bridging concept is common to all forms of LAN, of course. The concept of an ad hoc wireless LAN is shown in Figure 16-4.

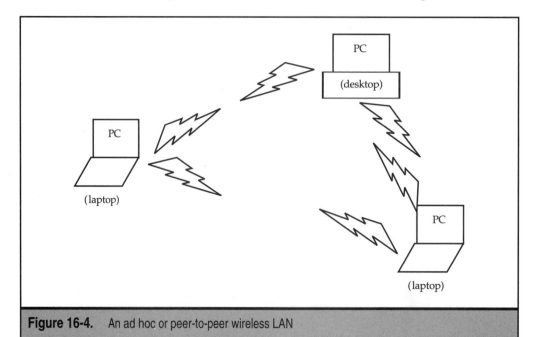

Figure 16-4.   An ad hoc or peer-to-peer wireless LAN

Each computer in the figure with a wireless interface (NIC) can communicate directly with all of the others. Ad hoc wireless LANs can be set up almost instantly and are a natural choice for meetings and other gatherings of PCs. But once the number of PCs grows beyond a certain point, this mesh of direct traffic becomes inefficient.

For more permanent wireless LANs, another type of wireless LAN can be set up. The second type of wireless LAN uses the concept of an *access point,* or *base station.* In this type of wireless LAN, the access point acts as a wired LAN hub. The hub provides the connectivity among the wireless computers. The hub can also connect through bridging to a wired LAN, allowing the wireless and wired portions of the LAN to communicate with each other to share resources and Internet connectivity without regard to which portion of the LAN they are on.

There are two types of access points defined in IEEE 802.11b: dedicated hardware access points (sometimes seen as HAPs) and software access points (sometimes SAPs). Generally, hardware access points offer more features and better overall performance, but this can vary from product to product. Software access points use special software that runs on a computer equipped with both a wireless NIC as in an ad hoc wireless LAN and the usual wired NIC (most likely Ethernet). The contrast between a hardware access point and a software access point is shown in Figure 16-5.

Many software access points are also software routers (not bridges) and so can include features that are not often found in the hardware access point configurations. For example, direct PPPoE support is common, as well as more flexible configuration options.

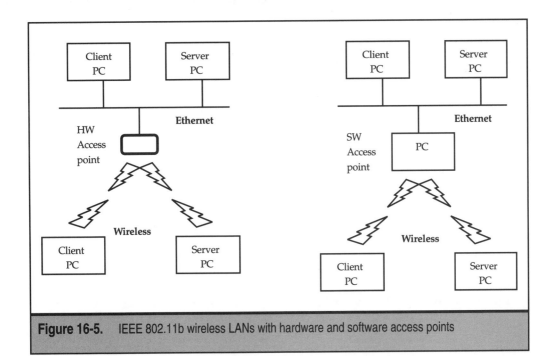

Figure 16-5. IEEE 802.11b wireless LANs with hardware and software access points

However, the software access points might not offer the full range of features defined by the IEEE 802.11b standard.

In theory, any hardware or software from any vendor that adheres to the IEEE 802.11b standard can interoperate. However, the standard itself is relatively new, so verification testing is still a requirement. This is especially true because the original IEEE 802.11 standard specified two different methods for wireless communications. These made up the *air interface* of the wireless devices. IEEE 802.11 wireless LANs could use either *frequency hopping (FH)* or *direct sequence spread spectrum (DSSS)* as an air interface. These are distinct methods and cannot interoperate. The newer IEEE 802.11b uses DSSS exclusively (with minor variations allowed).

Care is also necessary when it comes to operational speeds. The original IEEE 802.11 specification ran at 1 and 2 Mbps, much slower than the basic 10 Mbps Ethernet speed. IEEE 802.11b runs at either 5.5 Mbps or 11 Mbps (full Ethernet plus some overhead). Many vendors, in the interests of backward compatibility, support all speeds. But some vendors only support the higher speeds, and of course older equipment cannot run faster than 2 Mbps at all. And FH-type IEEE 802.11 equipment is essentially obsolete. There are also vendors who insist on "extending" a standard like IEEE 802.11b with their own unique features. Needless to say, this practice also complicates interoperability and might even prevent any form of meaningful interoperability among different vendors' IEEE 802.11b products.

Software access points might be more flexible than hardware access points in this regard. All that might be needed is to change the wireless NIC interface to upgrade an older wireless LAN to the new technology. Hardware access points usually must be replaced. For smaller LANs, software access points are common. Larger LANs usually use the hardware access point, but hardware solutions are becoming inexpensive enough to even be used in home network environments.

Also, a software access point does not limit the type or number of network interfaces. Hardware access points tend to be Ethernet, period. So if there happens to be a token ring LAN around, the software access point can be a good solution. In fact, a software access point can be used to provide connectivity between older, slower IEEE 802.11 networks and the newer IEEE 802.11b portion of the wireless LAN. A software access point is not limited to only two interfaces, although at some point the access point might become a real traffic bottleneck.

When software access points provide additional features such as shared Internet access, Web page caching, packet filtering, and so on, throughput is a concern. However, since the software access point is essentially just a PC, upgrading can be relatively painless. Software access points can be a real benefit to LAN administrators, but they can be a concern in some configurations.

When it comes to range, each access point has a finite range beyond which a client computer cannot communicate with the access point. The actual distance will vary depending on the environment. Vendors are fond of stating the maximum feasible distance, of course, and usually provide both indoor (a better indicator of real-world performance) and outdoor (a worse indicator) ranges to give a reasonable indication of performance. It is important to note that at the distance limits, connectivity might be maintained, but performance will drop, often dramatically, as the connection quality drops and the network tries to compensate, usually by slowing the air link down.

Typical indoor ranges are between 150–300 feet (about 75 to 100 meters) but can be much shorter in some older concrete and steel buildings. Support for longer distance is possible, but only at greatly reduced throughput rates. Outdoor ranges are sometimes quoted at 1,000 feet (more than 300 meters), but this depends on the environment as always. IEEE 802.11b LANs operate in the 2.4 GHz ISM (industrial, scientific, and medical) frequency range, so no license is required.

This basic operational range assumes direct client–to–access point communications. There are ways to extend the range of a wireless LAN by using more than one access point or using a wireless *relay point* (also called an *extension point*). And if a single area covered by the LAN is too large, multiple access points or extension points can be used. Actually, "extension points" are not defined in the IEEE 802.11b standard but have been developed by some vendors to aid in the construction of wireless LANs. This means that interoperability is a real concern.

A wired LAN can connect two or more access points. But when multiple access points are used, each wireless access point's coverage area should overlap its neighbors. This provides coverage without "dead spots." The idea of multiple and extension access points is shown in Figure 16-6.

Multiple extension points can be strung together to extend access to locations quite remote form the central access point(s).

The recommended capacity of any single access point is often limited to about 10, but vendor recommendations can vary. More expensive hardware access points can support up to 100 wireless users. More computers than recommended can be used, however, although this practice can cause reliability and performance to suffer. Even software access points have limits in terms of supported users.

It is possible for a wireless client computer to "roam" from one access point coverage area to another. The network should be able to maintain the connection by monitoring the strength of the signal as it fades from one access point and builds up in another access point coverage area. This is exactly the same method used by cellular phone networks to decide when a handoff is necessary. The new access point with the best signal quality is chosen. This process should be totally transparent to the user. However, some access points, especially software access points, require security authentication when a new user appears, and so the users might suddenly be confronted by an unexpected password dialog box. Not all access points will support roaming, especially less expensive models. Roaming is not an official feature of the IEEE 802.11 standard.

A wireless LAN can even be used to connect two or more wired LANs. This makes sense in a campus environment, where buildings are close by but permission to use facility or public thoroughfares between buildings is not attractive or possible. This type of configuration requires at least two access points, of course. Each of the access points acts

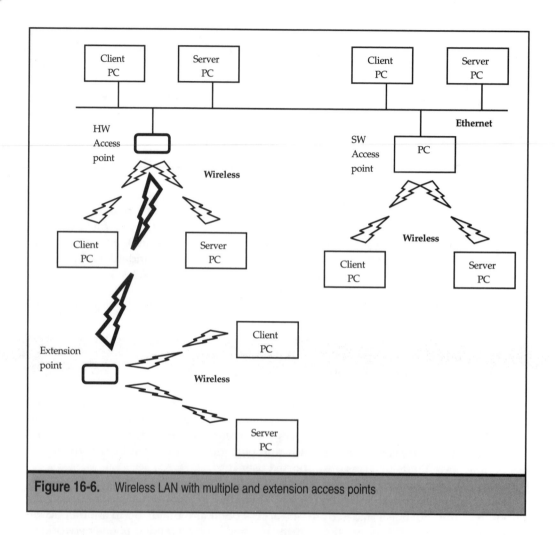

Figure 16-6.   Wireless LAN with multiple and extension access points

as a bridge (or router, depending on configuration and capability) connecting its own wired LAN to the wireless backbone. The wireless connection lets the access points communicate with each other, connecting the attached LANs. This is shown in Figure 16-7.

In the figure, both software and hardware access points are shown, but either type can be used on both ends of the wireless link. It should be noted that not all hardware access points have the capability to directly interconnect with another hardware access point. Many hardware access points expect communications with client PCs only. Again, it all depends on vendor capabilities.

Although wireless LANs are a natural for highly portable laptops, desktop PCs can use wireless LANs as well. It is estimated that the average office worker in the United States changes work location every two years. Put another way, 50 percent of the work force relocates

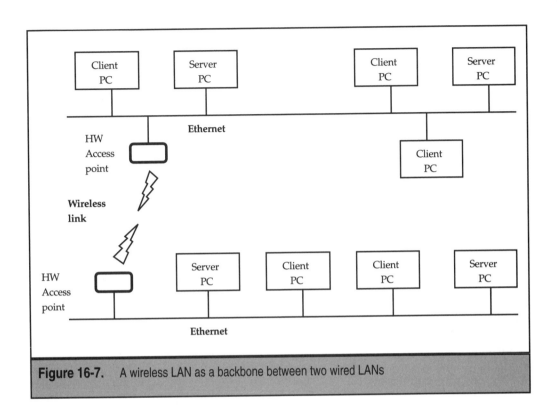

Figure 16-7. A wireless LAN as a backbone between two wired LANs

every year. With a wired LAN, these "adds, moves, and changes" can become an administrative nightmare as office kitchens become a maze of cubicles seemingly overnight. But this is not a problem for wireless LANs. Just pack up the PC and move it. And of course there are many advantages when using wireless LANs (and wireless networks in general) for temporary networks such as LANs built for conferences and exhibits.

Security is always a concern in wireless networking, and wireless LANs are no exception. An intruder does not even need physical access to the wired LAN in order to gain access to privileged information. But IEEE 802.11 information cannot even be received, much less decoded, by simple sniffers and scanners. It is not true, however, that wireless LANs are not at risk at all. Special equipment can be used to intercept and decode wireless LAN information, although this special equipment can be quite expensive. Moreover, all IEEE 802.11 wireless LANs have a function called *wireless equivalent privacy (WEP)*. This is a very good form of encryption that provides privacy at least as good as that achieved on a wired LAN. It is always a good idea to use WEP in a wireless LAN. It should also be noted that all normal virtual private network (VPN) procedures will function in exactly the same way over a wireless LAN as they do over a traditional wired LAN.

Early experiences with wireless LAN security have been a nightmare, however. According to one scenario, people set up a properly equipped laptop *outside* of a building with a wireless LAN. These people, who are in no way valid LAN users, can then get a network connection, obtain an IP address via DHCP, and access the Internet (and even check their e-mail) much faster than at home—and for free.

BLUETOOTH

Harold Bluetooth, king of Denmark from about 935 to 985, is noted for three things today. First, he introduced Christianity into a Denmark (and parts of Norway and Sweden) newly united by his father, Gorm the Old. Second, he was so fond of blueberries that they permanently stained his teeth, giving him his surname. Third, and of most interest here, Harold Bluetooth has had his unique name taken a millennium later for *Bluetooth,* a new wireless communications technology. The choice of name reflects on the prominence of Swedish equipment vendor Ericsson in proposing and promoting the Bluetooth technology.

Any relationship between the contemporary Bluetooth and blueberries is entirely conjectural. Bluetooth is a wireless technology that is neither WAN nor LAN, but something entirely new, really: the personal area network (PAN). PANs are intended to network the new breed of personal digital assistants (PDAs) and other small, hand-held digital devices such as portable game-playing units.

Handheld PDAs and PCs hold a lot of information in common. Schedules, address books, meeting notes, e-mail, and so on can dwell as easily on a PC as on a PDA. The problem is *synchronization:* taking information that has been updated on the PC or PDA and making sure it is the same on both units. Today, this synchronization of information is achieved mainly by a short cable running to the serial port on the back of the PC.

It would be nice if the PC and PDA could talk to each other wirelessly. It would be even nicer if someone holding a meeting could just send a copy of the presentation to the PDA of everyone in the room. And it would be nicest of all if this same technology could be made flexible enough to allow game players to play the same game, perhaps in groups of up to 10 or so, without the need for special attachment cables and the like between their hand-held game units. This is exactly what Bluetooth is for, and even more.

Bluetooth is designed to be more or less automatic. Walk into a room, and the Bluetooth devices automatically find each other. Whenever Bluetooth devices are brought into close proximity, they start networking: it's what they do. No need to turn Bluetooth on, even if the unit itself is powered off. Airlines have had an issue with this feature of Bluetooth, of course, and early prototypes had to have their batteries removed to eliminate any chance of the units interfering with aircraft navigational equipment on landing and takeoff.

Bluetooth, created in February 1998, has solid support among hardware and software vendors. Usually overhyped (Bluetooth will not be coming to your refrigerator soon) and sometimes even underhyped (Bluetooth *can* work with wireless LANs), the facts are that Bluetooth is not intended to be a full-blooded LAN replacement but can enhance the capabilities of LANs, both wired and wireless. Bluetooth is most effective when used as a replacement for ad hoc, temporary wireless LANs in meetings and small groups. Bluetooth, which usually runs at under 1 Mbps, just does not have the bandwidth needed for more than occasional use.

But Bluetooth is a good, low-power standard for both voice and data. Bluetooth-equipped pagers and cell phones can automatically switch to vibrate mode in restaurants or theaters, or turn the phones off on entering hospitals. In a palmtop or laptop, Bluetooth can link to the global positioning satellite (GPS) system to feed a display unit in a car without the need for configuration. Bluetooth is designed to be integrated into almost any portable device for as little as $5 per unit.

Bluetooth operates in the 2.4 GHz ISM (industrial, scientific, and medical) frequency range. Bluetooth uses frequency hopping at a typical rate of 1,600 times per second (a very fast hopping rate) to minimize interference to and from non-Bluetooth devices. The system has 78 possible channels spaced 1 MHz apart in almost every country except France (current French regulations limit Bluetooth to only 22 channels, but this might change).

There are three power classes for Bluetooth devices: 100 milliwatt, 2.5 milliwatt, and 1 milliwatt. The 100 milliwatt device range is 100 meters (a little more than 300 feet), the 2.5 milliwatt range is 10 meters (about 30 feet), and the 1 milliwatt range is only 10 centimeters (only a few inches). As with all wireless technologies, the ranges are approximate and can vary quite a bit depending on the local environment. Outdoors, some 2.5 milliwatt Bluetooth devices have been demonstrated to function over 100 meters.

The optimal throughput for Bluetooth devices is 1 Mbps. However, there is always some error-correcting overhead to add in and the need for retransmissions with some protocols and with high bit error rates. Throughput in the range of 700–800 Kbps should not be unreasonable to expect. Bluetooth operates in the same range as IEEE 802.11 wireless LANs, so the two technologies can and do interfere with each other, degrading the performance of both.

Bluetooth supports both voice and data traffic. The voice traffic, on up to three simultaneous but separate connections, has a much lower tolerance for interference than the data traffic, which can just keep trying until it gets through. To counter the interference from wireless LANs, just bring the Bluetooth devices closer together. A study done by equipment vendor Ericsson has shown that, even at 10 meters (30 feet or so), the probability of interference to Bluetooth voice is less than 10 percent. Data interference shows up as lower throughput. The same studies show that at 10 meters, throughput is reduced in about one quarter of the cases by 10 percent. In about another quarter of the cases, Bluetooth throughput ceased altogether. In the remaining cases, the results were somewhere in between.

The Ericsson study only measured the impact of IEEE 802.11b on Bluetooth. There is little data on the effect of Bluetooth on LAN throughput. Of the 79 Bluetooth channels, 17 channels are shared with IEEE 802.11b. The IEEE 802.11b group is working with the Bluetooth group (IEEE 802.15) to make sure that both technologies work well together.

Some of the frequency channels that Bluetooth uses to hop around on are also shared by microwave ovens. Bluetooth frequency hopping cycles through all 79 channels continuously, even if the channel is rendered useless by microwave noise. Any voice or data just vanishes when Bluetooth hops onto a noisy channel.

Bluetooth security is a combination of link-level and encryption security. The security is device-based and does not require the user to do anything at all to invoke the security. There are three levels of Bluetooth security:

▼ **Level 1**   No security at all.

■ **Level 2**   Service-level-enforced security. This security is established after the Bluetooth channel connection has been negotiated.

▲ **Level 3**   Link-level-enforced security. This requires the user to enter passwords and so on for services.

Level 2 security is recommended for most instances. Level 3 is the most secure but requires full encryption and constant user intervention. Level 3 security is not transparent and so makes Bluetooth more complex to use. Whatever the security level, Bluetooth devices are always vulnerable to denial of service attacks from flooding the channels with interference. But eavesdropping would be hard, since the devices use a pattern of 1,600 different channels from the 79-channel pool each and every second. Nevertheless, encryption above and beyond Bluetooth is advised end-to-end when privacy is a concern. There is nothing like IEEE 802.11b's WEP (wireless equivalent privacy) defined for Bluetooth.

There are two possible Bluetooth network topologies. These are called *piconet* and *scatternet*. Bluetooth has both point-to-point and point-to-multipoint modes of operation. When using a point-to-multipoint connection, the connection channel is shared among the Bluetooth units. These units form the piconet.

When two or more Bluetooth devices come into range, they automatically form a piconet. One unit becomes the master, while the other unit becomes the slave (the choice of terms is unfortunate, but defined in the standard). The master unit controls all of the traffic on the piconet, and both units frequency-hop together. The master gives all the slaves in a piconet its own clock-device ID and sets the hopping sequence for the group based on the master unit's device address. There can be up to seven active Bluetooth devices on a piconet. The general structure of a Bluetooth piconet is shown in Figure 16-8.

Piconet devices can be in one of five states. These are

▼ **Standby**   A Bluetooth device waiting to join a piconet.

■ **Inquire (also called Sniff)**   A Bluetooth device seeking other Bluetooth devices.

■ **Page**   A Bluetooth master seeking other Bluetooth devices.

■ **Connect**   A Bluetooth device active on a piconet in a master, slave, or simultaneous mode.

▲ **Park/hold**   A Bluetooth device in a low-power, but connected state.

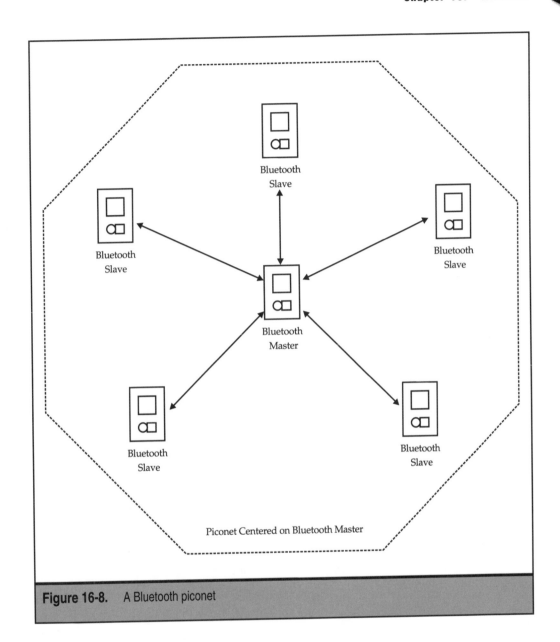

Figure 16-8.   A Bluetooth piconet

Bluetooth scatternets are formed when multiple masters exist within range of each other. In other words, a Bluetooth scatternet consists of multiple, overlapping piconets. A master unit might also be a slave on another piconet, and each piconet has a hopping sequence that is different within the 2.4 GHz frequency range. The masters do not have to worry much about hitting the same channel at the same time that another master does because of the differences in these hopping sequences.

There is no handoff of a slave in Bluetooth from one master to another. Bluetooth masters are not the same as IEEE 802.11b access points. This is a problem if, for example, a Bluetooth device is sending voice or data to another Bluetooth device and switches piconets. It takes a few seconds to reconnect, and anything sent in that time is lost. There are no plans to add a handoff to this ad hoc nature of Bluetooth, because this is what the more permanent IEEE 802.11b LANs are for.

Bluetooth interoperability should not be an issue, given the simple nature of Bluetooth operation. Users will get Bluetooth in the next few years inside of cell phones, PDAs, and perhaps even laptops, most likely if they want it or not. But Bluetooth is really a LAN extension that promotes connectivity, not bandwidth. The initial applications are likely to be a simple form of information sharing between PDAs and laptops and PCs, and ad hoc "conference room networks."

WIRELESS BROADBAND

Today, most users sitting in front of a PC to a certain extent take bandwidth for granted. This is especially true if the user is in an office with a 10 or 100 Mbps Ethernet, or at home with a DSL or cable modem (which usually have a 10 Mbps Ethernet interface even if the link bandwidth is considerably less). But wireless users rarely take bandwidth for granted. Usually, wireless Internet access is even slower than a dial-up link running at 56 Kbps or 33.6 Kbps. With wireless Internet access, even 9,600 bps is a luxury.

The wireless application protocol (WAP) was invented in large part to address this wireless bandwidth limitation. Is there any wireless technology (or technologies) that offers the same amount of bandwidth that users are accustomed to on their wired LANs? That is, beyond the modest ranges of wireless LANs?

As a matter of fact, there are at least two wireless technologies capable of delivering over a considerable distance the same kind of bandwidth enjoyed by DSL and cable modem users. These technologies are often grouped together as *wireless broadband* technologies. The term "broadband" as used here just means that the frequency ranges employed are much wider than the ranges traditionally used in wireless networks. The two technologies are the *multipoint, multichannel distribution system (MMDS)*, and the *local multichannel distribution system (LMDS)*.

MMDS

MMDS systems were first proposed as a form of wireless cable TV. It seems odd to position a service in this way, since the original free broadcast TV stations are essentially "wireless cable TV." But MMDS differs from broadcast TV as much as cable TV differs from broadcast TV. And MMDS is worthy of discussion because the MMDS service has been repositioned away from just being a TV and movie delivery platform and is now positioned as a very high speed adjunct system for cable TV or as a wireless broadband data service all on its own.

MMDS merges what are essentially TV channels from the Instructional Television Fixed Service (ITFS) and Multipoint Distribution Service (MDS). Some other channels were added so that MMDS has 33 analog channels, each 6 MHz wide, just as in broadcast and cable TV networks. The MMDS channels that came from IFTS have upstream bandwidth defined for user-to-network communications. ITFS was intended to be an experimental network for schools, but few schools ever used the ITFS channels. However, if there is a school near an MMDS network using one of the ITFS channels, the corresponding MMDS channels must be shut down and cannot be used.

The channels that make up MMDS in the United States are

▼ Two MDS channels between 2,150 and 2,162 MHz

■ Sixteen ITFS channels between 2,500 and 2,596 MHz

■ Eight MMDS channels between 2,596 and 2,644 MHz

▲ Four more ITFS channel interleaved with three more MMDS channels between 2,644 and 2,686 MHz

Outside the United States, MMDS runs with similar channel structures between 2 and 3 GHz.

For a while, MMDS was seen by the telephone companies in the United States as a way to compete with the cable TV companies for video services. Several trial MMDS networks were built and piloted by the telephone companies, but eventually abandoned. At MMDS frequencies, an unobstructed line of sight is not strictly needed from user (home antenna) to central tower, but it helps with signal quality. A pine tree was a serious obstacle to the signal. In hilly and heavily wooded areas, adequate *coverage* (the number of potential customers for the service) was hard to guarantee, especially as trees grew and buildings were erected. In most cases, only a small percentage of the people living within range of the signal could actually get the services. And 33 channels, once adequate for a TV-based service, were soon not enough to compete with newer cable TV architectures.

Outside of the United States, MMDS has been more successful. MMDS made sense for island nations and nations without a well-developed history of cable installation. This was especially true in nations with heavy rainfall or long rainy seasons. The 2.15–2.16 GHz band and the 2.5–2.686 GHz band that make up MMDS are not as susceptible to rain fade as other bands.

Lately, MMDS has been redefined as a high-speed service for IP packets and Internet traffic. In order to help in this process, the bands that make up MMDS have been reallocated for two-way communication. However, the channels are not paired in any way. Moreover, there are two competing technologies for how to use the MMDS bands. These are *frequency division duplexing (FDD)* and *time division duplexing (TDD)*. FDD uses two different frequencies for downstream and upstream traffic, while TDD uses the same frequency, but at different times, to "bounce" traffic back and forth between end points.

The basic MMDS architecture has a sectored (divided) cell site that has multiple "subscriber terminals" associated with each channel in every sector. A tricky issue is coordinating the MMDS service channels if there are preexisting ITFS and MDS operators in the area. The older systems are broadcast only, so in some cases there might be no upstream channels available for the MMDS operator to use.

MMDS can deliver up to 30 Mbps of data over a channel. Depending on the frequencies used, coverage can be to users 10 miles (about 16 km), 20 miles (32 km), or even 35 miles (56 km) away from the central tower. The high bandwidth available has made MMDS attractive for small and medium-sized businesses. Few homes require 30 Mbps at this time. Although MMDS has upstream channels available, most MMDS service providers use what is called *telco return* for traffic sent from user to network. That is, the user must still dial into an ISP (the MMDS service provider in most cases) at 56 Kbps or less to send anything at all *to* the Internet. Only downstream traffic *from* the Internet can flow at 30 Mbps. The basic organization of an MMDS network used for Internet access is shown in Figure 16-9.

The "MMDS modem" functions in almost exactly the same way as the cable modem. The upstream modem banks do not have to be owned and operated by the MMDS service

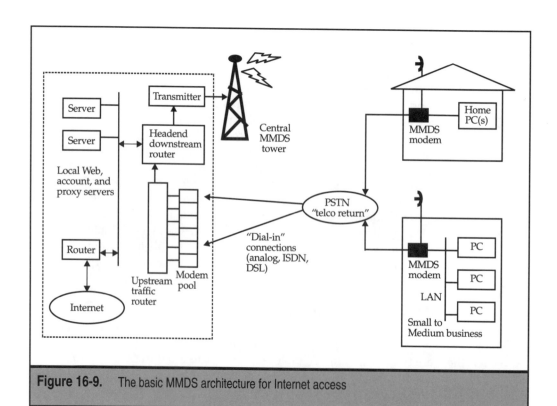

Figure 16-9. The basic MMDS architecture for Internet access

provider. In fact, many ISPs could be customers of the MMDS service provider, and the MMDS service provider need not be an ISP at all.

Each 6 MHz channel can support up to about 9,000 subscribers. Naturally, this all depends on user traffic patterns in the same way that cable modems depend on the same thing. When there are many low-traffic users, the number of users per channel can even be increased by using directional transmitters and more advanced equipment.

An ideal ISP location is not often the ideal MMDS head-end location, so the ISP and MMDS locations are split about 50 percent of the time, regardless of the relationship between ISP and MMDS operator. For areas located in the "shadow" of the central transmitter, a low-power broadband "booster" can be used when adequate line of sight is not possible. This type of arrangement is shown in Figure 16-10.

MMDS receivers must usually be installed with a signal level near the high end of the range, since signals tend to fade over time and not increase. Antennas can be mounted on the roof, or the side of a building, or even in a window, as long as the line of sight is adequate. The antennas vary widely in appearance, from traditional parabolas to foot-square flat arrays to dome-shaped designs.

The use of wireless returns for MMDS systems is relatively new. Current designs use the 2.4 GHz band upstream (the same ISM band used for wireless LANs). If the Internet service provider's office is closer than the downstream tower, this point of presence (POP) can be the target of the upstream transmissions, again with line of sight a concern. However, this requires separate sending and receiving equipment, each with its own alignment. But the upstream power requirement is considerably less.

Upstream speeds vary from 256 Kbps to 5.12 Mbps per user. Lower speeds support more users, of course. If two 6 MHz channels are combined into a 12 MHz band and split into 200 kHz "subchannels," then the data rate of each subchannel would be around 256 Kbps to

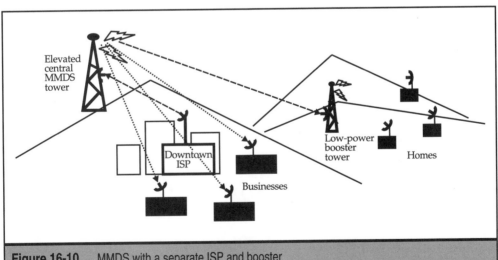

Figure 16-10.   MMDS with a separate ISP and booster

320 Kbps. These 60 user channels could support about 3,000 customers for data services. A full 6 MHz channel can deliver 30 Mbps, or 27 Mbps with forward error correction.

All in all, MMDS is a technology in transition. Faced with difficulties in the video services arena, the success of MMDS in the wireless Internet services arena is far from assured. Compared to other wireless systems, the basic MMDS channel speed of 30 Mbps is appreciable, but service providers have been cutting this speed drastically in the interest of supporting more users per channel and per MMDS system.

Recently, a lot of interest in the wireless broadband industry has shifted from MMDS to a related technology, LMDS.

LMDS

The *Local Multipoint Distribution System (LMDS)* is also a form of wireless broadband technology. If MMDS is a kind of "wireless cable TV," then LMDS might be characterized as a type of "cellular cable TV." That is, the larger, centralized transmitter towers characteristic of MMDS give way to smaller, distributed transmitter towers in LMDS.

In fact, LMDS is often seen without a transmitter tower at all. LMDS equipment has become inexpensive enough to deploy on a point-to-point basis between buildings in a metropolitan area. City governments in the United States have pioneered this use of LMDS, and LMDS is often the most cost-effective and quickest way to provide broadband speeds between two buildings located in congested downtown areas.

LMDS can be used as a video delivery system, and in fact the first LMDS test sites did offer exactly that type of service. But today LMDS is seen as a type of wireless access technology, especially for Internet access. LMDS is ideal for point-to-multipoint configurations where a central site connects multiple buildings. Traffic flow is determined by a system of queuing and multiplexing.

Although LMDS operates in the same microwave frequency range as point-to-point microwave systems, the small size and power of LMDS devices allow many more subscribers (really Mbps per square mile or kilometer) than any point-to-point microwave system. LMDS increases capacity by using a frequency reuse pattern similar to that used in cellular telephony, but based on signal polarization. But there is no handoff in LMDS, of course. LMDS can even be used as a perfectly adequate wireless local loop (WLL) technology (as can MMDS).

The small coverage area of an LMDS transmitter is sometimes a blessing. Service providers can start small and build up the network over time. And the equipment needed is quite inexpensive, especially compared to the cost of burying fiber.

LMDS networks consist of a physical transport network and a set of services delivered on this physical platform. The physical transport can handle both circuit-switched and packet-switched traffic. A series of base stations provide wireless access to users and provide the links to the central LMDS service provider office. Overall, the LMDS network is similar in layout to the cellular or PCS telephony network. Of course, there are no handoffs, as mentioned already, and different frequencies and speeds are used on the channels. This overall architecture for LMDS is shown in Figure 16-11.

The base station is usually configured as a hub with four sectors. There are multiple users assigned to each sector, as with MMDS. There is a lot of variability when it comes to the number of channels and the overall frequency plan for the network. It all depends on the type and number of users, the geographical area, and the capacity required.

A nice feature of LMDS is the ability to *oversubscribe*. The idea behind oversubscription (also called *overbooking*) is that bandwidth available can be assigned to many more customers than it should be able to support, because there are few times when all of the customers are using the bandwidth at the same time. For example, 12 Mbps might be assigned to a

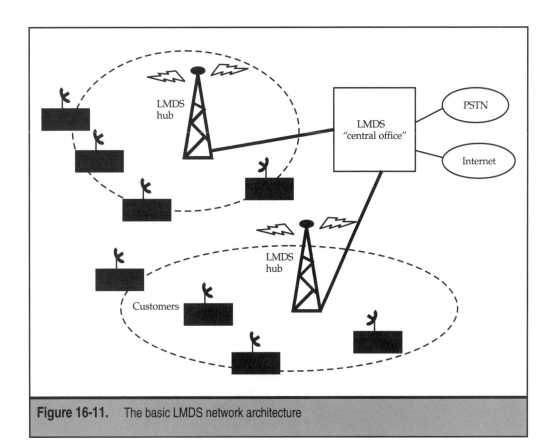

Figure 16-11. The basic LMDS network architecture

sector on an LMDS base station, but the users in that sector are paying for 24 Mbps in aggregate. There might be quality of service concerns, but this is at least possible with LMDS.

LMDS networks do not have to cover an entire geographical area. Services can be provided only where there are customers. LMDS is at its best when the LMDS channels serve a whole building. All of the users in the building share the channel without the need for each one to have a separate transmitter and receiver. Usually the LMDS equipment is located on the roof, and the individual customer sites are connected with fiber optic cable.

When it comes to services, the types of services being delivered over LMDS today include

▼   T1/E1 circuit replacement (1.5 or 2.0 Mbps)

■   Fractional T1/E1 (up to 1.5 or 2.0 Mbps in 256 Kbps increments)

■   LAN connectivity (typically 10 Mbps Ethernet speeds)

■   Frame relay/ATM network access (usually at T1/E1 speeds)

■   Voice and video services (e.g., teleconferencing)

■   Internet connectivity

■   Web services such as Web site hosting

■   Voice and faxing over IP

▲   ISDN (64 Kbps and 1.5 or 2.0 Mpbs)

There are multiple ways that an LMDS system can be deployed. Channels can be shared with time division duplexing (TDD), frequency division duplexing (FDD), or time division multiplexing (TDM). Various modulation methods can be used.

In the United States and Canada, LMDS uses the 24 GHz, 28 GHz, and 39 GHz bands. Spectrum is assigned in the 24 GHz band in pairs of 50 MHz blocks, so licenses can get a total of 100 MHz (sometimes it is possible to get 400 MHz). Subchannels have 10 MHz block allocations. The channels are paired for communication in each direction and "blocked" so that there is sure to be competition between two service providers in each area.

The 28 GHz band also follows the cellular practice of A and B blocks to ensure competition, and the subchannel structure is up to the service provider. The 39 GHz band has 50 MHz channel blocks, again paired for competition, and some 175 licenses will be given in each of 14 different 100 MHz blocks.

LMDS can also be used to extend the reach of a traditional hybrid fiber-coax (HFC) cable TV network. This use of LMDS is shown in Figure 16-12.

LMDS is not only for use in the United States. Outside the United States (and Canada), LMDS is known as Fixed Wireless Point to Multipoint (FWPMP).

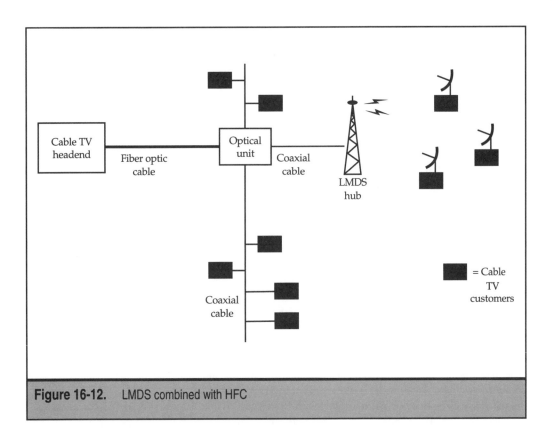

Figure 16-12.   LMDS combined with HFC

FWPMP

The Fixed Wireless Point to Multipoint (FWPMP) system is a European version of LMDS. The equipment and architectures are nearly identical, but the frequency allocations are more suitable for the European environment.

Oversubscription is possible in FWPMP, as with LMDS. There are WLL and cable TV extension applications. Generally, however, more restrictions are imposed on the FWPMP license than in the United States. For example, there is a minimum coverage requirement and a time line for universal services to be deployed (five years).

FWPMP operates in the 26 and 28 GHz bands (actually, 24.5–26.5 GHz and 27.5–29.5 GHz). The frequency blocks are multiples of 7 MHz, and channel allocations are usually in increments of 28 MHz. There is also a 38 GHz band; each of the bands, the 26 and 38 GHZ bands, have a channel block of 50 MHz and four channels 12.5 MHz wide, duplexed as expected.

FWPMP can also operate in the 3 GHz and 10 GHz bands, but these channels do not have the bandwidth that the higher frequencies offer.

Although private LMDS links can operate at SONET/SDH link speeds, most LMDS and FWPMP equipment is designed to run at the basic T1 (1.5 Mbps) and E1 (2.0 Mbps) rates. Multiple links can be and are provided to a single customer, however, and a large customer might have LMDS or FWPMP running at an aggregate speed of 4.5 Mbps or 6.0 Mbps.

SATELLITE SYSTEMS

Satellite systems form a special case of wireless communications. Satellites have been used for voice communications for many years, although in countries with highly developed communications infrastructures, the high delays are often unacceptable. This is also true for data services in some cases, but data satellite services such as DirecPC have been popular in spite of the delay.

The data service delay issue in satellite systems has been handled in the past in two ways. First, only the *downstream* (Internet to user) traffic has been sent over the satellite link. The upstream (user to Internet) traffic is usually handled over a traditional, dial-up link to the satellite ISP. This works well because the user device is usually a client PC, and the overwhelming proportion of the traffic flows not from client to server, but from server downstream to the client. Second, there is a lot of bandwidth on the downstream portion of the link, usually 400 Kbps of more. So even though it might take a few seconds for the Web page to start to appear, once it does the whole page is essentially there all as a whole (most Web pages are only about 50 to 100 kilobytes [KB] in size). There are exceptions, of course, and it all depends on how many independent connections have to be set up between client and server, but the overall effect is much better than with dial-up alone.

However, some service providers have begun to experiment with satellite systems that do not follow the normal geosynchronous orbits traditionally used by communications satellites. The geosynchronous orbits are the cause for the high satellite delays. Low earth orbit (LEO) and middle earth orbit (MEO) satellites will have much lower delays and make two-way communication with low-powered transmitters possible.

These systems have been promoted since the late 1990s, and yet they always seem to be a few years away. Rather than reproduce the system plans and schedules that always seem to slip further and further into the future, I refer the interested reader to the Internet for sources of information, especially regarding the Microsoft-backed system known as Teledesic.

FREE SPACE OPTICS

If wireless "wins," especially in the battle for the access portion of the network, then it seems that optical networking must some how "lose" this sector. But perhaps it is not quite correct to see wireless as a purely electrical networking scheme. Maybe there could be such a thing as a "wireless optical network," and as it turns out, there is.

The term for a wireless optical network currently in use is free space optics (FSO). Oddly enough, this discussion of FSO brings the theme of optical networking full circle back to Claude Chappe and his optical telegraph. There are significant differences between peering through a telescope at remote windmill-like semaphore arms and aiming a free space laser at a receiver, but there are important similarities as well. For one thing, weather now plays a significant role in the operation of the network link.

FSO is intended as an entry in the broadband access market. Until now, this segment has been the arena of DSL, cable modems, and wireless local loops. The ideas behind FSO have been around since the late 1980s, but these early systems left a lot to be desired in terms of distance and performance. The new systems have started to gain a lot of interest from ISPs, however.

The new interest in FSO stems from the fact that only a tiny percentage of the office buildings in the United States have fiber access. And running fiber to all of the office buildings without fiber will not only be expensive but take years. Many municipalities, fed up with years of trenching and repairing of streets and sidewalks, have even started to declare a moratorium on new fiber installation. So if an office is waiting for fiber to start using broadband speeds, they could be waiting for many years.

The idea behind FSO is simple enough. Lasers use bursts of light called *pulses* to send packets through the air in the terahertz (THz) spectrum range. So instead of waiting four to 12 months for a fiber link to a building and paying the service provider anywhere from $100,000 to $200,000, a customer can now be up and running in two or three days for only about $20,000 per building.

And FSO has plenty of speed, too. Links usually run at 155 Mbps or 622 Mbps, standard SONET/SDH link speeds. Some FSO equipment vendors are even claiming Gigabit Ethernet speeds as well. Neither DSL nor cable modems can match these numbers for now. Even wireless data MMDS and LMDS (or wireless broadband) systems cannot keep up with FSO speeds. FSO is generally useful only in urban areas, however. The reason is the considerable distance limitation between building and hub. But FSO fits nicely between wireless broadband and fiber optic cable in terms of speed and distance, as shown in Figure 16-13.

FSO is intended for last-mile access for small and mid-sized businesses. Even if another technology such as LMDS is used, FSO can be used to add capacity more or less on-the-fly to the other wireless link. FSO is also used as a backup or disaster recovery strategy to be used when the working fiber to a building is cut (few buildings have fiber, and even fewer are on metropolitan area rings). It is also possible that FSO can be installed and used for specific situations, such as when an optical link must be used to cross a river or highway and there is no other way to accomplish the task.

But FSO's greatest application success to date has been in the area of LAN or campus connectivity. This does not imply that universities are involved. For example, an FSO could be used to link a newsroom to the broadcasting station site, or to provide a link between two close buildings when the traffic between them threatens to overwhelm a backbone

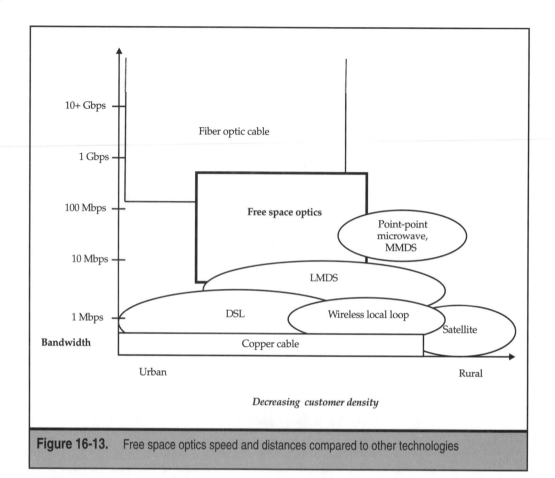

Figure 16-13.   Free space optics speed and distances compared to other technologies

LAN. The usual trade-off between distance and speed (i.e., to go farther, the link must go slower) is not a limitation to LAN or campus FSO systems, because of the short distances supported. However, when FSO is used as a last-mile access method, this trade-off between distance and bandwidth becomes important.

As a wireless local loop application, an FSO system requires a few basic pieces. There are usually several laser transmitters and receivers, or *laser terminals,* at each site or network node. These laser terminals can be configured to establish a basic point-to-point free space link, or a mesh topology, or a star topology (typically point to multipoint). The laser terminals can be installed on the roof of a building, or even inside windows, since the glass contributes very little to the overall signal loss. The optical signals are then sent between the laser terminals to or from *hubs* or *central nodes.* The hubs can be scattered throughout an urban area, since each laser terminal requires line of sight (LOS) to the central hub. The hubs themselves can be linked by FSO, a fiber ring, or even some other technology.

In a straight point-to-point configuration, current FSO equipment can support speeds from 155 Mbps all the way up to 10 Gbps and distances from 2 km (a little more than a mile) or 4 km (about 2.5 miles), always depending on speed. There is support for Gigabit Ethernet in some FSO equipment as well.

In a point-to-multipoint configuration, the same 155 Mbps to 10 Gbps speed range is supported. However, the distance drops to 1 km (about one-half mile) and 2 km. A full mesh drops the distances even further. Meshes currently only support 622 Mbps over 200 to 450 meters (about 600 to 1,400 feet).

The wavelengths used in the current crop of FSO products vary widely. Most use 780 nm lasers, while a few can operate in the 750–950 nm range. A couple even use the traditional SONET/SDH wavelength of 1,310 or the DWDM window at 1,550 nm.

The LOS issue has been an important one. What if a potential customer is too far to reach for the particular speed required? There are various ways to shorten the span with intermediate devices, but this is not an ideal solution. And reports of distance support over several kilometers might be achievable only under ideal weather conditions. Some skeptics maintain that services at any distances over a kilometer (or less!) will be greatly affected by atmospheric conditions.

FSO Advantages

The biggest FSO advantage, of course, is speed. But the cost effectiveness of this bandwidth is a big advantage as well. Running fiber to provide similar speeds can cost up to $70,000 or more per mile and involves right-of-way issues. Deployment of FSO has been estimated at about $20,000 per building, but most cost analyses like this come from the equipment vendors themselves. And the cost analysis only figured the cost for a mesh network with very short links up to 200 meters long.

A nice feature of FSO systems is the low start-up costs for the service. A laser terminal is only installed when needed. There are no massive capital outlays, and the return on investment can be measured in months, not years. Service providers can start with only a section of a city. Once two buildings have a laser terminal, one can be the "backbone" to all other buildings within line of sight of the first. Only when another customer is added does the equipment need to be deployed.

Security is an advantage for FSO, not a liability. The signal is hard to intercept, since the beam is narrow and invisible and not detectable by electronic techniques. The signals are all coded and encrypted, so even getting the signal is not enough. And it is hard to intercept the signal without blocking it. An eavesdropper would have to find the link, hang in the air to get it, and try not to disrupt the service at the same time. A formidable task, to say the least.

Right now, FSO systems operate in a frequency range that is totally unregulated, in contrast to MMDS and LMDS. However, this is mainly because there has been little traffic to date. If urban areas fill up with FSO traffic, regulation is sure to follow.

FSO Disadvantages

FSO systems share many of the disadvantages of many other line of sight wireless technologies such as MMDS or LMDS. All of the laser terminals require a clear path in a straight line to each other.

The distance support has also been a concern among potential customers and service providers. The shorter the distance supported, the more hubs are needed to cover a given area with the new service offering, which increases start-up costs. It is therefore tempting to run everything at the maximum supported distance, but engineering at the edge like this can make customers that know a few things about the technology nervous. Trials are one thing, but reliability and quality of service can only be demonstrated in the real world and over a period of time.

Not surprisingly, the biggest problem is weather. Certain atmospheric conditions make "seeing" at these laser wavelengths as difficult as seeing enough to drive a car in heavy rain or fog. Rain and snow can affect the signal, but fog is by far the worst offender. Fog is composed as many small beads of water suspended in the air. These small beads scatter the light signal and break it up entirely in many cases.

A few FSO equipment vendors have gone so far as to use a public band (no license required) microwave link as a backup. But the situation is far from ideal. Microwaves are more fog-resistant than optics, but microwaves suffer in heavy rains, which might alternate with heavy fog in some coastal areas. Current plans call for the microwave link, which has a much lower bandwidth than the optical link, to operate as backup only for a few minutes at a time, keeping the traffic flowing, but at a reduced rate. If the low-speed interlude is brief enough, the situation might be tolerable for users. It might be necessary to configure different speeds and distances in different cities not only based on geography, but on overall seasonal weather patterns, as some vendors claim to do now.

Weather is only the most important issue, not the only issue. Anything that blocks the path of the optical beam will disrupt services, if only for a moment. For instance, birds can be a real problem, as can very low flying helicopters. Outages should last only a few milliseconds in the case of birds, and the session is not even dropped for this brief period, although errors might cause resends. To offer some protection to the intruders, the lasers lower their power during the period of obstruction, then raise it again to full power when the path is again clear. Mesh architectures are best for dealing with these types of service disruptions, since a node can send or receive in several different directions. But meshes pay the heaviest price in terms of reach and speed.

A more subtle problem with FSO systems is the swaying of tall buildings. This occurs not only during heavy winds or storms, but all the time. A gentle roll is apparently quite common even in the most settled conditions, and vibrations due to heavy urban traffic, large trucks, and elevated trains can make a whole building shake. In order to keep the laser terminals precisely aligned, it is necessary to supply the device with sensitive auto-tracking equipment that constantly adjusts the LOS path between transmitter and receiver. In fact, one of the real reasons that FSO has only recently become affordable is due to the drop in cost of adding this auto-tracking feature.

Another real issue is access to the rooftop. Many customers are forbidden by the building owner or management company from fixing anything to the roof. Some municipalities have very strict local ordinances about the size and shape of anything projecting above the parapet of a building. Sometimes access is only secured after a considerable expense and signing a complicated contract or usage agreement. Naturally, equipment that can be placed in a window can avoid these problems. But this solution cannot usually be used to provide service to every potential customer in the building.

The final problem is just the fact that people are nervous about lasers. Hitting birds is one thing; potentially blinding helicopter pilots is another. But the lasers used are generally very low powered, in some cases less powerful than a classroom laser pointer.

FSO systems are new enough not to have to face an added problem that also affects all line-of-sight wireless systems. What if a tall building is built between laser terminals, blocking the line of sight? And consider the following potential conflict of interest scenario: A city government, always looking for ways to cut networking costs and constrained by budgets, is using a substantial FSO system in a major downtown area. However, government buildings such as the City Hall and courthouse are not usually very tall. What if someone wants to build an office tower in between? Well, the city government makes the zoning laws and approves all architectural plans. Is there a chance that the city government might block or delay plans for the office tower? And even if a rejection is entirely warranted, will not someone connected with the builder raise this conflict of interest issue?

FSO Architectures

There are three choices for the architecture of an FSO system, as mentioned earlier. This is not unusual for any technology, and FSO is no exception. Which architecture is chosen all depends on the customer's needs and the application requirements. The three choices are

▼ Point-to-point

■ Mesh

▲ Point-to-multipoint

Any FSO system can be built according to any one of these topologies. But it is even possible to combine all three in the same FSO network. Each has different benefits and drawbacks. In general, point-to-point FSO links can provide a higher bandwidth, dedicated connection. But this architecture does not scale well or cost-effectively. A mesh offers the best service redundancy for reliability purposes, but the distance covered by a mesh is greatly reduced. Finally, the point-to-multipoint (star) is less expensive but offers lower bandwidth than the point-to-point links.

Scalability is always an important issue when it comes to deploying a new service. A mesh or star makes it easiest to add nodes to an existing configuration. A mesh does not have to be a full mesh: it is usually enough to connect to three of four other network nodes. The general characteristics of FSO systems deployed in point-to-point, mesh, or point-to-multipoint architectures are shown in Figure 16-14. The perspective in the figure is looking down from above an urban area.

Some FSO vendors make equipment intended for a mesh architecture. A mesh can close on itself to form a ring, and "spurs" of point-to-point links can radiate out from the central mesh or ring. The cost is high for a mesh compared to the simpler star with

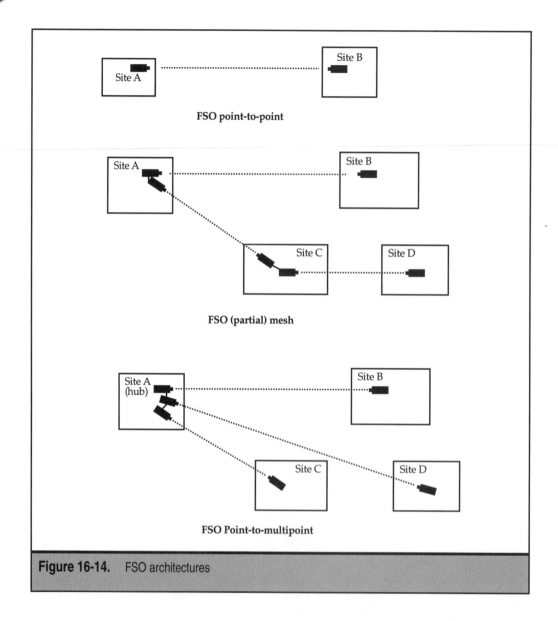

Figure 16-14.   FSO architectures

point-to-multipoint, but meshes do not suffer from single point-of-failure concerns. The need to reach the more distant central star node can limit speeds as well.

Even meshes are really just a series of point-to-point links. The laser terminal just has several different transmitters and receivers pointed in different directions. But no matter which architecture is used by a service provider, the bottom line is the number of nodes and the distances between them.

FSO and DWDM

So far, all of the FSO equipment vendors have used single wavelengths. However, *wireless DWDM* is coming to the FSO world also. The whole idea is to make provisioning of new customers in the same building easier, without the need to share bandwidth. There are fewer security concerns when different wavelengths are used as well. In many cases, DWDM support can be added to the laser terminal as needed.

WIRELESS AND OPTICAL NETWORKING

It should be noted that in every instance in this chapter that the word "wired" is mentioned, this could apply to an optical network. Every single place where optical networking has been applied to the Ethernet or voice networks or cable TV networks, the same considerations could be applied in this chapter. There is no reason that wireless technologies cannot be combined with optical networking in a number of different ways.

All things considered, wireless networking differs from optical networking in two areas. First, wireless networks have less bandwidth—much less—for users than optical networks. Second, wireless networks offer users the freedom and mobility than optical networks lack. So it makes sense to combine wireless networks with optical networks to increase wireless bandwidth on the backbone and at the same time allow more mobility to optical network users.

CHAPTER 17

Optical Networking Issues

As this book draws to a close, this is the time and place to review some of the most important aspects of optical networking, position optical networking in relation to networking as whole, and examine some of the unresolved issues surrounding optical networking. These issues are explored with regard to networks, equipment, and services.

Optical networks are quite different than the electrical-networks-with-optical-links that have been deployed up until now. The equipment used in an optical network is very different as well. And of course the services provided by optical networks can be very different than the services offered on "legacy" networks.

Yet the very differences that make optical networks unique might also make optical networks challenging. The impact of change is always a difficult thing to predict. Some changes seem so natural and desirable that the new methods and technologies sweep away everything that has gone before. For example, the arrival of the digital audio CD swept analog vinyl records into the dust bin quite rapidly. Analog audio cassettes, some observers claim, are kept alive only by the fact that auto makers do not charge extra to put cassette players in new cars. If all new cars had CD players, this line of reasoning goes, audio cassettes would follow vinyl records right out to the curb for pickup.

But some new methods and technologies, as attractive as they might seem at first glance, struggle for years to catch on. For instance, as attractive as DSL was and is for high-speed Internet access on copper local loops, for years the DSL industry was facing a game of catch-up with the cable modem industry.

WHAT CAN GO WRONG?

What could happen to make optical networking not like digital audio CDs in terms of marketability, but more like DSL? There are issues with optical networks themselves, with the equipment used in optical networks, and with the service provided by optical networks. The following lists of issues are not exhaustive, but they are all important.

Networks

Network issues are issues that should be resolved even before people begin buying boxes and start plugging them in to one another. These issues are basic ones and address aspects of optical networking overall.

What will happen to all of the "old" fibers that are not suited for optical networking? Older fibers, especially those installed before around 1990, might not be suitable for the multiple wavelengths that characterize optical networks with DWDM. And not even all of the "new" fibers can handle many wavelengths closely packed together. It is possible to replace these older fibers with newer ones, of course. But it is also possible to just leave the older fibers in place and add newer fibers more suited to DWDM and optical networking. This second approach seems much more likely.

What about the electrical repeaters and regenerators on fibers otherwise suitable for DWDM and optical networking? Even if an "old" fiber span seems fine for many wavelengths, the presence of even one electrical repeater or regenerator on the fibers makes the span unsuitable for DWDM. When trunks went digital, in many cases analog amplifiers were replaced where they stood with digital components. The longer spans possible between optical amplifiers make this relatively smooth, but labor-intensive, transition unlikely. This will just be another reason to install new fiber for DWDM and optical networking.

How should optical links be tested, repaired, and managed? It is tempting just to find a way to test, repair, and manage an optical network link and consider this issue closed. But there is much more to this issue than the mere *possibility* of optical link testing, repairing, and management. Standard test gear must evolve along with optical networks so that test gear does not become obsolete as the number of wavelengths grows. Repair personnel must be trained as well, and no one wants repair people trained for one method or vendor and yet useless for another. Standard optical network management is a key requirement also. Fortunately, this seems to be one area where everyone from customer to equipment vendor to service provider agrees that cooperation and standardization for interoperability is absolutely essential. The standard network management method for DWDM and optical networking might be based on TMN or MPLS or something else, but it will be firmly based on some common ground.

How should IP packets be encapsulated for sending across an optical network? This is a very crucial, yet sensitive issue. This book has pointed out that there are many ways that IP packets can make their way from OSI-RM Layer 3 to a stream of optical 0's and 1's. Naturally, if a network contains network nodes that consist mainly of ATM switches, it is in the best interests of that network that the IP packets be segmented into a stream of ATM cells at some point on their journey up and down the protocol stack. But if the standard and accepted way to put IP packets onto an optical network is just to put the IP packet into a Gigabit Ethernet frame and blast it out, there is not much room in the optical networking world for vendors and users of ATM gear.

Where do wireless networks fit in? Fiber has a lot of bandwidth and few errors, but it can't go everywhere. Wireless can go almost everywhere, but it has a lot of errors and little bandwidth compared to optical networks. What is the proper and natural relationship between the wireless and fiber optic portions of the network? As time goes on, it is tempting to take the position that optical networking is best for the backbone and wireless networking is best for the access portion of the network. However, as both optical and wireless networking evolve, this position might change in the near future. Some aspects of the recent evolution of wireless and optical networking are discussed in later sections of this chapter.

Equipment

Optical networking will be embodied in new devices. The new devices will have new types of interfaces, and capabilities not seen before. In many cases, the new devices will be made by new companies, and older companies will have to adapt themselves to supply the types of equipment that network operators are willing to buy.

What is the *minimum* that an optical networking device must do? This is always an important issue when it comes to new technology. As defined in this book, an optical networking device must at least be able to do some of the things that previously had to be done with electrical network node devices when optical networks were just optical links connecting electrical network nodes. But this definition is just one perspective. No "official" definition of an optical network really exists.

Is it enough just to replace electrical regenerators with optical amplifiers and call the result an "optical network"? Not by the definition used in this book. Only if the network nodes at the ends of the link were converted to DWDM devices would this configuration even begin to fit the minimum definition. DWDM is only a start on the road to optical networking today. However, if an optical cross-connect or switch were used to connect a series of these DWDM-with-optical-amplifier links, then the resulting network would fit the definition quite well. Beware of equipment vendors who suddenly find that their products were optical networking devices all along! Marketing can convert a simple on-button into a "convenient raised-surface action initiator."

How much of optical networking will be required on the premises? Users and customers can be very conservative. As has been pointed out earlier in this book, it should come as no surprise that Ethernet is so entrenched in the marketplace. Once a particular piece of customer premises equipment (CPE) has been cost-justified with meetings and memos and reports, someone somewhere is going to look silly if it becomes necessary to change direction soon afterward.

So when fiber access *does* come to the office, how much has to change? Is an optical cross-connect needed? Probably not. Is DWDM on the premises needed? Possibly, but not likely. There is no reason that an organization cannot simply keep on doing what it has been doing and let the service provider worry about the optical networking. But this does not mean that the customer or users should be oblivious to what the service provider is doing. For example, if the customer has an ATM switch as CPE, but the service provider is using Gigabit Ethernet over DWDM to access the building, then it is probably not a great idea to upgrade the ATM switch. It makes more sense for the customer to consider Gigabit Ethernet CPE.

What is the best optical networking premises equipment package? This issue is related to the preceding one. One of the major issues regarding optical networking, as has also been mentioned several times earlier in this book, is the best way to package IP packets for an optical network. There are many ways, of course. But as time goes on, it seems that the more IP-direct-to-some-form-of-Ethernet-frame there is, and the less "extras" such as ATM, the better. It is still too early to concede the entire optical access marketplace to Gigabit Ethernet or 10 Gigabit Ethernet, but that day could be fast approaching. Of course, there are few affordable, wavelength-agile Gigabit Ethernet or 10 Gigabit Ethernet CPE packages available. This too might change quickly, however.

Where are the end-to-end, turnkey optical network systems? There will always be organizations that will want to create their own private networks. And smaller service providers might not have the expertise in-house to create their own optical network system and service offering from scratch. Other organizations must step up to fill these roles in the near future. Either equipment vendors will add design and installation services, or larger service providers will add equipment procurement and installation to their menus. Companies with a tradition of expertise in the networking arena will step in here as well. However, one of the ongoing problems will be the rapid pace of change in the optical networking field. In many cases, an optical network might be obsolete the day it is installed.

How much will all this equipment cost?   This issue relates to the preceding one. If optical networking gear is priced very high, deployment can be slow. For instance, early ATM switches were very expensive, so there was pressure from the start to make IP routers "as good as ATM switches." By and large, this effort was successful. However, if optical networking gear is priced too low, there is little incentive for start-ups to stake their future on optical networking, and little reason for people selling higher-margin devices to shift emphasis to optical networking gear. In one case, a major SONET/SDH vendor also had the same sales representatives selling early DWDM devices at very low margins (DWDM was "priced to move"). But the sales goal was so hard to make with DWDM that the salespeople would mention DWDM only when pressed by customers. As a result, a potential DWDM leader lost valuable position in the new market.

Services

Money can be made from technology in one of two ways. The technology itself can be packaged and sold, like an automobile or a VCR. Or the service that the technology represents can be packaged and sold, like a limousine leasing service or a video rental store. The two service aspects might even be combined to some extent. The point is often made that customers do not buy technology, people buy services. Technology is always a little mysterious, but the things that a technology enables a person to do are as plain as day.

How will optical networking be marketed?   Okay, the optical network nodes and links are in place, the wavelengths are there, and the network is ready for...what? Just what do you do with an optical network? Initially, it appears that optical networks will do exactly what a lot of networks do today, just a lot more of it. There will be faster and better IP services, and faster and better IP routers and switches to go along with them. But this only scratches the surface of the potential of optical networks. Perhaps a wavelength could be set aside and used for *telepresence*, which has been positioned as the next generation of video conferencing systems. With telepresence, the conference table appears to extend right to the remote site. Half of the table is occupied by real people, and the other half is occupied by people at the remote site. But the illusion created with digital imaging and massive amounts of bandwidth is so good that it is "just like being there." This has been a dream of many organizations for a long time. The service will be very expensive. But it might still be less expensive than flying everyone around, especially when the lost productivity time spent in transit is factored in. This is, of course, only one example. But the same service could be repackaged for education or other applications.

Will there be any other form of networking besides optical networking?   This issue should more properly exclude wireless networking. It might be better to ask if there will be any other *wired* form of networking. For now, the answer appears to be, simply, no. All older forms of networking, from ATM to SONET/SDH and beyond, will simply port themselves right over to a wavelength on an optical network. This has been pointed out in many ways in many sections of this book. What lies beyond optical networking, if anything, is not known at present. However, several lines of research are mentioned at the end of this chapter.

How do you bill for a wavelength? Service providers make money by charging customers more than it costs the service provider to furnish the service and pocketing the difference. But optical networks might challenge and complicate this simple paradigm in several ways. What if a customer wants a wavelength leading to a new site? The cost to the service provider would be very high, naturally. Now consider a customer with several wavelengths leading to different sites. How much should it cost to add another wavelength to an already connected site? Arguably, much less that it cost for the first wavelength to a site. All that is needed for the added wavelength is another transmitter and receiver. Probably the best model for optical network billing is frame relay or ATM. There is a cost element for the physical link, usually quite high, and other cost elements for the virtual circuits that ride the link, usually a few dollars a month (some basic connectivity might even be free).

Service providers have also traditionally charged customers more for using more of something. The basis might vary from time (telephony) to quantity (thousands of packets), but there must be something measurable. However, a customer might pump a wavelength as many different speeds and using many different line codes, all of which is invisible to the service provider. Should a customer using a wavelength at (for example) 622 Mbps be charged the same as a customer using an adjacent DWDM wavelength at 10 Gbps? Why not? Dial-up modem users pay the same per unit time whether they are using a 56 Kbps modem or a modem from an old fax machine that runs at 9,600 bps. In fact, this is just another argument that the charge per wavelength should be low compared to the price element of the physical link. If the charge for a wavelength is low enough, few users will complain if the "modem pricing model" is used in optical networking, as appears will be the case.

How fast will optical networks and services be deployed? Speed of deployment is always a crucial measure of a technology's success. Cable modems took off and ran away with a considerable portion of the residential access market. DSL struggled to keep up. ISDN, as attractive as it was for many businesses, and some residential users, never seemed to come to an area or address when promised or needed. Potential ISDN users were forced to find other methods to gain high-speed access, and when the time came for ISDN, a considerable market segment was already using something else.

In the case of optical networks, the risk is that frustrated users might turn to wireless alternatives. It all depends on which technology will evolve to maturity faster. This issue requires a section of its own.

FASTER EVOLUTION: OPTICS OR WIRELESS?

Which technology will have the faster evolution: optics or wireless? This is a crucial factor, since technologies that grow and evolve tend to leave the technologies that are slow to adapt in the dust. Technologies like UNIX, IP, and Ethernet have always embodied the philosophy of "there's got to be more than one way to do this." Competing methods like DOS, ATM, and token ring were encumbered by concerns over backward compatibility, the "correct" way to do something, and the idea that something that works well does not need to be tampered with.

Many people are always "tampering" with optical networking and wireless networking. The evolution of cell phones has been astonishing, and no one can quite keep up with the advances in DWDM channel spacing, distance, and number of channels in experimental and production systems. Cost is always a factor. Expensive equipment always makes rapid evolution difficult. Early cell phones cost $700 or more. No one would buy a new one every year if the new cell phones were as costly. The PC evolved even faster on the way to hundred of megabits of memory and gigabyte hard drives. If PCs still cost $6–10,000 as they did in the early 1980s, no one would be able to buy a new PC every year or two.

Optical equipment is very expensive, and this might become a real concern in the near future. The resale value of technology is almost nil. An old 1989 analog cell phone that weighed 12 pounds is not only worthless, it's a joke. The same is true of the 1989 PC with (perhaps) 1 meg of memory and a 500-megabyte hard drive. Will anyone laugh at the optical networking gear of today 15 years from now? Perhaps, but only if the new optical systems cost a fraction of what they do today. Otherwise, the reaction would be a tendency to cry.

Right now, the rapid evolution of optical and wireless seems a dead heat. For every advance in digital signal processor (DSP) chips inside cell phones, there is an advance in lasers or fiber. But this might not continue to be true. Only time will tell if optical or wireless evolution—or both—are at the beginning of a cycle or near the end.

One of the things to watch is the number of companies and researchers in a given field. For example, not so long ago ATM was the technology of choice for many service providers. Users might still have IP, but only to feed massive ATM switches and backbones. But anyone that cared to look at the networking field in the early to mid-1990s could see storm clouds gathering for ATM. For every university student studying ATM, there were 10 studying IP. For every researcher looking at ways to make ATM better, there were 10 looking at ways to improve IP. For every ATM hardware or software vendor, it seemed there were 100 vendors making hardware or software geared toward IP networking. So whenever a network question came up in a real-world environment, there were many more IP voices in the room than ATM voices.

The best way to view the status of optical networking and wireless networking is probably not as some sort of competition, but as some sort of cooperation. The two technologies complement each other quite well. Fiber backbones have the massive bandwidths that service providers need. Wireless access has the freedom that wired solutions lack.

Wireless and optical networks might even come to share some networking techniques in common. One of these is code division multiple access (CDMA).

CDMA

One of the most exciting developments in the field of wireless communications is the evolution away from simple systems of channels sharing based on frequency division multiple access (FDMA) or time division multiple access (TDMA). The latest sharing technology is some form of *spread spectrum* technology employing code division multiple access (CDMA). CDMA is a direct sequence spread spectrum (DSSS) technology, as opposed to spread spectrum systems based on frequency hopping (FH). (There is also such a thing as *time hopping* spread spectrum, but this aspect of spread spectrum is seldom discussed in any detail.)

The following is not intended to be in any way a full tutorial on CDMA. Nevertheless, it is useful at this point to provide some of the details of just how CDMA works in order to understand what all the current excitement is about in the wireless community. There are several forms of CDMA, and any CDMA equipment vendor has to consider a number of factors to make operational decisions for the equipment being designed. This section is just a general introduction to the concept of CDMA; it does not necessarily imply that actual working CDMA systems function like this in detail.

The use of CDMA, especially wideband CDMA (W-CDMA), at least offers the promise of more bandwidth for wireless services. One reason that more bandwidth is available with CDMA is that CDMA does not chop up a spectrum allocation into bandwidth-limited channels, as do FDMA or TDMA. FDM allocates channels based on frequency, of course, and the channel can only support a stream of digits at a rate that can fit into the frequency range available. TDMA allocates channels based on a time slot, and no user can send bits faster than the rate established by the size of the time slot available, naturally.

With CDMA, it is possible to allow everyone to use the entire frequency range whenever they have bits to send. So there is no need to try to concentrate transmitter power inside a narrow channel. Power is *spread* across the *spectrum*, which is how spread spectrum got its name (actually, peak power is naturally lower the wider the range of frequencies generated by a transmitter, but this is not important for this discussion). The concepts behind spread spectrum have been around since World War II, but only recently have wireless components advanced to the point that such a concept as CDMA has become feasible. CDMA in World War II was explored for its *antijamming* features. This meant that CDMA signals could be freely intercepted by anyone, and yet the information content present was meaningful only to the sender and the receiver.

How is this possible? There are three key features of CDMA as a form of DSSS communications:

1. The signal sent occupies a bandwidth (frequency range) much greater than necessary to send the information. That is, a 4 KHz user signal might be sent in a 1.2 MHz CDMA bandwidth. This not only makes the signal immune to many jamming and interception techniques but makes it ideal for multiuser access, and at higher speeds than ever before.

2. The bandwidth is "spread" by use of a *code* (the "C" in CDMA) that is independent of the digital data sent. Now, modulation of a carrier will always spread the spectrum needed, but this is always seen as a sort of unwanted side effect (even in optical systems). It is the independence

of the code and data sent that distinguishes spread spectrum, and it is the size of the code chosen that spreads the spectrum out so widely.

3. The receiver synchronizes on the code used by the sender to recover the data content of the digital stream. The use of an independent code and this synchronization of sender and receiver are what allow multiple users to access the same frequency range at the same time.

The code used in CDMA is *pseudorandom*. This means that the code appears random but is actually highly predictable and repetitious, so that the receiver can easily synchronize on the pattern. The pseudorandom code is also frequently called *pseudonoise (PN)*, and CDMA is sometimes called *noise modulation*.

When used for voice systems, CDMA works directly on digital 64 Kbps voice channels. But CDMA has been applied to voice at 8, 16, 32, or 64 Kbps, mainly by "padding" the bit stream up to 64 Kbps. Usually some forward error correction (FEC) is added, doubling the actual bit rate to 128 Kpbs. But there is no need to limit CDMA input to voice or even 64 Kbps. W-CDMA increases the "channel" used in CDMA greatly, as high as 2 Mbps or more in some cases. But how do the individual users all sending on the same bandwidth at the same time not interfere with each other?

Simply because each user is employing a different PN code. The sender generates a different pseudorandom code for each channel and each connection. The information modulates this PN code, "spreading" the information. The resulting signal is the one that is sent. At the receiver, the same pseudorandom noise code is generated and matched to the arriving signal. The received signal is matched to the generated code sequence, recovering the data.

With CDMA, a digital stream of 0's and 1's is represented as a series of symbols called *chips*. The sequence of chips that make up the PN code for each user is unique, and this allows for overlap in time, frequency, and space for CDMA users. The number of code bits (chips) used for each information bit must be at least 10 but can be as high as 100 for commercial systems. Military systems, mostly concerned with security factors, can have PN codes up to 10,000 chips long. The number of code bits per information bit is called the *spreading factor* of the CDMA system. So spreading factors range from 10 to 100 in commercial CDMA systems. What this means is that the bandwidth used to send information is at least 10 times, and sometimes 100 times, greater than the bandwidth useful to the sender and receiver.

The trade-off for this apparently "wasted" bandwidth is in the lower peak power required for spread spectrum and the fact that CDMA allows multiple users to share the system bandwidth without elaborate control schemes that take time to set up or occupy otherwise usable channels.

The simplest way to modulate the information bits on a PN code sequence is just to invert the chips to indicate a 1 bit and to leave them alone to indicate a 0 bit. There are more elaborate schemes that can be used, such as *complex modulation*, but this simple inversion works fine and is widely used. The principles behind CDMA's use of pseudorandom noise codes, inversion modulation, and reception are shown in Figure 17-1.

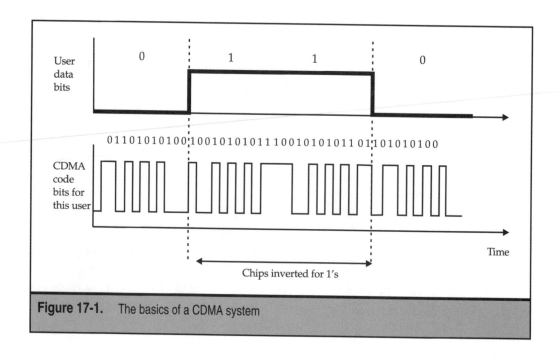

Figure 17-1. The basics of a CDMA system

Note that the signal can be corrupted in many places by other forms of noise and interference. This is natural, but noise tends to be concentrated into a narrow frequency range and at much higher power than the spread spectrum signal. So the noise will only cause the receiver to lose a few of the chips, preserving enough of the pattern to decide if the received bit should be a 0 or a 1. This is also shown in the figure.

CDMA works best with information that is bursty in nature. That is, periods of intense activity on a network between pairs of users are interspersed with longish idle periods. Fortunately, compressed and silence-suppressed voice, client/server traffic, and almost all other forms of digital information (with the possible exception of broadcast video) are quite bursty. CDMA does not limit bursty users to a narrow-channel bandwidth for sharing purposes. Users more or less freely compete for much wider bandwidth, and CDMA sorts it all out.

Optical CDMA

CDMA is just a technique for sharing a limited network resource (bandwidth) while requiring lower power and spreading the spectrum. But what works with wireless can also work with optical bandwidths as well. This brings us to optical CDMA, or OCDMA.

Why bother to apply CDMA to optical networks at all? DWDM systems are anything but limited by bandwidth. However, there are other concerns in optical DWDM systems that CDMA addresses nicely.

One issue that remains in optical networks is the fact that high-powered lasers tend to operate in the *nonlinear* regions of the fiber that carries the signal. Nonlinearities almost always are bad for the system, and the laser power must be carefully tuned to operate at the maximum power level without pushing the whole system into the nonlinear operating range where things just stop working altogether.

One of the features of CDMA is the ability to spread power over a wide range of frequencies (or wavelengths, which amounts to the same thing). Lower peak power means that there is less of a concern over nonlinearities in the optical system.

Interference (really just interactions of all sorts) between light at different, but closely packed, wavelengths is also a concern in optical networks. Optical CDMA bit streams would be much more resistive to these interactions for the same reason that wireless CDMA bit streams are much more reliable in the presence of noise. So OCDMA makes sense for a number of reasons.

Suppose four pseudorandom noise code patterns could be generated for each wavelength used on a DWDM link. Each bit stream could operate at 10 Gbps without much of a problem. If there are 32 wavelengths on the fiber, that total capacity of the link would be $32 \times 4 \times 10$ Gbps = 1.280 Terabits per second. The "spreading" would not be much at optical wavelengths, so the channel spacing of the wavelengths would still be 100 GHz (about 0.4 nm), or even 25 GHz (about 0.1 nm).

One concern with OCDMA is that this coding needs to be done electronically, introducing more electrical dependence into the optical system. However, it might be possible to perform this code conversion optically in the future.

Theoretically, it is possible to envision the optical equivalent of wideband CDMA (W-CDMA) as well. This would be broad or wide OCDMA (W-OCDMA?), of course. Optical signals could spread across the whole 1,300–1,600 nm window in the future.

OPTICAL COMPUTING

As has been pointed out several times previously in this book, the real promise of optical networking cannot be fulfilled as long as it is necessary to keep switching between optical and electrical forms of signals as information makes its way across a network from sender to receiver. Such electro-optical conversions add overhead, delay, and complexity to the network. A large part of the history of optical networking has been the development of newer methods and components that allow information, once converted to an optical format, to remain in optical form as long as possible on its journey.

So simple links of fiber had optical amplifiers added, getting rid of the need for electrical repeaters and regenerators. Optical cross-connects and switches acted on the very wavelengths that made up the signal, not their electrical counterparts. Optical routers will continue this trend, and optical transponders will allow wavelengths to shift and allow a series of wavelengths to form the path across the network.

But the weakest link in the optical network to date has been the inability to fabricate a truly optical computer. Why is this such a critical component of optical networking? Because networks really do three things, not just one. Everyone knows that networks exist mainly for the delivery of information across the network infrastructure. But in addition to this main *transfer* function, networks must have a way to *control* and *manage* this flow of optical signals.

These control and management functions are usually gathered under the label of operations, administration, and maintenance (OAM). Even such a simple task as redirecting the flow of light from one destination to another can be an OAM control function. One or more network nodes have to be instructed how to handle the light in the new way. Network management is most often mentioned in the context of alarm conditions and link or equipment failures, but management is really a much broader area of concern. Management can and should include such crucial network operations as capacity planning, reliability, and network performance.

The control and management tasks in an optical network today are done with a series of electronic integrated circuits (ICs) embedded in the network nodes and associated computers that make up the optical network. Naturally, it would be a real advantage for optical networking overall to make these control and management functions as optical in nature as the information signals passing through the devices. But this would require the development of an *optical IC* that could do everything an electronic IC can do, but using light instead of electricity.

Electronic ICs do two major things. First, they shuttle electrical signals around a very small component that has all of the characteristics that a complete computer has. So the IC has a memory area of the chip, and a central processor, and a unit to perform arithmetic and logical operations on the bits (which is why the IC is *integrated*: years ago, each of these components was a separate "box").

Second, an electronic IC must have a way to *store* the bits in small pieces of semiconductor material (such as silicon) until the components have had a chance to act on them. For example, consider the simple act of adding the bits in two different memory locations on the chip and placing the result in a third memory location. The electrical bits that tell the chip want to do are *fetched* into the processor portion of the IC, where the instruction sits for a while (probably a few billionths of second). Then the bits that have been sitting in the two memory locations referenced are gated to the *arithmetic and logic unit (ALU)* area of the chip, where they are held until ready to be acted upon. Once the bits have been added, the result is piped back to the memory section of the chip, where the result sits while the processor gets ready to perform the next instruction. (This is an extremely simplified version of what actually happens in an IC, but it is accurate enough for the purposes of this section.)

Now, fiber optics has already been used to form the waveguides on some IC chips that connect the IC components such as the processor and memory portions together. Without optical waveguides, electricity must follow what amounts to tiny wires *etched* onto the chip's semiconductor *substrate*. The problem is that flowing electricity always sets up additional magnetic fields and electrical currents that can actually disrupt the processes going on inside the chip itself. The IC *microprocessors* pick up these stray signals and the internal noise limits the size and performance of the IC. However, when optical waveguides are etched onto the substrate, flowing light does not have this effect on adjacent processes. Light can be shuttled around the chip much more efficiently than electricity.

So why not just make both the shuttling of signals around the IC chip *and* the components such as the processor and memory optical as well? This would be a giant step on the road to a completely optical computer to control and manage optical networks.

The problem is that once light arrives at an IC memory location or processor *register*, the light must wait around for something to happen. The light has to be *stored* in the IC for varying amounts of time, ranging from only a few nanoseconds for a processor to literally forever in a memory location. Light by its very nature wants to move, and move rapidly. In a billionth of a second (one *nanosecond*), left to its own devices, light will be already a foot (about 1/3 of a meter) from where it started, and far from the IC where it is needed. How can light be captured and stored in the optical equivalent of an IC?

Why not just confine the light with a system of simple, but very small, mirrors? That could be done, and is done in laser diodes, of course. But confining light inside a small space, unless done very carefully, will *change* the properties of the light (that's what the laser diode is all about). And since the light inside an IC controlling an optical network node represents information, changing the properties of the light, and hence the meaning of the information, is not a good idea at all. Some means must be found to store light in an IC size (or smaller) area without changing the "meaning" of the light.

This *trapping* of light in a small area without changing the light is a very active area of research today. Oddly, the most promising current approach to this idea of light trapping was invented independently by two researchers who had never met, but came to work within few miles of each other in the state of New Jersey.

In 1987, Eli Yablonovitch was working at what was then Bell Communications Research (Bellcore, now Telcordia) in Red Bank, New Jersey. Sajeev John worked not too far away on the faculty at Princeton University. Both had written a paper and submitted it to the same journal for consideration and possible publication. When the two met in the spring of that year, they decided on a name for their new idea: *photonic band gap*.

All the two had in 1987 was an idea. Today, the idea has become research into *photonic crystals*. Photonic crystals, like all crystals, have a regular *lattice* structure, meaning there is a pattern to their solid form that repeats itself over and over. This pattern looks like a cage to photons of light, and these tiny lattice cages can trap photons for a period of time.

The trick is to capture the light without absorbing or otherwise changing the essential properties of the light. Trapped light can be let out when needed, in the same way that electricity is let out when needed in an IC. The idea of a photonic *band gap* is based on the related phenomenon of the *semiconductor band gap*. Solid materials always have electrons that exist only in well-defined energy bands, or levels. This is also the reason that electrons can only orbit atoms at certain energy levels. Semiconductors are different than most ordinary solid materials in that semiconductors have very large "gaps" between the bands (energy levels) of electrons bound to atoms and the higher energy levels of electrons that flow through the material and carry electric current. That is why "semi"-conductors sometimes act as conductors and sometimes act as insulators, and semiconductor roles can be controlled very precisely. Pure silicon cannot support electrons at band-gap energies, but doping the silicon with impurities such as arsenic can create a semiconductor.

Photonic ICs would have information flows much faster, and with much more "bandwidth," than electrical ICs. Photonic ICs are potentially a thousand times or more faster than their 1 Gigahertz counterparts, into the Terahertz ranges. Even if it takes a while for photonic ICs to be used to construct a complete optical computer, the photonic ICs could be used on the line cards and boards and could drive 10 gigabit per second Ethernet links with ease. Throughput might be increased on such links by a hundred times or more once needless conversions from optical to electrical and back are eliminated. The Internet could potentially run 100 times faster over the same links that are in place today.

But before a photonic chip or chipset can be made, the optical equivalent of silicon is required. As silicon-based ICs trap electricity when and where needed, the photonic chip would trap light. There are gases that can slow light down to less than 100 miles per hour, but these methods require a roomful of exotic gases and control electronics. Photonic ICs would be as small as electrical ICs, if not smaller.

The long road to the photonic band gap started in 1958. Researcher Philip Anderson at Princeton showed that *electrons* could be trapped in a *disordered* array of atoms. Random arrays always bounced the free electron around and eventually returned the electron to where it started. Sajeev John asked if photons could be made to behave the same way.

The answer turned out to be yes, random materials could trap light. But it was not until 1997 that Diederik Wiersma and colleagues, and Ad Lagendijk used powered gallium arsenide to trap light. The grains in the powder were smaller than the wavelength of the laser light, and the light bounced around in loops without finding an exit. But it is hard to make an IC out of a powder. Some type of solid material would have to be found to do the same thing. But this would probably reintroduce some form of regular crystal structure. But how could the material be regular and random at the same time?

The answer turned out to be simple. Light always interacts, even with itself, whenever it can, although the term used to describe light interactions is *interference.* In normal usage, the term "interference" always has negative connotations, but when it comes to light, there are positive interactions (constructive interference) and negative interactions (destructive interference). When light travels through a crystal, some of the light reflects off each layer or plane of atoms while the rest of the light passes through, a phenomenon known as Bragg reflection. With enough layers, all of the light of a certain wavelength can be reflected. If the distance between the planes (layers) is exactly one half of the wavelength of the light, then the interference will be constructive. However, this applies only to light entering the structure at a certain angle. If light can be trapped no matter what angle it enters the lattice, this would capture the light as required.

Since this lattice would only work this way for a very narrow range of wavelengths, it would trap only those wavelengths inside this *photonic band gap.* Making the structure work for light at any angle would require not a series of flat crystal layers, but a series of spheres forming shells, one inside the other. This crystal structure is called a *face-centered cubic.* In the faced-centered cubic structure, which occurs in many natural crystals, not only is there one atom at each corner of the "sphere," but there is also one atom in the center of each cube's "face." This idea is shown in Figure 17-2.

Yablonovitch worked visually, with checkerboard-like doodles, while John worked with mathematics, but the outcome was the same. The next challenge was constructing the lattice, which normally has spacings of a few nanometers (nm), to have spacings that would work with light at about 700 nm or more. But the larger spacings would fill with air. And air, having a different index of refraction than the crystal material, might just

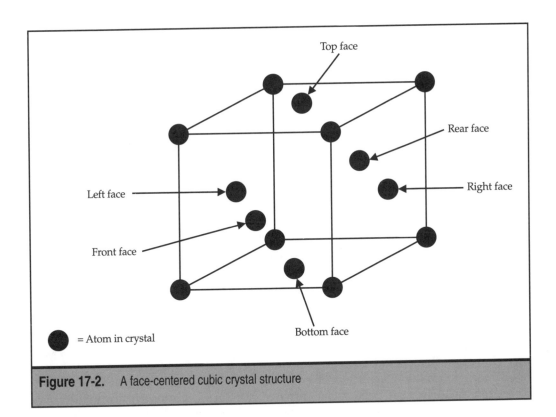

Figure 17-2.   A face-centered cubic crystal structure

make it seem that the spacings between layers were exactly one-half wavelength in all directions, which was exactly what was needed.

After years of experiments with trapping microwaves, which have wavelengths 100,000 longer than light, a few working prototypes were fabricated out of Plexiglas drilled with airholes. Work then turned to the creation of a diamond crystal stacked in a kind of crosshatch pattern with semiconductors forming "logs" and air in between. There are other architectures as well.

Of course, the photonic band gap cannot be *too* perfect. There has to be a way for the light to get in and a way for the light to get out when needed. In some cases, a photonic band gap film with a "hole" in one end can act as a laser, but a laser much smaller than current semiconductor lasers. Defects in the photonic band gap lattice can form waveguides for the light to follow. Punching a series of holes in just the right place of a photonic band gap device can make a waveguide that can even make right angle turns, something impossible to do with fiber optic cables.

And since the light inside the photonic crystal travels in air, the structure is much more attractive than the core-cladding all-glass structure of fiber optic cables. Glass tends to distort light in many ways, but photonic crystals are essentially hollow fibers. There are even prototype "coaxial fibers" with light confined between two concentric glass tubes by Bragg reflection. Such "fibers" could handle higher-powered light sources and much wider wavelength ranges than existing fibers. So photonic crystals have promise not only for optical ICs, but also for smaller lasers and newer types of fibers. The application of photonic crystals to optical networking has just begun.

INTO THE FUTURE...

Some new developments in optical networking have been areas of active research. These areas promise to keep the pace of evolution in optical networking very high. This section will mention just a few of the most active areas of research.

Dark Solitons

Solitons can be used to generate the pulses of light used to represent information on an optical link. Soliton DWDM technology promises to extend the reach of these new optical systems without requiring the reshaping of the optical signal due to the effects of dispersion. Solitons are a form of standing wave and when used in optical networks are essentially laser pulses inserted onto a fiber with intentional non-linearities. Usually such non-linearities are to be avoided, but solitons are pulses in the fiber core that keep their shape over long distances. Solitons do not disperse, but they do attenuate and need periodic amplification, a fact not usually played up. However, the fact that soliton pulses keep their shape over long distances is a real advantage. Think of solitons as a kind of "sonic boom" in light on the fiber. The problems with using solitons in general fiber systems have been threefold. First, solitons are hard to control precisely, mainly due to their non-linear characteristics. Second, solitons have to be relatively isolated in time, since two solitons placed close together will disrupt each other. Third, soliton generation has been very expensive in transmitters.

But *dark solitons* might be a way to get around these limitations. With dark solitons, the soliton is represented not by the presence of light in soliton form, but by the *absence* of light in soliton form. A dark soliton is a gap in a high-powered beam of light. The light, once on, is easier to control, and the "shade" needed for the dark soliton is simpler to build for the same reason that shutters are simpler than light bulbs. Dark solitons still need adequate spacing, but soliton spacing has been limited by the need to pulse the transmitter as well as the need to spread the solitons out.

Light Guiding Light

All fibers will tend to disperse a light pulse. This dispersion is almost always bad, although there are times that dispersion is used creatively to compensate for some other characteristic of the fiber optic cable. Overcoming dispersion often requires very precise control of the refractive index (RI) of the fiber optic cable, and light intensity affects the RI as well (known as the optical Kerr effect).

The idea behind *light guiding light* is that a high-powered beam of light can change the RI of any dispersive substance, even air, and form a kind of *waveguide* through the medium. Light guiding light is an important area of research in the field of free space optics (FSO).

ZBLAN Fiber

Fiber optic cables do not have to be made out of silicon glasses. New, experimental fibers are made from zirconium (chemical symbol Zr), barium (Ba), lanthanum (La), aluminum (Al), and sodium (Na). This is ZBLAN fiber, which has attenuation as low as 0.01 or 0.001 dB per kilometer. These fibers could easily cross the oceans without the need for amplification, a feat that has been a dream for some time now. (ZBLAN fiber signals can be amplified by the same rare-earth elements as current fibers, if necessary.)

ZBLAN fiber is not perfect, however. Lanthanum is far from common or inexpensive. ZBLAN fiber requires a larger core and different wavelengths than those currently used by optical networks. And the ZBLAN fluoride crystals that make up the fiber have to be coaxed into forming under "microgravity" conditions, as in the Space Shuttle or on a space station. Nevertheless, the promise is there.

This section has just scratched the surface of current optical networking research. New developments occur every day. But it does give a flavor for the range and depth of optical networking research.

PART VII

Reference

APPENDIX A

Fiber Optic Cables

By the end of Chapter 2, most of the basic components and properties of a single-fiber optic cable had been discussed. Details like structure (core and cladding), optical properties (refractive index and others), mode of operation (single mode and multimode), and so on, were discussed, but only from the perspective of an individual fiber optic strand. However, when deploying optical networks, it is extremely rare to run or work with a single strand of fiber optic cable. As mentioned at the end of Chapter 2, most fiber optic cables consist of many strands of fiber bundled together. Not only does this help when running many strands of fiber, but the cable construction helps to protect the fibers inside the cable jacket.

There are many types of fiber optic cables. This is because there are many types of environments in which the cables must perform. Fiber optic cables are made for a particular application, and there are literally hundreds of different types of fiber optic cables. In fact, it is possible to have a few miles or kilometers of special cable run off for a special project without too much of a problem. The details of cable construction are always open to negotiation with the cable manufacturer. Fortunately, they share many characteristics that make it at least possible to establish a workable number of categories, which we will explore in this appendix.

FIBERS NEED PROTECTION

The whole idea behind the concept of a fiber optic cable is to protect the fibers inside from environmental damage. Many of these threats have been discussed more fully in other parts of this book, but it cannot hurt to briefly mention them again here.

▼ **Stress** Fiber optic strands of a given cross section are stronger than steel. However, the small cross sections of a fiber optic cable make the cable vulnerable to tensile stress. Even if pulling very hard on a fiber cable does not actually break the cable, the stress can easily add to the attenuation of the installed cable.

■ **Bending** If a fiber is bent during installation beyond its designed bend radius (typically set at ten times the overall cable diameter), signal loss can result. Microbends are actual crimps in the cable and might occur, for example, when a heavy file cabinet is placed in error on a fiber cable during installation. This is not much of a problem with outdoor or undersea fiber optic cables.

■ **Environmental damage** Gnawing animals, such as rats, squirrels, gophers, and so on, will chew on cables whether they find them pleasant to the palate or not. The issue is just how much damage the gnawing will cause. The other major source of damage in this category is construction equipment. Ironically, in many cases a fiber optic cable break is caused by the installation of new cable along the same right of way by a fiber optic cable–laying machine.

■ **Installation damage** This was more of a problem in the early days of fiber. Some crews handled the fiber cable very roughly, while others tended to baby the cable as it was being installed. The result was there were many "good runs" that performed well and some "bad runs" that never seemed to work right. In those days, the fiber worked well *if* the fiber survived the installation process. However, even today, hanging cable for a long vertical core or pulling cable through a long conduit can damage the cable in spite of the best efforts on the part of the installation team.

▲ **Water**   Water is sometimes called the "universal solvent." Put something in water long enough, and water will get in. In water, fiber will gradually pick up hydroxyl ions that absorb light, and eventually the fiber stops working. Moreover, water has been known to cause "micro-cracking" in the glass, and this scatters the light. The micro-cracks also weaken the fiber. Waterproofing efforts in fiber optical cables often exceed those made for electrical cables.

BASIC FIBER CABLE ENVIRONMENTS

There are at least seven major types of environments where fiber optic cables are installed. There are fiber optic cables designed especially for each of these environments.

Long Distance Outdoor Buried Cable

The outdoor buried cable typically used for long-distance telecommunications have a large number of single mode fiber optic strands (up to 100). These cables also have extensive waterproofing, strengthening materials, and often have additional external armor.

Campus Range Outdoor Buried Cable

These cables are similar in construction to the long-distance cables, but usually are much lighter. They typically mix single mode and multimode fibers. The waterproofing is as good as the long distance cable, but there is less external armor and often less strengthening material. However, the conduit that the cable is placed in (often a two-inch steel pipe) makes up for the lack of cable protection.

Outdoor Aerial Cable

When fiber optic cable is run overhead, the main source of stress is the fact that the cable can only be supported at intervals, such as at utility poles. The typical interval between these poles is 75 to 100 feet (25 to 30 meters). These cables therefore need great tensile strength. These cables usually have a separate steel support cable that takes stress off of the fiber optic cable itself.

High-voltage Ground Wire Outdoor Aerial Cable

A popular place to run fiber optic cable today is inside the ground wire (earth wire) of a high voltage electrical transmission system. Power companies, as has been pointed out in this book, have rights of way that can be used for communication services. The ground wire is usually the top wire on the electrical transmission tower, and this give the fiber an added measure of protection from vandalism and other threats. These cables are often called Optical Ground Wire (OPGW) cables.

Undersea Cable

Many of the details about undersea fiber optic cables were investigated in Chapter 15. The major challenge for undersea cables is to keep the high-pressure sea water out of the cable. The need for elaborate waterproofing means that there are typically fewer fibers in an undersea cable than in a terrestrial long distance cable (perhaps 8 to 20 fibers). Undersea cables often need to carry electrical power for optical amplifiers as well.

Indoor Cable

Indoor fiber optic cables usually have only a few fibers, often only two, and these are almost always multimode fibers. There is little need for waterproofing or armor in an indoor, office environment, although there are exceptions such as steam tunnels and elevator shafts. The major threats are from rats gnawing the cable and installation damage. In addition, it is often required to use "low smoke" jackets so that the cables will not give off toxic fumes if they burn. Indoor cables longer than 1000 feet (300 meters) are rare. Indoor cables are very light and flexible, and are usually terminated with connectors and seldom spliced.

Jumper or Patch Cables

Short jumper or patch cables are used between optical network components in an office or equipment building environment. These fiber optic cables endure a relatively harsh environment compared to some other types of fiber optic cables. The cables are often just run across a floor where they are stepped on, chairs are rolled over them, and have heavy objects dropped on them. This activity often breaks the fibers.

BASIC FIBER OPTIC CABLE CONSTRUCTION

There are three basic ways that fiber optic strands can be packed inside a cable. These are:

▼ **Tight Buffered** In a tight buffer construction, there is a secondary coating of plastic encasing the fiber optic strand or strands. This is similar in concept to the construction used in electrical cable. The coated cables are about 1 mm in diameter. This type of cable is used for indoor applications. The number of fibers encased in the secondary coating is usually low, and the distances spanned by this type of cable are short (no more than campus application lengths).

■ **Loose Tube (Loose Buffer)** In a loose tube (also called loose buffer) construction, there is a small number of fibers run inside a plastic (usually polyvinyl chloride, or PVC) tube about 4 to 6 mm in diameter. There are typically two to eight fibers inside the tube. There is a lot of empty space in the tube for the fiber to move around in the "loose" construction. Since the fibers are loose, the fibers inside curl themselves into a helical twist. There is about five to ten percent more fiber inside a tube of a given length (e.g., 110 meters of fiber inside a 100 meter tube). This means that when the tube is bent or stressed, the fibers inside are not stressed or bent.

▲ **Gel Filled**   In most outdoor fiber optic cables today, the spaces inside the cable between the fiber strands are filled with a *gel*, a thick jelly-like substance. The gel adds many benefits to loose tube constructions; it also adds waterproofing characteristics to the cable, prevents the fibers from "chaffing" against each other when the cable is installed, and prevents microbends. All of this in addition to the stress relief that loose construction brings to the fiber.

The differences in the three basic fiber optic cable construction techniques are shown in Figure A-1.

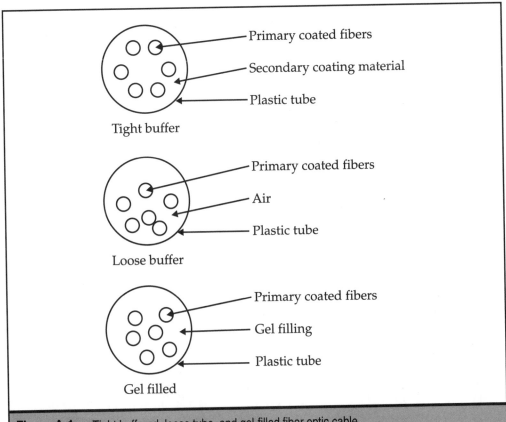

Figure A-1.   Tight buffered, loose tube, and gel-filled fiber optic cable

The precise nature of the gel used is an active area of research in the fiber optic cable construction industry. Petroleum jelly is inexpensive, and was used for years for this purpose, but exhibits wide ranges of viscosity (thickness) between extremes of temperature.

The viscosity or thickness of the gel used is very important. The gel must be thin enough to allow the fibers to move for stress-relief. It must also be thin enough so that it will flow easily and quickly during cable construction. On the other hand, the gel must be thick enough so that the fibers do not settle to the bottom of the cable and the gel does not settle if the cable is run vertically in a riser. The gel must also be thick enough at high temperatures so that a severed cable can be spliced on a hot day without the large amount of gel leaking out.

Today there are specially designed synthetic gels that display all of the needed characteristics. These gels have very stable viscosity over a wide temperature range.

THE FOUR MAJOR CABLE TYPES

There are four major types of fiber optic cable: outdoor cable, indoor cable, air blown fiber (ABF, used both indoor and outdoor), and undersea cable. Within each category, there is a lot of variation, so this section can give only the major characteristics of each type.

Outdoor Cable

A typical outdoor buried fiber optic cable has six gel filled tubes supported by other cable structures to add strength. The most important structure is a central steel cable, but there can also be steel wire outside armor. Some cables, especially those used for shorter distances, will use strong plastic instead of steel as a central support cable. In fact, in many cases the whole cable is non-metallic. The steel jacketing can be made of stainless steel, but this is very expensive and usually done only when the customer requests it, since stainless steel will not rust as ordinary steel does. In many areas, especially the tropics, an outer layer of nylon is usually added to discourage termites.

Not only does gel fill the fiber tubes, but the gel fills the spaces between the fiber tubes as well. This structure is shown in Figure A-2.

The cable in the figure has 6 fiber optic strands in each gel filled tube, so the entire cable contains 36 fibers. These can be used in pairs for inbound and outbound traffic, giving 18 fiber pairs for optical traffic. On the other hand, these fibers could all be used for inbound (or outbound) traffic and there could easily be another cable used to handle the traffic in the opposite direction.

There can be as many as 12 tubes and as many as 8 fibers per tube, giving 96 fibers in the total cable. If it is necessary to carry electrical power inside the same cable, copper wires can be run instead of the fibers in one or more of the tubes.

There is also a variation called *segmented-core* cable. This type of cable has a "backbone" or slotted core that looks like a small toothed wheel or gear. This is made of strong plastic and has a central core for a support cable. The individual fibers are placed in the slot between the spokes (often called channels) and the indentation is filled with gel. A segmented-core cable is shown in Figure A-3.

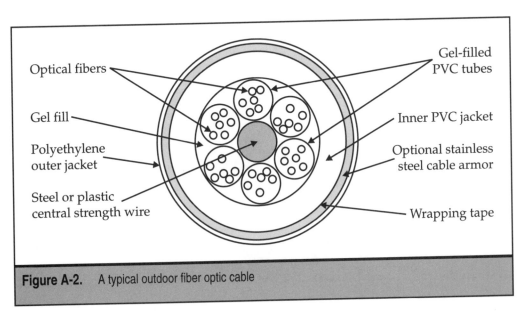

Figure A-2.   A typical outdoor fiber optic cable

There can be as many as 20–24 channels and as few as one fiber per channel. More typical segmented-core cables contain six channels and eight fibers per channel (48 fibers in total). Segmented-core cables can be lower in price than other types of outdoor cables.

Outdoor aerial fiber optic cables have stress problems, as pointed out earlier. It is tempting to create a very strong central cable to counter this stress. However, this just adds weight to the cable, and *increases* the stress at the points where the cable must be suspended. So a separate, but integrated, support wire is often added to the aerial cable as shown in Figure A-4.

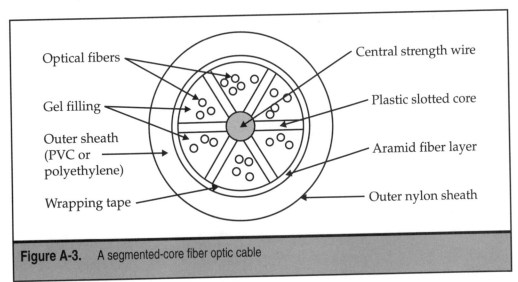

Figure A-3.   A segmented-core fiber optic cable

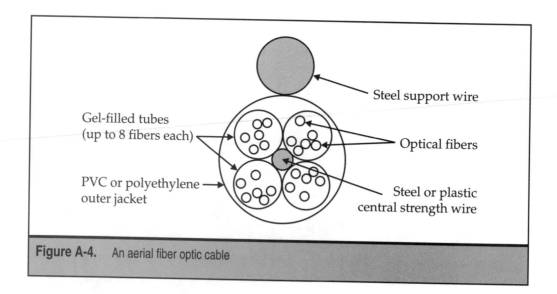

Figure A-4. An aerial fiber optic cable

The basic structure of the cable itself follows the typical gel-filled, outdoor cable. The steel support wire is also jacketed to protect it from rust.

In the United States in the early 1980s (the pioneer days for a lot of fiber cable types), an interesting variation was used for medium-distance outdoor cabling, up to about 6 miles (about 10 km) or so. This cable mounted multimode fibers as a ribbon cable, a very flat package about 5 mm across that often contained 12 fibers. This type of cable is shown in Figure A-5.

The fiber were held in place by glue and covered with two layers of mylar tape. The ribbons could be stacked to yield squareish structures with 144 fibers, and so on.

The great attraction of these cable types was their ease of termination. A gang connector could be added in one operation, saving time and money. This worked fine for multimode fiber, but could not be used with the precision that single mode fibers demanded. As single

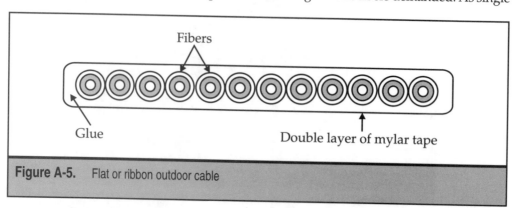

Figure A-5. Flat or ribbon outdoor cable

mode fibers replaced multimode fibers for telecommunications purposes, this system became obsolete and is no longer used.

Indoor Cables

Indoor fiber optic cables can consist of a single strand of fiber with some additional strength material and a PVC jacket. There is always a *secondary coating* of buffer material around the primary coating surrounding the fiber optic cladding. There is usually some strengthening material like Aramid fibers (not *optical* fibers, just fibers!) between the fiber optic strand cladding and the jacket. The whole package is about 2 mm in diameter. However, this single-strand structure is almost never used in the real world.

After all, it is more cost effective to build a cable with more than one fiber strand. And it saves on installation costs as well. Most applications require an inbound and outbound fiber anyway. So the simplest indoor fiber optic cable appears as shown in Figure A-6.

In the figure, two fibers are carried in a common sheath. The fibers are color coded and have a clear outer sheath.

If more than two fibers are carried this way, it is common to arrange the fibers around a central strength core. All of the fibers have their own secondary coating, strengthening material, and color coding. Cables like this are usually tight buffered for use in high-rise building cores. The tight buffering takes the stress of their own weight off of the fibers in a riser. This type of cable is shown in Figure A-7.

There can be up to 12 fibers in this type of tight buffered cable. It is also possible to have indoor ribbon cables, similar in size and shape to the outdoor ribbon cables discussed. These cables can be used under carpets in some cases, or in tight spaces where round cables cannot go. However, when used under carpets, the heavy foot traffic in some areas requires extra protection for the cable.

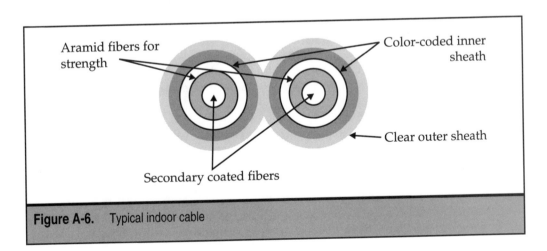

Figure A-6.   Typical indoor cable

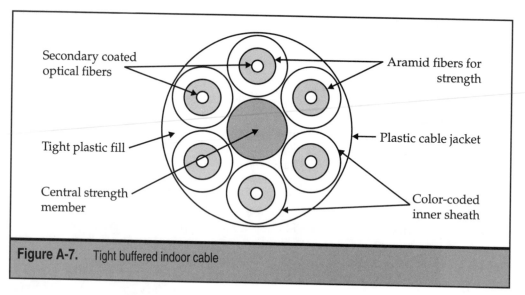

Secondary coated optical fibers

Aramid fibers for strength

Plastic cable jacket

Tight plastic fill

Central strength member

Color-coded inner sheath

Figure A-7. Tight buffered indoor cable

Air-Blown Fiber (ABF)

It is the repeated handling of fiber optic cables that causes damage during installation. In an air-blown fiber (ABF) installation, the main activity is handling the conduit used to "blow" the fiber cables through. Handling of the fiber is kept to a minimum. Very light fibers can and must be used for this purpose. The main propellant is usually compressed nitrogen, since compressed air can be corrosive to the plastic jackets used on fiber optic cables.

There are many different ABF systems. The most popular allow for the installation of 2 to 18 strands of fiber in a single conduit tube. Usually used indoors, ABF systems have been successfully deployed in metropolitan areas for runs up to 2 km (a little more than a mile).

The fiber bundles used for ABF systems are not only lightweight, but have aerodynamic properties. The fiber floats down the tube and there is little contact with the side walls of the conduit. The process is very fast, and fiber installation speeds are up to 50 meters (about 150 feet) per minute.

ABF systems were much more popular in the early 1980s than they are today. The idea behind ABF was that fiber designs and specifications were changing rapidly, yet fiber remained relatively expensive to purchase and install. An ABF system allows the older fiber to be drawn out, and the same system of conduit to be used to blow in new fiber. As fiber specifications have stabilized throughout the 1990s, ABF systems have become a luxury rather than a necessity. However, if optical networks and DWDM require massive changes to existing fiber infrastructures, ABF systems will enjoy a new round of popularity.

There are real advantages to ABF systems in addition to ease of fiber replacement. These are

▼ The conduit sections for ABF can be installed as straight runs, fibers blown segment by segment, and then the conduit joined later. This can be much more cost effective and more reliable than attempting to install fibers over very twisted paths all at once.

- The fiber blown in can be one unbroken length, without the need for splices or connectors that can introduce additional signal losses. This is especially nice when there are indoor and outdoor fiber segments. Usually, there would be a join between indoor and outdoor runs. ABF can cable a whole campus without numerous splices or connectors.

- The conduit tubes can have multiple cavities so that future fiber runs can be added. This not only saves money, but makes installation times much quicker and more flexible.

- The conduit tubes can remain empty until needed. It is even possible to withdraw a fiber from one tube and re-blow the fiber into another, previously empty tube. A company could, for example, plan to move a department in two years. ABF conduit can be run to both the present and future locations, and the active fiber withdrawn and redeployed when needed. Without ABF, this scenario becomes expensive and a real problem when it comes to scheduling.

- ▲ As mentioned, the invention of new fibers specifically for optical networking and DWDM makes any method that eases the installation of new fibers (and the retiring of older fibers) very, very attractive.

If nothing else, ABF systems will be a real advantage for any organization if and when single mode fiber replaces multimode fiber to the desktop.

Undersea Cables

Undersea cable construction and installation was discussed in detail in Chapter 15. There is no need to repeat this discussion in full, but some of the basics of undersea fiber cable designs are repeated here for completeness.

Undersea cables have fibers organized around a copper-jacketed steel cable called a *kingwire*. The whole assembly is wrapped in a protective sheath of nylon and hytrel (a kind of "plastic rubber") and hermetically sealed inside a copper sheath that is welded shut. This sheath not only protects the fibers but also can carry electricity to any powered components along the fiber path, such as optical repeaters. The assembly process is carried out in long, narrow buildings hundreds of feet in length. The copper-coated cable is further encased in liquid polyethylene plastic. High-strength steel cables are embedded in the plastic before it can harden. This structure is used for the mid-ocean portion of the cable, to keep them as light as possible as they hang off the back of a ship on their way to the ocean floor two miles (about 3 km) below.

Closer to shore, the cable is protected by another layer of galvanized steel cables. This portion of the cable can weigh 16 times more than the mid-ocean cables, and are much more costly. Yet the cables are surprisingly thin: the typical undersea cable is only about 1½ inches (38 mm) in diameter. Even when another layer of protection is added to the coastal portions, as is sometimes done, the resulting cable is about 3 inches (65 mm) in diameter. The structure of the modern undersea fiber optic cable is shown in Figure A-8.

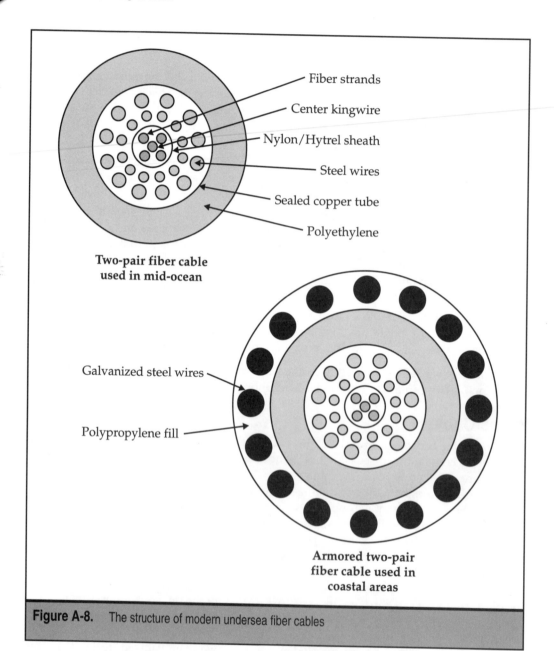

Fiber strands
Center kingwire
Nylon/Hytrel sheath
Steel wires
Sealed copper tube
Polyethylene

**Two-pair fiber cable
used in mid-ocean**

Galvanized steel wires

Polypropylene fill

**Armored two-pair
fiber cable used in
coastal areas**

Figure A-8. The structure of modern undersea fiber cables

The manufacturers and vendors of undersea fiber optic cable usually distinguish between "deep sea" and "coastal" (or "shallow water") cable types. In most cases, water is considered to be "deep" if the bottom is more than 1000 meters (about two-thirds of a mile, or 3000 feet) from the surface.

APPENDIX B

Acronym List

AAL5   ATM Adaptation Layer type 5

ABF   Air Blown Fiber

ABR   Available Bit Rate

ACELP   Algebraic Code-Excited Linear Predictive

A/D   Analog to Digital (conversion)

ADSL   Asymmetrical Digital Subscriber Line

ACSE   Association Control Service Element

ADM   Add-Drop Multiplexer

AIS   Alarm Indication Signal

AIS-L   Line level Alarm Indication Signal

AIS-V   Virtual tributary level Alarm Indication Signal

AIU   Access Interface Unit

ALU   Arithmetic and Logic Unit

AM   Amplitude Modulation, Administration Module

AMI   Alternate Mark Inversion

AMPS   Advanced Mobile Phone System

ANSI   American National Standards Institute

APD   Avalanche PhotoMode Detector (or PhotoDetector)

APS   Automatic Protection Switch

ARIB   Association of Radio Industries and Businesses

ARP   Address Resolution Protocol

ARPA   Advanced Research Project Agency

ASE   Amplified Spontaneous Emissions

ATM   Asynchronous Transfer Mode

ATSC   Advanced Television Systems Committee

AU   Administrative Unit

AUC   Authentication Center

AUI   Attachment Unit Interface

AU4-Nc   N AU4 signals concatenated

AU-n   Administrative Unit-level n (n = 3 or 4)

AWG   American Wire Gauge

B8ZS   Binary 8-Zero Substitution

B-DCS   Broadband Digital Crossconnect System

B-ISDN   Broadband ISDN (also, BISDN)

BER   Bit Error Rate (Ratio)

BFOC   Baseband Fiber Optic Connector

BIM   Byte Interleaved Multiplexer

BIP   Bit Interleaved Parity (check)

BITS   Building Integrated Timing Supply

BLSR   Bidirectional Line Switch Ring

BOC   Bell Operating Company

BPI   Baseline Privacy Interface

BRI   Basic Rate Interface

BSD   Berkeley Systems Distribution

BSC   Base Station Controller

BSS   Base System Station

BTS   Base Transceiver Station

C-n   Container-level n (n = 11, 12, 2, 3, or 4)

CAN   Campus Area Network

CBR   Constant Bit Rate

CC   Composited Clock

CCC   Clear Channel Capability

CCITT   International Telegraph and Telephone Consulative Committee

CCS   Common Channel Signaling

CD   Compact Disc

CDDI   Copper Distributed Data Interface

CDMA   Code Division Multiple Access

CDPD   Cellular Digital Packetized Data

CELP   Code-Excited Linear Predictive

CEPT-n   Conference of European Posts and Telecommunications = level n

CEPT-1   see E1

CEPT-2   see E2

CEPT-3   see E3

CEPT-4   see E4

CFI   Canonical Format Indicator

CLEC   Competitive Local Exchange Carrier

CLLI   Common Language Location Identifier

CLP   Cell Loss Priority

CNLP   Connectionless Network Layer Protocol

CMISE   Common Management Information Service Element

CMTS   Cable Modem Termination System

CNN   Cable News Network

CO   Central Office

CORBA   Common Object Request Broker Architecture

COT   Central Office Terminal

CPE   Customer Premises Equipment

CPM   Cross-Phase Modulation

CPU   Central Processor Unit

CRC   Cyclic Redundancy Check

CS   Convergence Sublayer

CSA   Carrier Serving Area

CSMA-CD   Carrier Sense Multiple Access with Collision Detection

CU   Channel Unit

CUG   Closed User Group

CV   Coding Violations

CV-P   Coding Violation = Path level

CV-S   Coding Violation = Section level

CV-V   Coding Violation = Virtual tributary level

DA   Destination Address

DAC   Dual Attached Concentrator

DACS   Digital Access Cross-connect System

DAS   Dual Attached Station

DAVIC   Digital Audio Video Council

dB   Decibel

DBR   Distributed Bragg Reflector

DCB   Digital Channel Bank

DCC   Data Communication Channel

DCOM   Distributed Common Object Model

DCS   Digital Cross-connect System

DDS   Digital Data Services

DF   Do Not Fragment

DFB   Distributed FeedBack laser

DHCP   Dynamic Host Configuration Protocol

DIU   Digital Interface Unit

DIX   DEC (Digital Equipment), Intel, and Xerox

DLC   Digital Loop Carrier

DLCI   Data Link Connection Identifier

DLPI   Data Link Provider Interface

DOCSIS   Data Over Cable Systems Interoperability Specification

DOD   Department of Defense

DNS   Domain Name System

DPSK   Differential Phase-Shift Keying

DQDB   Distributed Queue Dual Bus

DSF   Dispersion Shifted Fiber

DSL   Digital Subscriber Line

DSLAM   Digital Subscriber Line Access Multiplexer

DSP   Digital Signal Processor

DS-SMF   Dispersion Shifted Single Mode Fiber

DSSS   Direct Sequence Spread Spectrum

DS-0   Digital Signal-level 0 (64 kbps)

DS-1   Digital Signal-level 1 (1.544 Mbps)

DS-1C   Digital Signal-level 1C (3.152 Mbps)

DS-3   Digital Signal-level 3 (44.736 Mbps)

DS-3C   Digital Signal-level 3C (91.053 Mbps)

DSX-m   Digital Signal Cross-connect point for DSm signals

DS-n   Digital Signal-level n

DTV   Digital Television

DVB   Digital Video Broadcasting

DVB-RCCL   DVB Return Channels for Cable and LMDS

DWDM   Dense Wavelength Division Multiplexing

E1 (E-1)   ITU-T digital signal-level I (2.048 Mbps)

E2 (E-2)   ITU-T digital signal-level 2 (8.448 Mbps)

E3 (E-3)   ITU-T digital signal-level 3 (34.368 Mbps)

E4 (E-4)   ITU-T digital signal-level 4 (139.264 Mbps)

ECC   Embedded Communication Channel

EDFA   Erbium-Doped Fiber Amplifier

EIR   Equipment Identity Register

ELED   Edge Light Emitting Diode

E-O (E/O)   Electrical (signal) to optical (signal) conversion

EOC   Embedded Operations Channel

EPG   Electronic Program Guide

ES   Errored Seconds, Elementary Stream

ESCON   Enterprise System Connection

ESF   Extended Super Frame

ESP   Ethernet Service Provider

ESPN   Entertainment and Sports Programming Network

ETSI   European Telecommunications Standards Institute

FBG   Fiber Bragg Grating

FC   Fibre Channel

FCC   Federal Communications Commission

FCS   Frame Check Sequence

FDD   Frequency Division Duplexing

FDDI   Fiber Distributed Data Interface

FDM   Frequency Division Multiplexing

FDMA   Frequency Division Multiple Access

FEBE   Far End Block Error

FEC   Forward Error Correction

FERF   Far End Receive Failure (now RDI)

FH   Frequency Hopping

FFOL   FDDI Follow On LAN

FICON   Fiber Connection

FITL   Fiber in the Loop

FM   Frequency Modulation

FO   Fiber Optic

FOIRL   Fiber Optic Inter-Repeater Link

FOTS   Fiber Optic Transmission System

FP   Fabry-Perot

FSN   Full Service Network

FSO   Free Space Optics

FTF   Fiber to the Feeder

FTP   File Transfer Protocol

FTTC   Fiber to the Curb

FTTH   Fiber to the Home

FTTN   Fiber to the Neighborhood

FWHM   Full Width Half Maximum

FWM   Four Wave Mixing

FWPMP   Fixed Wireless Point to Multipoint

GBE   Gigabit Ethernet

GEO   Geosynchronous Earth Orbit

GHz   Gigahertz

GMII   Gigabit Media Independent Interface

GMSK   Global Mobile Shift Keying

GPRS   General Packet Radio Service

GPS   Global Positioning System

GRIN   GRaded INdex (fiber)

GSM   Global System for Mobile (Communications)

GUI   Graphical User Interface

HAP   Hardware Access Point

HBO   Home Box Office

HCDS   High-Capacity Digital Services

HDLC   High-level Data Link Control

HDSL   High-Speed Digital Subscriber Line

HDTV   High-Definition TV

HEC   Header Error Control

HFC   Hybrid Fiber-Coax

HLR   Home Location Register

HP   Homes Passed

HPPI   High-Performance Parallel Interface

HVC   High-order Virtual Container (VC-3/VC4)

IC   Integrated Circuit

ICMP   Internet Control Message Protocol

ID   Identification

IDLC   Integrated Digital Loop Carrier

IDSL   ISDN-type Digital Subscriber Line

IEEE   Institute of Electrical and Electronics Engineers

IETF   Internet Engineering Task Force

ILEC   Incumbent Local Exchange Carrier

IM   Inverse Multiplexer

IP   Internet Protocol

IPI   Intelligent Peripheral Interface

ISDB   Integrated Services of Digital Broadcasting

ISDN   Integrated Service Digital Network

ISM   Integrated (Intelligent) Synchronous Multiplexer

ISO   International Standards Organization

ISM   Industrial, Scientific, and Medical

ISP   Internet Service Provider

ITFS   Instructional Television Fixed Service

ITU   International Telecommunications Union

ITU-T   ITU—Telecommunications Standardization Sector (Formally CCITT)

IXC   Interexchange Carrier

L1   L-carrier, first level

L3   L-carrier, third level

L4   L-carrier, fourth level

L5   L-carrier, fifth level

L5   L-carrier, fifth level enhanced

LAN   Local Area Network

LANE   LAN Emulation

LAPD   Link Access Procedure - D channel

LATA   Local Access and Transport Area

LC   "Lucent Connector"

LCD   Liquid Crystal Display

LCM   Least Common Multiplier

LCV   Line Coding Violations

LDAP   Lightweight Directory Access Protocol

LDP   Label Distribution Protocol

LE   Local Exchange

LEAF   Large Effective Area Fiber

LEC   Local Exchange Carrier

LED   Light Emitting Mode (as a light source)

LEO   Low Earth Orbit

LER   Label Edge Router

LES   Line Errored Seconds

LLC   Logical Link Control

LMDS   Local Multipoint Distribution System

LOF   Loss Of Frame

LOH   Line Overhead

LOM   Loss Of Multiframe

LOP   Loss Of Pointer

LOP-P   Path level Loss Of Pointer

LOP-V   Loss Of Pointer = Virtual tributary level

LOS   Loss Of Signal, Line Of Sight

LSB   Least Significant Bit

LSES   Line Severe Errored Seconds

LSR   Label Switch Router

LTE   Line Termination Equipment (also MSTE)

LVC   Low-order Virtual Container (VC-I 1/VC-12/VC-2)

MAC   Media Access Control

MAF   Management Application Function

MAN   Metropolitan Area Network

MARS   Multicast Address Resolution Servers

MCF   Message Communications Function

MCNS   Multimedia Cable-Network System

MCVD   Modified Chemical Vapor Deposition

MD   Mediation Device

MDS   Multipoint Distribution Service

MDU   Multiple Dwelling Units

MEMS   MicroElectroMechanical Systems

MEO   Middle Earth Orbit

MF   More Fragments, Mediation Function

MGCP   Multimedia Gateway Control Protocol

MHz   Megahertz

MIC   Media Interface Connector

MII   Media Independent Interface

Mips   Millions of instructions per second

MMDS   Multipoint Multichannel Distribution System

MMF   MultiMode Fiber

MO   Managed Object

MP   MultiPurpose (connector)

MPEG   Motion Picture Experts Group

MPLS   Multiprotocol Label Switching

MPOA   Multiprotocol Over ATM

MSC   Mobile Switching Center

MSPR   Multiplex Section Shared Protection Ring

MS-AIS   Multiplex Section AIS

MSB   Most Significant Bit

MS-FERF   Multiplex Section FERF

MSOH   Multiplex Section Overhead

MSTE   Multiplex Section Terminating Equipment (also LTE)

MT   Media Termination

MTP   Message Transfer Part

MTSO   Mobile Telephone Switching Office

MTV   Music Television

N1   N-carrier, first level

N2   N-carrier, second level

N3   N-carrier, third level

N4   N-carrier, fourth level

NA   Numeric Aperture

NAMPS   Narrowband Advanced Mobile Phone System

NDF   New Data Flag

NDF-P   New Data Flag = Path level

NDF-V   New Data Flag = Virtual tributary level

NDIS   Network Driver Interface Specification

NE   Network Element (or Equipment)

NEF   Network Element Function

NEXT   Near End Crosstalk

NFL   National Football League

NFS   Network File System

NGDLC   Next Generation Digital Loop Carrier

NHRP   Next Hop Resolution Protocol

NIC   Network Interface Card

NLSS   Non–Locally Switched Special

NMT   Nordic Mobile Telephony

NNI   Network Node Interface

NOB   Network Operator Byte

NOS   Network Operating System

NPC   Network Parameter Control

NPI   Null Pointer Indication

NRM   Network Resource Management

NSF   National Science Foundation

NUT   Non-preemptive Unprotected Traffic

NZ-DSF   Non-zero Dispersion Shifted Fiber

OADM   Optical Add-Drop Multiplexer

OA&M   Operations, Administration, and Maintenance

OAM&P   Operations, Administration, Maintenance and Provisioning

OCDMA   Optical Code Division Multiple Access

OC-N   Optical Carrier for Nth level

ODCS   Optical Digital Cross-Connect System

ODI   Open Datalink Interface

ODN   Optical Distribution Network

O-E (O/E)   Optical (signal) to electrical (signal) conversion

OFS   Optical Fiber System

OFS   Out-of-Frame Second

OH   Overhead

OLC   Optical Loop Carrier

OLT   Optical Line Terminator

ONU   Optical Network Unit

OOF   Out-Of-Frame

OPGW   Optical Ground Wire

ORL   Optical Return Loss

OS   Operating System

OSS   Operations and Support System

OSF   Operations System Function

OSI   Open System Interconnection

OSI-RM   Open System Interconnection Reference Model

OSNR   Optical Signal-to-Noise Ratio

OSPF   Open Shortest Path First

OTDR   Optical Time Domain Reflectometer

OVD   Outside Vapor Deposition

OXC   Optical Cross-Connect

PAN   Personal Area Network

PARC   Palo Alto Research Center

PBX   Private Branch Exchange

PC   Personal Computer

PC   Protection Channel

PCI   Personal Computer Interface

PCM   Pulse Code Modulation

PCVD   Plasma-activated Chemical Vapor Deposition

PCS   Personal Communications Systems

PDA   Personal Digital Assistant

PDH   Plesiochronous Digital Hierarchy

PDU   Protocol Data Unit

PES   Packetized Elementary Stream

PIN   Positive-Intrinsic-Negative photodiode

PJ   Pointer Justification

PJC   Pointer Justification Count

PM   Performance Monitoring

PMD   Physical Media Dependent, Polarization Mode Dispersion

PMO   Present Method of Operations

POH   Path Overhead

PON   Passive Optical Network

POP   Point Of Presence

POS   Packet Over SONET(SDH)

POTS   Plain Old Telephone Service

PLM-P   Path level Path Label Mismatch

PLM-V   Virtual tributary Path Label Mismatch

PN Pseudo-Noise

ppm parts per million

PPP Point-to-Point Protocol

PPPoE Point-to-Point Protocol over Ethernet

PRC Primary Reference Clock

PRS Primary Reference Source (for clocks)

PS Protection Switching, Program Stream

PSC Protection Switching Count

PSD Protection Switching Duration

PSI Program Specific Information

PSTN Public Switched Telephone Network

PTE Path Terminating Equipment

PTI Payload Type Indicator

PTR Pointer

PUC Path User Channel

PVC Permanent Virtual Circuit, Polyvinyl Chloride

QA Q-Adapter

QAM Quadrature Amplitude Modulation

QOS Quality Of Service

QPSK Quadrature Phase-Shift Keying

RAID Redundant Array of Inexpensive Disks

RBOC Regional Bell Operating Company

RDI Remote Defect (Degrade) Indication (formerly FERF)

RDI-L Line level Remote Defect Indication

RDI-P Path level Remote Defect Indication

RDI-V   Remote Defect Indication = Virtual tributary

RECAP   Resonant-Cavity Photodetector

REI   Remote Error Indication

REI-L   Line level Remote Error Indication

REI-P   Remote Error Indication = Path level

REI-V   Remote Error Indication = Virtual tributary level

REDFA   Rare Earth Doped Fiber Amplifier

RELP   Residual Exciting Linear Predictive

RF   Radio Frequency

RFC   Request For Comment

RFI   Remote Failure Indication

RFI-V   Remote Failure Indication = Virtual tributary level

RI   Refractive Index

RJ   Recommended Jack

RM   Resource Management

ROSE   Remote Operations Service Element

RSM   Remote Switch Module

RSOH   Regenerator Section Overhead

RSTE   Regenerator Section Terminating Equipment (also STE)

RSVP   Resource Reservation Protocol

RT   Remote Terminal

Reg   Regenerator

Rx   Receive

SA   Source Address

SAC   Single Attached Concentrator

SAN   Storage (Server) Area Network

SAP   Software Access Point

SAR   Segmentation And Reassembly

SAS   Single Attached Station

SBCCS   Single Byte Command Code Sets

SBS   Stimulated Brillouin Scattering

SC   Subscriber (Square) Connector

SCCP   Signaling Connection Control Part

SCP   Service Control Point

SCSI   Small Computer Systems Interface

SCV   Section Coding Violations

SD   Signal Degrade

SDH   Synchronous Digital Hierarchy

SDSL   Symmetrical Digital Subscriber Line

SDTV   Standard Definition Television

SDV   Switched Digital Video

SECB   Severely Errored Cell Block

SEFS   Severe Errored Frame Seconds

SES   Section Errored Seconds

SF   Signal Fail

SF   Super Frame

SHIP   Silicon Hetero-Interface Photodetector

SIP   Session Initiation Protocol

SLC   Synchronous Line Carrier

SLED   Surface Light Emitting Diode

SLIP   Serial Line Interface Protocol

SLM   Single Longitudinal Mode

SLM   Synchronous Line Multiplexer

SM   Switching Module

SMDS   Switched Multimegabit Data Service

SMF   Single Mode Fiber

SMN   SDH Management Network

SMN   SONET Management Network

SMS   SDH Management Subnetwork

SMTP   Simple Mail Transfer Protocol

SNA   System Network Architecture

SNAP   Sub-Network Access Protocol

SNC   Sub-Network Connection

SNMP   Simple Network Management Protocol

SOA   Semiconductor Optical Amplifier

SOH   Section Overhead

SONET   Synchronous Optical Network

SPE   Synchronous Payload Envelope

SRS   Stimulated Raman Scattering

SS   Switching System

SSM   Synchronization Status Message

SSP   Service Switching Point

SS7   Signaling System 7

ST   Straight Tip (connector)

STE   Section Terminating Equipment (also RSTE)

STM   Synchronous Transfer Mode (see ATM)

STM   Synchronous Transport Module

STM-N   Synchronous Transport Module-level N (Nx 155.52 Mbps: N = 1, 4, 16, or 64)

STP   Shielded Twisted Pair, Spanning Tree Protocol, Signaling Transfer Point

STS   Synchronous Transport Signal

STS-N   Synchronous Transport Signal-level N (Nx 51.84 Mbps: N = 1, 3, 12, 48, or 192)

SVC   Switched Virtual Circuit

T1 (T-1)   A digital carrier system (1.544 Mbps) for DS-1 signal

T3 (T-3)   A digital carrier system (44.736 Mbps) for DS-3 signal

TACS   Total Access Communications System

TBD   To Be Determined

TC   Transmission Convergence

TCAP   Transaction Capability Application Part

TCM   Tandem Connection Monitoring

TCP   Transmission Control Protocol

TDD   Time Division Duplexing

TDM   Time-Division Multiplexing

TDMA   Time-Division Multiple Access

TEHO   Tail-End Hop Off

THz   Terahertz

TIA   Telecommunication Industry Association

TIM   Trace Identifier Mismatch

TIM-L   Line level Trace Identifier Mismatch

TIM-P   Path level Trace Identifier Mismatch

TIM-V   Virtual tributary level Trace Identifier Mismatch

TMN   Telecommunications Management Network

TOH   Transport Overhead (SOH + LOH)

TOS   Type Of Service

TS   Time Slot, Transport Stream

TSG   Timing Signal Generator

TTL   Time To Live

TU-n   Tributary Unit-level n (n= 11, 12, 2, or 3)

TUG-n   Tributary Unit Group n (n = 2 or 3)

TV   Television

Tx   Transmit

UAS   UnAvailable Second

UAT   UnAvailable Time

UBN   Unlimited Bandwidth Network

UDP   User Datagram Protocol

UE   User Equipment

ULP   Upper Layer Protocol

UMTS   Universal Mobile Telecommunications System

UNEQ   Unequipped

UNEQ-P   Unequipped at the Path level

UNEQ-V   Unequipped at the Virtual tributary level

UNI   User-Network Interface

UPC   Usage Parameter Control

UPSR   Unidirectional Path Switch Ring

UTP   Unshielded Twisted Pair

UV Ultraviolet

VAD Vapor Axial Deposition

VBR Variable Bit Rate

VC Virtual Channel, Voice Channel, Virtual Circuit

VC-n Virtual Container-level n (n= 11, 12, 2, 3 or 4)

VCC Virtual Channel Connection

VCELP Vector Code-Excited Linear Predictive

VCI Virtual Channel Identifier

VCSEL Vertical Cavity Surface Emitting Laser

VDSL Very-high-speed Digital Subscriber Line

VF Voice Frequency signal

VH1 Video Hits 1

VIA Virtual Interface Architecture

VLR Visitor Location Register

VoDSL Voice Over Digital Subscriber Line

VoIP Voice Over IP

VLAN Virtual LAN

VPI Virtual Path Identifier

VPN Virtual Private Network

VT Virtual Tributary

VTOA Voice and Telephony Over ATM

WAN Wide Area Network

WAP Wireless Application Protocol

W-CDMA Wideband Code Division Multiple Access

W-DCS Wideband Digital Cross-connect System

WDM   Wavelength Division Multiplexing

WEP   Wireless Equivalent Privacy

WLL   Wireless Local Loop

W-OCDMA   Wideband Optical Code Division Multiple Access

XMII   X-type Media Independent Interface

X.25   International standard packet-switching protocol at link and packet layers

ZBLAN   zirconium (chemical symbol Zr), barium (Ba), lanthanum (La), aluminum (Al), and sodium (Na)

APPENDIX C

References

BIBLIOGRAPHY

Abe, George, *Residential Broadband*. Cisco Press (Macmillan), Indianapolis, IN, 1997.

Benner, Alan F., *Fibre Channel*. McGraw-Hill, NY, 1996.

Bray, John, *The Communications Miracle*. Plenum, NY, 1995.

Cunningham, David G. and Lane, William G., *Gigabit Ethernet Networking*. Macmillan, Indianapolis, IN, 1999.

Dutton, Harry J. R., *Understanding Optical Communications*. Prentice-Hall, NY, 1998.

Feit, Dr. Sidnie, *TCP/IP*. McGraw-Hill, NY, 1999.

Goralski, Walter, *Introduction to ATM Networking*. McGraw-Hill, NY, 1995.

Goralski, Walter, *SONET, 2nd Edition*. McGraw-Hill, NY, 2000.

Harlow, Alvin F., *Old Wires and New Waves*. Arno Press, NY, 1971.

Hetch, Jeff, *City of Light: The Story of Fiber Optics*. Oxford, NY, 1999.

Laubach, Mark E., Farber, David J., and Dukes, Stephen D., *Delivering Internet Connections over Cable*. Wiley & Sons, NY, 2001.

Raskin, Donald, and Stoneback, Dean, *Broadband Return Systems for Hybrid Fiber/Coax Cable TV Networks*. Prentice-Hall, Saddle River, NJ, 1998.

Smith, Clint, *LMDS*. McGraw-Hill, NY, 2000.

Standage, Tom, *The Victorian Internet*. Berkley, NY, 1998.

Tunmann, Ernest, *Hybrid Fiber-optic Coaxial Networks*. Flatiron Publishing, NY, 1995.

MAGAZINE ARTICLES

Allen, Doug, The Second Coming of Free Space Optics. *Network Magazine*, March 2001, p. 55ff.

Blumenthal, Daniel J., Routing Packets with Light. *Scientific American*, January, 2001, p. 96ff.

Costa, Dan, DSL in Broadband Special Report. *PC Magazine*, February 6, 2001, p.147ff.

Kunzig, Robert, Trapping Light. *Discover*, April 2001, p. 72ff.

Mandell, Mel, 120 000 Leagues Under the Sea. *IEEE Spectrum*, April 2000, p.50ff.

Masud, Sam, Lighting Ethernet in the WAN. *Telecommunications*, September 2000, p. 26ff.

Shuchart, Steven J., The Bluetooth Invasion Begins. *Network Computing*, March 19, 2001, p. 101ff.

INDEX

 A

D

P

Y

 Z

INTERNATIONAL CONTACT INFORMATION

AUSTRALIA
McGraw-Hill Book Company Australia Pty. Ltd.
TEL +61-2-9417-9899
FAX +61-2-9417-5687
http://www.mcgraw-hill.com.au
books-it_sydney@mcgraw-hill.com

CANADA
McGraw-Hill Ryerson Ltd.
TEL +905-430-5000
FAX +905-430-5020
http://www.mcgrawhill.ca

**GREECE, MIDDLE EAST,
NORTHERN AFRICA**
McGraw-Hill Hellas
TEL +30-1-656-0990-3-4
FAX +30-1-654-5525

MEXICO (Also serving Latin America)
McGraw-Hill Interamericana Editores S.A. de C.V.
TEL +525-117-1583
FAX +525-117-1589
http://www.mcgraw-hill.com.mx
fernando_castellanos@mcgraw-hill.com

SINGAPORE (Serving Asia)
McGraw-Hill Book Company
TEL +65-863-1580
FAX +65-862-3354
http://www.mcgraw-hill.com.sg
mghasia@mcgraw-hill.com

SOUTH AFRICA
McGraw-Hill South Africa
TEL +27-11-622-7512
FAX +27-11-622-9045
robyn_swanepoel@mcgraw-hill.com

**UNITED KINGDOM & EUROPE
(Excluding Southern Europe)**
McGraw-Hill Education Europe
TEL +44-1-628-502500
FAX +44-1-628-770224
http://www.mcgraw-hill.co.uk
computing_neurope@mcgraw-hill.com

ALL OTHER INQUIRIES Contact:
Osborne/McGraw-Hill
TEL +1-510-549-6600
FAX +1-510-883-7600
http://www.osborne.com
omg_international@mcgraw-hill.com